RECENT PROGRESS IN COMPUTATIONAL AND APPLIED PDES

RECENT PROGRESS IN COMPUTATIONAL AND APPLIED PDES

Conference Proceedings for the International Conference Held in Zhangjiajie in July 2001

Edited by

TONY F. CHAN

University of California at Los Angeles
Los Angeles, California

YUNQING HUANG

Xiangtan University
Xiangtan City of Hunan, China

TAO TANG

Hong Kong Baptist University
Hong Kong, China

JINCHAO XU

Pennsylvania State University
University Park, Pennsylvania

and

LONG-AN YING

Peking University
Beijing, China

Springer Science+Business Media, LLC

Proceedings of the International Symposium on Computational and Applied PDEs, held July 1–7, 2001, in Zhangjiajie National Park, China
ISBN 978-1-4613-4929-7 ISBN 978-1-4615-0113-8 (eBook)
DOI 10.1007/978-1-4615-0113-8

Originally published by Kluwer Academic/Plenum Publishers, New York in 2002
Softcover reprint of the hardcover 1st edition 2002
http://www.wkap.com

10 9 8 7 6 5 4 3 2 1

A C.I.P. record for this book is available from the Library of Congress

Contents

Preface

The International Symposium on Computational & Applied PDEs was held at Zhangjiajie National Park of China from July 1-7, 2001. The main goal of this conference is to bring together computational, applied and pure mathematicians on different aspects of partial differential equations to exchange ideas and to promote collaboration. Indeed, it attracted a number of leading scientists in computational PDEs including Doug Arnold (Minnesota), Jim Bramble (Texas A & M), Achi Brandt (Weizmann), Franco Brezzi (Pavia), Tony Chan (UCLA), Shiyi Chen (John Hopkins), Qun Lin (Chinese Academy of Sciences), Mitch Luskin (Minnesota), Tom Manteuffel (Colorado), Peter Markowich (Vienna), Mary Wheeler (Texas Austin) and Jinchao Xu (Penn State); in applied and theoretical PDEs including Weinan E (Princeton), Shi Jin (Wisconsin), Daqian Li (Fudan) and Gang Tian (MIT). It also drew an international audience of size 100 from Austria, China, Germany, Hong Kong, Iseael, Italy, Singapore and the United States.

The conference was organized by Yunqing Huang of Xiangtan University, Jinchao Xu of Penn State University, and Tony Chan of UCLA through ICAM (Institute for Computational and Applied Mathematics) of Xiangtan university which was founded in January 1997 and directed by Jinchao Xu. The scientific committee of this conference consisted of Randy Bank of UCSD, Tony Chan of UCLA, K. C. Chang and Long-an Ying of Peking University, Qun Lin, Zhong-Ci Shi and Yaxiang Yuan of Chinese Academy of Sciences, Gang Tian of MIT, Mary Wheeler of UT Austin, Jinchao Xu of Penn State, and Yulin Zhou of Institute for Applied Physics and Computational Mathematics, Beijing.

The one-week conference featured 20 invited speakers, each of whom gave 45-minutes lectures, and about 40 other speakers. All invited talks were of high quality, covering several aspects of modern computational and theoretical partial differential equations. The conference site Zhangjiajie is the top one or two national park in China. It is located in Hunan Province. Participants visiting Zhangjiajie the first time were impressed by the beautiful view of the mountain, lake and canions. More details on the conference can be found in http://www.math.psu.edu/ccma/pde2001.

We would like to thank Professor Shucheng Li, Preseident of Xiangtan University, for his support and also for his participating the openning ceremony. Thanks also go to Professor Jiping Zhang, Dean of School of Mathematical Sciences at Peking University for his kind support to this conference. Moreover, the conference received considerable financial supports. The main grants were provided by the Institute for Computational and Applied Mathematics of Xiangtan University, School of Mathematical Sciences of Peking University, the Center for Computational Mathematics and Applications of Penn State University, the Science and Technology Department of Hunan Province, the National Science Foundation of China, and the State Key Basic Research Project "Large Scale Scientific Computing Research". We are grateful to all sponsors for their generous support.

These conference proceedings were refereed. We would like to thank all referees for their support. The performance of the meeting depended very much on many helpers, including Susan He, Xu Chen, Zhongbo Chen, Jianmei Yuan and Qishen Xiao of Xiangtan University and Rosemary Manning of Penn State University. We appreciate their assistance in making the conference organization a success. Finally, we thank Tammy Lam and Amy Lee of Hong Kong Baptist University for the considerable work they put into producing the final layout of this proceedings.

Editors:
T.F. Chan, UCLA
Y.-Q. Huang, XTU
T. Tang, HKBU
J.-C. Xu, Penn State
L.-A. Ying, PKU.

June 2002

List of participants

Shidfar Abdullah, shidfar@iust.ac.ir
Department of Mathematics, Iran University of Science and Technology

Douglas N. Arnold, dna@math.psu.edu; arnold@ima.umu.edu
Department of Mathematics, Penn State University

Steven F. Ashby, sfashby@llnl.gov
Center for Applied Scientific Computing, Lawrence Livermore National Laboratory

James H. Bramble, bramble@math.tamu.edu; barth@nas.nasa.gov
Mathematics Department, Texas A&M University

Achi Brandt, achi@wisdom.weizmann.ac.il
Department of Computer Science & Applied Mathematics, The Weizmann Institute of Science

Franco Brezzi, brezzi@ian.pv.cnr.it; brezzi@dragon.ian.pv.cnr.it
Dipartimento di Matematica, Universita' di Pavia

Qingdong Cai, caiqd@mech.pku.edu.cn
Department of Mechanical Engineering, Peking University

Xiaolin Cao, xiaolincao@163.com
School of Computer Science, National University of Defense Technology, China

Maria-Carme Calderer, mcc@math.psu.edu
Department of Mathematics, Penn State University

Tony Chan, chan@ipam.ucla.edu
Institute for Pure and Applied Mathematics (IPAM), UCLA

Chuanmiao Chen, cmchen@mail.hunnu.edu.cn
Department of Mathematics, Hunan Normal University, China

Jinru Chen, jrchen@pine.njnu.edu.cn
School of Mathematical Sciences, Nanjing Normal University, China

Shiyi Chen, syc@taylor.me.jhu.edu
Department of Mechanical Engineering, The Johns Hopkins University

Yixin Chen
Scientific Publishing Company of Hunan, China

Yanping Chen, ypchen@xtu.edu.cn
Department of Mathematics, Xiangtan University, China

Zhiming Chen, zmchen@lsec.cc.ac.cn
Institute of Computational Mathematics, Chinese Academy of Sciences

Yana Di Jbn58@water.pku.edu.cn
School of Mathematical Sciences, Peking University

Qiang Du, qdu@lsec.cc.ac.cn
Institute of Computational Mathematics, Chinese Academy of Sciences

Weinan E, weinan@math.princeton.edu; weinan@princeton.edu
Department of Mathematics, Princeton University

Eva Eggeling, eggeling@gmd.de; eva@fleuler.gmd.de
Institute of Algorithms and Scientific Computing (SCAI), GMD-German National Research Center for Information Technology

Shuguang Gong, gongsg@xtu.edu.cn
Department of Mathematics, Xiangtan University, China

Zhenting Hou
Department of Mathematics, Railway Area of Central South University, China

Yunqing Huang, huangyq@xtu.edu.cn
Department of Mathematics, Xiangtan University, China

Zhaohui Huang, zhhuang@amath6.amt.ac.cn

Jun Hu, hujunlya@263.net
Department of Mathematics, Xiangtan University, China

Qiya Hu, hqy@lsec.cc.ac.cn
Institute of Computational Mathematics, Chinese Academy of Sciences

Zhongxiao Jia, zxjia@dlut.edu.cn
Department of Applied Mathematics, Dalian University of Technology, China

Shi Jin, jin@math.wisc.edu
Department of Mathematics, University of Wisconsin-Madison

Daqian Li, dqli@fudan.edu.cn
Department of Mathematics, Fudan University

Qun Lin, qlin@staff.iss.ac.cn
Institute of Computational Mathematics, Chinese Academy of Sciences

Wenbing Liu, W.B.Liu@ukc.ac.uk
CBS, University of Kent

Mitchell Barry Luskin, luskin@math.umn.edu
School of Mathematics, University of Minnesota

Thomas A. Manteuffel, tmanteuf@colorado.edu
Department of Applied Mathematics, University of Colorado at Boulder

Meng Mao, maomeng@263.net
Department of Soil and Water Sciences, China Agricultural University

Luisa Donatella Marini, marini@ian.pv.cnr.it; marini@dragon.ian.pv.cnr.it
Dipartimento di Matematica and I.A.N.-C.N.R., Italy

Zeyao Mo, zy_mo@sina.com

Mo Mu, mamu@ust.hk
Department of Mathematics, Hong Kong University of Science & Technology

Joseph E. Pasciak, pasciak@math.tamu.edu
Department of Mathematics, Texas A&M University

Peter Markowich, peter.markowich@univie.ac.at
Institute of Mathematics, University of Vienna

Li Ren, renl@mx.cei.gov.cn
Department of Soil and Water Sciences, China Agricultural University

Xiumin Shao, shao@math03.math.ac.cn
Institute of Mathematics, Chinese Academy of Sciences,

Jie Shen, shen@math.psu.edu
Department of Mathematics, Penn State University

Ta'asan Shlomo, shlomo@andrew.cmu.edu; shlomo@sattva.math.cmu.edu
Department of Mathematics Sciences, Carnegie Mellon University

Shi Shu, shushi@xtu.edu.cn
Department of Mathematics, Xiangtan University, China

Jianzhong Su, su@uta.edu
Department of Mathematics, University of Texas at Arlington

Yi Sun, ysun@pku.edu.cn
School of Mathematical Sciences, Peking University

Tao Tang, ttang@math.hkbu.edu.hk
Department of Mathematics, Hong Kong Baptist University

Zhijun Tan, tanzhijun1221@sina.com
Department of Mathematics, Xiangtan University, China

Gang Tian, tian@math.mit.edu
Department of Mathematics, MIT

Lihe Wang, lwang@math.uiowa.edu
Department of Mathematics, University of Iowa

Ping Wang, pxw10@psu.edu
Department of Mathematics, Penn State University

Mary Wheeler, mfw@ticam.utexas.edu
Texas Institute for Computational and Applied Mathematics, University of Texas, Austin

Gabriel Christoph Wittum, wittum@iwr.uni-heidelberg.de
IFI, University Heidelberg

Haijun Wu, whj@lsec.cc.ac.cn
Morning Sight Center, Chinese Academy of Sciences

Yujiang Wu, myjaw@lzu.edu.cn
Department of Mathematics, Lanzhou University, China

Yuelong Xiao, Xiaoyl01@163.com
Department of Mathematics, Xiangtan University, China

Jie Xiao, Xiaojuan-xu@263.net
Department of Mathematics, Xiangtan University, China

Da Xu, daxu@hunnu.edu.cn
Department of Mathematics, Hunan Normal University, China

Jinchao Xu, xu@math.psu.edu
Department of Mathematics, Penn State University

Lung-an Ying, yingla@pku.edu.cn
School of Mathematical Sciences, Peking University

Harry Yserentant, harry@na.uni-tuebingen.de; yserentant@na.uni-tuebingen.de
Mathematisches Institut, Universität Tubingen

Haiyuan Yu
Department of Mathematics, Xiangtan University, China

Ju'e Yang, chrysan@china.com
Department of Mathematics, Xiangtan University, China

Ying Yang, y_y1976@263.net
Department of Mathematics, Xiangtan University, China

Jinping Zeng, blma@mail.hnu.net.cn; dhli@mail.hunu.edu.cn
Department of Applied Mathematics, Hunan University, China

Dakai Zhang, xiangshw@263.net ; zouaim@263.net
Department of Mathematics, Guizhou University, China

Deyue Zhang, zhangdeyue@263.net

Morning Sight Center, Chinese Academy of Sciences

Manping Zhang, zhangmanping@sina.com
Department of Mathematics, Xiangtan University, China

Pingwen Zhang, pzhang@math.pku.edu.cn; pzhang@pku.edu.cn
School of Mathematical Sciences, Peking University

Shan Zhao, asan@cz3.nus.edu.sg
Department of Computational Science, National University of Singapore

Weiying Zheng, zhengweiying@sina.com
Peking University

Zhoushun Zheng, zszheng@mail.csu.edu.cn
Department of Mathematics and Software, Central South University

Bin Zheng, bzheng@water.pku.edu.cn
School of Mathematical Sciences, Peking University

Huansong Zhou, hszhou@wipm.whcnc.ac.cn
Institute of Physics & Mathematics, Wuhan, Chinese Academy of Sciences

Yi Zhou, yizhou@fudan.ac.cn
Institute of Mathematics, Fudan University

SHIFT THEOREMS FOR THE BIHARMONIC DIRICHLET PROBLEM*

Constantin Bacuta
Dept. of Mathematics, The Pennsylvania State University
University Park, PA 16802, USA.
bacuta@math.psu.edu

James H. Bramble, Joseph E. Pasciak
Dept. of Mathematics, Texas A & M University
College Station, TX 77843, USA.
bramble@math.tamu.edu
pasciak@math.tamu.edu

Keywords: interpolation spaces, biharmonic operator, shift theorems

Abstract We consider the biharmonic Dirichlet problem on a polygonal domain. Regularity estimates in terms of Sobolev norms of fractional order are proved. The analysis is based on new interpolation results which generalizes Kellogg's method for solving subspace interpolation problems. The Fourier transform and the construction of extension operators to Sobolev spaces on R^2 are used in the proof of the interpolation theorem.

1. Introduction

Regularity estimates of the solutions of elliptic boundary value problems in terms of Sobolev-fractional norms are known as shift theorems or shift estimates. The shift estimates are significant in finite element theory.

The shift estimates for the Laplace operator with Dirichlet boundary conditions on nonsmooth domains are studied in [2], [12], [14] and [18]. On the question of shift theorems for the biharmonic problem on nonsmooth domains, there seems to be no work answering this question.

*This work was partially supported by the National Science Foundation under Grant DMS-9973328.

One way of proving shift results is by using the real method of interpolation of Lions and Peetre [3], [15] and [16]. The interpolation problems we are led to are of the following type. If X and Y are Sobolev spaces of integer order and X_K is a subspace of finite codimension of X then characterize the interpolation spaces between X_K and Y.

When X_K is of codimension one the problem was studied by Kellogg in some particular cases in [12]. The interpolation results presented in Section 2 give a natural formula connecting the norms on the intermediate subspaces $[X_K, Y]_s$ and $[X, Y]_s$. The main result of Section 2 is a theorem which provides sufficient conditions to compare the topologies on $[X_K, Y]_s$ and $[X, Y]_s$ and gives rise to an extension of Kellogg's method in proving shift estimates for more complicated boundary value problems.

In proving shift estimates for the biharmonic problem, we will follow Kellogg's approach in solving subspace interpolation problems on sector domains. The method involves reduction of the problem to subspace interpolation on Sobolev spaces defined on all of R^2. This reduction requires construction of "extension" and "restriction" operators connecting Sobolev spaces defined on sectors and Sobolev spaces defined on R^2. The method involves also finding the asymptotic expansion of the Fourier transform of certain singular functions. The remaining part of the paper is organized as follows. In Section 2 we prove a natural formula connecting the norms on the intermediate subspaces $[X_K, Y]_s$ and $[X, Y]_s$. The main result of the section is a theorem which provides sufficient conditions (the **(A1)** and **(A2)** conditions) to compare the topologies on $[X_K, Y]_s$ and $[X, Y]_s$. A new proof of the main subspace interpolation result presented in [12] and an extension to subspace interpolation of codimension greater than one are given in Section 3. The main result concerning shift estimates for the biharmonic Dirichlet problem is considered in Section 4.

2. Interpolation results

In this section we give some basic definitions and results concerning interpolation between Hilbert spaces and subspaces using the real method of interpolation of Lions and Peetre (see [15]).

2.1 Interpolation between Hilbert spaces

Let X, Y be separable Hilbert spaces with inner products $(\cdot,\cdot)_X$ and $(\cdot,\cdot)_Y$, respectively, and satisfying for some positive constant c,

$$\begin{cases} X \text{ is a dense subset of Y and} \\ \|u\|_Y \leq c\|u\|_X \quad \text{for all } u \in X, \end{cases} \tag{2.1}$$

where $\|u\|_X^2 = (u, u)_X$ and $\|u\|_Y^2 = (u, u)_Y$.

Let $D(S)$ denote the subset of X consisting of all elements u such that the antilinear form

$$v \to (u, v)_X, \ v \in X \tag{2.2}$$

is continuous in the topology induced by Y. For any u in $D(S)$ the antilinear form (2.2) can be extended to a continuous antilinear form on Y. Then by Riesz representation theorem, there exists an element Su in Y such that

$$(u, v)_X = (Su, v)_Y \quad \text{for all } v \in X. \tag{2.3}$$

In this way S is a well defined operator in Y, with domain $D(S)$. The next result illustrates the properties of S.

Proposition 2.1. *The domain $D(S)$ of the operator S is dense in X and consequently $D(S)$ is dense in Y. The operator $S : D(S) \subset Y \to Y$ is a bijective, self-adjoint and positive definite operator. The inverse operator $S^{-1} : Y \to D(S) \subset Y$ is a bounded symmetric positive definite operator and*

$$(S^{-1}z, u)_X = (z, u)_Y \quad \textit{for all } z \in Y, \ u \in X \tag{2.4}$$

If in addition X is compactly embedded in Y, then S^{-1} is a compact operator.

The interpolating space $[X, Y]_s$ for $s \in (0, 1)$ is defined using the K function, where for $u \in Y$ and $t > 0$,

$$K(t, u) := \inf_{u_0 \in X} (\|u_0\|_X^2 + t^2\|u - u_0\|_Y^2)^{1/2}.$$

Then $[X, Y]_s$ consists of all $u \in Y$ such that

$$\int_0^\infty t^{-(2s+1)} K(t, u)^2 \, dt < \infty.$$

The norm on $[X, Y]_s$ is defined by

$$\|u\|_{[X,Y]_s}^2 := \mathbf{c}_s^2 \int_0^\infty t^{-(2s+1)} K(t, u)^2 \, dt,$$

where

$$\mathbf{c}_s := \left(\int_0^\infty \frac{t^{1-2s}}{t^2 + 1} dt \right)^{-1/2} = \sqrt{\frac{2}{\pi} \sin(\pi s)}$$

By definition we take

$$[X,Y]_0 := X \text{ and } [X,Y]_1 := Y.$$

The next lemma provides the relation between $K(t,u)$ and the connecting operator S.

Lemma 2.1. *For all $u \in Y$ and $t > 0$,*

$$K(t,u)^2 = t^2 \left((I + t^2 S^{-1})^{-1} u, u\right)_Y.$$

Proof. Using the density of $D(S)$ in X, we have

$$K(t,u)^2 = \inf_{u_0 \in D(S)} (\|u_0\|_X^2 + t^2 \|u - u_0\|_Y^2)$$

Let $v = Su_0$. Then

$$K(t,u)^2 = \inf_{v \in Y} ((S^{-1}v, v)_Y + t^2 \|u - S^{-1}v\|_Y^2). \tag{2.5}$$

Solving the minimization problem (2.5) we obtain that the element v which gives the optimum satisfies

$$(I + t^2 S^{-1})v = t^2 u,$$

and

$$(S^{-1}v, v)_Y + t^2 \|u - S^{-1}v\|_Y^2 = t^2 \left((I + t^2 S^{-1})^{-1} u, u\right)_Y.$$

□

Remark 2.1. *Lemma 2.1 gives another expression for the norm on $[X,Y]_s$, namely:*

$$\|u\|^2_{[X,Y]_s} := c_s^2 \int_0^\infty t^{-2s+1} \left((I + t^2 S^{-1})^{-1} u, u\right)_Y dt. \tag{2.6}$$

In addition, by this new expression for the norm (see Definition 2.1 and Theorem 15.1 in [15]), it follows that the intermediate space $[X,Y]_s$ coincides topologically with the domain of the unbounded operator $S^{1/2(1-s)}$ equipped with the norm of the graph of the same operator. As a consequence we have that X is dense in $[X,Y]_s$ for any $s \in [0,1]$.

Lemma 2.2. *Let X_0, be a closed subspace of X and let Y_0, be a closed subspace of Y. Let X_0 and Y_0 be equipped with the topology and the geometry induced by X and Y respectively, and assume that the pair (X_0, Y_0) satisfies* (2.1). *Then, for $s \in [0,1]$,*

$$[X_0, Y_0]_s \subset [X,Y]_s \cap Y_0.$$

Proof. For any $u \in Y_0$ we have

$$K(t, u, X, Y) \leq K(t, u, X_0, Y_0).$$

Thus,

$$\|u_{[X,Y]_s}\| \leq \|u_{[X_0,Y_0]_s}\| \qquad \text{for all } u \in [X_0, Y_0]_s, \quad s \in [0,1], \tag{2.7}$$

which proves the lemma. □

2.2 Interpolation between subspaces of a Hilbert space

Let $\mathcal{K} = span\{\varphi_1, \dots, \varphi_n\}$ be a n-dimensional subspace of X and let $X_\mathcal{K}$ be the orthogonal complement of $\mathcal{K}$ in X in the $(\cdot,\cdot)_X$ inner product. We are interested in determining the interpolation spaces of $X_\mathcal{K}$ and Y, where on $X_\mathcal{K}$ we consider again the $(\cdot,\cdot)_X$ inner product. For certain spaces $X_\mathcal{K}$ and Y and $n = 1$, this problem was studied in [12]. To apply the interpolation results from the previous section we need to check that the density part of the condition (2.1) is satisfied for the pair $(X_\mathcal{K}, Y)$.

For $\varphi \in \mathcal{K}$, define the linear functional $\Lambda_\varphi : X \to C$, by

$$\Lambda_\varphi u := (u, \varphi)_X, \ u \in X.$$

Lemma 2.3. *The space $X_\mathcal{K}$ is dense in Y if and only if the following condition is satisfied:*

$$\begin{cases} \Lambda_\varphi \text{ is not bounded in the topology of } Y \\ \quad \text{for all } \varphi \in \mathcal{K}, \ \varphi \neq 0. \end{cases} \tag{2.8}$$

Proof. First let us assume that the condition (2.8) does not hold. Then for some $\varphi \in \mathcal{K}$ the functional L_φ is a bounded functional in the topology induced by Y. Thus, the kernel of L_φ is a closed subspace of X in the topology induced by Y. Since $X_\mathcal{K}$ is contained in $Ker(L_\varphi)$ it follows that

$$\overline{X_\mathcal{K}}^Y \subset \overline{Ker(L_\varphi)}^Y = Ker(L_\varphi).$$

Hence $X_\mathcal{K}$ fails to be dense in Y.

Conversely, assume that $X_\mathcal{K}$ is not dense in Y, then $Y_0 = \overline{X_\mathcal{K}}^Y$ is a proper closed subspace of Y. Let $y_0 \in Y$ be in the orthogonal complement of Y_0, and define the linear functional $\Psi : Y \to C$, by

$$\Psi u := (u, y_0)_Y, \ u \in Y.$$

Ψ is a continuous functional on Y. Let ψ be the restriction of Ψ to the space X. Then ψ is a continuous functional on X. By Riesz Representation Theorem, there is $v_0 \in X$ such that

$$(u, v_0)_X = (u, y_0)_Y, \qquad \text{for all } u \in X. \tag{2.9}$$

Let $P_{\mathcal{K}}$ be the X orthogonal projection onto $\mathcal{K}$ and take $u = (I - P_{\mathcal{K}})v_0$ in (2.9). Since $(I - P_{\mathcal{K}})v_0 \in X_{\mathcal{K}}$ we have $((I - P_{\mathcal{K}})v_0, y_0)_Y = 0$ and

$$0 = ((I - P_{\mathcal{K}})v_0, v_0)_X = ((I - P_{\mathcal{K}})v_0, (I - P_{\mathcal{K}})v_0)_X.$$

It follows that $v_0 = P_{\mathcal{K}}v_0 \in \mathcal{K}$ and, via (2.9), that $\psi = \Lambda_{v_0}$ is continuous in the topology of Y. This is exactly the opposite of (2.8) and the proof is completed. □

Remark 2.2. *The result still holds if we replace the finite dimensional subspace $\mathcal{K}$ with any closed subspace of X.*

For the next part of this section we assume that the condition (2.8) holds. By the above Lemma, the condition (2.1) is satisfied. It follows from the previous section that the operator $S_{\mathcal{K}} : D(S_{\mathcal{K}}) \subset Y \to Y$ defined by

$$(u, v)_X = (S_{\mathcal{K}}u, v)_Y \quad \text{for all } v \in X_{\mathcal{K}}, \tag{2.10}$$

has the same properties as S has. Consequently, the norm on the intermediate space $[X_{\mathcal{K}}, Y]_s$ is given by:

$$\|u\|^2_{[X_{\mathcal{K}},Y]_s} := \mathbf{c}_s^2 \int_0^\infty t^{-2s+1} \left((I + t^2 S_{\mathcal{K}}^{-1})^{-1}u, u\right)_Y \, dt. \tag{2.11}$$

Let $[X, Y]_{s,\mathcal{K}}$ denote the closure of $X_{\mathcal{K}}$ in $[X, Y]_s$. Our aim in this section is to determine sufficient conditions for φ_i's such that

$$[X_{\mathcal{K}}, Y]_s = [X, Y]_{s,\mathcal{K}}. \tag{2.12}$$

First, we note that the operators $S_{\mathcal{K}}$ and S are related by the following identity:

$$S_{\mathcal{K}}^{-1} = (I - Q_{\mathcal{K}})S^{-1}, \tag{2.13}$$

where $Q_{\mathcal{K}} : X \to \mathcal{K}$ is the orthogonal projection onto $\mathcal{K}$. The proof of (2.13) follows easily from the definitions of the operators involved.

Next, (2.13) leads to a formula relating the norms on $[X_{\mathcal{K}}, Y]_s$ and $[X, Y]_s$. Before deriving this formula in Theorem 2.1 , we introduce some notation. Let

$$(u, v)_{X,t} := \left((I + t^2 S^{-1})^{-1}u, v\right)_X \quad \text{for all } u, v \in X. \tag{2.14}$$

and denote by M_t the Gram matrix associated with the set of vectors $\{\varphi_1, \dots, \varphi_n\}$ in the $(\cdot, \cdot)_{X,t}$ inner product,i.e.,

$$(M_t)_{ij} := (\varphi_j, \varphi_i)_{X,t}, \; i, j \in \{1, \dots, n\}.$$

We may assume, without loss, that M_0 is the identity matrix.

Theorem 2.1. *Let u be arbitrary in $X_{\mathcal{K}}$. Then,*

$$\|u\|^2_{[X_{\mathcal{K}},Y]_s} = \|u\|^2_{[X,Y]_s} + \mathbf{c}_s^2 \int_0^\infty t^{-(2s+1)} \left\langle M_t^{-1} d, d \right\rangle \, dt, \qquad (2.15)$$

where $< \cdot, \cdot >$ is the inner product on $\mathbf{C}^n$ and d is the n-dimensional vector in $\mathbf{C}^n$ whose components are

$$d_i := (u, \varphi_i)_{X,t}, \; i = 1, \dots, n.$$

The proof of the of the theorem can be found in [2].

For $n = 1$, let $\mathcal{K} = span\{\varphi\}$ and denote $X_{\mathcal{K}}$ by X_φ. Then, for $u \in X_\varphi$, the formula (2.15) becomes

$$\|u\|^2{}_{[X_\varphi,Y]_s} = \|u\|^2{}_{[X,Y]_s} + \mathbf{c}_s^2 \int_0^\infty t^{-(2s+1)} \frac{|(u,\varphi)_{X,t}|^2}{(\varphi,\varphi)_{X,t}} \, dt. \qquad (2.16)$$

Next theorem gives sufficient conditions for (2.12) to be satisfied. Before we state the result we introduce the conditions:

(A.1) $[X_{\varphi_i}, Y]_s = [X, Y]_{s,\varphi_i}$ for $i = 1, \dots, n$.

(A.2) There exist $\delta > 0$ and $\gamma > 0$ such that

$$\sum_{i=1}^n |\alpha_i|^2 (\varphi_i, \varphi_i)_{X,t} \leq \gamma \langle M_t \alpha, \alpha \rangle$$
$$\text{for all } \alpha = (\alpha_1, \dots, \alpha_n)^{\mathbf{t}} \in \mathbf{C}^{\mathbf{n}}, \mathbf{t} \in (\delta, \infty).$$

In [2] we give the following result:

Theorem 2.2. *Assume that, for some $s \in (0,1)$, the conditions* **(A.1)** *and* **(A.2)** *hold. Then*

$$[X_{\mathcal{K}}, Y]_s = [X, Y]_{s,\mathcal{K}}.$$

For completness we include the proof.

Proof. Let s be fixed in $(0,1)$. Since $X_{\mathcal{K}}$ is dense in both these spaces, in order to prove (2.12) it is enough to find, for a fixed s, positive constants c_1 and c_2 such that

$$c_1\|u\|_{[X,Y]_s} \leq \|u\|_{[X_{\mathcal{K}},Y]_s} \leq c_2\|u\|_{[X,Y]_s} \quad \text{for all } u \in X_{\mathcal{K}}. \qquad (2.17)$$

The function under the integral sign in (2.15) is nonnegative, so the lower inequality of (2.17) is satisfied with $c_1 = 1$. For the upper part, we notice that, for $u \in X_{\mathcal{K}}$ and $w_{\mathcal{K}} := (I + t^2 S_{\mathcal{K}}^{-1})^{-1} u$

$$(w_{\mathcal{K}}, u)_Y = \left((I + t^2 S_{\mathcal{K}}^{-1})^{-1} u, u\right)_Y = (u, u)_Y - t^2 \left(S_{\mathcal{K}}^{-1}(I + t^2 S_{\mathcal{K}}^{-1})^{-1} u, u\right)_Y$$
$$\leq (u, u)_Y \leq c(s) \|u\|_{[X,Y]_s}^2$$

It was proved in [2] (Theorem 2.1) that

$$(w_{\mathcal{K}}, u)_Y = (w, u)_Y + t^{-2} \left\langle M_t^{-1} d, d \right\rangle. \tag{2.18}$$

Then, using (2.11), (2.18) and the above estimate, we have that for any positive number δ,

$$\|u\|_{[X_{\mathcal{K}},Y]_s}^2 \leq c(\delta, s) \|u\|_{[X,Y]_s}^2 + \int_\delta^\infty t^{-2s+1} (w_{\mathcal{K}}, u)_Y^2 \, dt$$
$$\leq c(\delta, s) \|u\|_{[X,Y]_s}^2 + \int_\delta^\infty t^{-2s+1} (w, u)_Y^2 \, dt$$
$$+ \int_\delta^\infty t^{-2s+1} \left\langle M_t^{-1} d, d \right\rangle \, dt.$$

Hence the upper inequality of (2.17) is satisfied if one can find a positive δ and $c = c(\delta)$ such that

$$\int_\delta^\infty t^{-2s+1} \left\langle M_t^{-1} d, d \right\rangle dt \leq c \|u\|_{[X,Y]_s}^2 \quad \text{for all } u \in X_{\mathcal{K}}. \tag{2.19}$$

From **(A.2)**, there exist $\delta > 0$ and $\gamma > 0$ such that

$$\left\langle M_t^{-1} \alpha, \alpha \right\rangle \leq \gamma \sum_{i=1}^n |\alpha_i|^2 \, (\varphi_i, \varphi_i)_{X,t}^{-1}$$

for all $\alpha = (\alpha_1, \ldots, \alpha_n)^{\mathrm{t}} \in \mathbf{C}^{\mathrm{n}}, t \in (\delta, \infty)$. In particular, for $\alpha_i = (u, \varphi_i)_{X,t}$, $i = 1, \ldots, n$, we obtain

$$\left\langle M_t^{-1} d, d \right\rangle \leq \gamma \sum_{i=1}^n \frac{|(u, \varphi_i)_{X,t}|^2}{(\varphi_i, \varphi_i)_{X,t}} \quad \text{for all } t \in (\delta, \infty), u \in X_{\mathcal{K}},$$

where $d = (d_1, \ldots, d_n)^t$. Thus, using the above estimate, (2.16) and **(A.1)** we have

$$\int_\delta^\infty t^{-2s+1} \langle M_t^{-1} d, d\rangle \, dt \leq \gamma \sum_{i=1}^n \int_\delta^\infty t^{-2s+1} \frac{|(u, \varphi_i)_{X,t}|^2}{(\varphi_i, \varphi_i)_{X,t}} dt$$
$$\leq \gamma \sum_{i=1}^n \int_0^\infty t^{-2s+1} \frac{|(u, \varphi_i)_{X,t}|^2}{(\varphi_i, \varphi_i)_{X,t}} dt$$
$$\leq \gamma c_s^{-2} \sum_{i=1}^n \|u\|^2_{[X_{\varphi_i}, Y]_s} \leq \gamma c_s^{-2} n \|u\|^2_{[X,Y]_s}$$

Finally, (2.19) holds, and the result is proved. □

Remark 2.3. *By Lemma 2.3, the space $X_{\mathcal{K}}$ is dense in $[X, Y]_s$ if and only if the functionals L_φ, $\varphi \in \mathcal{K}$ are not bounded in the topology induced by $[X, Y]_s$.*

3. Interpolation between subspaces of $H^\beta(R^N)$ and $H^\alpha(R^N)$.

In this section we give a simplified proof of the main interpolation result presented in [12]. An extension to the case when the subspace of interpolation has finite codimension bigger than one is also considered.

Let $\alpha \in R$ and let $H^\alpha(R^N)$ be defined by means of the Fourier transform. For a smooth function u with compact support in R^N, the Fourier transform $\hat{u}$ is defined by

$$\hat{u}(\xi) = (2\pi)^{-N/2} \int u(x) e^{-ix\xi} \, dx,$$

where the integral is taken over the whole R^N. For u and v smooth functions the
α -inner product is defined by

$$< u, v >_\alpha = \int (1 + |\xi|^2)^\alpha \, \hat{u}(\xi) \overline{\hat{v}(\xi)} \, d\xi.$$

The space $H^\alpha(R^N)$ is the closure of smooth functions in the norm induced by the α -inner product. For α, β real numbers ($\alpha < \beta$), and $s \in [0, 1]$ it is easy to check, using Remark 2.1, that

$$\Big[H^\beta(R^N), H^\alpha(R^N)\Big]_s = H^{s\alpha+(1-s)\beta}(R^N).$$

For $\varphi \in H^\beta(R^N)$, we are interested in determining the validity of the formula

$$\Big[H^\beta_\varphi(R^N), H^\alpha(R^N)\Big]_s = \Big[H^\beta(R^N), H^\alpha(R^N)\Big]_{s,\varphi}. \qquad (3.1)$$

For certain functions φ the problem is studied by Kellogg in [12]. Next, we give a new proof of Kellogg's result concerning (3.1) and extend it to the case when $H^{\beta}_{\varphi}(R^N)$ is replaced by a subspace of finite codimension. First, we consider the case when $0 = \alpha < \beta$. The operator S, associated with the pair $X = H^{\beta}(R^N)$, $Y = H^0(R^N) = L^2(R^N)$, is given by

$$\widehat{Su} = \mu^{2\beta}\hat{u}, \quad u \in D(S) = H^{2\beta}(R^N),$$

where $\mu(\xi) = (1+|\xi|^2)^{\frac{1}{2}}$, $\xi \in R^N$. For the remaining part of this chapter, H^{β} denotes the space $H^{\beta}(R^N)$ and $\hat{H}^{\beta}$ is the space $\{\hat{u}\,|u \in H^{\beta}\}$. For $\hat{u}, \hat{v} \in \hat{H}^{\beta}$, we define the inner product and the norm by

$$(\hat{u},\hat{v})_{\beta} = \int \mu^{2\beta}\hat{u}\bar{\hat{v}}\,d\zeta, \quad ||\hat{u}||_{\beta} = (\hat{u},\hat{u})_{\beta}^{1/2}.$$

To simplify the notation, we denote the the inner products $(\cdot,\cdot)_0$ and $<\cdot,\cdot>_0$ by $(\cdot,\cdot)$ and $<\cdot,\cdot>$, respectively. The norm $||\cdot||_0$ on H^0 or $\hat{H}^0$ is simply $||\cdot||$. Let $\phi \in \hat{H}^{\beta}$ be such that for some constants $\epsilon > 0$ and $c > 0$,

$$\begin{cases} |\phi(\xi) - b(\omega)\rho^{-\frac{N}{2}-2\beta+\alpha_0}| < c\rho^{-\frac{N}{2}-2\beta+\alpha_0-\epsilon} & \text{for all } \rho > 1 \\ 0 < \alpha_0 < \beta, \end{cases} \tag{3.2}$$

where $\rho \geq 0$ and $\omega \in S^{N-1}$ (the unit sphere of R^N) are the spherical coordinates of $\xi \in R^N$, and where $b(\omega)$ is a bounded measurable function on S^{N-1}, which is non zero on a set of positive measure.

Remark 3.1. *From the assumption* (3.2) *about ϕ and by using Lemma 2.3, we have that*

$$\hat{H}^{\beta}_{\phi} \text{ is dense in } \hat{H}^{\alpha} \text{ if and only if } \alpha \leq \alpha_0. \tag{3.3}$$

Theorem 3.1. *(Kellogg) Let $\varphi \in H^{\beta}$ be such that its Fourier transform ϕ satisfies* (3.2)*, and let $\theta_0 = \alpha_0/\beta$. Then*

$$\left[H^{\beta}_{\varphi}, H^0\right]_s = \left[H^{\beta}, H^0\right]_{s,\varphi}, \quad 0 \leq s \leq 1,\ 1-s \neq \theta_0, \tag{3.4}$$

Proof. From the way we defined $<\cdot,\cdot>_{\beta}$, (3.4) is equivalent to

$$\left[\hat{H}^{\beta}_{\phi}, \hat{H}^0\right]_s = \left[\hat{H}^{\beta}, \hat{H}^0\right]_{s,\varphi}, \quad 0 \leq s \leq 1,\ 1-s \neq \theta_0. \tag{3.5}$$

Following the proof of Theorem 2.2, we see that in order to prove (3.5), it is enough to verify (2.19) for some positive constants $c = c(s)$ and δ. Using (2.16), the problem reduces to

$$\int_{\delta}^{\infty} t^{-(2s+1)}\frac{|(\hat{u},\phi)_{X,t}|^2}{(\phi,\phi)_{X,t}}dt \leq c||\hat{u}||^2_{[X,Y]_s} \quad \text{for all } \hat{u} \in X_{\phi},$$

where $X = \hat{H}^{\beta}$ and $Y = \hat{H}^{0}$. Denoting $1 - s = \theta$ and $\Phi(t) = (\phi, \phi)_{X,t}$, this becomes

$$I := \int_{\delta}^{\infty} t^{2\theta-3} \frac{\left|\left(\frac{\mu^{4\beta}\hat{u}}{\mu^{2\beta}+t^2}, \phi\right)\right|^2}{\left(\frac{\mu^{4\beta}\phi}{\mu^{2\beta}+t^2}, \phi\right)} dt \leq c\|\hat{u}\|_{\theta\beta}^2 \quad \text{for all } \hat{u} \in \hat{H}_{\phi}^{\beta}. \tag{3.6}$$

Using (3.2) it is easy to see that, for a large enough $\delta \geq 1$

$$\left(\frac{\mu^{4\beta}\phi}{\mu^{2\beta}+t^2}, \phi\right) \geq ct^{2(\theta_0-1)} \qquad \text{for all } t \geq \delta, \tag{3.7}$$

and (3.2) also implies that

$$|\phi(\xi)| < c|\rho|^{-\frac{N}{2}-2\beta+\alpha_0} \quad \text{for} \quad |\xi| > 1. \tag{3.8}$$

Before we start estimating I, let us observe that by using spherical coordinates

$$\|\hat{u}\|_{\theta\beta}^2 = \int_0^{\infty} U^2(\rho)\, d\rho, \quad \hat{u} \in \hat{H}_{\phi}^{\beta}, \tag{3.9}$$

where

$$U(\rho) := \mu(\rho)^{\theta\beta} \rho^{\frac{N-1}{2}} \left(\int\limits_{|\xi|=1} |\hat{u}(\rho,\omega)|^2\, d\omega \right)^{1/2}, \quad \mu(\rho) = (1+\rho^2)^{1/2}.$$

First, we consider the case $0 < \theta < \theta_0$ and set $\theta_1 := \theta_0 - \theta$. For $\hat{u} \in \hat{H}_{\phi}^{\beta}$ we have

$$\left|\left(\frac{\mu^{4\beta}\hat{u}}{\mu^{2\beta}+t^2}, \phi\right)\right|^2 = t^4 \left|\left(\frac{\mu^{2\beta}\hat{u}}{\mu^{2\beta}+t^2}, \phi\right)\right|^2.$$

Thus, by this observation and (3.7) we get

$$I \leq c\int_{\delta}^{\infty} t^{3-2\theta_1} \left(\int \frac{\mu(\xi)^{2\beta}}{\mu(\xi)^{2\beta}+t^2} |\hat{u}(\xi)\phi(\xi)|\, d\xi \right)^2 dt.$$

Then,

$$I_1 = \int_{\delta}^{\infty} t^{3-2\theta_1} \left(\int\limits_{|\xi|<1} \frac{\mu(\xi)^{2\beta}}{\mu(\xi)^{2\beta}+t^2} |\hat{u}(\xi)\phi(\xi)|\, d\xi \right)^2 dt$$

$$\leq c\int_{\delta}^{\infty} \frac{t^{3-2\theta_1}}{t^4} \left(\int\limits_{|\xi|<1} |\hat{u}(\xi)\phi(\xi)|\, d\xi \right)^2 dt \leq c\int_{\delta}^{\infty} t^{-(1+2\theta_1)}\, dt\, \|\hat{u}\|^2\, \|\phi\|^2$$

$$\leq c(\theta)\|\hat{u}\|_{\theta\beta}^2.$$

On the other hand, by Fubini's theorem, we have

$$I_2 = \int_\delta^\infty t^{3-2\theta_1} \Bigg(\int\limits_{|\xi|>1} \frac{\mu(\xi)^{2\beta}}{\mu(\xi)^{2\beta} + t^2} |\hat{u}(\xi)\phi(\xi)| \, d\xi \Bigg)^2 dt$$

$$= \int_\delta^\infty t^{3-2\theta_1} \Bigg(\int\limits_{|\xi|>1} \frac{\mu(\xi)^{2\beta}}{\mu(\xi)^{2\beta} + t^2} |\hat{u}(\xi)\phi(\xi)| \, d\xi \Bigg) \Bigg(\int\limits_{|\eta|>1} \frac{\mu(\eta)^{2\beta}}{\mu(\eta)^{2\beta} + t^2} |\hat{u}(\eta)\phi(\eta)| \, d\eta \Bigg) dt$$

$$= \int\limits_{|\xi|>1} \int\limits_{|\eta|>1} |\hat{u}(\xi)\hat{u}(\eta)\phi(\xi)\phi(\eta)| \, \big(\mu(\xi)\mu(\eta)\big)^{2\beta} \int_\delta^\infty \frac{t^{3-2\theta_1}}{\big(\mu(\xi)^{2\beta} + t^2\big)\big(\mu(\eta)^{2\beta} + t^2\big)} \, dt \, d\eta \, d\xi.$$

To estimate the last integral we use the formula

$$\int_0^\infty \frac{t^{3-2\theta}}{(a+t^2)(b+t^2)} dt = \frac{1}{\mathbf{c}_\theta^2} \frac{a^{1-\theta} - b^{1-\theta}}{a-b}, \quad 0 < \theta < 2, \ \theta \neq 1, \ a, b > 0. \tag{3.10}$$

The integral can be calculated by standard complex analysis tools. If $a = b$, then the right side of the above identity is replaced by $\frac{1-\theta}{\mathbf{c}_\theta^2} a^{-\theta}$. Next, by using (3.10), (3.8) and spherical coordinates $\xi = (\rho, \omega)$, $\eta = (r, \rho)$, we obtain

$$I_2 \leq c(\theta) \int_1^\infty \int_1^\infty (\mu(r)\mu(\rho))^{2\beta - \beta\theta} (r\rho)^{-\frac{1}{2} - 2\beta + \alpha_0} R_{1-\theta_1}(\mu(r)^{2\beta}, \mu(\rho)^{2\beta}) \, U(r)U(\rho) \, d\rho \, dr,$$

where for $\alpha \in (0, 1)$, $x > 0$, $y > 0$, we denote

$$R_\alpha(x, y) = \begin{cases} \frac{x^\alpha - y^\alpha}{x - y}, & for \ x \neq y \\ \alpha x^{\alpha - 1}, & for \ x = y. \end{cases}$$

The function $x \to R_\alpha(x, y)$ is decreasing on $(0, \infty)$ for each $y \in (0, \infty)$ and it is symmetric with respect to x and y.

Using this observation, we get

$$I_2 \le c(\theta) \int_1^\infty \int_1^\infty (r\rho)^{-\frac{1}{2}+\beta\theta_1} R_{1-\theta_1}(r^{2\beta}, \rho^{2\beta})\, U(\rho)U(r)\; dr\, d\rho$$

$$\le c(\theta) \int_0^\infty \int_0^\infty K(r,\rho)U(r)U(\rho)\; dr\, d\rho,$$

where

$$K(r,\rho) = (r\rho)^{-\frac{1}{2}+\beta\theta_1} R_{1-\theta_1}(r^{2\beta}, \rho^{2\beta}). \tag{3.11}$$

In order to estimate the last integral, we apply the following lemma.

Lemma 3.1. *(Schur) Suppose $K(x,y)$ is nonnegative, symmetric and homogeneous of degree -1, and f, g are nonnegative measurable functions on $(0,\infty)$. Assume that*

$$k = \int_0^\infty K(1,x)x^{-\frac{1}{2}}\, dx \;<\; \infty.$$

Then

$$\int_0^\infty \int_0^\infty K(x,y)f(x)g(y)\; dx\, dy \le k\Big(\int_0^\infty f(x)^2\, dx\Big)^{\frac{1}{2}} \Big(\int_0^\infty g(y)^2\, dy\Big)^{\frac{1}{2}}. \tag{3.12}$$

We will prove this lemma later. For the moment, we see that the function $K(x,y)$, given by (3.11), is homogeneous of degree -1, and satisfies

$$k = \int_0^\infty K(x,1)x^{-\frac{1}{2}}\, dx \;<\; \infty.$$

Indeed

$$k = \int_0^\infty x^{-1+\beta\theta_1} \frac{x^{2\beta(1-\theta_1)}-1}{x^{2\beta}-1}\, dx$$

$$\overset{x^\beta = t}{=} \beta \int_0^\infty \frac{t^{1-\theta_1} - t^{\theta_1-1}}{t^2-1}\, dt \;<\; \infty,\; \textit{for } 0<\theta_1<1.$$

By Lemma 3.1,

$$I_2 \le c(\theta) \int_0^\infty U^2(\rho)\, d\rho \;\le\; c(\theta)\|\hat{u}\|^2_{\beta\theta}$$

and by combining the estimates I_1 and I_2, we obtain (3.6).

Let us consider now the case $\theta_0 < \theta < 1$, and let $\theta_1 = \theta - \theta_0$. Then, by using (3.7), we have

$$I \leq c\int_\delta^\infty t^{2\theta_1 - 1}\left(\int \frac{\mu(\xi)^{4\beta}}{\mu(\xi)^{2\beta}+t^2}|\hat{u}(\xi)\phi(\xi)|\,d\xi\right)^2 dt.$$

The remaining part of the proof is very similar to the proof of the first case. The theorem is proved. □

Proof of Lemma 3.1. By Fubini's theorem, it follows

$$\begin{aligned}
\int_0^\infty\int_0^\infty K(x,y)f(x)g(y)\,dx\,dy &= \int_0^\infty f(x)\left(\int_0^\infty K(x,y)g(y)\,dy\right)dx \\
= \int_0^\infty f(x)\int_0^\infty xK(x,xt)g(xt)\,dt\,dx &= \int_0^\infty f(x)\int_0^\infty K(1,t)g(xt)\,dt\,dx \\
= \int_0^\infty K(1,t)\int_0^\infty f(x)g(xt)\,dx\,dt & \\
\leq \int_0^\infty K(1,t)\left(\int_0^\infty f(x)^2\,dx\right)^{\frac{1}{2}}\left(\int_0^\infty g(xt)^2\,dx\right)^{\frac{1}{2}} dt & \\
\leq \int_0^\infty K(1,t)t^{-\frac{1}{2}}\,dt\left(\int_0^\infty f(x)^2\,dx\right)^{\frac{1}{2}}\left(\int_0^\infty g(x)^2\,dx\right)^{\frac{1}{2}}. &
\end{aligned}$$

Next we prepare for the generalization of the previous result.

Let $\phi_1, \phi_2, \ldots, \phi_n \in \hat{H}^\beta(R^N)$ such that for some constants $\epsilon > 0$ and $c > 0$ we have

$$\begin{cases} |\phi_i(\xi) - \tilde{\phi}_i(\xi)| < c\rho^{-\frac{N}{2}-2\beta+\alpha_i-\epsilon} \; for \; |\xi| > 1 \\ 0 < \alpha_i < \beta, \; i = 1, \ldots, n, \end{cases} \tag{3.13}$$

where

$$\tilde{\phi}_i(\xi) = b_i(\omega)\rho^{-\frac{N}{2}-2\beta+\alpha_i}, \; \xi = (\rho, \omega),$$

and $b_i(\cdot)$ is a bounded measurable function on S^{N-1}, which is non zero on a set of positive measure.

Define

$$\Phi_{ij}(t) = \left(\frac{\mu^{4\beta}\phi_i}{\mu^{2\beta}+t^2}, \phi_j\right), \quad \tilde{\phi}_{ij}(t) = \left(\frac{|\xi|^{4\beta}\tilde{\phi}_i}{|\xi|^{2\beta}+t^2}, \tilde{\phi}_j\right), \quad \theta_i = \frac{\alpha_i}{\beta},$$

$$[\tilde{\phi}_i, \tilde{\phi}_j] := \frac{1}{\beta}(b_i, b_j)_\sigma \int_0^\infty \frac{x^{\theta_i}x^{\theta_j}}{x(x^2+1)}\,dx, \; i, \; j = 1, 2, \ldots, n,$$

where $(\cdot,\cdot)_\sigma$ is the inner product on $L^2(S^{N-1})$.

Clearly, $[\cdot,\cdot]$ is an inner product on $span\{\tilde{\phi}_i \mid i = 1, 2, \ldots, n\}$.

Lemma 3.2. *With the above setting we have*

$$\tilde{\Phi}_{ij}(t) = [\tilde{\phi}_i, \tilde{\phi}_j] t^{\theta_i+\theta_j-2} \tag{3.14}$$

$$|\Phi_{ij}(t) - \tilde{\Phi}_{ij}(t)| \leq ct^{\theta_i+\theta_j-2-\eta}\ ,\ t > \delta, \tag{3.15}$$

for some constants $c > 0$, $\eta > 0$ *and* $\delta \geq 1$.

Proof. By using spherical coordinates, we have

$$\tilde{\Phi}_{ij}(t) = \int \frac{|\xi|^{4\beta}}{|\xi|^{2\beta}+t^2} \tilde{\phi}_i \bar{\tilde{\phi}}_j\, d\xi = \int_0^\infty \frac{\rho^{\alpha_i+\alpha_j-1}}{\rho^{2\beta}+t^2}\, d\rho \int_{|\xi|=1} b_i(\omega)\overline{b_j(\omega)}\, d\omega.$$

The change of variable $\rho^\beta = tx$ in the first integral completes the proof of (3.14). The proof of (3.15) is straightforward. □

Theorem 3.2. *Let* $\varphi_1, \varphi_2, \ldots, \varphi_n \in H^\beta$ *be such that the corresponding Fourier transforms* $\phi_1, \phi_2, \ldots, \phi_n$ *satisfy* (3.13) *and in addition, the functions* $\tilde{\phi}_1$, $\tilde{\phi}_2$, $\ldots$, $\tilde{\phi}_n$ *are linearly independent.*

Let $\mathcal{K} = span\{\varphi_1, \varphi_2, \ldots, \varphi_n\}$. *Then*

$$[H^\beta_{\mathcal{K}}, H^0]_s = [H^\beta, H^0]_{s,\mathcal{K}}\ ,\ (1-s)\beta \neq \alpha_i,\ for\ i = 1, 2, \ldots, n.$$

Proof. We apply the Theorem 2.2 for $X = H^\beta$, $Y = H^0$, $\mathcal{K} = span\{\varphi_1, \ldots, \varphi_n\}$ and s such that $(1-s)\beta \neq \alpha_i$, $i = 1, 2, \ldots, n$. By using the hypothesis (3.13) and Theorem 3.1, we get

$$[H^\beta_{\varphi_i}, H^0]_s = [H^\beta, H^0]_{s,\varphi_i},\ for\ i = 1, 2, \ldots, n.$$

So **(A1)** is satisfied. In order to verify the condition **(A2)**, we first observe that $(M_t)_{ij} = \Phi_{ij}(t)$. By denoting $D_t = diag(M_t)$, the condition **(A2)** can be written as follows:

There are $\delta > 0$ and $\gamma > 0$ such that

$$M_t - \gamma D_t \geq 0,\ \ for\ all\ t \in (\delta, \infty),$$

where for a square matrix A, $A \geq 0$ means that A is a nonnegative definite matrix. From the previous lemma we obtain the behavior of $(M_t)_{ij}$ for t large:

$$(M_t)_{ij} = \left([\tilde{\phi}_i, \tilde{\phi}_j] + f_{ij}(t)\right) t^{\theta_i-1} t^{\theta_j-1}$$

where $|f_{ij}(t)| < ct^{-\eta}$, for $t > \delta$. Denote $\tilde{M}_t$, $\tilde{M}$ the n x n matrices defined by

$$(\tilde{M}_t)_{ij} = [\tilde{\phi}_i, \tilde{\phi}_j] + f_{ij}(t),\ (\tilde{M})_{ij} = [\tilde{\phi}_i, \tilde{\phi}_j]$$

and let $\tilde{D}_t = diag\tilde{M}_t$, $\tilde{D} = diag\tilde{M}$. Next, for $z = (z_1, z_2, \ldots, z_n) \in C^n$, we have

$$\langle (M_t - \gamma D_t)z, z\rangle = \langle (\tilde{M}_t - \gamma \tilde{D}_t)z_t, z_t\rangle$$

where $< \cdot, \cdot >$ is the inner product on C^n and $(z_t)_i = z_i\, t^{\theta_i - 1}$, $i = 1, 2, \ldots, n$.

Hence, the condition **(A2)** is satisfied if one can find $\gamma > 0$, $\delta > 0$, such that

$$\tilde{M}_t - \gamma \tilde{D}_t \geq 0, \qquad \text{for all } t \in (\delta, \infty).$$

On the other hand, since $\tilde{\phi}_1, \tilde{\phi}_2, \ldots, \tilde{\phi}_n$ are linearly independent, $\tilde{M}$ is a symmetric positive definite matrix on C^n and

$$\lim_{\gamma \searrow 0, t \to \infty} (\tilde{M}_t - \gamma \tilde{D}_t) = \tilde{M}.$$

Therefore, there are $\gamma > 0$, $\delta > 0$ such that $\tilde{M}_t - \gamma \tilde{D}_t > 0$, for all $t \in (\delta, \infty)$, and **(A2)** holds. The result is proved by applying Theorem 2.2. □

The corresponding case of interpolation between subspaces of H^β of finite codimensions and H^α, where α, β are real numbers, $\alpha < \beta$, is a direct consequence of the previous theorem.

Let $\alpha < \beta$ and $\varphi_1, \varphi_2, \ldots, \varphi_n \in H^\beta$ be such that the corresponding Fourier transform $\phi_1, \phi_2, \ldots, \phi_n$ satisfy for some positive constants c and ϵ,

$$\begin{cases} |\phi_i(\xi) - \tilde{\phi}_i(\xi)| < c\rho^{-\frac{N}{2} - 2\beta + \gamma_i - \epsilon} \; for \; |\xi| > 1 \\ \alpha < \gamma_i < \beta, \; i = 1, \ldots, n, \end{cases} \tag{3.16}$$

where

$$\tilde{\phi}_i(\xi) = b_i(\omega)\rho^{-\frac{N}{2} - 2\beta + \gamma_i}, \; \xi = (\rho, \omega),$$

and $b_i(\cdot)$ is a bounded measurable function on S^{N-1}, which is non zero on a set of positive measure.

Theorem 3.3. *Let $\varphi_1, \varphi_2, \ldots, \varphi_n \in H^\beta$ be such that the corresponding Fourier transforms $\phi_1, \phi_2, \ldots, \phi_n$ satisfy* (3.16), *and in addition, the functions $\tilde{\phi}_1$, $\tilde{\phi}_2$, …, $\tilde{\phi}_n$ are linearly independent. Let $\L = span\{\varphi_1, \varphi_2, \ldots, \varphi_n\}$. Then*

$$[H^\beta_{\L}, H^\alpha]_s = [H^\beta, H^\alpha]_{s,\L}, \; s\alpha + (1-s)\beta \neq \gamma_i, \; for \; i = 1, 2, \ldots, n. \tag{3.17}$$

Furthermore, if $s\alpha + (1-s)\beta < \min\{\gamma_i, \; i = 1, 2, \ldots, n\}$, then

$$[H^\beta_{\L}, H^\alpha]_s = H^{s\alpha + (1-s)\beta}. \tag{3.18}$$

Proof. The first part follows from the main theorem 3.2 and the fact that $T : H^\alpha \to H^0$ defined by $\hat{Tu} = \mu^\alpha \hat{u},\ u \in H^\alpha$ is an isometry from H^α to $H^{\gamma-\alpha}$ for any $\gamma \in [\alpha, \beta]$.

Now let $s < \min\{\gamma_i,\ i = 1, 2, \ldots, n\}$. By the first part of the theorem, in order to prove (3.18) we need only to prove that $H^\beta_Ł$ is dense in $H^{s\alpha+(1-s)\beta}$. By Lemma 2.3, this is equivalent to proving that

$$\begin{cases} H^\beta \ni u \xrightarrow{\Lambda_\varphi} < u, \varphi >_\beta = (\hat{u}, \hat{\varphi})_\beta, \\ \textit{is not bounded in the topology of } H^{s\alpha+(1-s)\beta} \textit{ for all } \varphi \in Ł,\ \varphi \neq 0. \end{cases} \tag{3.19}$$

For a fixed $\varphi \in Ł$ we have $\hat{\varphi} = \sum_{i=1}^{n} c_i \phi_i$.

Since $\tilde{\phi}_1, \tilde{\phi}_2, \ldots, \tilde{\phi}_n$ are assumed to be linearly independent, φ fails to be a "good" function (better than φ_i, $i = 1, 2, \ldots, n$). More precisely, the asymptotic expansion at infinity of $\hat{\varphi}$ is of the same type (except maybe a different b-part) with one of the functions $\tilde{\phi}_1, \tilde{\phi}_2, \ldots, \tilde{\phi}_n$. Thus, it is enough to check (3.19) for $\varphi \in \{\varphi_1, \varphi_2, \ldots, \varphi_n\}$.

Assuming that Λ_{φ_i} is continuous, it implies that

$$(\hat{u}, \phi_i)_\beta = (\hat{u}, f_i)_{s\alpha+(1-s)\beta}, u \in H^\beta,$$

for a function $f_i \in \hat{H}^{s\alpha+(1-s)\beta}$.
Thus, by using the density of H^β in H^s, for $s < \beta$, we get that $f_i = \mu^{2\beta} \mu^{-2(s\alpha+(1-s)\beta)} \phi_i$.

On the other hand,

$$\begin{aligned} \int \mu^{2(s\alpha+(1-s)\beta)} |f_i|^2 \, d\xi &= \int \mu^{2\beta-2s\alpha+2s\beta} |\phi_i|^2 \, d\xi \\ \geq c \int_\delta^\infty & \rho^{2\beta-2s\alpha+2s\beta} \rho^{-N-4\beta+2\gamma_i} \rho^{N-1} d\rho \\ = c \int_\delta^\infty & \rho^{-1+2(\gamma_i-(s\alpha+(1-s)\beta))} d\rho = \infty \end{aligned}$$

for $s\alpha + (1 - s)\beta < \min\{\gamma_i,\ i = 1, 2, \ldots, n\}$. This completes the proof. □

4. Shift theorem for the Biharmonic operator on polygonal domains.

Let Ω be a polygonal domain in R^2 with boundary $\partial\Omega$. Let $\partial\Omega$ be the polygonal arc $P_1 P_2 \cdots P_m P_1$. At each point P_j, we denote the measure of the angle P_j (measured from inside Ω) by ω_j. Let $\omega := \max\{\omega_j : j = 1, 2, \ldots, m\}$.

We consider the biharmonic problem Given $f \in L^2(\Omega)$, find u such that

$$\begin{cases} \Delta^2 u = f \ \text{ in } \Omega, \\ u = 0 \ \text{ on } \partial\Omega, \\ \frac{\partial u}{\partial n} = 0 \ \text{ on } \partial\Omega. \end{cases} \tag{4.1}$$

Let $V = H_0^2(\Omega)$ and

$$a(u,v) := \sum_{1 \leq i,j \leq 2} \int_\Omega \frac{\partial^2 u}{\partial x_i \partial x_j} \frac{\partial^2 v}{\partial x_i \partial x_j}\, dx, \ u, v, \in V.$$

The bilinear form a defines a scalar product on V and the induced norm is equivalent to the standard norm on $H_0^2(\Omega)$. The variational form of (4.1) is : Find $u \in V$ such that

$$a(u,v) = \int_\Omega fv\, dx \quad \text{for all } v \in V. \tag{4.2}$$

Clearly, if u is a variational solution of (4.2), then one has $\Delta^2 u = f$ in the sense of distributions and because $u \in H_0^2(\Omega)$, the homogeneous boundary conditions are automatically fulfilled. As done in [2], the problem of deriving the shift estimate on Ω can be localized by a partition of unity so that only sectors domains or domains with smooth boundaries need to be considered. If Ω is a smooth domain, then it is known that the solution u of (4.2) satisfies

$$\|u\|_{H^4(\Omega)} \leq c\|f\|, \qquad \text{for all } f \in L^2(\Omega),$$

and

$$\|u\|_{H^2(\Omega)} \leq c\|f\|_{H^{-2}(\Omega)}, \qquad \text{for all } f \in H^{-2}(\Omega).$$

Interpolating these two inequalities yields

$$\|u\|_{2+2s} \leq c\|f\|_{-2+2s}, \qquad \text{for all } f \in H^{-2+2s}(\Omega), \ \ 0 \leq s \leq 1.$$

So we have the shift theorem for all $s \in [0,1]$. Let us consider the case of a sector domain. The threshold, s_0, below which the shift estimate for a polygonal domain holds is given, as in the Poisson problem, by the largest internal angle ω of the polygon. Thus, it is enough to consider the domain $S\omega$ defined by

$$S_\omega = \{(r,\theta),\ 0 < r < 1, -\omega/2 < \theta < \omega/2\}.$$

We associate to (4.1) and $\Omega = S_\omega$, the characteristic equation

$$\sin^2(z\omega) = z^2 \sin^2 \omega. \tag{4.3}$$

In order to simplify the exposition of the proof, we assume that

$$\sin\sqrt{\frac{\omega^2}{\sin\omega^2}-1}\neq\sqrt{1-\frac{\sin\omega^2}{\omega^2}} \tag{4.4}$$

and

$$Re z \neq 2 \text{ for any solution } z \text{ of (4.3).}$$

The restriction (4.4) assures that the equation (4.3) has only simple roots. Let $z_1, z_2, \dots, z_n$ be all the roots of (4.3) such that $0 < Re(z_j) < 2$. It is known (see [7], [10], [13], [17]) that the solution u of (4.2) can be written as

$$u = u_R + \sum_{j=1}^{n} k_j S_j, \tag{4.5}$$

where $u_R \in H^4(\Omega)$ and for $j = 1, 2, \dots, n$, we have $S_j(r,\theta) = r^{1+z_j} u_j(\theta)$, u_j is smooth function on $[-\omega/2, \omega/2]$ such that $u_j(-\omega/2) = u_j(\omega/2) = u_j'(-\omega/2) = u_j'(\omega/2) = 0$, $k_j = c_j \int_\Omega f\varphi_j\, dx$ and c_j is nonzero and depends only on ω. The function φ_j is called the dual singular function of the singular function S_j and $\varphi_j(r,\theta) = \eta(r)\, r^{1-z_j} u_j(\theta) - w_j$, where $w_j \in V$ is defined for a smooth truncation function η to be the solution of (4.2) with $f = \Delta^2(\eta(r)\, r^{1-z_j} u_j(\theta))$. In addition,

$$\|u_R\|_{H^4(\Omega)} \leq c\|f\|, \qquad \text{for all } f \in L^2(\Omega). \tag{4.6}$$

Next, we define $\mathcal{K} = span\{\varphi_1, \varphi_2, \dots, \varphi_n\}$. As a consequence of the expansion (4.5) and the estimate (4.6) we have

$$\|u\|_{H^4(\Omega)} \leq c\|f\|, \qquad \text{for all } f \in L^2(\Omega)_{\mathcal{K}}. \tag{4.7}$$

Combining (4.7) with the standard estimate

$$\|u\|_{H^2(\Omega)} \leq c\|f\|_{H^{-2}(\Omega)}, \qquad \text{for all } f \in H^{-2}(\Omega),$$

we obtain, via interpolation

$$\|u\|_{[H^4(\Omega),H^2(\Omega)]_{1-s}} \leq c\|f\|_{[L^2(\Omega)_{\mathcal{K}},H^{-2}(\Omega)]_{1-s}}, \; s \in [0,1]. \tag{4.8}$$

Let $s_0 = \min\{Re(z_j) \mid j = 1, 2, \dots, n\}$. Then, we have

Theorem 4.1. *If* $0 < 2s < s_0$ *and* $\Omega = S_\omega$, *then*

$$[L^2(\Omega)_{\mathcal{K}}, H^{-2}(\Omega)]_{1-s} = [L^2(\Omega), H^{-2}(\Omega)]_{1-s}. \tag{4.9}$$

Proof. First we prove that there are operators E and R such that

$$E : L^2(\Omega) \longrightarrow L^2(R),\; E : H_0^2(\Omega) \longrightarrow H^2(R^2),$$

$$R : L^2(R^2) \longrightarrow L^2(\Omega),\ R : H^2(R^2) \longrightarrow H_0^2(\Omega)$$

are bounded operators, and $REu = u$, for all $u \in L^2(\Omega)$.
Indeed, E can be taken to be the extension by zero operator.

To define R, let $\eta = \eta(r)$ be a smooth function on $(0, \infty)$ such that $\eta(r) \equiv 1$ for $0 < r \leq 1$ and $\eta(r) \equiv 0$ for $r > 2$. Define $\alpha = \frac{\omega}{2}$, $a = \frac{\alpha}{\pi - \alpha}$ and

$$g_1(\theta) = \frac{\alpha - \pi}{\alpha}\theta + \pi,\ g_2(\theta) = \frac{\pi - \alpha}{\alpha^2}(\alpha - \theta)^2 + \alpha,\ \theta \in [0, \alpha].$$

Note that $g_i(0) = \pi$ and $g_i(\alpha) = \alpha$, $i = 1, 2$. For a smooth function u defined on R^2 we define $Ru := u_3$, where

Step 1. $u_1 = \eta u$.
Step 2. $u_2(r, \theta) = u_1(r, \theta) + 3u_1(1/r, \theta) - 4u_1(1/2 + 1/(2r), \theta),$
$r < 1,\ \theta \in [0, 2\pi).$
Step 3. For $0 < r < 1$

$$u_3(r,\theta) = \begin{cases} u_2(r,\theta) + au_2(r, g_1(\theta)) - (1+a)u_2(r, g_2(\theta)), \\ \qquad\qquad 0 \leq \theta < \omega/2, \\ u_2(r,\theta) + au_2(r, -g_1(-\theta)) - (1+a)u_2(r, -g_2(-\theta)), \\ \qquad\qquad -\omega/2 < \theta < 0. \end{cases}$$

One can check that, for $u \in H_0^2(R^2)$, $u_3 \in H_0^2(\Omega)$ and $REu = u$. The operator R can be extended by density to $L^2(R^2)$. The extended operator R satisfies all the desired properties.

Next, let ϕ_j be the Fourier transform of $E\varphi_j$, $j = 1, \ldots, n$. Using asymptotic expansion of integrals theory presented in the Appendix 1, we have that the functions
$\{E\varphi_j,\ j = 1, \ldots, n\}$ satisfy for some positive constants c and ϵ,

$$\begin{cases} |\phi_j(\xi) - \tilde{\phi}_j(\xi)| < c\rho^{-1+(-2+s_j)-\epsilon}\ for\ |\xi| > 1 \\ -2 < -2 + s_i < 0,\ i = 1, \ldots, n, \end{cases} \tag{4.10}$$

where $s_j = Re(z_j)$ and

$$\tilde{\phi}_j(\xi) = b_i(\omega)\rho^{-1+(-2+s_j)},\ \xi = (\rho, \omega) \text{ in polar coordinates,}$$

and $b_j(\cdot)$ is a bounded measurable function on the unit circle, which is non zero on a set of positive measure. Thus, we have that the functions $\{E\varphi_j,\ j = 1, \ldots, n\}$ satisfy the hypothesis (3.16) of Theorem 3.3 with $N = 2$, $\beta = 0$, $\alpha = -2$ and $\gamma_j = -2 + s_j$, $j = 1, \ldots, n$. Denoting $Ł := span\{E\varphi_j,\ j = 1, \ldots, n\}$, by Theorem 3.3 applied with $1 - s$ instead of s, we have that

$$[L^2(R^2)_Ł, H^{-2}(R^2)]_{1-s} = [L^2(R^2), H^{-2}(R^2)]_{1-s} = H^{-2+2s}(R^2), \tag{4.11}$$

for $2s < s_0 := \min\{Re(z_j),\ j = 1, 2, \dots, n\}$.

Finally, using (4.11), the operators E, R and Lemma A.1 (adapted to the case when we work with subspaces of codimension $n > 1$), we conclude that (4.9) holds for $2s < s_0$.

From the estimate (4.8) and the interpolation result (4.9) we obtain

$$\|u\|_{2+2s} \leq c\|f\|_{-2+2s}, \qquad \text{for all } f \in H^{-2+s}(\Omega),\ \ 0 \leq 2s < s_0.$$

The above estimate still holds for the case when Ω is a polygonal domain and s_0 corresponds to the largest inner angle ω of the polygon. Figure 1 (see below) gives the graph of the function $\omega \to 2 + s_0(\omega)$ which represents the regularity threshold for the biharmonic problem in terms of the largest inner angle ω of the polygon. On the same graph we represent the the number of singular (dual singular) functions as function of $\omega \in (0, 2\pi)$. Note that if ω is bigger than 1.43π, which is an approximation for the solution in $(0, 2\pi)$ of the equation $\tan\omega = \omega$, the space $\mathcal{K}$ has the dimension six.

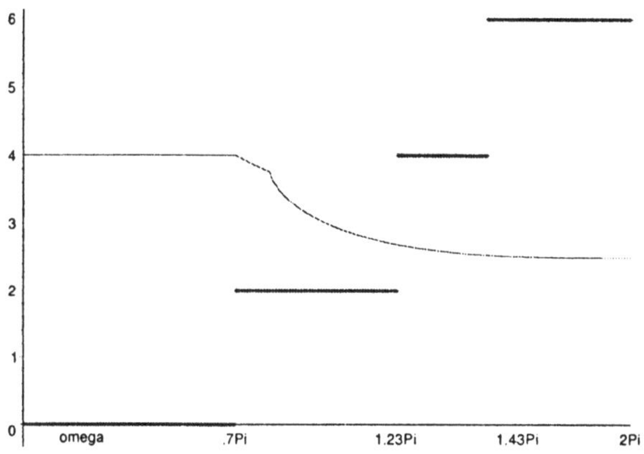

Figure 1. Regularity for the biharmonic problem.

Appendix: A. An interpolation result

Let $\Omega \subset \tilde{\Omega}$ be domains in R^2 and $V^1(\Omega)$, $V^1(\tilde{\Omega})$ be subspaces of $H^1(\Omega)$, $H^1(\tilde{\Omega})$, respectively. On $V^1(\Omega)$, $V^1(\tilde{\Omega})$ we consider inner products such that the induced norms are equivalent with the standard norms on $H^1(\Omega)$, $H^1(\tilde{\Omega})$, respectively. In addition, we assume that $V^1(\Omega)$, $V^1(\tilde{\Omega})$ are dense in $L^2(\Omega)$, $L^2(\tilde{\Omega})$, respectively. Let's denote the duals of $V^1(\Omega)$,

$V^1(\widetilde{\Omega})$ by $V^{-1}(\Omega)$, $V^{-1}(\widetilde{\Omega})$, respectively. We suppose that there are linear operators E and R such that

$$E : L^2(\Omega) \to L^2(\widetilde{\Omega}), \quad E : V^1(\Omega) \to V^1(\widetilde{\Omega}) \text{ are bounded operators,} \tag{A.1}$$

$$R : L^2(\widetilde{\Omega}) \to L^2(\Omega), \quad R : V^1(\widetilde{\Omega}) \to V^1(\Omega), \text{ are bounded operators,} \tag{A.2}$$

$$REu = u \qquad \text{for all } u \in L^2(\Omega). \tag{A.3}$$

Let $\psi \in L^2(\Omega)$, $\widetilde{\psi} = E\psi \in L^2(\widetilde{\Omega})$ and $\theta \in (0,1)$ be such that

$$L^2(\Omega)_\psi := \{u \in L^2(\Omega) : (u, \psi) = 0\} \text{ is dense in } [L^2(\Omega), V^{-1}(\Omega)]_\theta, \tag{A.4}$$

$$L^2(\widetilde{\Omega})_{\widetilde{\psi}} := \{u \in L^2(\widetilde{\Omega}) : (u, \widetilde{\psi}) = 0\} \text{ is dense in } V^{-1}(\widetilde{\Omega}), \tag{A.5}$$

$$[L^2(\widetilde{\Omega})_{\widetilde{\psi}}, V^{-1}(\widetilde{\Omega})]_\theta = [L^2(\widetilde{\Omega}), V^{-1}(\widetilde{\Omega})]_\theta. \tag{A.6}$$

Lemma A.1. *Using the above setting, assume that* (A.1)-(A.6) *are satisfied. Then,*

$$[L^2(\Omega)_\psi, V^{-1}(\Omega)]_\theta = [L^2(\Omega), V^{-1}(\Omega)]_\theta. \tag{A.7}$$

Proof. Using the duality , from (A.1)-(A.3) we obtain linear operators E^*, R^* such that

$$E^* : L^2(\widetilde{\Omega}) \to L^2(\Omega), \quad E^* : V^{-1}(\widetilde{\Omega}) \to V^{-1}(\Omega), \text{ are bounded operators,} \tag{A.8}$$

$$R^* : L^2(\Omega) \to L^2(\widetilde{\Omega}), \quad R^* : V^{-1}(\Omega) \to V^{-1}(\widetilde{\Omega}) \text{ are bounded operators,} \tag{A.9}$$

$$E^* R^* u = u \qquad \text{for all } u \in L^2(\Omega), \tag{A.10}$$

$$E^* \text{ maps } L^2(\widetilde{\Omega})_{\widetilde{\psi}} \text{ to } L^2(\Omega)_\psi, \tag{A.11}$$

$$R^* \text{ maps } L^2(\Omega)_\psi \text{ to } L^2(\widetilde{\Omega})_{\widetilde{\psi}}. \tag{A.12}$$

From (A.8) and (A.11), by interpolation, we obtain

$$\|E^* v_{[L^2(\Omega)_\psi, V^{-1}(\Omega)]_\theta}\| \le c \|v_{[L^2(\widetilde{\Omega})_{\widetilde{\psi}}, V^{-1}(\widetilde{\Omega})]_\theta}\| \qquad \text{for all } v \in L^2(\widetilde{\Omega})_{\widetilde{\psi}}. \tag{A.13}$$

For $u \in L^2(\Omega)_\psi$, let $v := R^* u$. Then, using (A.12), we have that $v \in L^2(\widetilde{\Omega})_{\widetilde{\psi}}$. Taking $v := R^* u$ in (A.13) and using (A.10), we get

$$\|u_{[L^2(\Omega)_\psi, V^{-1}(\Omega)]_\theta}\| \le c \|R^* u_{[L^2(\widetilde{\Omega})_{\widetilde{\psi}}, V^{-1}(\widetilde{\Omega})]_\theta}\| \qquad \text{for all } u \in L^2(\Omega)_\psi. \tag{A.14}$$

Also, from the hypothesis (A.6), we deduce that

$$\|R^* u_{[L^2(\widetilde{\Omega})_{\widetilde{\psi}}, V^{-1}(\widetilde{\Omega})]_\theta}\| \le c \|R^* u_{[L^2(\widetilde{\Omega}), V^{-1}(\widetilde{\Omega})]_\theta}\| \qquad \text{for all } u \in L^2(\Omega)_\psi. \tag{A.15}$$

From (A.9), again by interpolation, we have in particular

$$\|R^* u_{[L^2(\tilde{\Omega}),V^{-1}(\tilde{\Omega})]_\theta}\| \le c\|u_{[L^2(\Omega),V^{-1}(\Omega)]_\theta}\| \quad \text{for all } u \in L^2(\Omega)_\psi. \tag{A.16}$$

Combining (A.14)-(A.16), it follows that

$$\|u_{[L^2(\Omega)_\psi,V^{-1}(\Omega)]_\theta}\| \le c\|u_{[L^2(\Omega),V^{-1}(\Omega)]_\theta}\| \quad \text{for all } u \in L^2(\Omega)_\psi. \tag{A.17}$$

The reverse inequality of (A.17) holds because $L^2(\Omega)_\psi$ is a closed subspace of $L^2(\Omega)$. Thus, the two norms in (A.17) are equivalent for $u \in L^2(\Omega)_\psi$. From the assumption (A.4), $L^2(\Omega)_\psi$ is dense in both spaces appearing in (A.7). Therefore, we obtain (A.7). □

Remark A.1. *The proof does not change if we consider $\Omega \subset \tilde{\Omega}$ to be domains in R^N and H^1 is replaced by any other Sobolev space of positive integer order k.*

Appendix: B. Asymptotic expansion for the Fourier integrals

For a more general presentation of asymptotic expansion of functions defined by integrals see [4], [8], [19].

Integrals of the form

$$\int_a^b e^{ixt} f(t)\, dt,$$

are called Fourier integrals. We shall present the asymptotic behavior as $x \to \infty$ of the Fourier integrals for a particular type of function f. If ϕ and ψ are two real functions defined on the interval $I = (0, \infty)$ and ψ is a strictly positive function on I, we write $\phi = O(\psi)$ as $x \to \infty$ if ϕ/ψ is bounded on an interval $I = (\delta, \infty)$ for a positive δ, and $\phi = o(\psi)$ as $x \to \infty$ if $\lim\limits_{x\to\infty} \phi/\psi = 0$.

Theorem B.2. *Let ϕ be a continuously differentiable function on the interval $[a,b]$ and $\lambda \in (0,1)$.*

a) If $\phi(b) = 0$ then

$$\int_a^b e^{ixt}(t-a)^{\lambda-1}\phi(t)\, dt = -\Gamma(\lambda)\phi(a)e^{\frac{\pi}{2}i(\lambda-2)}e^{ixa}x^{-\lambda} + O(x^{-1}).$$

b) If $\phi(a) = 0$ then

$$\int_a^b e^{ixt}(b-t)^{\lambda-1}\phi(t)\, dt = \Gamma(\lambda)\phi(b)e^{-\frac{\pi}{2}i\lambda}e^{ixb}x^{-\lambda} + O(x^{-1}).$$

Here Γ is the Euler's gamma function.

Remark B.2. *The result holds for $\lambda = 1$ provided $O(x^{-1})$ is replaced by $o(x^{-1})$ in the above formulas.*

The proof of Theorem B.2 can be found in [8] Section 2.8.

Next we study the asymptotic behavior of the Fourier transforms of the dual singular functions which appear in Section 4. To this end, let $\eta = \eta(r)$ be a smooth real function on $[0, \infty)$ such that $\eta(r) \equiv 0$ for $r > 3/4$ and let $u = u(\theta)$ be a sufficiently smooth real function on $[0, 2\pi]$. For any non-zero $s \in (-1, 1)$ we define

$$u(x) = \eta(r)r^s u(\theta), \quad x = (r, \theta) \in R^2,$$

and

$$\Phi(\rho,\omega) = 2\pi\bar{\hat{u}}(\xi) = \int_{R^2} e^{ix\cdot\xi}u(x)\,dx, \quad \xi = (\rho,\omega) \in R^2,$$

where (r,θ) and (ρ,ω) are the polar coordinates of x and ξ, respectively. One can easily see that

$$\Phi(\rho,\omega) = \int_0^1\int_0^{2\pi} \eta(r)r^{1+s}u(\theta)e^{ir\rho\cos(\theta-\omega)}\,d\theta dr. \tag{B.1}$$

To study the asymptotic behavior of Φ for large ρ, we use the technique of [12] to reduce the double integral to a single integral. For a fixed ω, we consider the line $r\cos(\theta-\omega) = t$ in the x plane and denote by $l(t,\omega)$ the intersection of this line with the unit disk. Next, in the (r,t) variables the integral (B.1) becomes:

$$\Phi(\rho,\omega) = \int_{-1}^1 g(t)e^{it\rho}\,dt, \tag{B.2}$$

where

$$g(t) = \int_{l(t,\omega)} \frac{\eta(r)r^{1+s}}{\sqrt{r^2-t^2}}u(\theta)\,dr,$$

$\theta = \omega + \cos^{-1}(t/r)$, if $\theta \in [\omega,\omega+\pi]$ and $\theta = \omega - cos^{-1}(t/r)$, if $\theta \in [\omega-\pi,\omega]$. The function g is continuous differentiable on $[-1,1]$ and $g(-1) = g(1) = 0$. Thus, from (B.2) we have

$$\Phi(\rho,\omega) = \frac{i}{\rho}\int_{-1}^1 g'(t)e^{it\rho}\,dt \tag{B.3}$$

The function g can be described as

$$g(t) = \int_{|t|}^1 \frac{\eta(r)r^{1+s}}{\sqrt{r^2-t^2}}u(\omega+\cos^{-1}(t/r))\,dr + \int_{|t|}^1 \frac{\eta(r)r^{1+s}}{\sqrt{r^2-t^2}}u(\omega-\cos^{-1}(t/r))dr,$$

and the integral in (B.3) can be split in $\int_{-1}^0 + \int_0^1$. Thus, the function Φ is defined by a sum of four integrals. We will use Theorem B.2 in order to find the asymptotic behavior as $\rho \to \infty$ of each of the integrals. We shall present the estimate for only one of them.

Let $s \in (-1,0)$ be fixed and let h be the function defined by

$$h(t) = \int_t^1 \frac{\eta(r)r^{1+s}}{\sqrt{r^2-t^2}}u(\theta)\,dt,$$

where $\theta = \omega + cos^{-1}(t/r)$. We apply Theorem B.2 for the integral

$$\int_0^1 h'(t)e^{it\rho}\,dt. \tag{B.4}$$

To compute $h'(t)$ (by Leibnitz's formula) we set $x = r - t$ to rewrite h as

$$h(t) = \int_{x=0}^{1-t} \frac{\eta(x+t)(x+t)^{1+s}}{\sqrt{x}\sqrt{x+2t}}u(\theta)\,dx.$$

This leads to

$$h'(t) = \int_0^{1-t} \left[\left(\frac{(1+s)\eta(x+t)(x+t)^s}{\sqrt{x}\sqrt{x+2t}} + \frac{\eta'(x+t)(x+t)^{1+s}}{\sqrt{x}\sqrt{x+2t}} \right.\right.$$
$$\left.\left. - \frac{\eta(x+t)(x+t)^{1+s}}{\sqrt{x}(x+2t)^{3/2}} \right) u(\theta) - \frac{\eta(x+t)(x+t)^s}{x+2t} u'(\theta) \right] dx$$

Going back to the r variable, via the change $r = x + t$, we get

$$h'(t) = \int_t^1 \left[\left(\frac{(1+s)\eta(r)r^s}{\sqrt{r^2-t^2}} + \frac{\eta'(r)r^{1+s}}{\sqrt{r^2-t^2}} - \frac{\eta(r)r^{1+s}}{\sqrt{r^2-t^2}(r+t)} \right) u(\theta) \right.$$
$$\left. - \frac{\eta(r)r^s}{r+t} u'(\theta) \right] dr$$

A new change of variable $r = yt$ leads to the fact that $h'(t) = t^s\phi(t)$, where the function ϕ is continuous differentiable on $[0,1]$, $\phi(0)$ is in general not zero and $\phi(1) = 0$. According to Theorem B.2 (with $\lambda = 1 + s$) we have that

$$\int_0^1 h'(t)e^{it\rho}\,dt = b_1(\omega)\rho^{-1-s} + O(\rho^{-1}), \tag{B.5}$$

where the constant in the term $O(\rho^{-1})$ is bounded uniformly in ω. Therefore, from (B.2) and (B.5), for the case $s \in (-1, 0)$ we obtain that

$$\Phi(\rho, \omega) = b(\omega)\rho^{-2-s} + O(\rho^{-2}), \tag{B.6}$$

where the constant in the term $O(\rho^{-2})$ is bounded uniformly in ω. By Remark B.2, (B.6) holds for $s = 0$ provided $O(\rho^{-2})$ is replaced be $o(\rho^{-2})$. The case $s \in (0, 1)$ can be treated in a similar way. Since $h'(1) = 0$, one can easily see that in fact we have $g'(1) = 0$ and $g'(-1) = 0$. Then, from (B.2) we get

$$\Phi(\rho, \omega) = \frac{-1}{\rho^2} \int_{-1}^1 g''(t)e^{it\rho}\,dt. \tag{B.7}$$

All the considerations for g used in the case $s \in (-1, 0)$ can be reproduced in the case $s \in (0, 1)$ for the functions g' in order to get

$$\Phi(\rho, \omega) = b(\omega)\rho^{-2-s} + O(\rho^{-3}), \tag{B.8}$$

where the constant in the term $O(\rho^{-3})$ is bounded uniformly in ω.

References

[1] C. Bacuta, J. H. Bramble, J. Pasciak. New interpolation results and applications to finite element methods for elliptic boundary value problems. To appear.

[2] C.Bacuta, J. H. Bramble and J. Pasciak. Using finite element tools in proving shift theorems for elliptic boundary value problems. To appear in "Numerical Linear Algebra with Applications"..

[3] C. Bennett and R. Sharpley. *Interpolation of Operators*. Academic Press, New-York, 1988.

[4] N. Bleistein and R. Handelsman. *Asymptotic expansions of integrals*. Holt, Rinehart and Winston, New York, 1975.

[5] S. Brenner and L.R. Scott. *The Mathematical Theory of Finite Element Methods*. Springer-Verlag, New York, 1994.

[6] P. G. Ciarlet. *The Finite Element Method for Elliptic Problems*. North Holland, Amsterdam, 1978.

[7] M. Dauge. *Elliptic Boundary Value Problems on Corner Domains*. Lecture Notes in Mathematics 1341. Springer-Verlag, Berlin, 1988.

[8] A. Erdelyi. *Asymptotic Expansions*. Dover Publications, Inc., New York, 1956.

[9] V. Girault and P.A. Raviart. *Finite Element Methods for Navier-Stokes Equations*. Springer-Verlag, Berlin, 1986.

[10] P. Grisvard. *Elliptic Problems in Nonsmooth Domains*. Pitman, Boston, 1985.

[11] P. Grisvard. *Singularities in Boundary Value Problems*. Masson, Paris, 1992.

[12] R. B. Kellogg . Interpolation between subspaces of a Hilbert space ,Technical note BN-719. Institute for Fluid Dynamics and Applied Mathematics, University of Maryland, College Park, 1971.

[13] V. Kondratiev. *Boundary value problems for elliptic equations in domains with conical or angular points. Trans. Moscow Math. Soc.*, 16:227-313, 1967.

[14] V. A. Kozlov, V. G. Mazya and J. Rossmann. *Elliptic Boundary Value Problems in Domains with Point Singularities*. American Mathematical Society, Mathematical Surveys and Monographs, vol. 52, 1997.

[15] J. L. Lions and E. Magenes. *Non-homogeneous Boundary Value Problems and Applications, I*. Springer-Verlag, New York, 1972.

[16] J. L. Lions and P. Peetre. Sur une classe d'espaces d'interpolation. *Institut des Hautes Etudes Scientifique. Publ.Math.*, 19:5-68, 1964.

[17] S. A. Nazarov and B. A. Plamenevsky. *Elliptic Problems in Domains with Piecewise Smooth Boundaries*. Expositions in Mathematics, vol. 13, de Gruyter, New York, 1994.

[18] J. Nečas . *Les Methodes Directes en Theorie des Equations Elliptiques*. Academia, Prague, 1967.

[19] F. W. Olver. *Asymptotics and Special Functions*. Academic Press, New York, 1974.

NUMERICAL L METHODS FOR SCHRÖDINGER EQUATIONS *

Weizhu Bao
Department of Computational Science,
National University of Singapore, Singapore 117543
bao@cz3.nus.edu.sg

Shi Jin
Department of Mathematics, University of Wisconsin-Madison,
Madison, WI 53706, USA
jin@math.wisc.edu

Peter A. Markowich
Institute of Mathematics, University of Vienna,
Boltzmanngasse 9, A-1090 Vienna, Austria
peter.markowich@univie.ac.at

Abstract In this note we review the time-splitting spectral method, recently studied by the authors, for linear [2] and nonlinear [3] Schrödinger equations (NLS) in the semiclassical regimes, where the Planck constant ε is small. The time-splitting spectral method under study is unconditionally stable and conserves the position density. Moreover it is gauge invariant and time reversible when the corresponding Schrödinger equation is. Numerical tests are presented for linear, for weak/strong focusing/defocusing nonlinearities, for the Gross-Pitaevskii equation and for current-relaxed quantum hydrodynamics. The tests are geared towards understanding admissible meshing strategies for obtaining 'correct' physical observables in the semi-classical regimes. Furthermore, comparisons be-

*This research was supported by the International Erwin Schrödinger Institute in Vienna. W.B. acknowledges support in part by the National University of Singapore grant No. R-151-000-016-112. S.J. acknowledges support in part by NSF grant No. DMS-0196106. P.A.M. acknowledges support from the EU-funded TMR network 'Asymptotic Methods in kinetic Theory' and from his WITTGENSTEIN-AWARD 2000, funded by the Austrian National Science Fund FWF.

tween the solutions of the nonlinear Schrödinger equation and its hydrodynamic semiclassical limit are presented.

Keywords: Time-splitting spectral method, Schrödinger equation, semi-classical regime, admissible meshing strategy, physical observable

1. Introduction

Many problems of solid state physics require the solution of the following linear/nonlinear Schrödinger equation (NLS) with a small (scaled) Planck constant ε $(0 < \varepsilon \ll 1)$:

$$i\varepsilon u_t^\varepsilon + \frac{\varepsilon^2}{2}\,\Delta u^\varepsilon - V(\mathbf{x})u^\varepsilon - f(|u^\varepsilon|^2)u^\varepsilon - \varepsilon\tau\,\arg(u^\varepsilon)u^\varepsilon = 0, \quad (1.1)$$

$$u^\varepsilon(\mathbf{x}, t=0) = u_0^\varepsilon(\mathbf{x}), \qquad \mathbf{x} \in \mathbb{R}^d. \quad (1.2)$$

In this equation, $V = V(\mathbf{x})$ is a given real-valued electrostatic potential, f is a real-valued smooth function, $\tau \geq 0$ is a constant relaxation rate, $u^\varepsilon = u^\varepsilon(\mathbf{x}, t)$ is the wave function and $\arg(u^\varepsilon)$ is defined (up to an additive constant) as:

$$\varepsilon\,\arg(u^\varepsilon(\mathbf{x}, t)) = S^\varepsilon(\mathbf{x}, t), \qquad \nabla S^\varepsilon = J^\varepsilon/\rho^\varepsilon, \quad \text{when } \rho^\varepsilon \neq 0, \quad (1.3)$$

where ρ^ε (the position density) and J^ε (the current density) are primary physical quantities and can be computed from the wave function u^ε:

$$\rho^\varepsilon(\mathbf{x}, t) = |u^\varepsilon(\mathbf{x}, t)|^2 \quad (1.4)$$

and

$$J^\varepsilon(\mathbf{x}, t) = \varepsilon\,\mathrm{Im}(\overline{u^\varepsilon(\mathbf{x}, t)}\,\nabla u^\varepsilon(\mathbf{x}, t)), \quad (1.5)$$

where "—" denotes complex conjugation. We assume that $|u_0^\varepsilon|$ decays to zero sufficiently fast when $|\mathbf{x}| \to \infty$.

The general form of (1.1) covers many linear/nonlinear Schrödinger equations arising in various different applications. For example, when $f \equiv 0$ and $\tau = 0$, (1.1) reduces to the linear Schrödinger equation; when $V \equiv 0$, $f(\rho) = \beta_\varepsilon\,\rho$ and $\tau = 0$, it is the cubic nonlinear Schrödinger equation (called the focusing NLS if $\beta_\varepsilon < 0$ and the defocusing NLS if $\beta_\varepsilon > 0$; when $V(x) = \frac{\omega}{2}|\mathbf{x}|^2$ with $\omega > 0$ a constant, $f(\rho) = \rho$ and $\tau = 0$, it is related to Bose-Einstein condensation (Gross-Pitaevskii equation, cf. [6]). For $\tau > 0$ the equation can be rewritten as current-relaxed quantum hydrodynamical system for ρ^ε and J^ε (cf. [14, 16]). It is well known that the equation (1.1) propagates oscillations in space and time, preventing u^ε from converging strongly as $\varepsilon \to 0$. On the other hand, the weak convergence of u^ε is, for example, not sufficient for passing to the limit in the quadratic macroscopic densities (1.4)–(1.5). The analysis of the so-called semiclassical limit is a mathematically complex issue.

Much progress has been made recently in understanding semiclassical limits of the linear Schödinger equation, particularly by the introduction of tools from microlocal analysis, such as defect measures [8], H-measures [22], and Wigner measures [7, 9, 18]. These techniques have provided powerful technical tools to exploit properties of the linear Schrödinger equation in the semiclassical limit regime, allowing the passage to the limit $\varepsilon \longrightarrow 0$ in the macroscopic densities by revealing an underlying kinetic structure. These techniques have not been successfully extended to the semiclassical limit of the (cubically) nonlinear Schrödinger equation, which was solved in the one-dimensional defocusing nonlinearity using techniques of inverse scattering [12, 13]. Thus numerical study of the semiclassical limit of linear/nonlinear Schrödinger equation is a very important and interesting problem. The numerical experiments may shed some lights on analytical understanding of the semiclassical behavior of the Schrödinger equations.

The oscillatory nature of the solutions of the Schrödinger equation with small ε provides severe numerical burdens. Even for stable numerical approximations (or under mesh size and time step restrictions which guarantee stability) the oscillations may very well pollute the solution in such a way that the quadratic macroscopic quantities and other physical observables come out completely wrong unless the spatial-temporal oscillations are fully resolved numerically, i.e., using many grid points per wave length of $O(\varepsilon)$. In [19, 20], Markowich at. el. studied the finite difference approximation of the linear Schrödinger equation with small ε. Their results show that, for the best combination of the time and space discretizations, one needs the following constraints in order to guarantee good approximations to all (smooth) observables for ε small: mesh size $h = o(\varepsilon)$ and time step $k = o(\varepsilon)$. Failure to satisfy these conditions leads to wrong numerical observables.

Recently, we syetematically studied the time-splitting spectral method for linear [2] and nonlinear [3] Schrödinger equation in the semiclassical regimes (1.1), (1.2). The method under study is unconditionally stable and conserves the position density. Moreover it is gauge invariant and time reversible when the corresponding Schrödinger equation is. For the linear Schrödinger equation, we have proved a uniform L^2-approximation of the wave-function for $k = o(\varepsilon)$ and $h = O(\varepsilon)$. Our extensive numerical experiments suggest the following meshing strategies for obtaining 'correct' physical observables: k independent of ε and $h = O(\varepsilon)$ for linear case [2]; $k = O(\varepsilon)$ and $h = O(\varepsilon)$ for defocusing nonlinearities and weak $O(\varepsilon)$ focusing nonlinearities; $k = o(\varepsilon)$ and $h = O(\varepsilon)$ for strong $O(1)$ focusing nonlinearities (when the Krasny Filter [15] is applied). Furthermore, comparisons between the solution of the nonlinear Schrödinger equation and its hydrodynamic semiclassical limit are presented [3].

The note is organized as follows. In section 2 we present the formal semi-classical limit of the NLS (1.1). In section 3 we review time-splitting spectral method for the NLS (1.1), (1.2) in 1-d. In section 4 we report numerical results for the NLS.

2. Formal semiclassical limit

Suppose that the initial datum u_0^ε in (1.2) is rapidly oscillating on the scale ε, given in WKB form:

$$u_0^\varepsilon(\mathbf{x}) = A_0(\mathbf{x}) \exp\left(\frac{i}{\varepsilon} S_0(\mathbf{x})\right), \qquad \mathbf{x} \in \mathbb{R}^d, \tag{2.1}$$

where the amplitude A_0 is assumed to decay to zero sufficiently fast as $|\mathbf{x}| \to \infty$ and the phase S_0 are smooth real-valued functions. Plugging the radial-representation of the wave-function

$$u^\varepsilon(\mathbf{x},t) = A^\varepsilon(\mathbf{x},t) \exp\left(\frac{i}{\varepsilon} S^\varepsilon(\mathbf{x},t)\right) = \sqrt{\rho^\varepsilon(\mathbf{x},t)} \exp\left(\frac{i}{\varepsilon} S^\varepsilon(\mathbf{x},t)\right) \tag{2.2}$$

into (1.1), one obtains the following quantum hydrodynamic form of the Schrödinger equation for $\rho^\varepsilon = |A^\varepsilon|^2$, $J^\varepsilon = \rho^\varepsilon \nabla S^\varepsilon$ [17]

$$\rho_t^\varepsilon + \operatorname{div} J^\varepsilon = 0, \tag{2.3}$$

$$J_t^\varepsilon + \operatorname{div}\left(\frac{J^\varepsilon \otimes J^\varepsilon}{\rho^\varepsilon}\right) + \nabla P(\rho^\varepsilon) + \rho^\varepsilon \nabla V + \tau J^\varepsilon = \frac{\varepsilon^2}{4} \operatorname{div}(\rho^\varepsilon \nabla^2 \log \rho^\varepsilon);$$

with initial data

$$\rho^\varepsilon(\mathbf{x},0) = |A_0(\mathbf{x})|^2, \qquad J^\varepsilon(\mathbf{x},0) = |A_0(\mathbf{x})|^2 \, \nabla S_0(\mathbf{x}), \tag{2.4}$$

(see Grenier [11], Jüngel [14], Lin and Li [16] for mathematical analyses of this system). Here the hydrodynamic pressure $P(\rho)$ is related to the nonlinear potential $f(\rho)$ by

$$P(\rho) = \rho f(\rho) - \int_0^\rho f(s)\, ds, \tag{2.5}$$

i.e. f' is the enthalpy. Letting $\varepsilon \to 0+$, one obtains formally the following Euler system

$$\rho_t + \operatorname{div} J = 0, \tag{2.6}$$

$$J_t + \operatorname{div}\left(\frac{J \otimes J}{\rho}\right) + \nabla P(\rho) + \rho \nabla V + \tau J = 0. \tag{2.7}$$

Note that $\frac{1}{\tau}$ is the actual relaxation time. In the case $f' > 0$ we expect (2.6), (2.7) to be the 'rigorous' semiclassical limit of (1.1) as long as caustics do not occur, i.e. in the pre-breaking regime. After caustics the dispersive behavior of the NLS takes over and (2.6), (2.7) is not correct any more. Note that the nonlinear Schrödinger equation (1.1) is time reversible iff $\tau = 0$, i.e. iff no current relaxation occurs.

3. Time-splitting spectral method

In this section we review the time-splitting spectral method for the NLS (1.1), (1.2) of periodic problems. For nonperiodic problems, Bao [1] recently proposed a time-splitting Chebyshev-spectral method. For simplicity of notation we shall introduce the method in one space dimension ($d = 1$). Generalizations to $d > 1$ are straightforward for tensor product grids and the results remain valid without modifications. For $d = 1$, the problem becomes

$$i\varepsilon u_t^\varepsilon + \frac{\varepsilon^2}{2} u_{xx}^\varepsilon - V(x)u^\varepsilon - f(|u^\varepsilon|^2)u^\varepsilon - \varepsilon\tau \arg(u^\varepsilon)u^\varepsilon = 0, \tag{3.1}$$

$$u^\varepsilon(x, t = 0) = u_0^\varepsilon(x), \ a \le x \le b, \tag{3.2}$$

$$u^\varepsilon(a, t) = u^\varepsilon(b, t), \qquad u_x^\varepsilon(a, t) = u_x^\varepsilon(b, t), \ t > 0. \tag{3.3}$$

3.1 Time-splitting spectral approximations

We choose the spatial mesh size $h = \Delta x > 0$ with $h = (b - a)/M$ for M an even positive integer, the time step $k = \Delta t > 0$ and let the grid points and the time step be

$$x_j := a + j\, h, \qquad t_n := n\, k, \qquad j = 0, 1, \cdots, M, \qquad n = 0, 1, 2, \cdots$$

Let $U_j^{\varepsilon,n}$ be the approximation of $u^\varepsilon(x_j, t_n)$ and $U^{\varepsilon,n}$ be the solution vector at time $t = t_n = nk$ with components $U_j^{\varepsilon,n}$.

The first-order time-splitting spectral method (SP1). From time $t = t_n$ to $t = t_{n+1}$, the NLS equation (3.1) is solved in two steps. One solves first

$$i\varepsilon u_t^\varepsilon + \frac{\varepsilon^2}{2} u_{xx}^\varepsilon = 0, \tag{3.4}$$

for one time step (of length k), followed by solving

$$i\varepsilon u_t^\varepsilon - V(x)u^\varepsilon - f(|u^\varepsilon|^2)u^\varepsilon - \varepsilon\tau \arg(u^\varepsilon)u^\varepsilon = 0, \tag{3.5}$$

for the same time step. Equation (3.4) will be discretized in space by the Fourier spectral method and integrated in time *exactly*. For $t \in [t_n, t_{n+1}]$, the ODE (3.5) leaves $|u^\varepsilon|$ invariant in t:

$$\begin{aligned} \frac{\partial}{\partial t}\left(|u^\varepsilon|^2\right) &= -\frac{2}{\varepsilon}\,\mathrm{Re}\left(i\left(V(x) + f(|u^\varepsilon|^2) + \varepsilon\tau \arg(u^\varepsilon)\right)|u^\varepsilon|^2\right) \\ &= 0 \end{aligned} \tag{3.6}$$

(since V and f are real valued) and therefore (3.5) becomes

$$i\varepsilon u_t^\varepsilon - V(x)u^\varepsilon - f(|u^\varepsilon(x, t_n)|^2)u^\varepsilon - \varepsilon\tau \arg(u^\varepsilon)u^\varepsilon = 0. \tag{3.7}$$

If $\tau = 0$, (3.7) can be integrated *exactly* and the solution is given by

$$u^\varepsilon(x,t) = e^{-i(V(x)+f(|u^\varepsilon(x,t_n)|^2))(t-t_n)/\varepsilon}\, u^\varepsilon(x,t_n),\ t \in [t_n, t_{n+1}]. \quad (3.8)$$

If $\tau \neq 0$, since $|u^\varepsilon|$ remains invariant for the ODE (3.7), we set

$$u^\varepsilon(x,t) = |u^\varepsilon(x,t_n)| \exp\left(\frac{i}{\varepsilon} S^\varepsilon(x,t)\right), \qquad t \in [t_n, t_{n+1}], \quad (3.9)$$

where S^ε is a function to be determined. Plugging (3.9) into (3.7), using (1.4) and (1.5), one obtains

$$S_t^\varepsilon(x,t) + \tau S^\varepsilon(x,t) + V(x) + f(|u^\varepsilon(x,t_n)|^2) = 0, \quad (3.10)$$
$$J^\varepsilon(x,t) = \rho^\varepsilon(x,t_n) S_x^\varepsilon(x,t). \quad (3.11)$$

Differentiating the equation (3.10) with respect to x gives

$$J_t^\varepsilon(x,t) + \tau J^\varepsilon(x,t) + \left[V_x(x) + f(|u^\varepsilon(x,t_n)|^2)_x\right] \rho(x,t_n) = 0. \quad (3.12)$$

Solving this ODE, one obtains

$$\begin{aligned} J^\varepsilon(x,t) &= -\left[V_x(x) + f(|u^\varepsilon(x,t_n)|^2)_x\right] \rho(x,t_n) \frac{1 - e^{-\tau(t-t_n)}}{\tau} \\ &\quad + e^{-\tau(t-t_n)} J^\varepsilon(x,t_n). \end{aligned} \quad (3.13)$$

Substituting (3.12) into (3.11), and using (1.4) and (1.5), we find

$$\begin{aligned} S_x^\varepsilon(x,t) &= \frac{J^\varepsilon(x,t)}{\rho^\varepsilon(x,t_n)} = e^{-\tau(t-t_n)} \varepsilon \,\mathrm{Im}\left(\frac{u_x^\varepsilon(x,t_n)}{u^\varepsilon(x,t_n)}\right) \\ &\quad - \left[V_x(x) + f(|u^\varepsilon(x,t_n)|^2)_x\right] \frac{1 - e^{-\tau(t-t_n)}}{\tau}, \end{aligned} \quad (3.14)$$

and set

$$\begin{aligned} \frac{S^\varepsilon(x,t)}{\varepsilon} &= e^{-\tau(t-t_n)} \int_a^x \mathrm{Im}\left(\frac{u_v^\varepsilon(v,t_n)}{u^\varepsilon(v,t_n)}\right) dv \\ &\quad - \left[V(x) + f(|u^\varepsilon(x,t_n)|^2)\right] \frac{1 - e^{-\tau(t-t_n)}}{\varepsilon\tau} \\ &\equiv S^\varepsilon(u^\varepsilon,t) - \left[V(x) + f(|u^\varepsilon(x,t_n)|^2)\right] \frac{1 - e^{\tau(t_n-t)}}{\varepsilon\tau}. \end{aligned} \quad (3.15)$$

Plugging (3.14) into (3.9), one obtains the solution of (3.5) in the case $\tau \neq 0$. Notice that $S^\varepsilon(x,t)$ is determined up to a constant and the choice of the constant does not affect the observables.

The detailed method is given by:

$$U_j^{\varepsilon,*} = \frac{1}{M} \sum_{l=-M/2}^{M/2-1} e^{-i\varepsilon\mu_l^2 k/2}\, \widehat{U}_l^{\varepsilon,n}\, e^{i\mu_l(x_j-a)}, \quad j = 0,1,2,\cdots, M-1,$$

$$U_j^{\varepsilon,n+1} = \begin{cases} e^{-i(V(x_j)+f(|U_j^{\varepsilon,*}|^2))k/\varepsilon}\, U_j^{\varepsilon,*}, & \text{if } \tau = 0, \\ e^{-i(V(x_j)+f(|U_j^{\varepsilon,*}|^2))(1-e^{-k\tau})/\varepsilon\tau + iS_j^\varepsilon(U^{\varepsilon,*},k)}\, |U_j^{\varepsilon,*}|, & \text{if } \tau \neq 0; \end{cases}$$

where $\widehat{U}_l^{\varepsilon,n}$, the Fourier coefficients of $U^{\varepsilon,n}$, are defined as

$$\mu_l = \frac{2\pi l}{b-a}, \quad \widehat{U}_l^{\varepsilon,n} = \sum_{j=0}^{M-1} U_j^{\varepsilon,n}\, e^{-i\mu_l(x_j-a)}, \; l = -\frac{M}{2}, \cdots, \frac{M}{2}-1, \tag{3.16}$$

with

$$U_j^{\varepsilon,0} = u^\varepsilon(x_j,0) = u_0^\varepsilon(x_j), \qquad j = 0,1,2,\cdots, M \tag{3.17}$$

and $\mathcal{S}^\varepsilon(U,k)$ is an approximation $\mathcal{S}^\varepsilon(u^\varepsilon,t)$ in (3.14). Here we use the composite trapezoidal rule to obtain $\mathcal{S}^\varepsilon$ numerically:

$$\mathcal{S}_j^\varepsilon(U,t) = e^{-\tau t} \sum_{l=1}^{j} \frac{h}{2} \operatorname{Im}\left(\frac{D_x^s U|_{x=x_{l-1}}}{U_{l-1}} + \frac{D_x^s U|_{x=x_l}}{U_l} \right), \tag{3.18}$$

$$j = 1,2,\cdots, M, \qquad S_0(U,k) = 0.$$

with D_x^s the spectral approximation of ∂_x:

$$D_x^s U|_{x=x_j} = \frac{1}{M} \sum_{l=-M/2}^{M/2-1} i\mu_l\, \widehat{U}_l\, e^{i\mu_l(x_j-a)}, \tag{3.19}$$

with

$$\widehat{U}_l = \sum_{j=0}^{M-1} U_j\, e^{-i\mu_l(x_j-a)}, \; l = -\frac{M}{2}, \cdots, \frac{M}{2}-1. \tag{3.20}$$

Note that the only time discretization error of this method is the splitting error, which is $O(k)$ for any fixed $\varepsilon > 0$. For future reference we define the trigonometric interpolant of a function f on the grid $\{x_0, x_1, \cdots, x_M\}$:

$$f_I(x) = \frac{1}{M} \sum_{l=-M/2}^{M/2-1} \widehat{f}_l\, e^{i\mu_l(x-a)}, \quad \widehat{f}_l = \sum_{j=0}^{M-1} f(x_j) e^{-i\mu_l(x_j-a)}. \tag{3.21}$$

The Strang splitting spectral method (SP2). From time $t = t_n$ to $t = t_{n+1}$, we combine the split steps via the standard Strang splitting:

$$
\begin{aligned}
U_j^{\varepsilon,*} &= \sum_{l=-M/2}^{M/2-1} e^{-i\varepsilon\mu_l^2 k/4}\, \widehat{U}_l^{\varepsilon,n}\, e^{i\mu_l(x_j-a)}, \quad j=0,1,2,\cdots,M-1,\\
U_j^{\varepsilon,**} &= \begin{cases} e^{-i(V(x_j)+f(|U_j^{\varepsilon,*}|^2))k/2\varepsilon}\, U_j^{\varepsilon,*}, & \text{if } \tau=0,\\[2mm] e^{-i(V(x_j)+f(|U_j^{\varepsilon,*}|^2))(1-e^{-k\tau})/2\varepsilon\tau + iS_j^{\varepsilon}(U^{\varepsilon,*},k/2)}\, |U_j^{\varepsilon,*}|, & \text{if } \tau\neq 0,\end{cases}\\
\widehat{U}_l^{\varepsilon,n+1} &= \frac{1}{M} e^{-i\varepsilon\mu_l^2 k/4} \sum_{j=0}^{M-1} U_j^{\varepsilon,**}\, e^{-i\mu_l(x_j-a)},
\end{aligned}
$$

where $\widehat{U}_l^{\varepsilon,n}$, the Fourier coefficients of the numerical solution $U^{\varepsilon,n}$ at time $t = t_n$ $(n = 0,1,2,\cdots)$. In this algorithm, we need to do a Fourier transformation (i.e. (3.16) with $n = 0$) for the initial value $U_j^{\varepsilon,0} = u_0^{\varepsilon}(x_j)_{j=0}^{M}$ before time marching and do an inverse Fourier transformation after all time marching to get the solution at the final time step. Again, the overall time discretization error comes solely from the splitting, which is now $O(k^2)$ for fixed $\varepsilon > 0$.

Our numerical experiments [3] show that, when ε is small, SP1 and SP2 work very well for all considered cases except the strong O(1) focusing nonlinearity, i.e. $f(\rho) = -\beta\rho$ and $\beta = O(1) > 0$. In this case, due to the modulational instability (see [4]), the numerical solution is stable but qualitatively wrong for small ε due to the accumulation of round-off errors. Therefore, we apply the Krasny filter [15] to the solution at each time step (see also [2] for similar applications). That is, we set to zero all the Fourier modes of the numerical solution whose magnitudes are below a certain filter level. This filter is applied only for the strong $O(1)$ focusing nonlinearity. It is no need to be used in all other cases.

The schemes SP1 and SP2 are time reversible, just as the IVP for the NLS, if $\tau = 0$. Also, a main advantage of the time-splitting methods is their gauge invariance, when $\tau = 0$ in (3.1), just as it holds for the NLS itself. If a constant α is added to the potential V, then the discrete wave functions $U_j^{\varepsilon,n+1}$ obtained from SP1 and SP2 get multiplied by the phase factor $e^{-i\alpha(n+1)k/\varepsilon}$, which leaves the discrete quadratic observables unchanged. This property does not hold for finite difference scheme.

3.2 Stability and error estimates in the linear case

Let $U = (U_0, \cdots, U_{M-1})^T$ and let $\|\cdot\|_{L^2}$ and $\|\cdot\|_{l^2}$ be the usual L^2-norm and discrete l^2-norm respectively on the interval (a, b), i.e.

$$\|u\|_{L^2} = \sqrt{\int_a^b |u(x)|^2\, dx}, \qquad \|U\|_{l^2} = \sqrt{\frac{b-a}{M}\sum_{j=0}^{M-1} |U_j|^2}. \tag{3.22}$$

For the *stability* of the time-splitting spectral approximations SP1 and SP2, we havethe following lemma [2, 3], which shows that the total charge is conserved.

Lemma 3.1. *The time-splitting spectral schemes (SP1) and (SP2) are unconditionally stable. In fact, for every mesh size $h > 0$ and time step $k > 0$,*

$$\|U^{\varepsilon,n}\|_{l^2} = \|U^{\varepsilon,0}\|_{l^2} = \|U_0^{\varepsilon}\|_{l^2}, \qquad n = 1, 2, \cdots \tag{3.23}$$

For constant potential $V(x) \equiv V =$ constant, $f \equiv 0$ and $\tau = 0$ in (1.1), we have the following error estimate for both SP1 and SP2:

Theorem 3.1. *Let u^{ε} be the exact solution of (3.1), (3.3), let $V =$ constant and $u_I^{\varepsilon,n}$ be the trigonometric interpolant of $U^{\varepsilon,n} = (U_j^{\varepsilon,n})_{j=0}^{M-1}$ as obtained from SP1 or SP2. Under assumption (A) [2], we have for all integers $m \geq 1$*

$$\left\|u_I^{\varepsilon,n} - u^{\varepsilon}(t_n)\right\|_{L^2} \leq D\, C_m \left(\frac{h}{\varepsilon(b-a)}\right)^m, \tag{3.24}$$

where $D > 0$ is a constant.

For variable potential $V = V(x)$, $f \equiv 0$ and $\tau = 0$, we have the following error estimate for SP1:

Theorem 3.2. *Let $u^{\varepsilon} = u^{\varepsilon}(x,t)$ be the exact solution of (3.1), (3.3) and $U^{\varepsilon,n}$ be the discrete approximation SP1. Under assumption (B) [2], and assuming $k = O(\varepsilon)$, $h = O(\varepsilon)$, we have for all positive integers $m \geq 1$ and $t_n \in [0, T]$:*

$$\left\|u^{\varepsilon}(t_n) - u_I^{\varepsilon,n}\right\|_{L^2} \leq G_m \frac{T}{k}\left(\frac{h}{\varepsilon(b-a)}\right)^m + \frac{CTk}{\varepsilon}, \tag{3.25}$$

where C is a positive constant independent of ε, h, k and m and G_m is independent of ε, h, k.

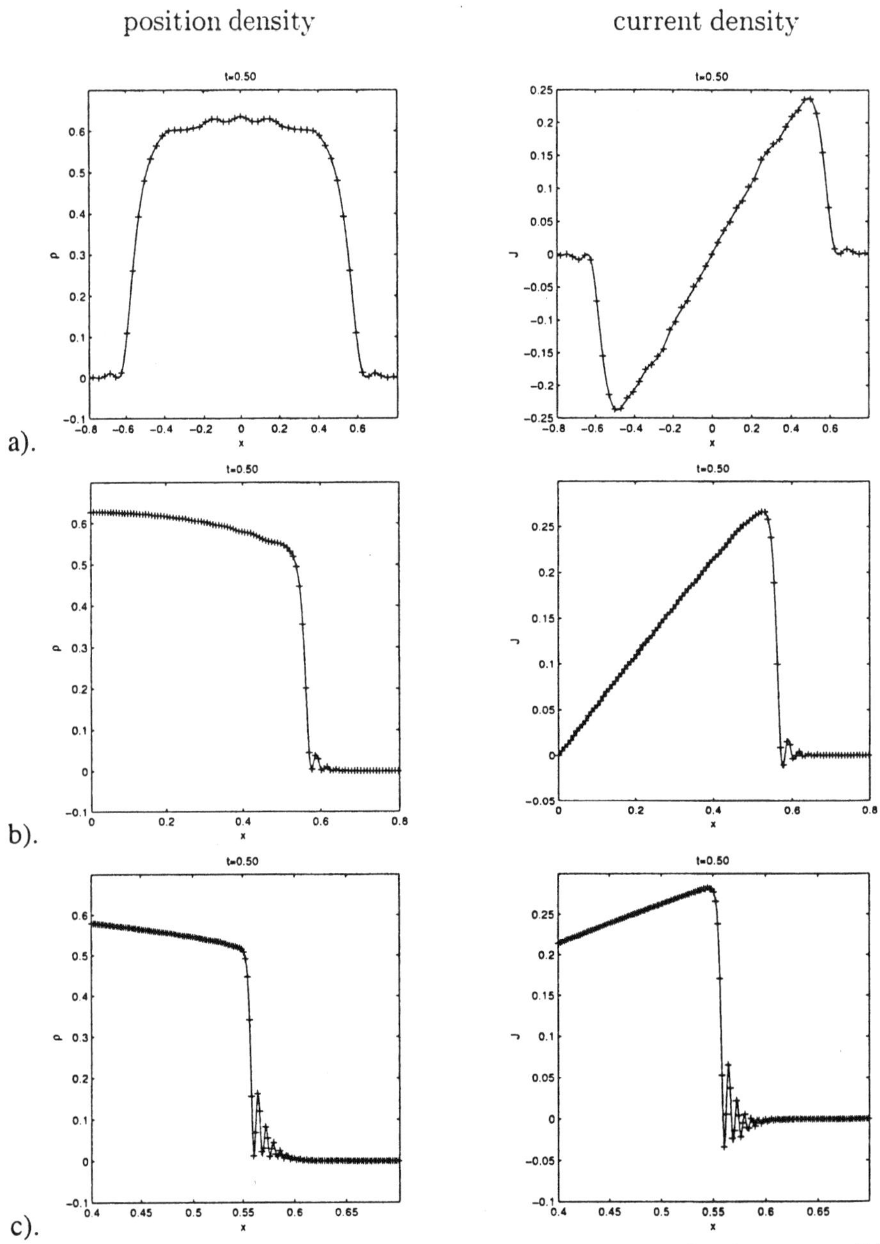

Figure 1: Numerical solutions at $t = 0.5$ in the Example for the strong $O(1)$ defocusing nonlinearity by using SP2. $V(x) \equiv 0$, $f(\rho) = \rho$, $\tau = 0$. '—': 'exact' solution, '+ + +': numerical solution. a). $\varepsilon = 0.04$, $k = 0.01$, $h = \frac{1}{32}$, b). $\varepsilon = 0.01$, $k = 0.0025$, $h = \frac{1}{128}$, c). $\varepsilon = 0.0025$, $k = 0.000625$, $h = \frac{1}{512}$. Here $h = O(\varepsilon), k = O(\varepsilon)$.

4. Numerical examples

Here we consider an example of strong $O(1)$ defocusing nonlinearity,

$$V(x) \equiv 0, \qquad f(\rho) = \rho, \qquad \tau = 0.$$

In our computations, the initial condition (3.2) is always chosen in the classical WKB form:

$$u^\varepsilon(\mathbf{x}, t = 0) = u_0^\varepsilon(\mathbf{x}) = A_0(\mathbf{x})\, e^{iS_0(\mathbf{x})/\varepsilon} = \sqrt{\rho_0(\mathbf{x})}\, e^{iS_0(\mathbf{x})/\varepsilon}, \qquad (4.1)$$

with A_0 and S_0 independent of ε, real valued, regular and with $A_0(\mathbf{x})$ decaying to zero sufficiently fast as $|\mathbf{x}| \to \infty$. We compute with SP2 on the interval $[-8, 8]$, which is large enough for the computations such that the periodic boundary conditions do not introduce a significant (aliasing) error relative to the whole space problem. The initial condition is taken as

$$A_0(x) = \begin{cases} 1 - |x|, & |x| < 1, \\ 0, & \text{otherwise}; \end{cases} \quad S_0(x) = -\ln(e^x + e^{-x}),\ x \in \mathbb{R}. \quad (4.2)$$

To test the numerical method, for each fixed ε, we compute the numerical solution with a very fine mesh, e.g. $h = \frac{1}{4096}$, and a very small time step, e.g. $k = 0.00001$, as the reference 'exact' solution u^ε. Figure 1 shows the numerical results at $t = 0.5$ with $\varepsilon = 0.04$, $k = 0.01$, $h = \frac{1}{32}$; $\varepsilon = 0.01$, $k = 0.0025$, $h = \frac{1}{128}$; $\varepsilon = 0.0025$, $k = 0.000625$, $h = \frac{1}{512}$, corresponding to the meshing strategy $h = O(\varepsilon)$ and $k = O(\varepsilon)$.

Figure 1 seems to suggest the following meshing strategy in order to guarantee good approximations of observables for defocusing nonlinearity: $h = O(\varepsilon)$ and $k = O(\varepsilon)$.

For more numerical experiments on various Schrödinger equations see [2, 3].

References

[1] Bao, W. (2001). *Time-splitting Chebyshev-spectral approximations for (non)linear Schrödinger equation*, preprint.

[2] Bao, W., Jin, Shi, and Markowich, P.A. (2001). *On time-splitting spectral approximations for the Schrödinger equation in the semiclassical regime*, J. Comput. Phys., to appear.

[3] Bao, W., Jin, Shi, and Markowich, P.A. (2001). *Numerical study of time-splitting spectral discretizations of nonlinear Schrödinger equations in the semi-classical regimes*, SIAM J. Sci. Comput., submitted.

[4] Bronski, J.C., and McLaughlin, D.W., (1994). *Semiclassical behavior in the NLS equation: optical shocks - focusing instabilities*, Singular Limits of Dispersive Waves, Plenum Press, New York and London.

[5] Ceniceros, H.D., and Tian, F.R. A numerical study of the semi-classical limit of the focusing nonlinear Schrödinger equation, Phys. Lett. A., to appear.

[6] Gardiner, S.A., Jaksch, D., Dum, R., Cirac, J.I., and Zollar, P. (2000). *Nonlinear matter wave dynamics with a chaotic potential*, Phys. Rev. A, 62, pp. 023612-1:21.

[7] Gasser, I., and Markowich, P.A., (1997).*Quantum hydrodynamics, Wigner transforms and the classical limit*, Asymptotic Analysis 14, pp. 97-116.

[8] Gérard, P., (1991). it Microlocal defect measures, Comm. PDE. 16, pp. 1761-1794.

[9] Gérard, P., Markowich, P.A., Mauser, N.J., and Poupaud, F., (1997). *Homogenization limits and Wigner transforms*, Comm. Pure Appl. Math. 50, pp. 321-377.

[10] Gottlieb, D., and Orszag, S.A., (1977). *Numerical Analysis of Spectral Methods*, SIAM, Philadelphia.

[11] Grenier, E. (1998). *Semiclassical limit of the nonlinear Schrödinger equation in small time*, Proc. Amer. Math. Soc., 126, pp. 523-530.

[12] Jin, Shan, Levermore, C.D., and McLaughlin, D.W., (1999). *The semiclassical limit of the defocusing NLS hierarchy*, Comm. Pure Appl. Math. LII, pp. 613-654.

[13] Jin, Shan, Levermore, C.D., and McLaughlin, D.W., (1994). *The behavior of solutions of the NLS equation in the semiclassical limit*, Singular Limits of Dispersive Waves, Plenum Press, New York and London.

[14] Jüngel, A. (2001). *Quasi-hydrodynamic semiconductor equations*, Progress in Nonlinear Differential Equations and Its Applications, Birkhäuser, Basel.

[15] Krasny, R. (1986). *A study of singularity formulation in a vortex sheet by the point-vortex approximation*, J. Fluid Mech., 167, pp. 65-93.

[16] Lin, C.K., and Li, H. (2001). *Semiclassical limit and well-posedness of Schödinger-Poisson and quantum hydrodynamics*, preprint.

[17] Laudau, and Lifschitz (1977). *Quantum Mechanics: non-relativistic theory*, Pergamon Press, New York.

[18] Markowich, P.A., Mauser, N.J., and Poupaud, F., (1994). *A Wigner function approach to semiclassical limits: electrons in a periodic potential*, J. Math. Phys. 35, pp. 1066-1094.

[19] Markowich, P.A., Pietra, P., and Pohl, C., (1999). *Numerical approximation of quadratic observables of Schrödinger-type equations in the semi-classical limit*, Numer. Math. 81, pp. 595-630.

[20] Markowich, P.A., Pietra, P., Pohl, C., and Stimming, H.P., (2000). *A Wigner-Measure Analysis of the Dufort-Frankel scheme for the Schrödinger equation*, preprint.

[21] Miller, P.D., Kamvissis, S., (1998). *On the semiclassical limit of the focusing nonlinear Schrödinger equation*, Phys. Letters A, 247, pp. 75-86.

[22] Tartar, L., (1990). *H-measures: a new approach for studying homogenization, oscillations and concentration effects in partial differential equations*, Proc. Roy. Soc. Edinburgh Sect. A 115, pp. 193-230.

INVERSE DOPING PROBLEMS FOR SEMICONDUCTOR DEVICES

Martin Burger*, Heinz W. Engl
Institut für Industriemathematik, Johannes Kepler Universität Linz, Altenbergerstr. 69, A-4040 Linz, Austria
burger@indmath.uni-linz.ac.at, http://www.sfb.013.uni-linz.ac.at/~martin
engl@indmath.uni-linz.ac.at, http://www.indmath.uni-linz.ac.at/

Peter A. Markowich †
Institut für Mathematik, Universität Wien, Boltzmanngasse 9, A-1090 Vienna, Austria
peter.markowich@univie.ac.at, http://mailbox.univie.ac.at/peter.markowich/

Abstract This paper is devoted to a class of inverse problems arising in the testing of semiconductor devices, namely the identification of doping profiles from indirect measurements of the current or the voltage on a contact. In mathematical terms, this can be modeled by an inverse source problem for the drift-diffusion equations, which are a coupled system of elliptic or parabolic partial differential equations.

We discuss these inverse problems in a stationary and a transient setting and compare these two cases with respect to their mathematical properties. In particular, we discuss the identifiability of doping profiles in the model problem of the unipolar drift-diffusion system. Finally, we investigate the important special case of a piecewise constant doping profile, where the aim is to identify the p-n junctions, i.e., the curves between regions where the doping profile takes positive and negative values.

1. Introduction

Due to their tremendous impact on modern electronics, the mathematical modeling of semiconductor devices has developed well in the last fifty years, since Van Roosbroeck (cf.[20]) first formulated the fundamental semiconductor

*Supported by the Austrian National Science Foundation FWF under project grants SFB F 013 / 08 and P 13478-INF
†Supported by the Austrian National Science Foundation through the Wittgenstein Award 2000

device equations (see Section 2 for an overview). For a detailed expositions concerning the modeling, analysis and simulation of semiconductor devices we refer to the monographs [12, 14, 17] and for an overview of recent advances and hierarchies of models we refer to [9].

Although of increasing technological importance, optimal design and identification problems related to semiconductor devices seem to be poorly understood so far. Only recently, there was some effort in optimizing the performance of devices (cf. e.g. [8, 18, 19]) and in identifying relevant material properties (cf. [4, 10]). The position-dependent function $C = C(x)$ to be identified or optimized is the *doping profile*, which is the density difference of ionized donors and acceptors. In some cases (e.g. for the p-n diode discussed below), it may be assumed that the doping profile is piecewise constant over the device; the interesting quantities are then the curves or surfaces between the subdomains where the doping is constant. These curves are usually called *pn-junctions*, when they separate subdomains where the doping profile takes positive and negative values, respectively. In the most important doping technique of silicon devices, *ion implantation*, it is only possible to obtain a rough estimate of the doping profile by process modeling (cf. e.g. [17] for further details). In order to determine the real doping profile, reconstruction methods from indirect data have to be used. We shall use the notion *inverse doping problem* introduced in [4] for the identification of the doping profile in general.

2. Stationary and Transient Semiconductor Equations

In the following we review the drift-diffusion (DD) model for semiconductor devices, both in the stationary and transient case. The drift-diffusion model is a coupled system of nonlinear partial differential equations for the electrostatic potential V, the electron density n (≥ 0) and the hole density p (≥ 0) , which is solved in a domain $\Omega \subset \mathbf{R}^d$ ($d = 1, 2, 3$) representing the semiconductor device and in a time interval $[0, T]$ in the transient case.

2.1 The Transient DD-Model

The drift-diffusion equations in the *transient case* are given by (cf. [14])

$$0 = \operatorname{div}(\epsilon_s \nabla V) - q(n - p - C) \qquad \text{in } \Omega \times (0, T)$$

$$\frac{\partial n}{\partial t} = \operatorname{div}(D_n \nabla n - \mu_n n \nabla V) \qquad \text{in } \Omega \times (0, T)$$

$$\frac{\partial p}{\partial t} = \operatorname{div}(D_p \nabla p + \mu_p p \nabla V) \qquad \text{in } \Omega \times (0, T)$$

where ϵ_s denotes the semiconductor permittivity, q the elementary charge, μ_n and μ_p are the electron and hole mobility, D_n and D_p are the electron and hole diffusion coefficients. R denotes the *recombination-generation rate*, which generally depends on n and p. We assume that R is of the standard form

$$R = F(n, p, x)(np - n_i^2), \tag{2.1}$$

where F is a nonnegative smooth function, which holds e.g. for the frequently used *Shockley-Read-Hall* rate

$$R_{SRH} = \frac{np - n_i^2}{\tau_p(n + n_i) + \tau_n(p + n_i)}.$$

The parameters ϵ_S and q are positive (dimensional) constants and $\mu_{n/p}$ and $D_{n/p}$ are modeled by positive functions.

This system is supplemented by homogeneous Neumann boundary conditions on a part $\partial\Omega_N$ (open in $\partial\Omega$) of the boundary. On the remaining part $\partial\Omega_D$ (with positive $(d-1)$-dimensional Lebesgue-measure), the following Dirichlet conditions are imposed:

$$\begin{aligned}
V(x,t) &= V_D(x,t) = U(x,t) + V_{bi}(x) = U(x) + U_T \ln\left(\frac{n_D(x)}{n_i}\right) \\
n(x,t) &= n_D(x) = \frac{1}{2}\left(C(x) + \sqrt{C(x)^2 + 4n_i^2}\right) \\
p(x,t) &= p_D(x) = \frac{1}{2}\left(-C(x) + \sqrt{C(x)^2 + 4n_i^2}\right)
\end{aligned}$$

on $\partial\Omega_D \times (0, T)$, where n_i is the intrinsic carrier density, U_T (≥ 0) the thermal voltage and U is the applied potential. Moreover, the initial conditions

$$n(x, 0) = n_0(x) \geq 0, \qquad p(x, 0) = p_0(x) \geq 0 \qquad \text{in } \Omega \tag{2.2}$$

have to be supplied.

2.2 The Stationary DD-Model

The *stationary drift-diffusion model* is obtained from the transient case by setting

$$\frac{\partial n}{\partial t} = \frac{\partial p}{\partial t} = 0$$

and omitting the initial conditions. Using the Einstein relations, which are standard assumptions about the mobilities and diffusion coefficients of the form

$$D_n = \mu_n U_T, \qquad D_p = \mu_p U_T, \tag{2.3}$$

we may transform the system using the so-called *Slotboom variables* u and v defined by

$$n = C_0 \delta^2 e^{V/U_T} u, \qquad p = C_0 \delta^2 e^{-V/U_T} v, \tag{2.4}$$

where $\delta^2 = \frac{n_i}{C_0}$ with a typical value C_0 for the doping profile, which is also scaled by C_0. Rescaling all quantities to non-dimensional analogues (cf. [14] for further details) we obtain the system

$$\lambda^2 \Delta V = \delta^2(e^V u - e^{-V} v) - C \qquad \text{in } \Omega \tag{2.5}$$

$$\mathrm{div} J_n = \delta^4 Q(u, v, V, x)(uv - 1) \qquad \text{in } \Omega \tag{2.6}$$

$$\mathrm{div} J_p = -\delta^4 Q(u, v, V, x)(uv - 1) \qquad \text{in } \Omega \tag{2.7}$$

$$J_n = \mu_n \delta^2 e^V \nabla u \qquad \text{in } \Omega \tag{2.8}$$

$$J_p = -\mu_p \delta^2 e^{-V} \nabla v \qquad \text{in } \Omega \tag{2.9}$$

where λ^2 is a positive constant and Q is defined via the relation $F(n, p, x) = Q(u, v, V, x)$. The new variables J_n and J_p are the scaled electron and hole current densities; the above mixed formulation seems to be natural for cases where one is interested in these quantities, since it contains them explicitly. Unless specified otherwise, we shall set δ to 1 in the sequel.

The Dirichlet boundary conditions can be written as

$$V = U + V_{bi} = U + \ln\left(\frac{1}{2\delta^2}(C + \sqrt{C^2 + 4\delta^2})\right) \qquad \text{on } \partial\Omega_D \tag{2.10}$$

$$u = e^{-U} \qquad \text{on } \partial\Omega_D \tag{2.11}$$

$$v = e^{U} \qquad \text{on } \partial\Omega_D. \tag{2.12}$$

On the remaining part $\partial\Omega_N = \partial\Omega - \partial\Omega_D$, the homogeneous Neumann conditions can be formulated in terms of J_n and J_p, i.e.,

$$\frac{\partial V}{\partial \nu} = J_n.\nu = J_p.\nu = 0 \qquad \text{on } \partial\Omega_N \tag{2.13}$$

We note that the mobilities μ_n and μ_p generally depend on the electric field strength, i.e, on $|\nabla V|$ in a realistic model. Such a dependence could be incorporated in our subsequent analysis. However, since the technical details one has to deal with in this general case do not contribute to the understanding of inverse doping problems and their solution, we will assume that μ_n and μ_p are positive constants in the following.Also, for the sake of simplicity we shall henceforth use $F \equiv 0$, i.e., no recombination-generation.

3. Available Data

In a typical experiment, measurements are always taken on the boundary of the device, more precisely on a contact $\Gamma_1 \subset \partial\Omega_D$. In the following we will therefore use the notation

$$\Sigma_1 := \begin{cases} \Gamma_1 & \text{in the stationary case} \\ \Gamma_1 \times (0, T) & \text{in the transient case} \end{cases}$$

For general semiconductor devices, two different types of data can be measured, namely:

- *Voltage-Current Data* (denoted by I_U) are given by measurements of the normal component of the current density $J := (J_n + J_p)$ on Σ_1, i.e.,

$$I_U := (J_n + J_p)|_{\Sigma_1} \tag{3.1}$$

  for all applied voltages $U \in \mathcal{U}$, where $\mathcal{U}$ is an appropriate class of functions on $\partial\Omega_D$ in the stationary and on $\partial\Omega_D \times (0,T)$ in the transient case.

- *Capacitance Data* (denoted by Q_Φ) around the voltage U are measurements of the variation of the electric flux in normal outward direction ($\frac{\partial V}{\partial \nu}$ on Σ_1) with respect to the voltage Φ, i.e.,

$$Q_\Phi := \lim_{s\to 0} s^{-1}\left(\frac{\partial V^{s\Phi+U}}{\partial \nu} - \frac{\partial V^U}{\partial \nu}\right)|_{\Sigma_1} \tag{3.2}$$

  for all voltages $\Phi \in \mathcal{U}$, where V^Φ denotes the solution of the Poisson equation with $U = \Phi$ and $\mathcal{U}$ is as above. For simplicity we assume that $U \equiv 0$ in the following, i.e., we are interested in capacitance data around equilibrium.

Using well-posedness and regularity results for the solutions of the stationary and transient DD-model, one can show that for given doping profile C, both current and capacitance are well-defined outputs for appropriate choices of the applied voltage U. In the stationary case, "appropriate" means smoothness (e.g. $U \in H^{\frac{3}{2}}(\partial\Omega_D)$) and we assume (for linearized stability of the DD system) smallness of U (cf. [4]), since hysteresis might occur for large voltages (cf. [14] for examples of non-unique solutions). In the transient case, a smallness assumption on U is not necessary, which is due to the fact the transient DD-model and its linearization are invertible for arbitrarily large applied voltage (cf. [13]).

The computation of the current consists of solving the DD-equations and evaluation of a trace type operator, which can be realized numerically by standard tools. The computation of the capacitance is more involved, since it requires the solution of the DD-model and its linearization. At equilibrium, i.e., $U \equiv 0$, the electron and hole density are given by $n^0 = e^V, p^0 = e^{-V}$ with the corresponding Slotboom variables $u^0 = v^0 = 1$. The potential V^0 solves the Poisson equation

$$\lambda^2 \Delta V^0 = \delta^2 (e^{V^0} - e^{-V^0}) - C \tag{3.3}$$

subject to the boundary conditions given above with $U \equiv 0$ (note that this holds both in the transient as in the stationary case). The capacitance can now be computed by a linearization of the drift-diffusion model with respect to the applied voltage, i.e., it is given by

$$Q_\Phi = \frac{\partial \hat{V}}{\partial \nu}|_{\Sigma_1}, \tag{3.4}$$

where in the transient case $(\hat{V}, \hat{n}, \hat{p})$ solves the linearized equations (given here after scaling)

$$\begin{aligned} 0 &= \lambda^2 \Delta \hat{V} - q(\hat{n} - \hat{p}) && (3.5)\\ \frac{\partial \hat{n}}{\partial t} &= \operatorname{div}\left(\mu_n(\nabla \hat{n} - \hat{n}\nabla V^0 - e^{V^0}\nabla \hat{V})\right) && (3.6)\\ \frac{\partial \hat{p}}{\partial t} &= \operatorname{div}\left(\mu_p(\nabla \hat{p} + \hat{p}\nabla V^0 + e^{-V^0}\nabla \hat{V})\right) && (3.7) \end{aligned}$$

in $\Omega \times (0, T)$, subject to homogenous initial conditions, homogeneous Neumann boundary conditions on $\partial\Omega_N$, and the Dirichlet boundary conditions

$$\hat{V} = \Phi, \qquad \hat{n} = \hat{p} = 0, \tag{3.8}$$

on $\partial\Omega_D$. Again, in the stationary case, the boundary conditions remain the same and the corresponding differential equations are obtained from the transient ones by setting $\frac{\partial n}{\partial t} = \frac{\partial p}{\partial t} = 0$.

The above formulation of the data set is a rather general one, in specific applications one usually has to deal with some of the following choices for the function class $\mathcal{U}$:

- *Full data:* here $\mathcal{U}$ denotes a linear function space of admissible applied voltages, e.g., $\mathcal{U} = H^{\frac{3}{2}}(\partial\Omega_D)$ in the stationary case. This case is a mathematical idealization of a situation with a very large number of measurements. For full data, the identification problem has many analogies to the important field of impedance tomography (cf. [6, 11]).

- *Parameterized data set:* in this case $\mathcal{U}$ is a special function class that can be parametrized using parameters $s_j \in (-S, S)$, $j = 1, \dots, m$, with some $S \in \mathbf{R}_+$. Of particular importance is the case where U is piecewise constant on some disjoint sets $\Gamma_j \subset \partial\Omega_D$, which represent different ohmic contacts. The parameter s_j denotes the voltage applied on the j-th contact.

- *Finite number of measurements:* here $\mathcal{U}$ consists of a finite number N of functions U_j, $j = 1, \dots, N$ on $\partial\Omega_D$ (and possible in the time interval $(0, T)$).

A frequently appearing special case is the one with a *single measurement*, i.e., the preceding case with $N = 1$.

An immediate observation for all cases of data is that the amount of available data is much larger in the transient case than in the stationary case. Together with the simpler mathematical analysis, this clearly makes the transient case favourable. However, under practical conditions it is not always possible to obtain meaningful transient measurements, since the time variation only occurs in a small initial time layer. Therefore one has to use either the stationary or the transient model dependent on the specific application.

For the sake of simplicity, we restrict our attention here to the case of a finite number of applied potentials, with measured current or capacitance (or both of them). Under the standard conditions on the applied potential U and the domain Ω one can show that current and capacitance are well-defined on a contact $\Gamma \subset \partial\Omega_D$.

4. Identification of Doping Profiles

In the following we discuss some mathematical problems concerned with the identification of spatially varying doping profiles. The domain of admissible doping profiles C is given by

$$\mathcal{D} := \left\{ C \in L^2(\Omega) \mid \underline{C} \leq C \leq \overline{C} \text{ a.e. in } \Omega \right\} \tag{4.1}$$

for some constants $\underline{C}, \overline{C} \in \mathbf{R}$.

All the above cases can be transformed to the standard form for an inverse problem, namely the nonlinear operator equation

$$F(C) = Y^\delta, \tag{4.2}$$

where F stands for the parameter-to-output map

$$\begin{array}{rccc} F: & \mathcal{D} & \rightarrow & L^2(\Sigma_1)^N \\ & C & \mapsto & (I_{U_j})_{j=1,\ldots,N} \end{array} \tag{4.3}$$

for current measurements, and

$$\begin{array}{rccc} F: & \mathcal{D} & \rightarrow & L^2(\Sigma_1)^N \\ & C & \mapsto & (Q_{\Phi_j})_{j=1,\ldots,N} \end{array} \tag{4.4}$$

for capacitance measurements. The right-hand side Y^δ represents noisy current or capacitance data, and we assume that the data error is bounded by δ, i.e.,

$$\|Y^\delta - Y\|_{L^2(\Sigma_1)^N} \leq \delta \tag{4.5}$$

for the exact data Y.

We are now able to state the following result on the parameter-to-output operator, for a proof in the stationary case we refer to [4] and in the transient case to [5]:

Theorem 1. *The parameter-to-output map F is well-defined by* (4.3) *respectively* (4.4) *and Frèchet-differentiable on $\mathcal{D}$.*

The well-definedness and differentiability of the operator F enables the application of *iterative regularization methods* for the solution of the identification problem, such as the Landweber iteration

$$C_{k+1} = C_k - \omega F'(C_k)^*(F(C_k) - Y^\delta), \tag{4.6}$$

with appropriate damping parameter $\omega \in \mathbf{R}_+$, or Newton-type methods, e.g., the *Levenberg-Marquardt method*

$$C_{k+1} = C_k - (F'(C_k)^* F'(C_k) + \alpha_k)^{-1} F'(C_k)^*(F(C_k) - Y^\delta), \tag{4.7}$$

where $(\alpha_k)_{k \in \mathbf{N}}$ is a sequence of positive real numbers. We mention that the evaluation of the directional derivative $F'(C)\hat{C}$ and of the adjoint $F'(C_k)^*\hat{Y}$ require the solution of a linear system similar to the original drift-diffusion equations, which causes a high effort in the numerical solution of the identification problem. We refer to Burger et al [4] for the iterative regularization of the inverse doping problem (4.2) and to [7] for a unified overview of iterative regularization methods.

In the remaining part of this section we focus on a fundamental question in the identification of parameters from indirect measurements:

Do the data determine the doping profile uniquely, respectively which set of data is suffficient for uniquely determining the doping profile ?

The mathematical equivalent of this question is called *identifiabilty* (cf. [1]) and means to investigate the injectivity of the parameter-to-output map F, which is a difficult task for inverse doping problems as we shall see below. Therefore we restrict our attention to the special case of unipolarity, where a rigorous analysis can be carried out. The unipolar drift diffusion equations arise from the original DD-system by setting $p \equiv 0$ in Ω. We start this discussion in the stationary case, where a large amount of data is necessary in order to ensure identifiability. In the transient case we shall show that even a single measurement of capacitance and current can determine the doping profile uniquely.

4.1 Stationary Inverse Doping

We start our investigation of stationary inverse doping problems with the spatially one-dimensional case, i.e., $\Omega = (0, L)$. We assume without restriction of generality that the voltage is applied at $x = 0$ and the measurements of current and capacitance are taken at $x = L$. Even if we are able to measure both, a single measurement consists only of two real numbers, which can clearly not suffice to identify the doping profile as a function of the spatial variable x. Full data in this case means to measure current and capacitance as a function of the applied voltage $U \in (-r, r)$ with appropriate $r \in \mathbf{R}^+$, respectively the variation of the voltage $\Phi \in \mathbf{R}$.

At a first glance it seems possible to identify the doping profile from full data, since the data are now a function of one variable and, roughly speaking, of the same dimensionality as the parameter to be identified. However, the information content in the data is much smaller, which can be seen very easily for capacitance measurements. Here the map $\Phi \mapsto Q_\Phi$ is affinely linear because it only consists in solving the linearized drift-diffusion equations and evaluating a linear trace operator. Since an affinely linear function of one variable can be characterized by two real numbers, the information content is the same as if one would measure the capacitance at only two different values $\Phi \in \mathbf{R}$ and hence cannot suffice to determine the doping profile uniquely. For current measurements a similar argument holds, since it can be shown that the current behaves around $U = 0$ like an exponential function of U (cf. [14]).

In two spatial dimensions, the situation is different. Here a single measurement of current or capacitance is given by a function of one spatial variable over the contact Γ_1. Of course, a single measurement is again not sufficient for identifiability of the doping profile, which is a function over the two-dimensional domain Ω, but by exploiting similarities to electrical impedance tomography (cf. [6, 15]) we may argue that full data are sufficient, which we shall rigourously prove in a special case below.

As a starting point for the analysis we investigate the unipolar drift-diffusion system around equilibrium, i.e., its linearization with respect to the voltage at $U = 0$. Since $p = 0$, this implies $u \equiv 1$, $v \equiv 0$ and the linearization $\hat{v}$ solves the elliptic differential equation

$$\operatorname{div}\,(e^{V^0} \nabla \hat{u}) = 0 \qquad \text{in } \Omega \tag{4.8}$$

subject to the boundary conditions

$$\hat{u} = \Phi \quad \text{on } \partial\Omega_D, \qquad \frac{\partial \hat{u}}{\partial \nu} = 0 \quad \text{on } \partial\Omega_N. \tag{4.9}$$

The function V^0 is the solution of the Poisson equation at equilibrium, i.e.,

$$\Delta V^0 = e^{V^0} - C \qquad \text{in } \Omega \tag{4.10}$$

with the boundary conditions

$$V^0 = V_{bi} \quad \text{on } \partial\Omega_D, \qquad \frac{\partial V^0}{\partial \nu} = 0 \quad \text{on } \partial\Omega_N. \tag{4.11}$$

The output current in this case can be identified after rescaling with the Neumann boundary data of $\hat{u}$, i.e.,

$$I_\Phi := \frac{\partial \hat{u}}{\partial \nu}|_{\Gamma_1}. \tag{4.12}$$

From the standard theory of elliptic differential equations one may conclude that for a domain Ω with regular boundary, there is a one-to-one relation between functions $V^0 \in H^2(\Omega)$ satisfying the Poisson equation and potentials $C \in L^2(\Omega)$. This motivates the investigation of the identifiability of $V^0 \in H^2(\Omega)$ in (4.8) directly, since for known potential V^0 the doping profile is determined uniquely by (4.10). The identifiability of the potential V^0 in (4.8), respectively the identifiability of the conductivity $a = e^{V^0}$ has been investigated by Nachman [15] with a positive answer only in the case of full data, which leads to the following result:

Theorem 2. *Let $\Omega \subset \mathbf{R}^2$ be a bounded Lipschitz domain, $\Gamma_1 = \partial\Omega_D = \partial\Omega$, and let for two doping profiles C_1 and C_2 in $\mathcal{D}$ denote their output currents by I^1_Φ and I^2_Φ obtained from linearization around equilibrium. Then the equality*

$$I^1_\Phi = I^2_\Phi, \qquad \forall\, \Phi \in H^{\frac{3}{2}}(\partial\Omega)$$

implies $C_1 = C_2$.

4.2 Transient Inverse Doping

We have seen in the previous section that the identification in the stationary case makes no sense for spatial dimension one. In the transient case, the situation is different, because a second dimension is added via the time variable. If we measure current and capacitance over a time interval $(0,T)$, the dimensionality of the data is the same as of the doping profile, namely that of functions over an interval. Since there are many examples of parabolic identification problems, where a measurement on the boundary over some time interval determines a spatially distributed parameter uniquely (cf. [11] and the references therein), it seems reasonable that one can identify the doping profile from a single transient measurement.

For a rigorous justification of the identifiability, we consider the model problem of the unipolar transient drift-diffusion equations in $\Omega = (0,L)$. After appropriate scaling (setting $\mu_n = 1$), they are given by

$$0 = \lambda^2 \frac{\partial^2 V}{\partial x^2} - n + C \qquad \text{in } (0,L)\times[0,T] \tag{4.13}$$

$$\frac{\partial n}{\partial t} = \frac{\partial}{\partial x}\Big(\frac{\partial n}{\partial x} - n\frac{\partial V}{\partial x}\Big) \qquad \text{in } (0,L)\times[0,T], \tag{4.14}$$

subject to appropriate Dirichlet boundary conditions for n and V at $x = 0$ and $x = L$, and the initial condition

$$n(x,0) = n_0(x) \geq 0. \tag{4.15}$$

We assume that the potential $U = U(t)$ is applied at $x = 0$, i.e.,

$$V(0,t) = U(t) + V_D(0), \qquad V(1,t) = V_D(1). \tag{4.16}$$

Introducing the antiderivatives N and D determined by

$$\frac{\partial N}{\partial x} = n, \quad N(L,.) = 0, \qquad D_x = C, \quad D(L) = 0 \tag{4.17}$$

for all $t \in [0,T]$, we can integrate (4.13) to

$$\lambda^2 \frac{\partial V}{\partial x} = N - D + \alpha, \tag{4.18}$$

where $\alpha = \alpha(t)$. Thus, the system (4.13), (4.14) can be reduced to the single equation

$$\frac{\partial N}{\partial t} = J + \beta = \frac{\partial^2 N}{\partial x^2} - \lambda^{-2} \frac{\partial N}{\partial x}(N - D) - \frac{\partial N}{\partial x}\alpha + \beta, \tag{4.19}$$

where α and β are functions of time only, which have to be determined from boundary data. Using the boundary values of N and D at $x = L$, we deduce $\alpha(t) = \lambda^2 \frac{\partial V}{\partial x}(L,t)$. Moreover, the current J at $x = 1$ determines β, since

$$0 = \frac{\partial N}{\partial t}(L,t) = J(L,t) + \beta(t).$$

I.e., under the knowledge of $\frac{\partial V}{\partial x}(L,t)$ and $J(L,t)$, the reconstruction of the doping profile C in the unipolar case reduces to the identification of the spatially varying source D in the parabolic equation (4.19) from the overposed boundary data for $\frac{\partial N}{\partial x}$ at $x = 0, L$ and N at $x = L$. Using results for parabolic inverse problems based on Carleman estimates, we may derive the following result (cf. [5]):

Theorem 3. *Let $C \in C^{1,1}(\Omega)$, $\frac{dU}{dt}(s) \neq 0$ for some $s \in (0,T)$ and assume that $\frac{\partial V}{\partial x}(L,t)$ and*

$$J(L,t) = \frac{\partial n}{\partial x}(L,T) - n(L,T)\frac{\partial V}{\partial x}(L,t)$$

are known in a finite time interval $(0,T)$ and $\frac{\partial V}{\partial x}(L,.) \in C^{1,1}(0,T)$. Then the solution (n,V,C) of the unipolar identification problem is uniquely determined.

In two spatial dimensions, the situation is similar, since we want to identify a function on $\Omega \subset \mathbf{R}^2$ from a measurement on $\Gamma_1 \times (0,T) \subset \mathbf{R}^2$. However, there are no rigorous results on the multi-dimensional inverse doping profile yet. Nonetheless, there is hope to derive uniqueness results at least in special cases, which raises important problems for future research.

5. Identification of P-N Junctions

Finally, we consider the case, where the doping profile is a piecewise constant function of position. We assume that there exists a decomposition $\overline{\Omega} = \overline{\Omega_P} \cup \overline{\Omega_N}$ and values $C_+ \in \mathbf{R}_+$, $C_- \in \mathbf{R}_-$, such that

$$C \equiv C_+ \quad \text{in } \Omega_P, \qquad C \equiv C_- \quad \text{in } \Omega_N. \tag{5.1}$$

Both Ω_N and Ω_P shall only consist of a finite number of connected components. Under this a-priori information it seems now reasonable to consider also the stationary case with a finite number of measurements.

In the case of one spatial dimension, we only have to identify a finite number of points $x_j \in (0, L)$ that mark the location of the p-n junction. This seems possible from voltage and capacitance measurements if the number of junctions is not too large. E.g., for a p-n diode, which is a device where the Ω_N and the Ω_P region consist of only one connected component each, one only seeks the location of one junction, i.e., a single real value in the interval $(0, L)$, which seems reasonable to be determined from a single measurement of the current or the capacitance. We will rigorously prove the identifiability of the p-n junction for a unipolar p-n diode in the following section.

In two spatial dimensions one may argue again that a single measurement on a contact is sufficient for identifiability of the p-n diode, which is now a curve, i.e., a function of one variable (the arclength parameter). We will investigate the identification in a special case, namely a p-n diode with zero space charge and low injection below; for this problem the identification of the p-n junction reduces to an inverse boundary problem for the Laplace equation and one can rigorously prove identifiability. However, it is well-known that such inverse problems are *severely ill-posed*, i.e., measurement errors are amplified dramatically. Typically, stability estimates for the unknown boundaries in such problems are only of logarithmic type with respect to the data error δ (cf. e.g. [2]).

5.1 A Unipolar P-N Diode in $\mathbf{R}^1$

In the following we investigate a simple identification problem for a one-dimensional p-n diode ($\Omega = (0, L)$) in the unipolar case. The linearization around equilibrium ($U = 0$, as a system for the equilibrium potential V^0 and the perturbation $\hat{u}$ of u) in Slotboom variables reads

$$\lambda^2 \frac{\partial^2 V^0}{\partial x^2} = e^{V^0} - C \qquad \frac{\partial}{\partial x}\left(e^{V^0} \frac{\partial \hat{u}}{\partial x}\right) = 0, \tag{5.2}$$

in Ω, with Dirichlet boundary conditions for V^0 and $\hat{u}$ at $x = 0$ and $x = L$. If we are given the linearized current $J = e^{V^0}\frac{\partial \hat{u}}{\partial x}$ at $x = 0$, we may conclude that

$$\frac{\partial \hat{u}}{\partial x} = e^{-V^0} J \qquad \text{in } \Omega.$$

With the given Dirichlet boundary values $u(0)$ and $u(L)$ we can reduce the identification of the p-n junction to the identification of the p-n junction in the equilibrium Poisson equation with the additional condition

$$\hat{u}(L) - \hat{u}(0) = J \int_0^L e^{-V^0(x)}\, dx. \tag{5.3}$$

For the solution of this identification problem we can now prove an identifiability result:

Theorem 4. *Let $C_+ > 0$ and $C_- \leq 0$ be given and let C_1 and C_2 be two doping profiles satisfying*

$$C_i = \begin{cases} C_+ & \text{for } x < p_i \\ C_- & \text{for } x > p_i \end{cases} \tag{5.4}$$

for $p_i \in (0, L)$, $i = 1, 2$. Denote by $(V_i^0, \hat{u}_i)$ the corresponding solutions of (5.2) with doping profile C_i. If the output currents $J_i = e^{V_i^0(0)}\frac{\partial \hat{u}_i}{\partial x}(0)$ are equal, then $C_1 = C_2$ in Ω.

Proof. We have seen above that we may equivalently consider the identification of C in the Poisson equation with the additional integral condition (5.3). Let in the following w.r.o.g. $p_1 \leq p_2$ and set $w := V_1^0 - V_2^0$, then by the mean value theorem we may deduce the existence of bounded functions a and b such that w satisfies

$$\begin{aligned} -\lambda^2 \frac{\partial^2 w}{\partial x^2} + e^a w = C_1 - C_2 = (C_+ - C_-)1|_{(p_1, p_2)} &\geq 0, \\ \int_0^L e^{b(x)} w(x)\, dx &= 0. \end{aligned}$$

Moreover, w satisfies homogeneous Dirichlet boundary conditions at $x = 0$ and $x = L$. Using the maximum principle for elliptic differential equations we deduce that $w \geq 0$ in Ω and hence, the above integral identity can only hold for $w \equiv 0$, which implies also $\chi(p_1, p_2) \equiv 0$. Thus, we have $p_1 = p_2$ and consequently $C_1 \equiv C_2$. □

5.2 A P-N Diode with Zero Space Charge and Low Injection

The case of zero space charge and low injection means to let first tend $\lambda \to 0$ and then $\delta \to 0$ in the stationary drift-diffusion equations. It has been shown by Schmeiser [16] that the arising limiting problem for u and v can be solved explicitely and identifiability can be shown in two space dimensions (cf . [4]) using tools from the theory of harmonic functions.

In [4], numerical test have been performed for the identification of the p-n junction in this special case. The data where generated by numerically solving the stationary drift-diffusion equations with appropriate parameter choices (cf. [3] for details on the numerical scheme employed) and subsequently evaluating the current over the contact. The inverse problem was solved using an iterative regularization method for the simplified model. The results for two different values of $C_0 = C_+ = |C_-|$, which can be interpreted as two different noise levels in the current measurements, are shown in Figure 1. One observes that the quality of the reconstruction improves with increasing C_0, which is related to the better approximation of the reduced equation to the original drift-diffusion model. The results obtained indicate that the doping profile is not only identifiable, but can be reconstructed with reasonable precision also from noisy data obtained in practice.

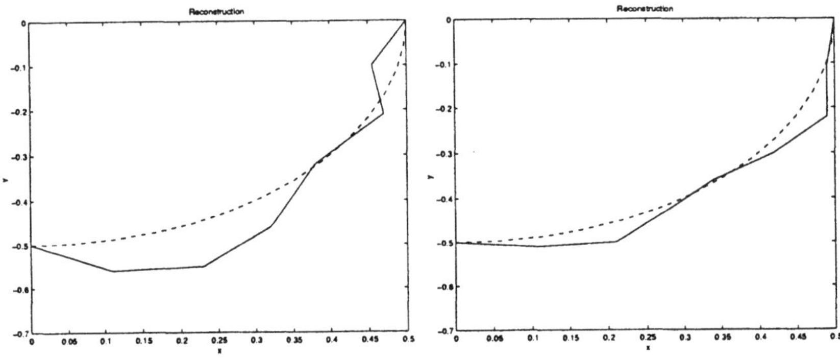

Figure 1. Reconstruction using data from the full drift-diffusion model for $C_0 = 10^{20}m^{-3}$ (left) and $C_0 = 10^{21}m^{-3}$ (right) compared to the exact junction (dotted).

References

[1] H.T.Banks, K.Kunisch, *Estimation Techniques for Distributed Parameter Systems* (Birkhäuser, Basel, Boston, 1989).

[2] E.Beretta, S.Vessella, *Stable determination of boundaries from Cauchy data*, SIAM J. Math. Anal. **30** (1998), 220-232.

[3] F.Brezzi, L.D.Marini, P.Pietra *Two-dimensional exponential fitting and applications in drift-diffusion models*, SIAM J. Numer. Anal. **26** (1989), 1342-1355.

[4] M.Burger, H.W.Engl, P.Markowich, P.Pietra, *Identification of doping profiles in semiconductor devices*, Inverse Problems (2001), to appear.

[5] M.Burger, P.Markowich, *Identification of doping profiles from transient measurements*, in preparation.

[6] M.Cheney, D.Isaacson, J.C. Newell, *Electrical impedance tomography*, SIAM Review **41** (1999), 85-101.

[7] H.W.Engl, O.Scherzer, *Convergence rate results for iterative methods for solving nonlinear ill-posed problems*, in D.Colton, H.W. Engl, A.Louis, J.McLauglin, W.Rundell, eds., *Surveys on Solution Methods for Inverse Problems* (Springer, Vienna, 2000).

[8] M.Hinze, R.Pinnau, *Optimal control of the drift-diffusion model for semiconductor devices* , Math. Models and Methods in Applied Sciences (2001), to appear.

[9] A.Jüngel, *Quasi-hydrodynamic Semiconductor Equations*, Progress in Nonlinear Differential Equations (Birkhäuser, Boston, Basel, 2001).

[10] N.Khalil, *ULSI Characterization with Technology Computer-Aided Design* (PhD-Thesis, Technical University Vienna, 1995).

[11] V.Isakov, *Inverse Problems for Partial Differential Equations* (Springer, New York, 1998).

[12] P.A.Markowich, *The Stationary Semiconductor Device Equations* (Springer, Wien, New York, 1986).

[13] P.A.Markowich, C.A.Ringhofer, *Stability of the linearized transient semiconductor device equations*, Z. Angew. Math. Mech. **67** (1987), 319-332.

[14] P.A.Markowich, C.A.Ringhofer, C.Schmeiser, *Semiconductor Equations* (Springer, Wien, New York, 1990).

[15] A.I.Nachman, *Global uniqueness for a two-dimensional inverse boundary value problem*, Annals of Mathematics **143** (1996), 71-96.

[16] C.Schmeiser, *Voltage-current characteristics of multi-dimensional semiconductor devices*, Quarterly of Appl. Math. **4** (1991), 753-772.

[17] S.Selberherr, *Analysis and Simulation of Semiconductor Devices* (Springer, Wien, New York, 1984).

[18] M.Stockinger, R.Strasser, R.Plasun, A.Wild, S.Selberherr, *A qualitative study on optimized MOSFET doping profiles*, Proc. of SISPAD 98 (Springer, 1998), 77-80.

[19] M.Stockinger, R.Strasser, R.Plasun, A.Wild, S.Selberherr, *Closed-loop MOSFET doping profile optimization for portable systems*, Proceedings Intl. Conf. on Modeling and Simulation of Microsystems, Semiconductors, Sensors, and Actuators (San Juan, 1999), 411-414.

[20] W.R. Van Roosbroeck, *Theory of flow of electrons and holes in germanium and other semiconductors*, Bell Syst. Tech. J. **29** (1950), 560-607.

SUPERCONDUCTIVITY ANALOGIES, FERROELECTRICITY AND FLOW DEFECTS IN LIQUID CRYSTALS

M. Carme Calderer*
School of Mathematics, University of Minnesota
Minneapolis, MN 55455
mcc@math.psu.edu

Abstract Liquid crystal phases present a large variety of physical phenomena often leading to interesting applications. Here we address issues involving nematic and smectic phases, from the point of view of modeling and analysis. These include superconductivity analogies of the phase transition from nematic to smectic A*, ferroelectricity of the smectic C* and flow defects in some special regimes. Molecular chirality plays an important role in such behaviors.

1. Introduction

The theory of liquid crystals and liquid crystalline polymers spans a wide spectrum from the rheology of complex fluids to optics and defects. All aspects of the theory are highly linked with experiment, placing liquid crystal research in an interesting position; it is driven by both theory and experiment and by both the abstract and the applied.

This article addresses mathematical and physical issues concerning the nematic, smectic A and smectic C phases of liquid crystals. Equilibrium as well as flow problems are considered.

Liquid crystal molecules are rod-like (on the scale of 20 A° in length and 5 A° in diameter) and in the nematic phase tend to follow a preferred direction of alignment. The smectic phases present an additional one–dimensional positional ordering as layer structures. Smectic A molecules tend to align perpendicularly to the layers, while those of the smectic C phase are at a preferred

*This work has been partially supported by a grant from the National Science Foundation, contract number DMS-9704714.
I would like to thank the organizers of the International Symposium on Computational and Applied PDEs and the community of Zhangjiajie, for the kind welcome and hospitality during the week of July 1-7, 2001.

Recent Progress in Computational and Applied PDES
Edited by Tony Chan *et al.*, Kluwer Academic/Plenum Publishers, 2002

(temperature dependent) angle different than $\frac{\pi}{2}$. The phase transition to the Smectic C takes place upon lowering the temperature of the smectic A liquid crystal sample. (Some liquid crystals may experience a transition directly from nematic to smectic C). Some liquid crystals are made of chiral molecules and exhibit a natural twist with a well determined pitch (wave number), τ, depending on the material and temperature. (Conventional notation assigns a "*" symbol to chiral phases.)

A survey on modeling is followed by *case studies* emphasizing connections to applications and to other problems of soft matter physics. We present one topic from the static theory dealing with the phase transition between the chiral nematic and the smectic A*, emphasizing the analogy with the conductor to superconductor transition; we also illustrate how ferroelectricity arises in smectic liquid crystals. From the point of view of dynamics, we present a flow problem of nematic liquid crystals to illustrate how the competition between elastic and viscous torques causes defects to develop.

The average perpendicular alignment of molecules with respect to the layers in the smectic A* phase results in no spontaneous polarization. This is not the case with the smectic C* that may present significant polarization leading to seizable ferroelectric effects, and to multiple ferroelectric (and antiferroelectric) phases. Such phases often present complex interactions with electric fields allowing for important optic applications. We show that the (de Gennes) model yields the two phases associated to the Clark-Lagerwall effect, that allows for ferroelectric switching devices.

Chirality is an issue at the center of our studies of smectic phases. This is a molecular property that expands many fields, since chiral molecules are widely encountered in nature and may yield many liquid crystal phases. They are extensively studied by biologists, chemists and biophysicists [16], and there is still much theoretical work to be done on the wealth of experimental data. Chiral smectic liquid crystals exhibit the analogue of the Meissner phase in superconductors, in which chirality is excluded from the bulk (just as the magnetic field is expelled in a superconductor). The constraints imposed by topology and geometry can eliminate chirality as well.

The last section of the paper deals with the interaction between molecular alignment and flow in nematic liquid crystals. Liquid crystal flow is genuinely non-Newtonian presenting complex defect structures. The development of defects and texture is known to pose difficulties in the processing and subsequent uses of polymeric liquid crystal materials. Experimentally, it is found that the defect density in flow increases with the Ericksen number. This is a dimensionless quantity that measures the ratio of the viscous torque with respect to the elastic one, and together with the Reynolds number and the interface number (ratio of elastic surface stress with respect to bulk ones) fully characterizes flow regimes. The wealth of available experimental results for shear flow geometries proves very useful in probing and analyzing mathematical models. I present the case study of such a flow, both non-steady as well as steady, with the purpose of obtaining quantitative information on how the defect density depends on the dimensionless parameters of the flow, showing, in particular, that the number of defects increases with the Ericksen number, $\mathcal{E}$. Experimental results also indicate that for very large values of $\mathcal{E}$, the flow behaves nearly isotropically, yielding the Newtonian viscosity and vanishing of the normal stress differences. We address the latter issues by obtaining the effective flow configurations at the limit of very large $\mathcal{E}$. This yields the limiting configurations in terms of Young measures associated with sequences of weak solutions [10]. Moreover, we show that the resulting effective equations correspond to the Newtonian flow and obtain additional relations for Young measures representing a remnant microstructure of ordered states. Numerical simulations of such a problem support the analytic as well as the experimental results [1].

2. Free energy functions of smectic materials

In the Oseen–Frank theory, static configurations of uniaxial nematic liquid crystals are given by unit vector fields $\mathbf{n}$ (director fields) that minimize the total energy $\mathcal{F} = \int_\Omega F_{\rm N}(\mathbf{n})\,d\mathbf{x}$, with

$$F_{\rm N} = K_1|\nabla\cdot\mathbf{n}|^2 + K_2|\mathbf{n}\cdot(\nabla\times\mathbf{n}) + \tau|^2 + K_3|\mathbf{n}\times(\nabla\times\mathbf{n})|^2$$
$$+(K_2 - K_4)\big((\mathrm{tr}(\nabla\mathbf{n}))^2 - (\nabla\cdot\mathbf{n})^2\big), \tag{1}$$

where K_i, $i = 1\,,2\,,3$ correspond to the splay, twist and bend elasticity constants, respectively; they indicate the energetic expense of a configuration in terms of these three pure distortions. The scalar τ represents the chiral pitch of the helical structures of the cholesteric phases. If $\tau = 0$ then any constant vector $\overline{\mathbf{n}}$ with $|\overline{\mathbf{n}}| = 1$ describes an undistorted equilibrium configuration for F_N. If $\tau \neq 0$ (chiral liquid crystals) the special twist configuration

$$\mathbf{n}_\tau(\mathbf{x}) = (\cos(\tau z), \sin(\tau z), 0) \tag{2}$$

makes the fundamental energy contributions of splay, twist, and bend in (1) vanish. The fourth term in $F_{\rm N}$ is a null-Lagrangian; its integral is determined by $\mathbf{n}$ on $\partial\Omega$.

The Oseen-Frank theory for $\mathbf{n}$ can be regarded as a special case of the Landau–de Gennes theory involving the traceless, symmetric *order parameter tensor*, Q, as introduced through averaging of molecular directions. Biaxial optics corresponds to the case of two distinct eigenvalues of Q, whereas the case with two equal, nonzero eigenvalues is associated with uniaxiality. In the latter case, the eigenvector corresponding to the common eigenvalue is, in fact, the director $\mathbf{n}$ of the Oseen-Frank theory, with the eigenvalue, s corresponding to the variable degree of orientation, uniaxial order parameter. The latter takes values on the interval $(-\frac{1}{2}, 1)$, with $s = 0$, representing the isotropic phase, $s = 1$ perfect alignment; $s = -\frac{1}{2}$ indicates a degenerate limiting case with molecules placed on a plane perpendicular to the director. The free energy in terms of the order tensor $\mathbf{Q}$ in the Landau-de Gennes form is quadratic in $\nabla\mathbf{Q}$, with additional scalar polynomial terms on $\mathbf{Q}$. According to Ericksen [15], in the uniaxial case, the free energy is given by

$$F_S = K|\nabla s|^2 + s^2 F_N + \nu f(s), \tag{3}$$

where the double-well potential $f(s)$ is parameterized by the concentration or the temperature.

In uniform smectic A materials the layers have the director field $\mathbf{n}$ as their normal and are commonly between one and two molecular lengths thick ($\sim$ 30 $A°$). The number q is the *layer number*, and $\frac{2\pi}{q}$ is the *layer thickness*.

In the case of an undistorted (non-chiral) nematic, $\mathbf{n} = \overline{\mathbf{n}}$ = constant and the liquid crystal has a locally uniform molecular mass density ρ_0. For the case of an undistorted smectic A material, $\mathbf{n} = \overline{\mathbf{n}}$ and the mass density modulates causing layering; moreover the density's Fourier modes near $\{\pm q\}$ dominate. In this case the mass density is taken as $\delta(\mathbf{x}) = \rho_0 + \rho_1 \cos(q\overline{\mathbf{n}} \cdot \mathbf{x})$, where the amplitude ρ_1 indicates the intensity of the smectic layering.

In [dG] de Gennes introduced a complex wave function, $\Psi(\mathbf{x})$, to describe smectic layering. For the undistorted case, with $\mathbf{n} = \overline{\mathbf{n}}$ and $\delta(\mathbf{x})$ as above, de Gennes set

$$\Psi(\mathbf{x}) = \rho_1 e^{iq\overline{\mathbf{n}}\cdot\mathbf{x}}. \tag{4}$$

In general he identified $|\Psi(\mathbf{x})|$ with the amplitude of modulation, $\nabla \arg(\Psi(\mathbf{x}))$ with the normal to the layer structure, and $|\nabla \arg(\Psi(\mathbf{x}))|$ with the layer number at $\mathbf{x}$.

The free energy associated with the smectic layering is given by the following covariant form proposed by de Gennes and it involves the *complex order parameter* Ψ ([14, 12]):

$$F_C = \left((C_{\|} - C_{\perp})\mathbf{n} \otimes \mathbf{n} + \mathrm{C}_{\perp}\mathbf{I}\right) : (\nabla - \mathrm{i}q\mathbf{n})\Psi \otimes (\nabla + \mathrm{i}q\mathbf{n})\Psi * \\ + r|\Psi|^2 + \frac{g}{2}|\Psi|^4 + F_N. \tag{5}$$

The smectic A* case corresponds to $C_{\|} = C_{\perp} = 1$, in which case (5) yields

$$F_A = |(i\nabla + q\mathbf{n})\Psi|^2 + r|\Psi|^2 + \frac{g}{2}|\Psi|^4 + F_N. \tag{6}$$

with $g > 0$ and $r = T - T_{NA}$; here T denotes the (constant) temperature of the material and T_{NA} is the nematic–smectic transition temperature at $\tau = 0$.

Writing $\Psi = \rho e^{i\Phi}$, the first term in (6) can be rewritten as $|(i\nabla + q\mathbf{n})\Psi|^2 = |\nabla\rho|^2 + \rho^2|\nabla\Phi - q\mathbf{n}|^2$. Note that this term vanishes in the case of uniform smectic layering (1.5).

The smectic C model corresponds to the case $C_{||} \neq C_{\perp}$ in (5); such a quantity is related to the $\beta \neq 0$ angle of the C-phase. Moreover, when the phase is also chiral, successive smectic C* layers show a gradual change in the direction of tilt, and such that the director precesses about the normal axis to the layer, lying on the surface of a cone of angle 2β. This creates a helical structure in the chiral smectic C* that results in a spontaneous molecular polarization; the polarization vector **P** is perpendicular to the molecule and contained in the layer plane. Therefore, all possible directions for the vector are tangent to the circle of intersection of the cone with the plane.

3. Smectic A* phases and the superconductivity analog

The topics discussed in this section are join work with Bauman, Phillips and Liu [26]. Related issues on defect structures in smectic materials are presented in several articles by Calderer and co-authors ([4], [5], [3], [6]). In this section we study minimizers to the Landau-de Gennes energy for chiral smectic A liquid crystals (smectic A^*):

$$\mathfrak{F} = \int_{\Omega} F_{\mathrm{A}}\, d\mathbf{x}, \tag{7}$$

where Ω is the domain occupied by the liquid crystal. Such an energy is appropriate for describing the behavior near the transition temperature between nematic and smectic A phases. The leading mechanism underlying all of the observed phenomena is the competition between the tendency of the molecules to form layers in the smectic phase and the helical twist preferred by nematic chiral structures. This fact also becomes the central mathematical issue of the analysis in this work.

We establish the existence of minimizers within a general class of admissible fields that assumes physically natural boundary conditions. We then study the nature of minimizers for parameter values near the transition regime. Three parameters are distinguished: q, τ and $r = T - T_{NA}$. The latter measures the temperature of the material relative to T_{NA} which denotes the transition temperature for a nonchiral material ($\tau = 0$). We find two curves $r = \underline{r}(q\tau)$ and $r = \overline{r}(q\tau)$, in the $q\tau - r$ plane where

$$\underline{r}(q\tau) < \overline{r}(q\tau) < 0 \qquad \text{for} \quad q\tau > 0. \tag{8}$$

These curves bound the transition region provided $\frac{q}{\tau}$ and the Frank constants, K_2 and K_3 are sufficiently large. We prove that there is a constant $\overline{\lambda} \geq 1$ so that if $q\tau \leq \frac{q^2}{\overline{\lambda}}$, and $(q\tau, r)$ is such that $r < \underline{r}(q\tau)$ then minimizers will be in the smectic A^* phase while if $r > \overline{r}(q\tau)$ the minimizers will be chiral nematic (N^*).

Minimizing configurations $(\Psi, \mathbf{n})$ for $\mathfrak{F}$ for which $\Psi \equiv 0$ are called *nematic* (N) if $\tau = 0$ and *chiral nematic* (N^*) if $\tau \neq 0$. Minimizers for which $\Psi \not\equiv 0$ are denoted *smectic A* if $\tau = 0$ and *smectic* A^* if $\tau \neq 0$.

If $T \geq T_{NA}$ ($r \geq 0$) it is clear that $\Psi \equiv 0$ minimizes F_A and that minimizers for $\mathfrak{F}$ in general will be nematic. If however $T < T_{NA}$, having $\Psi \not\equiv 0$ may be energetically favorable. An example of this is the nonchiral case where $\tau = 0$ and $\mathbf{n} = \overline{\mathbf{n}}$=constant. Then F_A is minimized by the configuration with uniform smectic layers determined by $\Psi = \rho_1 e^{iq\overline{\mathbf{n}}\cdot\mathbf{x}}$ where $\rho_1 = (-\frac{r}{g})^{1/2}$. This is an example of quasi-static nucleation of the smectic A phase, as r decreases, with its onset at $r = 0$ ($T = T_{NA}$).

One purpose of the work is to justify the phase diagram. Specifically, we want to find for what parameter ranges such the chiral nematic is the minimizer, and for what ranges is a layer structure (smectic A*). In the former case, we seek to characterize the minimizer in terms of the basic twist (2).

3.1 Existence of Energy Minimizers

We consider a bounded simply connected domain Ω in $\mathbb{R}^3$ which is contained between two parallel plates, $\Omega \subset \{\mathbf{x} = (x, y, z) : |z| < L\}$, for some fixed L. We investigate minimizers for $\mathfrak{F}$ in an admissible set $\mathcal{A}$ consisting of configurations $(\Psi, \mathbf{n}) \in \mathcal{W}^{1,2}(\Omega; \mathbf{C}) \times \mathbf{W}^{1,2}(\Omega; \mathbf{S}^2)$ such that $\mathbf{n}(x, y, \pm L) \cdot \mathbf{e}_3 = 0$ for all $(x, y, \pm L) \in \partial\Omega$ if $\partial\Omega \cap \{\mathbf{x} : z = \pm L\}$ is nonempty. In particular, configurations of the form $(\Psi, \mathbf{n}_\tau)$ for d as above are in $\mathcal{A}$ for arbitrary $\tau \geq 0$.

We assume that g is fixed,

$$g > 0,\ q \geq 0, \text{ and } \tau \geq 0. \tag{9}$$

We also make the following assumptions on the Frank constants: There exist fixed positive constants c_0 and c_1 so that

$$c_1 \geq K_2 + K_4 \geq c_0, K_1 \geq K_2 + K_4,\ K_3 \geq K_2 + K_4, \text{ and } 0 \geq K_4. \tag{10}$$

These inequalities are analogous to those derived by Ericksen to ensure the positivity of the Oseen-Frank energy in the case without chirality and without taking boundary conditions into account. Such inequalities were applied by Kinderlehrer, Lin and Hardt to prove existence of minimizers of the nematic problem.

Theorem 1. *Theorem Let $\mathfrak{F}$ be as in (1.1) with F_A given by (6), and such that (9) and (10) hold. Then there exists a minimizer for $\mathfrak{F}$ in $\mathcal{A}$.*

3.2 Classification of minimizers

In order to classify the solutions, additional hypotheses need to be place on the Frank constants. Specifically, we will require that K_2 and K_3 blow up near the transition temperature. This is a very realistic hypothesis that corresponds to the pre-transition effects observed in second order transitions. We show that that for K_2 and K_3 large the director field of the minimizer approaches the pure twist.

Theorem 2. *Theorem Let* $0 \leq q \leq q_0$, $0 \leq \tau \leq q_0$. $|r| \leq r_0$, *and assume that* K_1, K_2, *and* K_4 *satisfy (10) with* $K_2 = K_3$. *Assume further that* $(\tilde{\Psi}, \mathbf{n}_\tau) \in \mathcal{A}$ *for* $0 \leq \tau \leq q_0$. *Then given* $\varepsilon > 0$ *there exists a constant* $\Pi = \Pi(\varepsilon, q_0, r_0)$ *so that if* $K_2 \geq \Pi$ *and* $(\Psi, \mathbf{n})$ *minimizes* $\mathfrak{F}$ *in* $\mathcal{A}$ *then*

$$\|\mathbf{n}(\mathbf{x}) - Q\mathbf{n}_\tau(Q^t\mathbf{x})\|_{4;\Omega} < \varepsilon$$

for some $Q \in SO(3)$.

We observe that the pure twist referred to in the previous theorem may involve the case that $\Psi \equiv 0$ (nematic) as well as that with $\Psi \neq 0$ (smectic A). We discuss the issue in the next two subsections. We seek conditions on the parameters that guarantee that the solution is either nematic or smectic A. A distinctive feature of the nematic–smectic A phase transition is the relative magnitudes of the different Frank constants. In particular, experiments show that K_2 and K_3 are large relative to K_1 near the transition temperature. (See [13], pg.515.) We prove that for K_2 and K_3 sufficiently large, the director of a minimizer is close to (a rotation of) $\mathbf{n}_\tau$. Because of this feature the helical field, (1.3), is especially significant. The parameter τ is the helix's pitch. In many smectic A^* materials the layer number is large relative to the helical twist; typically $\frac{q}{\tau} > 100$. (See [27].) Here we assume $\frac{q}{\tau} >> 1$.

3.3 The pure twist as energy minimizer at high temperature

In this section and in the next we examine the effect of the chiral parameter τ and the wave number q on the type of phase taken on by a minimizer. We recall that a solution with $\Psi \equiv 0$ corresponds to the nematic phase, with or without chirality according to $\tau \neq 0$ or $\tau = 0$, respectively.

Here we prove that for $\tau \neq 0$, assuming $\frac{q}{\tau}$ and K_2 sufficiently large, minimizers are nematic for temperatures $r \geq \overline{r}(q\tau)$ where $\overline{r}$ has the form $\overline{r}(q\tau) = -\overline{\beta}\min(q\tau, (q\tau)^2)$ for some $\overline{\beta} > 0$. In particular, since $\overline{r} < 0$ for $\tau > 0$, we see that the nematic regime extends below $T_{NA}(r = 0)$ if chirality is present. We shall make two estimates in determining $\overline{r}$ based on whether $q\tau$ is large or small. The fact that $\overline{r}$ changes from linear to quadratic as $q\tau$ decreases is due to the fact that Ω is bounded.

To be definite we consider a domain representing liquid crystal confined between two parallel plates. Consider a bounded simply connected domain confined between two planes, $\Omega \subset \{\mathbf{x}\colon |z| \le L\}$, where $\partial\Omega$ is assumed to be locally a piecewise-$C^{2,\alpha}$ surface for some $\alpha > 0$. We take the admissible set to be

$$\begin{aligned} \mathcal{A} := \{(\Psi, \mathbf{n}) \in \mathcal{W}^{1,2}(\Omega) \times \mathbf{W}^{1,2}(\Omega; \mathbf{S}^2) : \\ \mathbf{n}(x, y, \pm L) \cdot \mathbf{e}_3 = 0 \text{for} (x, y, \pm L) \in \partial\Omega\}. \end{aligned} \tag{11}$$

Thus the top and bottom plates are physically treated so that the director on these surfaces in $\partial\Omega$ is forced to lie parallel to them. Note that $(\Psi, \mathbf{n}_\tau) \in \mathcal{A}$ for any $\Psi \in \mathcal{W}^{1,2}(\Omega)$ and any τ. As such the results from sections 3.1 and 3.2 are applicable here.

Remark. For $\mathcal{A}$ as in (11), if $(\Psi, \mathbf{n}) \in \mathcal{A}$ then $(0, \mathbf{n}) \in \mathcal{A}$ as well. Now $r \ge 0$ implies that $F_A \ge 0$ and we see that $\mathcal{F}(0, \mathbf{n}) \le \mathcal{F}(\Psi, \mathbf{n})$ with equality if and only if $\Psi \equiv 0$ in Ω. It follows that if $r \ge 0$ then minimizers in $\mathcal{A}$ are always nematic. Due to this, we shall assume that $r < 0$.

Theorem 3. *There are positive constants $\overline{\lambda}$ and $\overline{\beta}$, depending on Ω, and a function $\overline{K}(q, \Omega)$ so that if $K_2 > \overline{K}$, $q > \overline{\lambda}\tau$, and $(\Psi, \mathbf{n})$ is a minimizer for $\mathcal{F}$ in $\mathcal{A}$ then*

$$r \ge \overline{r}(q\tau) = -\overline{\beta}(\min(q\tau, (q\tau)^2) \tag{12}$$

implies $\Psi \equiv 0$ in Ω.

3.4 The layering energy minimizer at low temperature

In this section we estimate the transition regime from below by a curve $\underline{r} = \underline{r}(q\tau)$ valid for $\frac{q}{\tau}$ and K_2 large. If $(q\tau, r)$ is such that $r < \underline{r}(q\tau)$ then minimizers for $\mathfrak{F}$ in $\mathcal{A}$ are smectic, i.e., $\Psi \not\equiv 0$ in Ω. We determine $\underline{r}$ as follows. If $(0, \mathbf{n}') \in \mathcal{A}$ is a minimizer then necessarily $\frac{d^2}{ds^2}\mathfrak{F}(s\Upsilon, \mathbf{n}')|_{s=0} \ge 0$ for all $\Upsilon \in \mathcal{W}^{1,2}(\Omega)$, i.e.,

$$\int_\Omega (|(i\nabla + q\mathbf{n}')\Upsilon|^2 + r|\Upsilon|^2) d\mathbf{x} \ge 0.$$

We determine $\underline{r} = \underline{r}(q\tau)$ so that

$$\int_\Omega |(i\nabla + q\mathbf{n}')\tilde{\Upsilon}|^2 d\mathbf{x} < -\underline{r} \int_\Omega |\tilde{\Upsilon}|^2 d\mathbf{x}$$

for some $\tilde{\Upsilon} \in \mathcal{W}^{1,2}(\Omega)$. This implies $r > \underline{r}$. The structure of $\underline{r}(q\tau)$ depends on the magnitude of $q\tau$. We find that $\underline{r}$ is linear for $q\tau \ge 1$ and quadratic for $q\tau < 1$. This is the same qualitative structure as that of $\overline{r}$.

Theorem 4. *There is a positive constant* $\underline{\beta}$, *depending on* Ω, *and a constant* $\underline{K}(q,\Omega)$ *so that if* $K_2 \geq \underline{K}$, $q \geq \tau$, *and* $(\bar{\Psi}, \mathbf{n})$ *minimizes* $\mathfrak{F}$ *in* $\mathcal{A}$ *then*

$$r \leq \underline{r}(q\tau) = -\underline{\beta}\min(q\tau, (q\tau)^2)$$

implies $\Psi \not\equiv 0$ *in* Ω.

4. Ferroelectricity in Liquid Crystals

4.1 Flexoelectric Nematic

Spontaneous polarization in liquid crystals was discovered by R. Meyer in certain nematic materials that he labeled as flexoelectric [25]. In flexoelectric nematics, splay and bend deformations polarize the material, and, viceversa, an electric field will induce a change of alignment. (This effect is analogous to piezoelectricity of solids). The polarization vector is of the form

$$\mathbf{P} = \mathrm{e}_1(\nabla\cdot\mathbf{n})\mathbf{n} + \mathrm{e}_3\mathbf{n}\cdot\nabla\mathbf{n}, \tag{13}$$

where e_1 and e_3 denote flexoelectric coefficients.

The total energy of the liquid crystal in the presence of an electric field $\mathbf{E}$ is now

$$\mathfrak{F} = \int_\Omega F_{\mathrm{N}} + F_{\mathrm{Elec}}\, d\mathbf{x} \tag{14}$$

$$F_{\mathrm{Elec}} = -\frac{1}{2}\mathbf{D}\cdot\mathbf{E}, \quad \mathbf{D} = \varepsilon\mathbf{E} + \mathbf{P}, \tag{15}$$

$$\varepsilon = I + \varepsilon_a\mathbf{n}\otimes\mathbf{n}, \tag{16}$$

$\mathbf{P}$ denotes the polarization vector, ε is the susceptibility tensor with the scalar ε_a representing the dielectric anisotropy. Since the total energy may be unbounded from below due to the term F_{Elec}, we characterize equilibrium configurations as critical points of $\mathfrak{F}$ subject to constraints

$$\nabla\times\mathbf{E} = 0, \quad \nabla\cdot\mathbf{D} = 0. \tag{17}$$

We assume that Ω is bounded with sufficiently smooth boundary $\partial\Omega$, and that $\bar{\mathbf{n}}\colon \partial\Omega \to \mathrm{S}^2$ is Lipschitz and $\bar{\phi} \in \mathrm{H}^{\frac{1}{2}}(\partial\Omega)$. The following are sufficient conditions for the existence of a critical point $\big(\mathbf{n} \in H^1(\Omega;\mathrm{S}^2), \mathbf{E} = -\nabla\phi \in L^2(\Omega), \phi = \bar{\phi}$ and $\mathbf{n} = \bar{\mathbf{n}}$ on $\partial\Omega\big)$ of (13)-(17):

$$K_1 - K_2 - K_3 > 0, K_3 > K_2 + K_4 > C \text{ and } K_4 < 0,$$

$$C = \frac{8\pi}{\varepsilon_0}\Big(1 + 2\pi(|\mathrm{e}_{11}| + |\mathrm{e}_{33}|)\Big)(|\mathrm{e}_{11}| + |\mathrm{e}_{33}|)^2$$

$$+2\pi(|\mathrm{e}_{11}| + |\mathrm{e}_{33}|).$$

4.2 Smectic C*

In the case of smectic C* liquid crystals the polarization vector field is a direct consequence of the nonzero angle between the director and the layer normal,

$$\mathbf{P} = \mathbf{c}(\mathbf{n} \times \nabla \arg(\Psi)), \tag{18}$$

where c is a temperature-dependent material constant. In electro-optical applications of liquid crystals, it is necessary to explore the connection between chirality and ferroelectricity of the smectic C* phase. In particular, a bulk smectic C* sample, free to develop its helical structure, will not show ferroelectric behavior since the spontaneous polarization will average to zero over one pitch . Clark and Lagerwall [11] proposed a way to suppress the helix, and developed the surface stabilized ferroelectric liquid crystal (SSFLC). In a SSFLC device, the helix is suppressed by using a cell gap smaller than the helical pitch. Moreover, interaction forces between the liquid crystal and the bounding plates unwind the intrinsic helix. Symmetry arguments show that the boundary condition also causes the molecular orientation for each layer to be the same and the material exhibits ferroelectric behavior. The director is favored to lie in the plane of the bounding plates.

Because of this condition and the fact that the director is constrained to be at a certain angle from the normal to the layer (i.e. to lie on the intersection of a cone and the bounding plate), two stable states are found. The polarization vector, therefore, must be normal to the bounding plates, and its two states are in opposite directions. Electro-optical effects are achieved by applying an electric field that induces changes in the director orientation. Since the polarization vector is coupled to the director, it is also switched between the two stable states by the electric field. The Clark-Lagerwall effect in SSFLCs shows a much more rapid response to the externally applied electric field than in nematic liquid crystals because it is interacting with the sizeable spontaneous polarization rather than with an induced polarization.

Now, we can formulate the analogous problem to the flexoelectric one but for the smectic C*, i.e., find the critical points of (14)-(16), (5), (18) and (17).

Theorem 5. *Suppose that (9), (10) and* $C_{\|} \geq C_{\perp} > 0$ *hold. Then the problem (5), (14)-(16) and (18) admits a critical point in* $A = \{\mathbf{n}, \Psi \in H^1 : |\mathbf{n}| = 1, \mathbf{n} = \bar{\mathbf{n}}, \Psi = \bar{\Psi}$ *on* $\partial\Omega\}$,

Remark. For $\Omega = \{(x, y, z) : -d < x < d\}$, and for a given constant field $\mathbf{E} = E\mathbf{e}_1$, there is a positive number $d_0 = d_0(q, E, C, C_a)$ such that, for $0 < d \leq d_0$, there exist two critical points $\left(\mathbf{n}_{f\pm}, \Psi_{f\pm} = \rho e^{kz}\right)$ of the energy, with the following properties: $\mathbf{n}_{f\pm} = (0, n^y_{f\pm}, n^z_{f\pm}), n^y_{f+} = -n^y_{f-}, n^z_{f+} = n^z_{f-}$, with ρ, k and $\mathbf{n}_{f\pm}$ constant. Although such solutions seem to correspond to two unwound ferroelectric states, their stability remains to be investigated.

5. Structures in Nematic Liquid Crystal Flow

The goal of this section is to analyze liquid crystal flow that admits a large density of defects. We study a prototype problem following the model developed by Ericksen ([15]) that involves the uniaxial order parameter in addition to the director field. It turns out that the presence of the order parameter in the model allows for the description of defects and non-Newtonian phenomena not predicted by the Leslie-Ericksen system. In addition to $\mathbf{n}$ and s, the fields to describe such a flow include the velocity field $\mathbf{v}$, the pressure p. The discussion presented in this section is based on work by Calderer and co-authors ([8], [9], [24], [7], [1], [10]).

We refer to L, V, K and η as typical length, velocity, elasticity constant and viscosity of the flow, respectively. Such quantities determine the three nondimensional parameters that characterize liquid crystal flow, namely, the Reynolds number $\mathcal{R} = \frac{\rho_0 LV}{\eta}$, the Ericksen number $\mathcal{E} = \eta\frac{VL}{K}$, and the Interface number $\mathcal{I} = L^2\frac{\nu}{K}$, ν as in (3). The condition of $\mathcal{E}$ being large renders rigorous the notion of *fast flow* and it is a physically relevant condition for polymeric materials. The quantity $\mathcal{I}$ is associated with the free energy that is required in order to maintain defects in the flow configuration. Flow of polymeric liquid crystals often presents large values of $\mathcal{E}$. This, in turn, is responsible for the presence of defects in the flow region, and complex Non–Newtonian phenomena ([18], [19], [28]).

The governing equations for the variables $\mathbf{v}$, $\mathbf{n}$ and s correspond to balance of linear and generalized momenta [15]:

$$\rho_0\dot{\mathbf{v}} = \nabla\cdot\sigma, \tag{19}$$

$$\beta_2(s)\dot{s} = \nabla\cdot(\frac{\partial\mathcal{F}}{\partial\nabla s}) - \frac{\partial\mathcal{F}}{\partial s} - \beta_3(s)\mathbf{n}\cdot\mathbf{A}\mathbf{n}, \tag{20}$$

$$\begin{aligned}\gamma_1(s)\dot{\mathbf{n}}\times\mathbf{n} = \nabla\cdot(\frac{\partial\mathcal{F}}{\partial\nabla\mathbf{n}})\times\mathbf{n} - \frac{\partial\mathcal{F}}{\partial\mathbf{n}}\times\mathbf{n}+\\ \gamma_1(s)\Omega\mathbf{n}\times\mathbf{n} - \gamma_2(s)\mathbf{A}\mathbf{n}\times\mathbf{n}.\end{aligned} \tag{21}$$

$$\sigma = -pI - \nabla\mathbf{n}^T\frac{\mathcal{F}}{\nabla\mathbf{n}} - \nabla s\otimes\frac{\mathcal{F}}{\nabla s} + \hat{\sigma}, \tag{22}$$

$$\begin{aligned}\hat{\sigma} = (\beta_1\dot{s} + \alpha_1\mathbf{n}\cdot\mathbf{A}\mathbf{n})\mathbf{n}\otimes\mathbf{n} + \alpha_2\dot{\mathbf{N}}\otimes\mathbf{n}+\\ \alpha_3\mathbf{n}\otimes\dot{\mathbf{N}} + \alpha_4\mathbf{A} + \alpha_5\mathbf{A}\mathbf{n}\otimes\mathbf{n} + \alpha_6\mathbf{n}\otimes\mathbf{A}\mathbf{n},\end{aligned} \tag{23}$$

$$2\mathbf{A} = \nabla\mathbf{v} + (\nabla\mathbf{v})^T, \quad 2\Omega = \nabla\mathbf{v} - (\nabla\mathbf{v})^T, \tag{24}$$

$$\dot{\mathbf{N}} = \dot{\mathbf{n}} - \Omega\mathbf{n}, \tag{25}$$

$p(\mathbf{x}, t)$ is the pressure, σ denotes the stress tensor and $\rho_0 > 0$ the density. The Leslie coefficients are constitutive functions $\alpha_i(s)$, $\beta_i(s)$ and $\gamma_i(s)$ on the interval $(-\frac{1}{2}, 1)$ satisfying restrictions imposed by the second law of thermodynamics and Parodi's relations (([15]), equation (4.18)). Such inequalities guarantee that the system (19)-(25) is dissipative (increasing entropy), implying that the total free energy $E(t) = \int_\Omega \frac{1}{2}\rho\mathbf{v}\cdot\mathbf{v} + \mathfrak{F}\, d\mathbf{x}$, stored by the nematic flow under isothermal conditions, satisfies

$$\frac{dE(t)}{dt} = -\int_\Omega \Delta(\mathbf{x})\, d\mathbf{x} \tag{26}$$

$$\Delta = \alpha_4 \mathrm{tr}\mathbf{A}^2 + (\alpha_5+\alpha_6)\mathbf{n}\cdot\mathbf{A}^2\mathbf{n} + \alpha_1(\mathbf{n}\cdot\mathbf{A}\mathbf{n})^2 +$$

$$\gamma_1\dot{\mathbf{N}}\cdot\dot{\mathbf{N}} + 2\gamma_2\dot{\mathbf{N}}\cdot\mathbf{A}\mathbf{n} + \beta_2\dot{s}^2 + 2\beta_1\dot{s}\mathbf{n}\cdot\mathbf{A}\mathbf{n}, \tag{27}$$

$$\Delta \geq 0. \tag{28}$$

For the Leslie–Ericksen system, Lin and Liu ([20], [22] and [23]) show that the dissipative relation yields uniform bounds of norms of weak solutions of the governing system, resulting in global existence and regularity of weak solutions, and, in some cases, even the existence of classical solutions of initial boundary value problems.

We let the flow domain $\Omega = \{(x,y,z) : x \in (-1,1)\}$. We consider plane flow $\mathbf{v} = (0,0,v(x))$ with director field configurations

$$\mathbf{n} = (\sin\phi(x), 0, \cos\phi(x)). \tag{29}$$

To be specific, we consider Poiseuille boundary conditions for our problem (with small modifications, the results hold for shear flow as well):

$$\frac{\partial p}{\partial z} = 1, \quad v(-1) = 0 = v(1), \quad \text{and}$$

$$s(-1) = s_{-1}, \quad s(1) = s_1, \quad \phi(-1) = \phi_{-1}, \quad \phi(1) = \phi_1. \tag{30}$$

We also prescribe initial data, s_0, ϕ_0 and v_0, with $s_0 \in (-\frac{1}{2}, 1)$, and denote $\mu = \mathcal{E}^{-1}$. The governing system for Poiseuille flow is (see [8], [9] for the derivation):

$$s_t = \mu\left(a_1 s'' - a_2 s(\phi')^2\right) - \mathcal{J}^{-1} f'(s) - g_3(s) v' \sin\phi\cos\phi \tag{31}$$

$$\gamma_1(s)\phi_t = \left(g_1(s)\sin^2\phi - g_2(s)\cos^2\phi\right)v' + \mu a_2(s^2\phi')', \tag{32}$$

$$v_t = \frac{1}{\mathcal{R}}(v' g(s,\phi))' - 1, \quad \text{with} \tag{33}$$

$$g_1(s) = \frac{1}{2}(-\gamma_1+\gamma_2), \quad g_2(s) = \frac{1}{2}(\gamma_1+\gamma_2), \quad g_3(s) = \beta_3, \tag{34}$$

$$g(s,\phi) = \frac{1}{2}\alpha_4 + \alpha_1\sin^2\phi\cos^2\phi + \frac{1}{2}(\alpha_5-\alpha_2)\sin^2\phi + \tag{35}$$

$$\frac{1}{2}(\alpha_3+\alpha_6)\cos^2\phi. \tag{36}$$

$$a_i = \frac{k_i}{K_0}, \quad i = 1,2, \quad K_0 \equiv \max\{k_1, k_2\} \tag{37}$$

With the help of the maximum principle, we show that the system (31)-(37) satisfies the following dissipative relation.

Theorem 6. *If s, ϕ and v are smooth solutions of the system (31)-(37) then they satisfy the following dissipative relation provided $\mathcal{R}$ is small enough:*

$$\frac{1}{2}\frac{d}{dt}\int_{-1}^{1}\{a_1|s'|^2 + a_2 s^2|\phi'|^2 + |\mathrm{v}|^2\}\,dx +$$
$$\int_{-1}^{1}\{\tilde{\mu}_1|a_1 s'' - a_2 s|\phi'|^2|^2 + \tilde{a}_2^2\frac{\mu}{\gamma_1}|(s^2\phi')'|^2 + \frac{1}{\mathcal{R}}\tilde{g}(s,\phi)|v'|^2\}\,dx$$
$$\leq M(T), \tag{38}$$

for any $0 < T < \infty$. Here $M(T)$ depends only on T, and $\tilde{\mu}_1$, $\tilde{a}_2$ and $\tilde{g}$ are positive constants.

We apply the Galerkin method to prove existence of weak solutions. For a given positive integer m, we consider the finite dimensional approximation of v of the form

$$v_m(x,t) = \sum_{j=0}^{m} c_{jm}(t)\Phi_j(x). \tag{39}$$

The conditions of v vanishing on the boundary, allow for the following choice:

$$\Phi_j(x) = \cos\frac{\pi}{2}(2j+1)x. \tag{40}$$

We define the orthogonal projection operator P_m as follows:

$$P_m : L^2(-1,1) \longrightarrow \mathrm{Span}\{\Phi_1, \ldots, \Phi_m\}. \tag{41}$$

We look at the solutions of the following approximated system (the analogous approximation is used in [21]):

$$s_{m_t} = \mu\big(a_1 s_m{}'' - a_2 s_m(\phi_m{}')^2\big) - \mathcal{J}^{-1}f'(s_m)$$
$$-g_3(s_m)v_m' \sin\phi_m \cos\phi_m, \tag{42}$$
$$(\gamma_1(s_m) + \epsilon_1^2)\phi_{m_t} = \big(g_1(s_m)\sin^2\phi_m - g_2(s_m)\cos^2\phi_m\big)v_m'$$
$$+\mu a_2((s_m{}^2 + \epsilon_1^2)\phi_m')', \tag{43}$$
$$v_{m_t} = \frac{1}{\mathcal{R}}P_m\big((v_m{}'g(s_m,\phi_m))' - 1\big). \tag{44}$$

Theorem 7. *For any integer $m > 0$, $\epsilon_1 > 0$ and $T > 0$, there exists a unique solution such that $v_m \in L^\infty(0,T; L^2(-1,1)) \cap L^2(0,T; H^1(-1,1))$, $s_m\phi'_m \in L^\infty(0,T; L^2(-1,1))$, and $s_m \in L^\infty(0,T; H^1(-1,1))$. Moreover, this solution satisfies the relations $\big(a_1 s_m{}'' - a_2 s_m|\phi'_m|^2\big) \in L^2(0,T; L^2(-1,1))$,* *and* $((s_m^2+\epsilon_1^2)\,\phi'_m)'\cdot(\gamma_1(s_m)+\epsilon_1^2)^{-1/2}\in L^2(0,T;L^2(-1,1))$. *Furthermore for any* $m>0$ *the solutions satisfy the dissipative relation of 6.*

Now we multiply equation (31) by s to get the new equation

$$\frac{1}{2}(s^2)_t = \mu[\frac{a_1}{2}((s^2)'' - 2(s')^2) - a_2 s^2(\phi')^2] \tag{45}$$

$$-\mathcal{J}^{-1} s f'(s) - s g_3(s) v' \sin\phi\cos\phi. \tag{46}$$

Remark. We notice that equation (31) is equivalent to (46) when $s\neq 0$. On the other hand, $s=0$ also satisfies the original equation (31).

Passing to the limit as $m\to\infty$ and $\epsilon_1\to 0$ in the fields of Theorem7, yields:

Theorem 8. *There exists a global weak solution of the governing systems (46), (32)-(33) together with initial and boundary conditions.*

5.1 Stationary flow

We discuss properties of steady state flow with large Ericksen number. This is a prevalent feature of polymeric liquid crystals in processing conditions. One relevant issue about such a flow is the large number of defects present in the region ([29], [2], [17]). In our analyses, we obtain configurations with arrays of line defects parallel to the flow. In each of such lines, s vanishes (i.e., the material becomes locally isotropic), and ϕ experiences a jump from $-\frac{\pi}{4}$ to $\frac{\pi}{4}$ across it. The governing equations are singularly perturbed by a parameter $\mu=\mathcal{E}^{-1}$, become singular at points where $s=0$ and are highly nonlinear.

First, let us consider aligning regimes, i.e., those that satisfy $|\frac{\gamma_1}{\gamma_2}|\leq 1$. Although the oscillatory and singular behavior of solutions is still more prevalent in non-aligning regimes, we restrict the present description to the former case. The proof of singular oscillatory solutions to the boundary value problem proceeds through the following steps. Let $\mathrm{I}=(-1,1)$.

Turning points and the saddle point property. The turning points $x\in\mathrm{I}$ of the governing equations belong to one of the solution branches $S=0$ and $S_T(\cdot,\pm1)$. The branches S_T are defined by the algebraic relations,

$$0 = f'(s(x)) - \frac{e}{2}\mathcal{J}g^{-1}(s,\phi(s))g_3(s)h(s(x))x, e=\pm 1 \tag{47}$$

with $h(s)=(1-\frac{\gamma_1{}^2}{\gamma_2{}^2}(s))^{\frac{1}{2}}$. In such branches, ϕ takes the following values:

$$\sin\phi = \pm\sqrt{\frac{1}{2}(1+\frac{\gamma_1}{\gamma_2}(s))}, \tag{48}$$

respectively. (Note that ϕ is undefined when $s=0$.) Equations (47) and (48) result from setting $\mu=0$ in the steady state equations. $S=0$ and $S_T(\cdot,\pm1)$, are invariant manifolds of the governing equations and satisfy the saddle point property.

The oscillatory property. The solutions of the steady state problem oscillate between the branches $S_T(x, e), e = \pm 1$, through $S = 0$, with the number of oscillations $N = O(\mu^{-\frac{1}{2}})$, for small $\mu > 0$. Moreover, the graph of ϕ presents either a discontinuity or a cusp at points where $s = 0$.

Young measures and effective equations. The high density of defects and oscillations through the isotropic state for small μ suggest to address the problem from the point of view of finding effective quantities and flow equations. We characterize such configurations in terms of Young measures associated with sequences of weak solutions. The resulting effective equations correspond to the Newtonian flow together with additional relations for Young measures representing a remnant microstructure of ordered states. The analysis covers aligning as well as non-aligning regimes.

If $\{s_\mu, \phi_\mu\}$ is a sequence of functions such that $s_\mu \to 0$ uniformly,

$$g(s_\mu, \phi_\mu) \to \frac{1}{2}\alpha_4(0),$$

uniformly on I as μ goes to zero. Up to a subsequence, velocity v_μ tends to a limit v_0 in $L^2(I)$, and weakly in $W^{1,2}(I)$. Hence, the product $g(s_\mu, \phi_\mu)v'_\mu$ of a strongly convergent sequence g and a weakly convergent sequence v'_μ converges to $\frac{1}{2}\alpha_4(0)v'_0$ weakly in $L^2(I)$. This implies that v_0 satisfies the effective equation

$$\eta_{\text{eff}} v''_0 = 1, \tag{49}$$

which is the equation of the Newtonian Poiseuille flow with effective viscosity $\eta_{\text{eff}} = \frac{1}{2}\alpha_4(0)$.

Remarks.

1 When the Ericksen number is large and the Reynolds number is on the order of 1, the typical viscosity is much larger than the typical elasticity. It is natural to expect that alignment of the molecules will be destroyed by the diffusion, so that liquid crystal flow is that of an isotropic liquid with a constant viscosity. If that were the case, equation (4.2) would be the only effective equation of the limiting flow. Rigorous analysis suggests, however, that one should also consider Young measure generated by the sequence ϕ_μ. This measure is nontrivial and it describes a remnant of ordered states compatible with the boundary conditions. Our analysis justifies such a conjecture.

2 We observe that the governing system is shift-invariant; indeed if (s, ϕ) is a solution, then $(s, \phi + k\pi)$ is also a solution for any integer k. Starting

with an increasing $\tilde{\phi}$, we can split the interval I into subintervals on which $k\pi \leq \tilde{\phi} \leq (k+1)\pi$ and then define a ϕ by shifting appropriately on each subinterval. The resulting ϕ will be bounded, rapidly oscillating and discontinuous, and hence not an element of $W^{1,2}(I)$. We refer to such functions generalized solutions, to distinguish them from weak solutions. This behavior of ϕ is associated with the presence of multiple defects. (We point out that such generalized solutions ϕ can also be obtained from asymptotic analysis of the governing system for small μ, when neglecting the internal layer components of the solution field).

The following notation allows us to summarize the previous discussion in Theorem 9.

$$G_1(s,\phi,x) = \frac{1}{2}\beta_1(s)g^{-1}(s,\phi,x)x\sin 2\phi + \frac{1}{\mathcal{J}}f'(s), \quad (50)$$

$$G_2(s,\phi,x) = \frac{1}{2\eta_0}(\gamma_1(s) + \gamma_2(s)\cos 2\phi)g^{-1}(s,\phi,x)x, \quad (51)$$

$$A = (a_2/a_1)^{1/2}. \quad (52)$$

Theorem 9. *Let $\tilde{s}_\mu, \tilde{\phi}_\mu$ be a sequence of weak solutions satisfying the a priori estimates*

$$\| \tilde{s}' \|_{L^2(I)} \leq C,$$

$$\| \tilde{s}\tilde{\phi}' \|_{L^2(I)} \leq C,$$

with C independent of μ. Let s_μ, ϕ_μ be a corresponding sequence of generalized solutions. Then, up to a subsequence,
i) $s_\mu \to 0$ uniformly on I,
ii) The sequence ϕ_μ generates a Young measure ν_x satisfying momentum relations

$$\int_{-1}^{1}\int_{-\pi/2}^{\pi/2} (G_1(0,z,x)\sin Az - \frac{1}{A}\frac{G_2}{s}(0,z,x)\cos Az)d\nu_x(z)h(x)dx = 0, \quad (53)$$

$$\int_{-1}^{1}\int_{-\pi/2}^{\pi/2} (G_1(0,z,x)\cos Az + \frac{1}{A}\frac{G_2}{s}(0,z,x)\sin Az)d\nu_x(z)h(x)dx = 0;$$

iii) The sequence $s(\tilde{\phi}'_\mu)^2$ converges to a measure ρ in the sense of distributions. Moreover,

$$\rho = \int_{-\pi/2}^{\pi/2} G_1(0,z,x)d\nu_x(z). \quad (54)$$

Numerical simulations. We now include the graphs of $(s(x), \phi(x))$, $x \in (-1, 1)$ obtained in a numerical simulations of such flow [1] with Ericksen number 10^3 ($\mu = 10^{-3}$). The boundary conditions are $s(-1) = .7 = s(1)$ and $\phi(-1) = 0 = \phi(1)$.

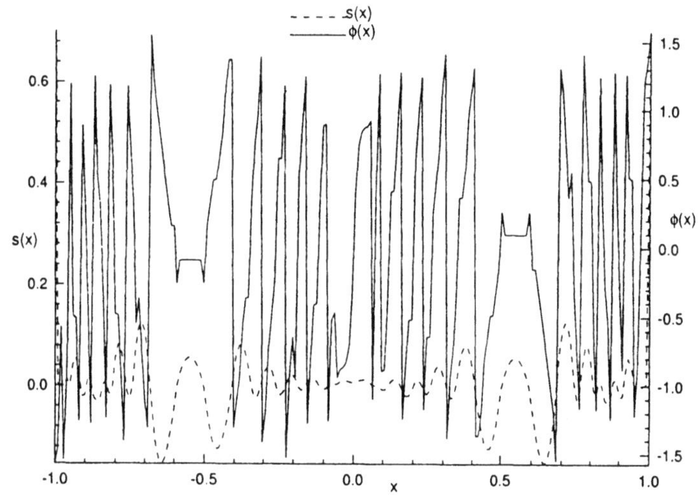

References

[1] S. Muzumder B. Mukherjee and M.C. Calderer. Poiseuille flow of liquid crystals: highly oscillatory regimes. *Journal of Non Newtonian Fluid Mechanics*, 99:37–55, 2001.

[2] G.C. Berry. Rheological properties of nematic solutions of rodlike polymers. *Mol. Cryst. Liq. Cryst.*, 165:333–360, 1988.

[3] M. C. Calderer and C. Liu. Mathematical developments in the study of smectic a liquid crystals. *Int. J. Engr. Sci. Mech.*, 38:1113–1128, 2000.

[4] M. C. Calderer, C. Liu, and K. Voss. Radial configurations of smectic a materials and focal conics. *Physica D*, 124:11–22, 1998.

[5] M. C. Calderer, C. Liu, and K. Voss. Smectic a liquid crystal configurations with interface defects. *Mathematical Methods in the Applied Sciences*, 52:473–489, 2001.

[6] M. C. Calderer and P. Palffy-Muhoray. Ericksen's mar and modeling of the smectic a –nematic phase transition. *SIAM J. Appl. Math.*, 60:1073–1098, 2000.

[7] M.C. Calderer and C. Liu. Liquid crystal flow: Dynamic and static configurations. *SIAM Journal of Applied Mathematics*, 60:1925–1949, 2000.

[8] M.C. Calderer and B. Mukherjee. Chevron patterns in liquid crystal flows. *Physica D*, 98:201–224, 1996.

[9] M.C. Calderer and B. Mukherjee. On poiseuille flow of liquid crystals. *Liquid Crystals*, 22:121–135, 1997.

[10] M.C. Calderer and A. Panchencko. Young measures and order-disorder transition in steady flow of liquid crystals. *SIAM Jour. Appl. Math. (submitted)*, 2001.

[11] N. A. Clark and S. T. Lagerwall. Submicrosecond bistable electro-optic switching in liquid crystals. *Applied Physics Letters*, 36:899–901, 1980.

[12] P. G. de Gennes and J. Prost. *The Physics of Liquid Crystals*. Oxford University Press, Oxford, 1993.

[13] P. G. de Gennes and J. Prost. *The Physics of Liquid Crystals*. Oxford University Press, Oxford, 1993.

[14] P.G. de Gennes. An analogy between superconductivity and smectics a. *Solid State Commun.*, 58(10):753–756, 1972.

[15] J.L. Ericksen. Liquid crystals with variable degree of orientation. *Arch. Rational Mech. Anal.*, 113:97–120, 1991.

[16] S. Fraden and R.D. Kamien. Self-assembly in vivo. *Biophysical Journal*, 78:2189–2190, 2000.

[17] P.L. Maffettone G. Marrucci. Description of the liquid crystalline phase of rodlike polymers at high shear rates. *American Chemical Society*, 22:4076–4082, 1989.

[18] P. Palffy-Muhoray J.T. Gleeson and W. Van Saarloos. Propagation of excitations induced by shear flow in nematic liquid crystals. *Physical Review A*, 44:2588–2595, 1991.

[19] R.G. Larson and D.W. Mead. Development of orientation and texture during shearing of liquid-crystalline polymers. *Liquid Crystals*, 12:751–768, 1992.

[20] F.-H. Lin and C. Liu. Nonparabolic dissipative systems modeling the flow of liquid crystals. *Comm. Pure Appl. Math.*, 48:501–537, 1995.

[21] F. H. Lin and C. Liu. Nonparabolic dissipative systems modeling the flow of liquid crystals. *Comm.Pur.Appl.Math.*, 48:501–537, 1995.

[22] F.-H. Lin and C. Liu. Partial regularity of the nonlinear dissipative systems modeling the flow of liquid crystals. *Discrete Contin. Dynam. Systems*, 2(1):1–22, 1996.

[23] F.-H. Lin and C. Liu. Existence of solution for erickse-leslie system. *Arch. Rat. Mech. Anal.*, 154(2):135–156, 2000.

[24] M.C. Calderer. Dissipation, surface energy and defects. *European Journal on Applied Mathematics*, 8:301–310, 1997.

[25] R.B. Meyer. The flexoelectric effect in nematic liquid crystals. *Phys. Rev. Lett.*, 22:918–922, 1969.

[26] C. Liu P. Bauman, M. C. Calderer and D. Phillips. The phase transition between chiral nematic and smectic a* liquid crystals. *Arch. Rat. Mech. Anal.(submitted)*, 2002.

[27] S.R. Renn and T.C. Lubensky. Abrikosov dislocation lattice in a model of the cholesteric - to smectic-a transition. *Phys. Rev. A*, 38:2132–2147, 1988.

[28] J. Wahl and F. Fisher. Elastic and viscosity constants of nematic liquid crystals from a new optical method. *Molecular Crystals and Liquid Crystals*, 22:359–373, 1973.

[29] K.F. Wissbrun. Rheology of rod-like polymers in the liquid crystalline state. *Journal of Rheology*, 25:619–662, 1981.

BAYESIAN INPAINTING BASED ON GEOMETRIC IMAGE MODELS

Tony F. Chan *
Department of Mathematics, UCLA, Los Angeles, CA 90095, USA
chan@math.ucla.edu

Jianhong Shen
School of Mathematics, University of Minnesota, Minneapolis, MN 55455, USA
jhshen@math.umn.edu

Abstract Image inpainting is an image restoration problem, with wide applications in image processing, vision analysis, and the movie industry. This paper surveys and summarizes all the recent inpainting models based on the Bayesian and variational principle. A unified view is developed around the central topic of geometric image models. We also discuss their associated Euler-Lagrange PDE's and numerical implementation. A few open problems are proposed.

Keywords: Inpainting, interpolation, Bayesian, curve model, image model, Euclidean invariance, elastica, bounded variation, Mumford-Shah, curvature, Γ-convergence, numerical PDE.

1. Introduction

The word "inpainting" is an artistic synonym for "image interpolation," as frequently used among museum restoration artists, who manually remove cracks from degraded ancient paintings by following as faithfully as possible the original intention of their creators [EM76, Wal85]. A mathematical illustration is depicted in Figure 1.

As fine art museums go digital, all real paintings are scanned into computers. No doubt, digital inpainting provides the safest way to restore those degraded ancient paintings, simply by trying computer codes and softwares on the digital copies. Unlike the manual inpainting process which applies directly to the

*Research supported by grants from NSF under grant number DMS-9626755 and from ONR under N00014-96-1-0277.

original, digital inpainting brings tremendous freedom in making errors or improving results progressively, with no risk of destroying the original precious painting on the canvas, which is often unique in the entire world.

But the application of digital inpainting goes far beyond on-line art museums. In on-line real estate business, for example, a potential customer may ask: I do not like the palm tree and bush in the front yard; what will the house look like without them? If the tree and bush are too close to the house, it is impossible to capture by ordinary cameras an unblocked overview of the entire house. However, treating the tree and bush as an inpainting domain, it is very hopeful that a (clever) inpainting scheme can get rid of them in a realistic fashion.

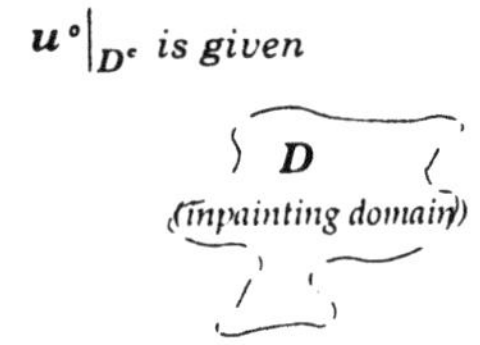

Figure 1. For a typical inpainting problem, the image is missing on an inpainting domain D, and the available part $u^0|_{D^c}$ is often noisy. D can be disconnected.

Ever since the original work of Bertalmio et al. [BSCB00], digital inpainting has found wide applications in image processing, vision analysis, and the movie industry. Recent examples include: automatic scratch removal in digital photos and old films [BSCB00, CS01a], text erasing [BBC+01, BSCB00, CS01a, CS01c], special effects such as object disappearance and wire removal for movie production [BSCB00, CS01c], disocclusion [MM98], zooming and super-resolution [BBC+01, CS01a, TAYW01], lossy perceptual image coding [CS01a], and removal of the laser dazzling effect [CCBT01], and so on. On the other hand, in the engineering literature, there also have been many earlier works closely related to inpainting, which include image interpolation [AKR97, KMFR95a, KMFR95b], image replacement [IP97, WL00], and error concealment [JCL94, KS93] in the communication technology.

As scattered as the applications are, the methods for inpainting related problems have also been very diversified, ranging from nonlinear filtering method, wavelets and spectral method, and statistical method (especially for textures), etc.

The most recent approach to non-texture inpainting is based on the PDE method and Calculus of Variations, and can be classified into two categories. The first class is based on the simulation of micro-inpainting mechanisms. It includes the axiomatic approach of Caselles, Morel and Sbert [CMS98], the transport process [BSCB00] (the first high-order PDE model), the diffusion process [CS01c], and their combination [BBS01, CS01b]. The second category includes all variational models simulating the unique macro-inpainting mechanism: "best guess," or the Bayesian framework [GG84, KR96, Mum94b]. The latter includes the total variation model [CS01d, CS01a, RO94, ROF92], the functionalized elastica model [CKS01, MM98], the value-and-direction joint model [BBC+01], the active contour model based on Mumford and Shah's segmentation [TAYW01], and the inpainting scheme based on the Mumford-Shah-Euler image model [ES01].

The current paper surveys and summarizes this latter category. The main goal is to develop a systematic approach and mathematical foundation for all these previously scattered works, so that the survey can serve as a fresh starting point, rather than a concluding chapter, for further research on this challenging topic.

The philosophy behind Bayesian inpainting is quite simple and intuitive (see Figure 1): the way we human inpainters inpaint an incomplete picture mostly relies on two factors — how we read the existing part of the picture $u^0|_{\Omega\backslash D}$ (i.e. *data model*), and what class of images we believe the original good picture u belongs to (i.e. *image prior model*). (For example, if it is known that we are inpainting an image of a kitchen with tomatoes, peppers, and apples, we have the *a priori* preference of smooth shapes and the green and red colors.) In the Bayesian language, a balanced optimal guess is to maximize the posterior probability $\text{Prob}(u|u^0)$ (MAP) given by

$$\text{Prob}(u|u^0) = \frac{\text{Prob}(u^0|u) \cdot \text{Prob}(u)}{\text{Prob}(u^0)}. \tag{1}$$

Once an image u^0 is given, the denominator is a fixed constant. Thus we are to maximize the product of the data probability and the image probability.

For inpainting, the data model is usually simple as illustrated in Figure 1: the available part $u^0|_{\Omega\backslash D}$ is the restriction of the original good image u on $\Omega\backslash D$, polluted independently by white noise n, i.e.

$$u^0\big|_{\Omega\backslash D} = u\big|_{\Omega\backslash D} + n.$$

On the other hand, since there is no data available on the inpainting domain D, the task of reconstructing the image on D solely falls on the image model. This makes a good image model more crucial for inpainting than for any other classical restoration problems such as denoising, deblurring, and segmentation [CS01a, ES01].

Image models can be learned from image data banks based on filtering, parametric or non-parametric estimation, and the entropy method (See Zhu, Wu and Mumford [ZM97, ZWM97] for examples). Such statistical approach is especially important for inpainting or synthesizing images with rich textures [IP97, WL00].

On the other hand, in most inpainting problems, the inpainting domain often "erases" some perceptually important geometric information of the image, edges, for example. To reconstruct geometry, it is necessary that the image model well resolves the geometry *a priori*. Most conventional probabilistic models lack such feature. Fortunately, "energy" forms do exist in the literature, which have been explicitly motivated by geometry. Well-known examples include the TV (total variation) model of Rudin, Osher and Fatemi [ROF92] and the object-edge model of Mumford and Shah [MS89]. The link between probabilistic image models and such geometric image models, as Mumford pointed out [Mum94b], is formally made through Gibbs' formula in statistical mechanics [Gib02]:

$$\text{Prob}(u) = \frac{1}{Z}\exp(-\beta E[u]),$$

where $E[u]$ is the energy of u (e.g., the total variation of u), β denotes the inverse absolute temperature, and Z the partition function. (Working with energy also frees one from laboring on the definability of the partition function Z, which is generally a highly non-trivial mathematical issue.) The Bayesian formula (1) is re-expressed in the energy form by

$$E[u|u^0] = E[u^0|u] + E[u] + \text{const.},$$

where the constant can be dropped safely as far as energy minimization is concerned.

The organization goes as follows. Section 2 starts with an axiomatic approach for curve models, which, to our best knowledge, is new. The latter half of the section explains two approaches for constructing geometric image models from curve models:

(i) through direct functionalization based on the level-sets; and

(ii) by having a curve model embedded as an edge model in the object-edge primitive image model.

These approaches unify the four geometric image models appearing in the recent inpainting literature:

(1) the TV image model of Rudin, Osher and Fatemi [ROF92, RO94], first applied to inpainting modeling by Chan and Shen [CS01a];

(2) the functionalized elastica image model as proposed and studied by Masnou and Morel [MM98], and Chan, Kang, and Shen [CKS01];

(3) the Mumford-Shah image model [MS89] applied to inpainting by Tsai, Yezzi, and Willsky [TAYW01], Chan and Shen [CS01a], and Esedoglu and Shen [ES01]; and

(4) the Mumford-Shah-Euler image model designed for image inpainting by Esedoglu and Shen [ES01].

Section 3 explains all the recent inpainting schemes based on these geometric image models. We discuss the associated Euler-Lagrange PDE's, their geometric meaning, and robust ways of numerical implementation. Digital examples are given for each inpainting model. Conclusion and open problems are written into Section 4.

Throughout the paper, Ω denotes the entire image domain, D the missing inpainting domain, u^0 the available part of the image on $\Omega \backslash D$, and u the targeted inpainting restoration. The standardized symbols ∇, $\nabla\cdot$ and Δ represent the gradient, divergence, and Laplacian operators separately. For any multi-variable function or functional $F(X, Y)$, the symbol $F(X|Y)$ still means $F(X, Y)$, but emphasizing that Y is fixed as known. This is to imitate the symbol of conditional probability or expectation appearing in the Bayesian formula (but without probabilistic normalization).

2. Geometric Image Models

2.1 Curve models

Geometry plays a crucial role in visual perception and image understanding, including classification and pattern recognition. The most important geometry for image analysis is hidden in edges. From David Marr's classical work on primal sketch [MH80] to David Donoho's *geometric* wavelets analysis [Don00], edges always stay in the heart of many issues: image coding and compression, image restoration, segmentation and tracking, just to name a few.

Therefore it is of fundamental significance to understand how to mathematically model edges and curves.

>From the Bayesian point of view, this is to establish a probability distribution Prob(Γ) over "all" curves. An instant example coming to mind is the Brownian motion and its Wiener measure [KS97]. The problem is that Brownian paths are parameterized curves (by "time") and are even almost surely no-where differentiable. For image analysis, however, edges are intrinsic (1-D) manifolds and their regularity is an important visual cue.

According to the previously stated Gibbs' formulation, we are to look for a suitable energy form $E[\Gamma]$. It is always convenient to first start with its digital version.

In digital image processing, Freeman's *Chain Coding* [Fre61] is a popular data structure for representing object borders and edges. The underlying idea is to represent a 1-D curve Γ by a chain of ordered sample points

$$x_0, x_1, \cdots, x_N,$$

dense enough to ensure reasonable approximation. Working directly with such chains of finite length, we need to define appropriate energy forms

$$E[x_0, x_1, \cdots, x_N].$$

To the best of our knowledge, the following axiomatic approach is new in the literature. We shall naturally construct two of the most useful planer curve models: the length energy and Euler's elastica energy.

Axiom 1. Euclidean invariance.

Let $Q \in O(2)$ (conventionally called a rotation, though including all reflections), and $c \in R^2$ an arbitrary point. Euclidean invariance consists of two parts: the rotational invariance

$$E[Qx_0, Qx_1, \cdots, Qx_N] = E[x_0, x_1, \cdots, x_N],$$

and the translation invariance

$$E[x_0 + c, x_1 + c, \cdots, x_N + c] = E[x_0, x_1, \cdots, x_N].$$

Axiom 2. Reversal invariance.

It requires that

$$E[x_0, \cdots, x_N] = E[x_N, \cdots, x_0],$$

which means that the energy does not depend on the orientation of the curve.

Axiom 3. p-point accumulation ($p = 2, 3, \cdots$).

This is fundamentally a rule on *locality*. A p-point accumulative energy satisfies the accumulation law:

$$E[x_0, \cdots, x_n, x_{n+1}] = E[x_0, \cdots, x_n] + E[x_{n-p+2}, \cdots, x_{n+1}],$$

for all $n \geq p - 2$. Through cascade, we easily establish that

Proposition 1. *Suppose E is p-point accumulative. Then for any* $N \geq p - 1$,

$$E[x_0, \cdots, x_N] = \sum_{n=0}^{N-p+1} E[x_n, \cdots, x_{n+p-1}]$$

Thus, for example, a 2-point accumulative energy must be in the form of

$$E[x_0, \cdots, x_N] = E[x_0, x_1] + E[x_1, x_2] + \cdots + E[x_{N-1}, x_N];$$

and a 3-point accumulative energy satisfies

$$E[x_0, \cdots, x_N] = E[x_0, x_1, x_2] + E[x_1, x_2, x_3] + \cdots + E[x_{N-2}, x_{N-1}, x_N].$$

Generally, a p-point accumulative energy E is completely determined by its fundamental form

$$E[x_0, \cdots, x_{p-1}].$$

In what follows, we study the cases of $p = 2$ and $p = 3$.

2.1.1 2-point accumulative energy and the *length* .

Proposition 2. *A Euclidean invariant 2-point accumulative energy must be in the form of*

$$E[x_0, \cdots, x_N] = \sum_{n=0}^{N-1} f(|x_{n+1} - x_n|),$$

for some non-negative function $f(s)$.

Proof. We only need to show that

$$E[x_0, x_1] = f(|x_1 - x_0|).$$

Translation invariance leads to

$$E[x_0, x_1] = E[0, x_1 - x_0] = F(x_1 - x_0),$$

with $F(x) = E[0, x]$. Then rotational invariance further implies that

$$F(Qx) \equiv F(x), \quad Q \in O(2), x \in R^2.$$

Thus if we define $f(s) = F((s, 0))$, then $F(x) = f(|x|)$. □

If in addition, we impose

Axiom 4. Linear additivity:
For any $\alpha \in (0, 1)$, and $x_1 = \alpha x_0 + (1 - \alpha) x_2$,

$$E[x_0, x_2] = E[x_0, x_1] + E[x_1, x_2].$$

Then it is easy to show that the energy is unique up to a multiplicative constant.

Theorem 1. *A Euclidean invariant 2-point accumulative energy E with linear additivity must be the* length *energy, i.e.,*

$$E[x_0, \cdots, x_N] = c \sum_{n=0}^{N-1} |x_n - x_{n+1}|,$$

for some fixed positive constant c.

For a summable curve Γ, as $N \to \infty$, and the sampling size

$$\max_{0 \le n \le N-1} |x_n - x_{n+1}|$$

tends to zero, such digital energy converges to the length.

2.1.2 3-point accumulative energy and the curvature.

To determine the fundamental form $E[x_0, x_1, x_2]$, first recall Frobenius' classical theorem [COS+98]. The three points x_0, x_1, x_2 live in $R^6 = R^2 \times R^2 \times R^2$, and the dimension of a Euclidean orbit is 3: 1 from the rotation group, and 2 from the translation group. Therefore, Frobenius' theorem applied to the Euclidean invariance gives

Proposition 3. *One can find exactly three independent joint invariants: I_1, I_2, and I_3, such that $E[x_0, x_1, x_2]$ is a function of them.*

Define

$$a = |x_1 - x_0|, \quad b = |x_2 - x_1|, \quad c = |x_2 - x_0|.$$

Then the ordered triple (a, b, c) is apparently Euclidean invariant, and two chains $[x_0, x_1, x_2]$ and $[y_0, y_1, y_2]$ are Euclidean congruent if and only if they share the same (a, b, c). Thus there must exist a non-negative function $F(a, b, c)$ such that

$$E[x_0, x_1, x_2] = F(a, b, c).$$

Define the two elementary symmetric functions of a and b:

$$A_1 = \frac{a+b}{2} \quad \text{and} \quad B_1 = ab.$$

The reversal invariance implies the symmetry of F with respect to a and b. Thus E has to be a function of A_1, B_1, and c:

$$E[x_0, x_1, x_2] = f(A_1, B_1, c).$$

Let s denote the half perimeter of the triangle (x_1, x_2, x_3):

$$s = A_1 + \frac{c}{2},$$

and Δ its area:

$$\Delta = \sqrt{s(s-a)(s-b)(s-c)} = \sqrt{s(s-c)(s^2 - 2A_1 s + B_1)}.$$

Then we can define the digital curvature at x_1 [COS+98, COT96, Bou00] by

$$\kappa_1 = 4\frac{\Delta}{B_1 c} = \frac{\sin\theta_1}{c/2},$$

where θ_1 is the angle facing the side $[x_0, x_2]$. It is shown by Calabi, Olver, and Tannenbaum [COT96] that for a generic smooth curve and a fixed point x_1 on it, as $a, b \to 0$,

$$\kappa_1 = \kappa(x_1) + O(|b-a|) + O(a^2 + b^2),$$

where $\kappa(x_1)$ is the curvature at x_1.

Now it is easy to see that κ_1, A_1, B_1 is a complete set of joint invariants for both the Euclidean and reversal invariances, and

$$E[x_0, x_1, x_2] = g(\kappa_1, A_1, B_1).$$

Therefore, we have proved

Theorem 2. *A 3-point accumulative energy E with both Euclidean and reversal invariances must be in the form of*

$$E[x_0, \cdots, x_N] = \sum_{n=1}^{N-1} g(\kappa_n, A_n, B_n).$$

Further notice that, as the sampling size $a, b = O(h), h \to 0$ at a fixed point $x_1 \in \Gamma$,

$$\kappa_1 = O(1), \quad A_1 = O(h), \quad B_1 = O(h^2).$$

Applying Taylor expansion to g for the infinitesimals A_1 and B_1 (assuming that g is smooth), we have

$$g(\kappa_1, A_1, B_1) = g_1(\kappa_1)A_1 + g_2(\kappa_1)B_1 + \cdots,$$

for some functions $g_1, g_2, \cdots$. In the linear integration theory, by neglecting all high order (≥ 2) infinitesimals, we end up with

$$g(\kappa_1, A_1, B_1) = g_1(\kappa_1)A_1.$$

Therefore we have derived,

Corollary 1. *Following Theorem 2, in addition, suppose that for any smooth summable simple curve* Γ*, and its Chain Coding approximation* $[x_0, x_1, \cdots x_N]$ *with the size*

$$h = \max_{0 \le n \le N-1} |x_n - x_{n+1}|$$

tending to zero, $E[x_0, \cdots x_N]$ *converges. Then as far as the limit is concerned, there is only one class of such energy, which is given by*

$$E[x_0, \cdots, x_N] = \sum_{n=1}^{N-1} f(\kappa_n) A_n.$$

As $h \to 0$*, it converges to*

$$E[\Gamma] = \int_\Gamma f(\kappa) ds,$$

where ds *is the length element.*

For instance, if we take $f(\kappa) = \alpha + \beta\kappa^2$ for two fixed weights α and β, the resulting energy is called the elastica energy [Mum94a], which was first studied by Euler in modeling the shape of a torsion free thin rod in 1744. If $\beta = 0$, the elastica energy degenerates to the length energy.

2.2 Image models via functionalizing curve models

Once a curve model $E[\Gamma]$ is established, it can be "lifted" to an image model by direct functionalization and the level-set approach.

Let $u(x)$ be an image defined on a domain $\Omega \subset R^2$. For the moment assume that u is smooth so that almost surely for each gray value λ, the level-set

$$\Gamma_\lambda = \{x \in \Omega : u(x) = \lambda\}$$

is a smooth 1-D manifold. Let $w(\lambda)$ be an appropriate non-negative weight function. Then based on a given *curve* model $E[\Gamma]$, we can construct an *image* model:

$$E[u] = \int_{-\infty}^{\infty} E[\Gamma_\lambda] w(\lambda) d\lambda.$$

Conventionally $w(\lambda)$ is set to 1 to reflect human perceptual sensitivity. Suppose we have a bundle of level-sets whose gray values are concentrated over $[\lambda, \lambda + \Delta\lambda]$. If $\Delta\lambda$ is small, then the image appears smooth over the region made of these level-sets, and is thus less sensitive to perception. The energy assigned to such bundles should be small accordingly. On the other hand, if $\Delta\lambda$ is large, for example in the situation when the bundle contains a sharp edge, then the level-sets carry important visual information and the associated energy should be large. Therefore, the Lebesgue measure $d\lambda$ is already perceptually motivated and $w(\lambda)$ can be set to 1, which we shall assume in the following.

Suppose we take the length energy in Theorem 1 as the curve model, then the resulting image model

$$E[u] = \int_{-\infty}^{\infty} \text{length}(\Gamma_\lambda) d\lambda$$

is exactly Rudin-Osher-Fatemi's TV model [ROF92, RO94]:

$$E[u] = \int_{\Omega} |\nabla u| dx.$$

This is because for a smooth image u, along any level-set Γ_λ,

$$d\lambda = |\nabla u| d\sigma, \quad \text{length}(\Gamma_\lambda) = \int_{\Gamma_\lambda} ds,$$

with ds and $d\sigma$ denoting the arc lengthes of the level-sets and gradient flows, which are orthogonal to each other, and thus

$$ds d\sigma = dx$$

is the area element. Therefore,

$$E[u] = \int_{-\infty}^{\infty} \int_{\Gamma_\lambda} |\nabla u| d\sigma ds = \int_{\Omega} |\nabla u| dx.$$

The above derivation is in a formal level and can be rigorously established based on the theory of BV functions [Giu84], where the length of a level-set is replaced by the perimeter of its associated region, the Sobolev gradient norm by the TV radon measure. Then the lifting process is exactly the famous co-area formula.

Similarly, suppose we take the curvature curve model in Corollary 1, then the lifted image model becomes

$$E[u] = \int_{\Omega} f\left(|\nabla \cdot [\frac{\nabla u}{|\nabla u|}]| \right) |\nabla u| dx.$$

Especially, if $f(s) = \alpha + \beta s^2$, it is called the elastica image model [CKS01, MM98].

2.3 Image models with embedded edge models

The second approach to construct image models from curve models is based on the object-edge primary model, as proposed by Mumford-Shah [MS89]. In such image models, the curve model is embedded to weigh the energy from the edges, i.e., abrupt jumps in images.

For example, the classical Mumford-Shah image model employs the length curve model:

$$E[u,\Gamma] = \int_{\Omega\setminus\Gamma} |\nabla u|^2 dx + \alpha\,\text{length}(\Gamma).$$

Here Γ denotes edge collection. Unlike the TV image model, once the singular set Γ is singled out, for the rest of the image domain, Sobolev smoothness can be legally imposed.

Mumford-Shah image model has been very successful in image segmentation and denoising. For image inpainting, as Esedoglu-Shen discussed in [ES01], it is intrinsically insufficient. Therefore a new image model called Mumford-Shah-Euler is proposed in [ES01] based on the elastica curve model:

$$E[u,\Gamma] = \int_{\Omega\setminus\Gamma} |\nabla u|^2 dx + \int_\Gamma (\alpha + \beta\kappa^2) ds.$$

We now start to discuss how to carry out inpainting based on these image models.

3. Inpainting Models and Their PDE's

In this section, we survey all the recent inpainting schemes based on the geometric image models mentioned above. We shall describe the PDE forms for all the variational models, and their digital realization based on the numerical PDE method. Digital examples are demonstrated for each inpainting scheme.

3.1 The TV inpainting

In [CS01a], we first touched on the Bayesian idea for the inpainting problem, as an alternative to the PDE approach invented by Bertalmio et al. [BSCB00] based on the transport mechanism. The image model employed in [CS01a] is the well-known Rudin-Osher-Fatemi's TV image model, as first proposed for the denoising and deblurring application [ROF92, RO94].

The TV inpainting model is to minimize the posterior energy

$$J_{\text{tv}}[u|u^0, D] = \int_\Omega |\nabla u| dx + \frac{\lambda}{2}\int_{\Omega\setminus D} (u-u^0)^2 dx. \tag{2}$$

Define

$$\lambda_D(x) = \lambda \cdot 1_{\Omega\setminus D}(x).$$

Then the steepest descent equation for the energy is

$$\frac{\partial u}{\partial t} = \nabla \cdot \left[\frac{\nabla u}{|\nabla u|} \right] + \lambda_D(x)(u^0 - u), \tag{3}$$

which is a diffusion-reaction type of nonlinear equation. To justify the drop of the boundary integral coming from the variational process, the associated boundary condition along $\partial\Omega$ is adiabatic: $\partial u / \partial \vec{\nu} = 0$, where $\vec{\nu}$ denotes the normal direction of the boundary.

The diffusion is anisotropic to respect sharp edges as in the Perona-Malik diffusion [PM90] since the diffusivity coefficient $1/|\nabla u|$ becomes small where u has sharp jumps. The reaction term has u^0 as its attractor to keep the solution close to the given noisy image. But notice that on the inpainting domain D, the equation is a pure diffusion process, which of course originally comes from the TV image model.

In [CKS01], the existence of a minimizer of J_{tv} in the BV space is established based on the direct method of Calculus of Variation. The uniqueness, however, is generally not guaranteed. An example is given in [CKS01]. Non-uniqueness, from the vision point of view, reflects the uncertainty of human visual perception in certain situations, and thus should not be cursed in terms of faithful modeling.

For the digital realization of model (3), the degenerate diffusion coefficient $1/|\nabla u|$ is always conditioned to

$$\frac{1}{|\nabla u|_a}, \quad |\nabla u|_a = \sqrt{a^2 + |\nabla u|^2},$$

for some small positive constant a. From the energy point of view, it amounts to the minimization of the modified J_{tv}:

$$J_{\text{tv}}^a[u] = \int_\Omega |\nabla u|_a + \frac{\lambda}{2} \int_{\Omega \setminus D} (u - u^0)^2 dx. \tag{4}$$

This energy form connects image inpainting to the classical problem of *non-parametric minimal surfaces* [Giu84]. In fact, in the (x, y, z) space, the first term of $J_{\text{tv}}^a[u]$, up to the multiplicative constant a, is exactly the area of the surface

$$z = z(x, y) = u(x, y)/a.$$

In the case when the available part u^0 is noise-free, we have

$$\lambda = \infty, \quad z|_{\Omega \setminus D} = z^0|_{\Omega \setminus D}.$$

Thus we end up with the exact minimal surface problem on the inpainting domain D:

$$\min \int_D \sqrt{1 + |\nabla z|^2} dx \quad \text{with} \quad z = z^0 \text{ along } \partial D.$$

Here along the boundary, $z\big|_{\partial D}$ is understood as the trace from the interior. Since this Dirichlet problem might not be solvable for general inpainting domains D (see [Giu84] for example), as far as inpainting is concerned, we may formulate a weaker version even for the noise-free case:

$$\min \int_D \sqrt{1+|\nabla z|^2} + \frac{\mu}{2} \int_{\partial D} (z - z^0)^2 dH_1,$$

where μ is a large positive weight and dH_1 the 1-dimensional Hausdorff measure of ∂D. Then the existence of a minimum can be easily established based on the direct method.

Compared with all other variational inpainting schemes, the TV model has the lowest complexity and easiest digital implementation. It works remarkably well for more local inpainting problems such as digital zoom-in (Figure 2) and text removal [CS01a]. But for large-scale inpainting problems, the TV inpainting model suffers from its origin in the length curve energy. One major drawback is its failure to realize the Connectivity Principle in visual perception as discussed in [CS01c].

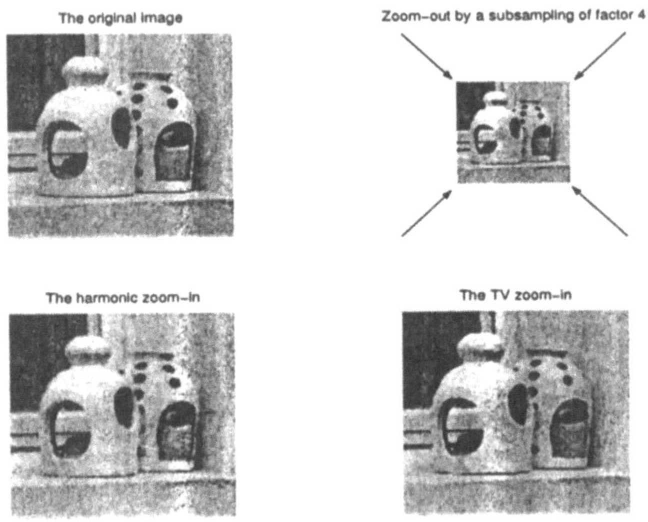

Figure 2. Digital zoom-in based on the TV inpainting scheme, as compared with that based on the harmonic inpainting scheme, i.e., that based on the Sobolev image model: $E[u] = \int_\Omega |\nabla u|^2 dx$. Notice that the TV gives much sharper boundary reconstruction.

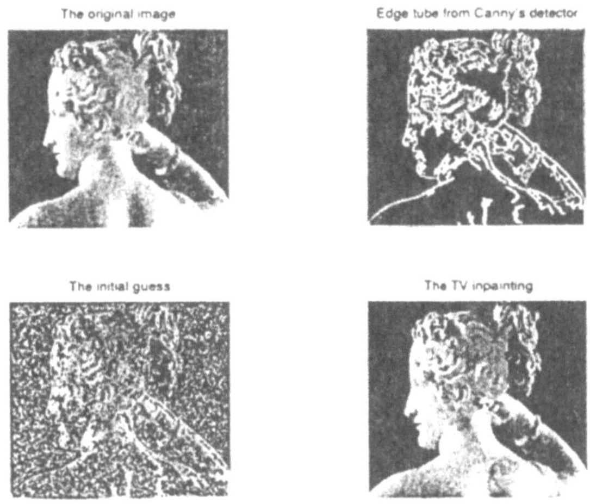

Figure 3. TV inpainting applied to the primal-sketch based image decoding.

3.2 The elastica inpainting

In [CKS01], Chan, Kang, and Shen proposed to improve the TV inpainting model by using the elastica image model

$$E[u] = \int_\Omega (\alpha + \beta\kappa^2)dx, \quad \kappa = \nabla \cdot \left[\frac{\nabla u}{|\nabla u|}\right].$$

The elastica inpainting model is thus to minimize the posterior energy

$$J_e[u|u^0, D] = \int_\Omega \phi(\kappa)dx + \frac{\lambda}{2}\int_{\Omega\setminus D}(u - u^0)^2 dx, \tag{5}$$

where $\phi(s) = \alpha + \beta s^2$.

By Calculus of Variation, it is shown in [CKS01] that the steepest descent equation is given by

$$\frac{\partial u}{\partial t} = \nabla \cdot \vec{V} + \lambda_D(x)(u^0 - u), \tag{6}$$

$$\vec{V} = \phi(\kappa)\vec{n} - \frac{\vec{t}}{|\nabla u|}\frac{\partial(\phi'(\kappa)|\nabla u|)}{\partial \vec{t}}. \tag{7}$$

Here $\vec{n}, \vec{t}$ are the normal and tangent directions:

$$\vec{n} = \frac{\nabla u}{|\nabla u|}, \quad \vec{t} = \vec{n}^\perp, \quad \frac{\partial}{\partial \vec{t}} = \vec{t} \cdot \nabla.$$

Notice that the coupling of $\vec{t}$ and $\partial/\partial\vec{t}$ in (7) makes it safe to take any direction of $\vec{n}^{\perp}$ for $\vec{t}$. The natural boundary conditions along $\partial\Omega$ are

$$\frac{\partial u}{\partial \vec{\nu}} = 0 \quad \text{and} \quad \frac{\partial(\phi'(\kappa)|\nabla u|)}{\partial \vec{\nu}} = 0.$$

The vector field $\vec{V}$ is called the flux of the elastica energy. Its decomposition in the natural orthogonal frame $(\vec{n}, \vec{t})$ in (7) has significant meaning in terms of micro-inpainting mechanisms.

(i) The normal flow $\phi(k)\vec{n}$ carries the feature of an important inpainting scheme invented earlier by Chan and Shen called CDD (*curvature driven diffusion*) [CS01c]. CDD was discovered in looking for micro mechanisms that can realize the Connectivity Principle in visual perception [CS01c, Kan79, NMS93].

(ii) The tangential component can be written as

$$\vec{V}_t = -\left(\frac{1}{|\nabla u|^2}\frac{\partial(\phi'(\kappa)|\nabla u|)}{\partial \vec{t}}\right)\nabla^{\perp} u,$$

and its divergence is

$$\nabla \cdot \vec{V}_t = \nabla^{\perp} u \cdot \nabla\left(\frac{-1}{|\nabla u|^2}\frac{\partial(\phi'(\kappa)|\nabla u|)}{\partial \vec{t}}\right)$$

since $\nabla^{\perp} u$ is divergence free. Define the smoothness measure

$$L_\phi = \frac{-1}{|\nabla u|^2}\frac{\partial(\phi'(\kappa)|\nabla u|)}{\partial \vec{t}}.$$

Then the tangent component is in the form of the transport inpainting mechanism as originally invented by Bertalmio et al. [BSCB00].

Pure transport can lead to shocks as in traffic models, while pure curvature driven diffusion (CDD) is only motivated by the Connectivity Principle in vision research and lacks theoretical support. The elastica inpainting PDE (6) combines their strength and also offers a theoretical framework.

For the numerical realization of the model, we mention the following aspects. More detail can be found in [CKS01]. Two digital examples are illustrated in Figure 4.

(a) To accelerate the convergence of the steepest descent marching (6) toward its equilibrium solution, one can adopt the Marquina-Osher method [MO99] by adding a non-negative "time correcting factor" $T(u, |\nabla u|)$:

$$\frac{\partial u}{\partial t} = T \cdot \left(\nabla \cdot \vec{V} + \lambda_D(x)(u^0 - u)\right).$$

For instance, take $T = |\nabla u|$. As shown in [MO99], such simple technique can substantially improve the numerical marching size and speed up the convergence.

(b) As in the TV inpainting model, for the computation of κ and $\vec{V}$, $1/|\nabla u|$ is always conditioned to $1/|\nabla u|_a$ to avoid a zero denominator.

(c) To more efficiently denoise and propagate sharp edges, classical numerical techniques from computational fluid dynamics (CFD) can be very useful, including those originally designed for capturing shocks. Techniques adopted in [CKS01] are the *upwind scheme* and the *min-mod* scheme [OR90].

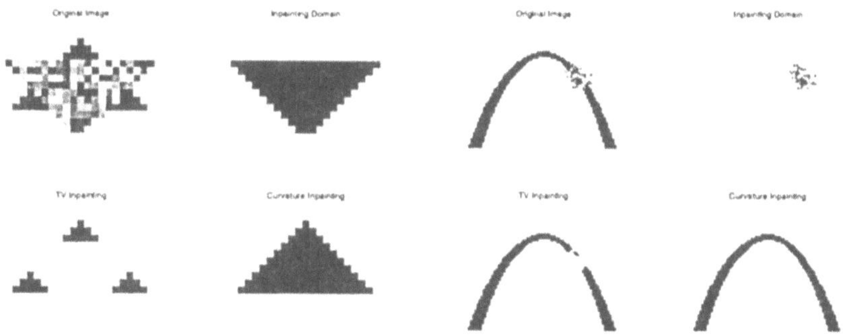

Figure 4. Two examples of elastica inpainting, as compared with TV inpainting. In the case of large aspect ratios [CS01c], the TV inpainting model fails to comply to the Connectivity Principle.

3.3 Inpainting via Mumford-Shah image model

The idea of applying the Mumford-Shah image model to inpainting and image interpolation first appeared in Tsai, Yezzi, and Willsky [TAYW01], and Chan and Shen [CS01a], and has been recently studied again by Esedoglu and Shen [ES01] based on the Γ-convergence theory.

The model is to minimize the posterior energy

$$J_{\mathrm{ms}}[u, \Gamma | u^0, D] = \frac{\gamma}{2} \int_{\Omega \setminus \Gamma} |\nabla u|^2 dx + \alpha \, \mathrm{length}(\Gamma) + \frac{\lambda}{2} \int_{\Omega \setminus D} (u - u^0)^2 dx, \tag{8}$$

where γ, α, and λ are positive weights. Notice that if D is empty, i.e. there is no spatially missing domain, then the model is exactly the classical Mumford-Shah denoising and segmentation model [MS89]. Also, notice that unlike the previous two models, it outputs two objects: the completed and cleaned image u, and its associated edge collection Γ.

For a given edge layout Γ, variation of $J_{\text{ms}}[u|\Gamma, u^0, D]$ gives

$$\gamma\Delta u + \lambda_D(x)(u^0 - u) = 0 \quad \text{on } \Omega\backslash\Gamma, \tag{9}$$

with the natural adiabatic condition $\partial u/\partial\vec{\nu} = 0$ along both Γ and $\partial\Omega$.

Denote the solution to the elliptic equation (9) by u_Γ. Then the steepest descent infinitesimal move of Γ for $J_{\text{ms}}[\Gamma|u_\Gamma, u^0, D]$ is given by

$$\frac{dx}{dt} = \left(\alpha\kappa + \left[\frac{\gamma}{2}|\nabla u_\Gamma|^2 + \frac{\lambda_D}{2}(u_\Gamma - u^0)^2\right]_\Gamma\right)\vec{n}. \tag{10}$$

Here $x \in \Gamma$ is an edge pixel and $\vec{n}$ the normal direction at x. The symbol $[g]_\Gamma$ denotes the jump of a scalar field $g(x)$ across Γ:

$$[g]_\Gamma(x) = \lim_{\sigma\to 0^+}(g(x+\sigma\vec{n}) - g(x-\sigma\vec{n})).$$

The sign of the curvature κ and the direction of the normal $\vec{n}$ are coupled so that $\kappa\vec{n}$ points to curvature center of Γ at x.

Note that the curve evolution equation (10) is a combination of the *mean curvature motion* [ES91]

$$dx/dt = \alpha\kappa\vec{n}$$

and a field-driven motion specified by the second term. The field-driven motion attracts the curve toward the expected edge set, while the mean curvature motion makes sure that the curve does not develop ripples and stays smooth.

Like the TV inpainting model, inpainting based on the Mumford-Shah image model is of second order. But the extra complexity comes from its free boundary nature. In [CS01a, TAYW01], the level-set method of Osher and Sethian [OS88] is proposed.

In the most recent work by Esedoglu and Shen [ES01], a simpler numerical scheme is developed based on the Γ-convergence theory of Ambrosio and Tortorelli [AT90, AT92].

In the Γ-convergence theory, the 1-dimensional edge Γ is approximately represented by its associated signature function

$$z : \Omega \to [0, 1],$$

which is nearly 1 almost everywhere except on a narrow (specified by a small parameter ϵ) tubular neighborhood of Γ, where it is close to 0. The posterior energy $J_{\text{ms}}[u, \Gamma|u^0, D]$ is approximated by:

$$\begin{aligned} J_\epsilon[u, z|u^0, D] =& \frac{1}{2}\int_\Omega \lambda_D(x)(u-u^0)^2 dx + \frac{\gamma}{2}\int_\Omega z^2|\nabla u|^2 dx \\ &+ \alpha\int_\Omega \left(\epsilon|\nabla z|^2 + \frac{(1-z)^2}{4\epsilon}\right) dx. \end{aligned} \tag{11}$$

Taking variation on u and z separately yields the Euler-Lagrange system:

$$\lambda_D(x)(u - u^0) - \gamma\nabla \cdot (z^2 \nabla u) = 0 \tag{12}$$

$$(\gamma|\nabla u|^2)z + \alpha\left(-2\epsilon\Delta z + \frac{z-1}{2\epsilon}\right) = 0, \tag{13}$$

with the natural adiabatic boundary conditions along $\partial\Omega$ (due to the boundary integrals coming from integration-by-parts):

$$\frac{\partial u}{\partial \vec{\nu}} = 0, \qquad \frac{\partial z}{\partial \vec{\nu}} = 0.$$

Define two elliptic operators acting on u and z separately:

$$L_z = -\nabla \cdot z^2 \nabla + \lambda_D/\gamma \tag{14}$$

$$M_u = (1 + 2(\epsilon\gamma/\alpha)|\nabla u|^2) - 4\epsilon^2\Delta. \tag{15}$$

Then the Euler-Lagrange system (12) and (13) is simply written as

$$L_z u = (\lambda_D/\gamma)u^0 \qquad \text{and} \qquad M_u z = 1. \tag{16}$$

This coupled system can be solved easily by any efficient elliptic solver and an iterative scheme. Two digital examples are included in Figure 5 and 6.

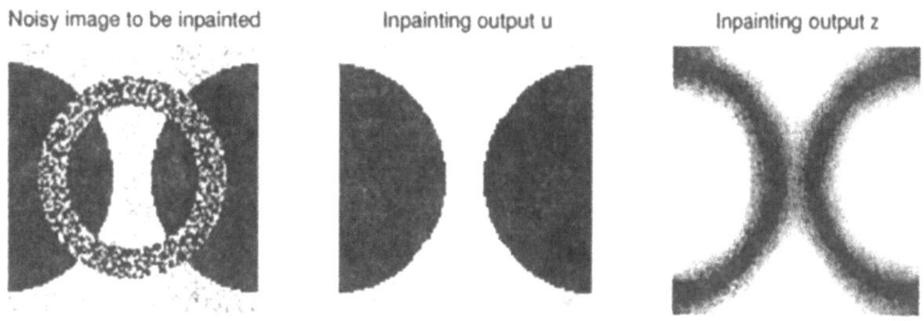

Figure 5. Inpainting based on the Γ-convergence approximation (11) and its associated elliptic system (16).

3.4 Inpainting via Mumford-Shah-Euler image model

Like the TV image model, the Mumford-Shah image model is insufficient for large-scale image inpainting problems due to the embedded length curve energy. To improve, Esedoglu and Shen [ES01] recently proposed the inpainting scheme based on the Mumford-Shah-Euler image model.

In this model, the posterior energy to be minimized is

$$J\text{mse}[u, \Gamma|u^0, D] = \frac{\gamma}{2}\int_{\Omega\setminus\Gamma} |\nabla u|^2 dx + \int_\Gamma (\alpha + \beta\kappa^2)ds + \frac{\lambda}{2}\int_{\Omega\setminus D}(u - u^0)^2 dx, \tag{17}$$

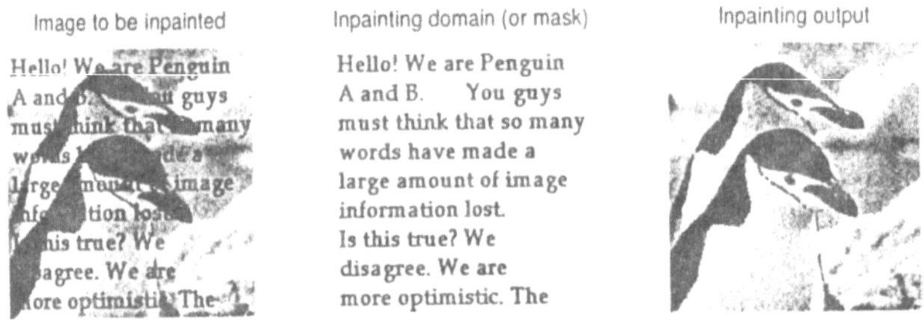

Figure 6. Text erasing by inpainting based on the Mumford-Shah image model.

where the length energy in J_{ms} has been upgraded to Euler's elastica energy.

As in the previous inpainting model, for a given edge layout Γ, the Euler-Lagrange equation for $J_{\text{mse}}[u|\Gamma, u^0, D]$ is

$$\gamma \Delta u + \lambda_D(x)(u^0 - u) = 0, \qquad x \in \Omega \backslash \Gamma, \tag{18}$$

with the adiabatic condition along Γ and $\partial\Omega$: $\partial u / \partial \vec{\nu} = 0$.

For the solution u_Γ to this equation, the infinitesimal steepest descent move of Γ is given by [CKS01, Mum94a, LS84]:

$$\frac{dx}{dt} = \alpha\kappa - \beta\left(2\frac{d^2\kappa}{ds^2} + \kappa^3\right) + \left[\frac{\gamma}{2}|\nabla u_\Gamma|^2 + \frac{\lambda_D}{2}(u_\Gamma - u^0)^2\right]_\Gamma. \tag{19}$$

The meaning of the symbols is the same as in the previous section.

The digital implementation of this 4th order nonlinear evolutionary equation is highly non-trivial. The challenge lies in finding an effective numerical representation of the 1-dimensional object Γ, and robust ways to compute its geometry, i.e., the curvature and its differentials.

In Esedoglu and Shen [ES01], the equation is numerically implemented based on the Γ-convergence theory of De Giorgi [Gio61]. As for the previous Mumford-Shah image model, Γ-convergence approximation leads to simple elliptic systems that can be solved efficiently in computation.

De Giorgi [Gio61] proposed to approximate Euler's elastica curve model

$$e(\Gamma) = \int_\Gamma (\alpha + \beta\kappa^2) ds,$$

by an elliptic integral of the signature z (the two constants α and β may vary):

$$E_\epsilon[z] = \alpha \int_\Omega \left(\epsilon|\nabla z|^2 + \frac{W(z)}{4\epsilon}\right) dx + \frac{\beta}{\epsilon}\int_\Omega \left(2\epsilon\Delta z - \frac{W'(z)}{4\epsilon}\right)^2 dx, \tag{20}$$

where $W(z)$ can be the symmetric double-well function

$$W(z) = (1 - z^2)^2 = (z + 1)^2(z - 1)^2. \tag{21}$$

Unlike the choice of $W(z) = (1 - z)^2$ for the Mumford-Shah image model, here the edge layout Γ is embedded as the zero level-set of z. Asymptotically, as $\epsilon \to 0^+$, a boundary layer grows to realize the sharp transition between the two well states $z = 1$ and $z = -1$.

Then the original posterior energy J_{mse} on u and Γ can be replaced by an elliptic energy on u and z:

$$J_\epsilon[u, z|u^0, D] = \frac{\gamma}{2}\int_\Omega z^2|\nabla u|^2 dx + E_\epsilon[z] + \frac{1}{2}\int_\Omega \lambda_D(u - u^0)^2 dx. \tag{22}$$

For a given edge signature z, variation on u in $J_\epsilon[u|z, u^0, D]$ gives

$$\lambda_D(u - u^0) - \gamma\nabla \cdot (z^2\nabla u) = 0, \tag{23}$$

with the adiabatic boundary condition $\partial u/\partial\vec{\nu} = 0$ along $\partial\Omega$. For the solution u, the steepest decent marching of z for $J_\epsilon[z|u, u^0, D]$ is given by

$$\frac{\partial z}{\partial t} = -\gamma|\nabla u|^2 z + \alpha g + \frac{\beta W''(z)}{2\epsilon^2} g - 4\beta\Delta g, \tag{24}$$

$$g = 2\epsilon\Delta z - \frac{W'(z)}{4\epsilon}, \tag{25}$$

again with the Neumann adiabatic conditions along the boundary $\partial\Omega$:

$$\frac{\partial z}{\partial\vec{\nu}} = 0, \qquad \text{and} \qquad \frac{\partial g}{\partial\vec{\nu}} = 0.$$

Eq. (24) is of fourth-order for z, with the leading head $-8\epsilon\beta\Delta^2 z$. Thus, to ensure stability, an explicit marching scheme would require $\Delta t = O((\Delta x)^4/\epsilon\beta)$. There are a couple of ways to stably increase the marching size. First, as inspired by Marquina and Osher [MO99], one can add a time correcting factor (as in Section 3.2):

$$\frac{\partial z}{\partial t} = T(\nabla z, g|u)\left(-\gamma|\nabla u|^2 z + \alpha g + \frac{\beta W''(z)}{2\epsilon^2} g - 4\beta\Delta g\right),$$

where $T(\nabla z, g, |u)$ is a suitable positive scalar, for example, $T = |\nabla z|$ [MO99].

The second alternative is to turn to implicit or semi-implicit schemes. Eq. (24) can be rearranged to

$$\frac{\partial z}{\partial t} + \gamma|\nabla u|^2 z - 2\alpha\epsilon\Delta z + 8\beta\epsilon\Delta^2 z = -\frac{\alpha}{4\epsilon}W'(z) + \frac{\beta W''(z)}{2\epsilon^2} g + \frac{\beta}{\epsilon}\Delta W'(z), \tag{26}$$

Or simply

$$\frac{\partial z}{\partial t} + L_u z = f(z),$$

where L_u denotes the positive definite elliptic operator (u-dependent)

$$L_u = \gamma |\nabla u|^2 - 2\alpha\epsilon\Delta + 8\beta\epsilon\Delta^2,$$

and $f(z)$ the entire right hand side of (26). Then a semi-implicit scheme can be designed as: at each discrete time step n,

$$(1 + \Delta t L_u) z^{(n+1)} = z^{(n)} + \Delta t f(z^{(n)}),$$

where the positive definite operator $1 + \Delta t L_u$ is numerically inverted based on many fast solvers [GO92, Str93]. A digital example is given in Figure 7.

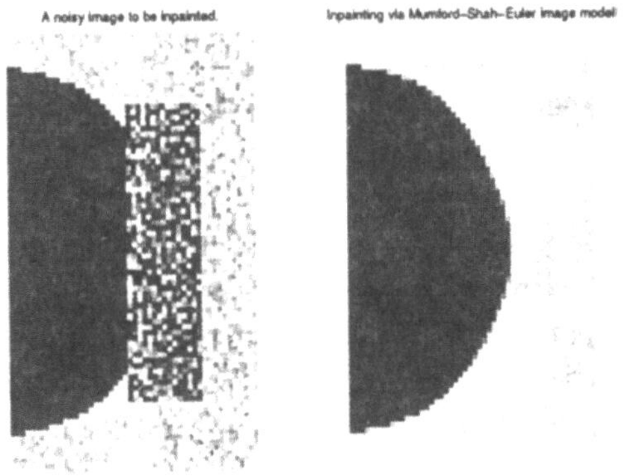

Figure 7. Inpainting based on the Mumford-Shah-Euler image model can satisfactorily restore a smooth edge as expected.

4. Conclusion and open problems

In this paper, we have surveyed all the recent inpainting models based on the combination of the Bayesian principle and geometric image models. As for the classical denoising, deblurring, and segmentation applications, the Bayesian framework has proven to be very effective in designing and improving general inpainting models.

We have explained that the fundamental ingredient for a geometric image model is the associated or embedded curve model. Based on some natural axioms such as the Euclidean invariance and reversal invariance, we have been able to understand the general structure of geometric curve model on the 2-dimensional image domain. We have described two general ways for "lifting" a curve model to a geometric image model.

We have observed that conventional first-order geometric image models, such as the TV model and Mumford-Shah model, function very well for classical denoising, deblurring, and segmentation problems, as well as for inpainting problems with a more local nature (such as zoom-in and text erasing). But for large-scale inpainting problems, they are insufficient for reconstructing perceptually meaningful outputs. Therefore, high order geometric image models, such as the elastica model and the Mumford-Shah-Euler model, become necessary for more general inpainting applications. The tradeoff is that high order inpainting models are computationally much more challenging.

We have described all the Euler-Lagrange PDE's associated to these models, their geometric meaning, and their digital realization based on techniques from numerical PDE's and the Γ-convergence theory.

Finally we post three interesting open problems.

(i) **Video inpainting.** Video inpainting has profound application in the movie industry, surveillance analysis, and dynamic vision analysis. The first open problem is: how to integrate the extra dimension of "time" into the spatial inpainting techniques? And how to define geometric prior models for spatial-temporal images?

(ii) **Texture inpainting.** Textures by definition are image patterns with rich statistical features. Geometric image models can well describe the boundaries of different texture patches, but are apparently insufficient for inpainting the textures themselves. Therefore, the second open problem is: how to integrate geometric image models and statistical texture models? And how to *grow* textures through texture synthesis without creating artificial boundaries?

(iii) **Fast and efficient digital realization.** Throughout this survey, numerical PDE has been a core computational tool for all the geometric inpainting models. The third open problem concerns fast and efficient digital implementation of the associated PDE's, especially for the high order ones. There are a number of non-trivial questions that wait to be answered: How to develop discretization schemes that respect geometry, the curvature and its differentials, for examples? How to speed up convergence based on various numerical techniques such as the multigrid method and the multiresolution decomposition? since speed is always highly concerned in applications. Finally, the energies of high order inpainting models, such as the elastica and the Mumford-Shah-Euler models, often have many local energy wells. It is thus another important issue to develop numerical schemes that can efficiently avoid being trapped in local energy wells, as in Molecular Dynamics [MW97].

Acknowledgments

We would like to thank Professor Sapiro's group for first introducing us to the topic. We would like to acknowledge the support from both the Institute of Mathematics and its Applications (IMA) at University of Minnesota and the Institute of Pure and Applied Mathematics (IPAM) at UCLA during this project. We are very grateful for the help from Professors Stan Osher, Luminita Vese, Sang Ha Kang, Li-Tien Cheng, Jean-Michel Morel and Simon Masnou, Willard Miller, Peter Olver, Fadil Santosa, Bob Gulliver, Dan Kersten, Selim Esedoglu, and Rachid Deriche.

References

[AKR97] S. Armstrong, A. Kokaram, and P.J.W. Rayner. Nonlinear interpolation of missing data using min-max functions. *IEEE Int. Conf. Nonlinear Signal and Image Processings*, 1997.

[AT90] L. Ambrosio and V. M. Tortorelli. Approximation of functionals depending on jumps by elliptic functionals via Γ-convergence. *Comm. Pure Appl. Math.*, 43:999–1036, 1990.

[AT92] L. Ambrosio and V. M. Tortorelli. On the approximation of free discontinuity problems. *Boll. Un. Mat. Ital.*, 6-B:105–123, 1992.

[BBC+01] C. Ballester, M. Bertalmio, V. Caselles, G. Sapiro, and J. Verdera. Filling-in by joint interpolation of vector fields and grey levels. *IEEE Trans. Image Process.*, 10(8):1200–1211, 2001.

[BBS01] M. Bertalmio, A. L. Bertozzi, and G. Sapiro. Navier-Stokes, fluid dynamics, and image and video inpainting. IMA Preprint 1772 at: www.ima.umn.edu/preprints/jun01, Juun, 2001.

[Bou00] M. Boutin. Numerically invariant signature curves. *Int. J. Comp. Vision*, 40(3):235–248, 2000.

[BSCB00] M. Bertalmio, G. Sapiro, V. Caselles, and C. Ballester. Image inpainting. Computer Graphics, *SIGGRAPH 2000*, July, 2000.

[CCBT01] L. Chanas, J. P. Cocquerez, and J. Blanc-Talon. Highly degraded sequences restoration and inpainting. Preprint, 2001.

[CKS01] T. F. Chan, S.-H. Kang, and J. Shen. Euler's elastica and curvature based inpaintings. *SIAM J. Appl. Math.*, submitted. Available at UCLA CAM Report 2001-12 at: www.math.ucla.edu/~imagers, 2001.

[CMS98] V. Caselles, J.-M. Morel, and C. Sbert. An axiomatic approach to image interpolation. *IEEE Trans. Image Processing*, 7(3):376–386, 1998.

[COS+98] E. Calabi, P. J. Olver, C. Shakiban, A. Tannenbaum, and S. Haker. Differential and numerically invariant signature curves applied to object recognition. *Int. J. Comp. Vision*, 26(2):107–135, 1998.

[COT96] E. Calabi, P. J. Olver, and A. Tannenbaum. Affine geometry, curve flows, and invariant numerical approximations. *Adv. Math.*, 124(1):154–196, 1996.

[CS01a] T. F. Chan and J. Shen. Mathematical models for local non-texture inpaintings. *SIAM J. Appl. Math.*, in press, 2001.

[CS01b] T. F. Chan and J. Shen. Morphologically invariant PDE inpaintings. UCLA CAM Report 2001-15 at: www.math.ucla.edu/~imagers; submitted to *IEEE Trans. Image Process.*, 2001.

[CS01c] T. F. Chan and J. Shen. Non texture inpainting by curvature driven diffusions (CDD). *J. Visual Comm. Image Rep.*, to appear, 2001.

[CS01d] T. F. Chan and J. Shen. Variational restoration of non-flat image features: models and algorithms. *SIAM J. Appl. Math.*, 61(4):1338–1361, 2001.

[Don00] D. L. Donoho. Curvelets. Invited talk at workshop on *Wavelets, Statistics, and Image Processing*, Geogia Inst. Tech., 1999; Invited talk at MSRI workshop on *Mathematics of Imaging*; 1999. Beamlets, Invited talk at IMA workshop on *Image Analysis and Low Level Vision*, 2000.

[EM76] G. Emile-Male. *The Restorer's Handbook of Easel Painting*. Van Nostrand Reinhold, New York, 1976.

[ES91] L. C. Evans and J. Spruck. Motion of level sets by mean curvature. *J. Diff. Geom.*, 33(3):635–681, 1991.

[ES01] S. Esedoglu and J. Shen. Digital inpainting based on the Mumford-Shah-Euler image model. UCLA CAM Report 2001-24 at: www.math.ucla.edu/~imagers; submitted to *European J. Appl. Math.*, 2001.

[Fre61] H. Freeman. On the encoding of arbitrary geometric configuration. *IRE Transactions on Electronic Computers*, EC-10(2):260–268, 1961.

[GG84] S. Geman and D. Geman. Stochastic relaxation, Gibbs distributions, and the Bayesian restoration of images. *IEEE Trans. Pattern Anal. Machine Intell.*, 6:721–741, 1984.

[Gib02] W. Gibbs. *Elementary Principles of Statistical Mechanics*. Yale University Press, 1902.

[Gio61] E. De Giorgi. Frontiere orientate di misura minima. *Sem. Mat. Scuola Norm. Sup. Pisa*, 1960-61.

[Giu84] E. Giusti. *Minimal Surfaces and Functions of Bounded Variation*. Birkhäuser, Boston, 1984.

[GO92] G. H. Golub and J. M. Ortega. *Scientific Computing and Differential Equations*. Academic Press, 1992.

[IP97] H. Igehy and L. Pereira. Image replacement through texture synthesis. *Proceedings of IEEE Int. Conf. Image Processing*, 1997.

[JCL94] K.-H. Jung, J.-H. Chang, and C. W. Lee. Error concealment technique using data for block-based image coding. *SPIE*, 2308:1466–1477, 1994.

[Kan79] G. Kanizsa. *Organization in Vision*. Praeger, New York, 1979.

[KMFR95a] A. C. Kokaram, R. D. Morris, W. J. Fitzgerald, and P. J. W. Rayner. Detection of missing data in image sequences. *IEEE Trans. Image Process.*, 11(4):1496–1508, 1995.

[KMFR95b] A. C. Kokaram, R. D. Morris, W. J. Fitzgerald, and P. J. W. Rayner. Interpolation of missing data in image sequences. *IEEE Trans. Image Process.*, 11(4):1509–1519, 1995.

[KR96] D. C. Knill and W. Richards. *Perception as Bayesian Inference*. Cambridge Univ. Press, 1996.

[KS93] W. Kwok and H. Sun. Multidirectional interpolation for spatial error concealment. *IEEE Trans. Consumer Electronics*, 39(3), 1993.

[KS97] I. Karatzas and S. E. Shreve. *Brownian motion and stochastic calculus*. Springer, New York, 1997.

[LS84] J. Langer and D. A. Singer. The total squared curvature of closed curves. *J. Diff. Geom.*, 20:1–22, 1984.

[MH80] D. Marr and E. Hildreth. Theory of edge detection. *Proc. Royal Soc. London*, B:207: 187–217, 1980.

[MM98] S. Masnou and J.-M. Morel. Level-lines based disocclusion. *Proceedings of 5th IEEE Int'l Conf. on Image Process., Chicago*, 3:259–263, 1998.

[MO99] A. Marquina and S. Osher. *Lecture Notes in Computer Science*, volume 1682, chapter "A new time dependent model based on level set motion for nonlinear deblurring and noise removal", pages 429–434. 1999.

[MS89] D. Mumford and J. Shah. Optimal approximations by piecewise smooth functions and associated variational problems. *Comm. Pure Applied. Math.*, XLII:577–685, 1989.

[Mum94a] D. Mumford. Elastica and computer vision. In C. L. Bajaj, editor, *Algebraic Geometry and its Applications*, pages 491–506. Springer-Verlag, New York, 1994.

[Mum94b] D. Mumford. *Geometry Driven Diffusion in Computer Vision*, chapter "The Bayesian rationale for energy functionals", pages 141–153. Kluwer Academic, 1994.

[MW97] J. J. Moré and Z. Wu. Issues in large-scale global molecular optimization. In L. T. Biegler, T. F. Coleman, A. R. Conn, and F. N. Santosa, editors, *Large-Scale Optimization with Applications*, pages 99–121. Springer, New York, 1997.

[NMS93] M. Nitzberg, D. Mumford, and T. Shiota. *Filtering, Segmentation, and Depth*. Lecture Notes in Comp. Sci., Vol. 662. Springer-Verlag, Berlin, 1993.

[OR90] S. Osher and L. Rudin. Feature-oriented image enhancement using shock filters. *SIAM J. Num. Anal.*, 27(4):919–940, 1990.

[OS88] S. Osher and J. A. Sethian. Fronts propagating with curvature-dependent speed: Algorithms based on Hamilton-Jacobi formulations. *J. Comput. Phys.*, 79(12), 1988.

[PM90] P. Perona and J. Malik. Scale-space and edge detection using anisotropic diffusion. *IEEE Trans. Pattern Anal. Machine Intell.*, 12:629–639, 1990.

[RO94] L. Rudin and S. Osher. Total variation based image restoration with free local constraints. *Proc. 1st IEEE ICIP*, 1:31–35, 1994.

[ROF92] L. Rudin, S. Osher, and E. Fatemi. Nonlinear total variation based noise removal algorithms. *Physica D*, 60:259–268, 1992.

[Str93] G. Strang. *Introduction to Applied Mathematics*. Wellesley-Cambridge Press, MA, 1993.

[TAYW01] A. Tsai, Jr. A. Yezzi, and A. S. Willsky. Curve evolution implementation of the Mumford-Shah functional for image segmentation, denoising, interpolation and magnification. *IEEE Trans. Image Process.*, 10(8):1169–1186, 2001.

[Wal85] S. Walden. *The Ravished Image*. St. Martin's Press, New York, 1985.

[WL00] L.-Y. Wei and M. Levoy. Fast texture synthesis using tree-structured vector quantization. Preprint, Computer Science, Stanford University, 2000; Also in *Proceedings of SIGGRAPH*, 2000.

[ZM97] S. C. Zhu and D. Mumford. Prior learning and Gibbs reaction-diffusion. *IEEE Trans. on PAMI*, 19(11):1236–1250, 1997.

[ZWM97] S. C. Zhu, Y. N. Wu, and D. Mumford. Minimax entropy principle and its applications to texture modeling. *Neural Computation*, 9:1627–1660, 1997.

A COMBINATION OF ALGEBRAIC MULTIGRID ALGORITHMS WITH THE CONJUGATE GRADIENT TECHNIQUE

Qianshun Chang and Zhaohui Huang
Academy of Mathematics and System Sciences, Chinese Academy of Sciences, Beijing 100080, People's Republic of China
cqs@amath8.amt.ac.cn
zhhuang@amath6.amt.ac.cn

Abstract In this paper, conjugate gradient acceleration of algebraic multigrid methods (AMGCG) is described. Theoretical analysis and numerical experiments demonstrate that iterant recombination increases the efficiency and robustness of algebraic multigrid methods. To judge the performance of AMGCG, we not only focus on its convergence behavior, but also take both computing times and memory requirement into account.

Keywords: algebraic multigrid, preconditioned conjugate gradient, iterant recombination, convergence

1. Introduction

Multigrid method (MG) is an nearly optimal-order iterative method for large sparse systems that arise from discretizations of partial differential equations. In this method, the process of relaxation possessing certain smoothing properties alternates with a process of coarse-grid approximation [1, 4, 5, 12, 16].

Although MG is a very efficient way and its scope of applications is also broader and broader, it is not always easy to choose the optimal components for difficult problems and therefore the acceleration of MG has become more and more popular during the last years to further increase the efficiency and robustness of standard MG methods. Krylov subspace methods e.g. conjugate gradient (CG) acceleration provides a powerful tool for speeding up the convergence of multigrid methods, instead of trying to optimize the interplay between the various multigrid components [14, 17].

Algebraic multigrid (AMG) methods are automatic procedures for coarsening the set of equations, relying exclusively on its algebraic relations. AMG is widely employed for solving discretized partial differential equations on unstructured grids, or even many types of discrete systems not arising from differential equations [2, 3, 6, 7, 8, 9, 10, 11, 13, 15].

However, it is also not trivial to design AMG components. Especially, its interpolation will hardly ever be optimal. In this paper, by CG acceleration, we can put less effort into the expensive setup phase and use AMG as preconditioner and therefore the efficiency of AMG is also enhanced largely. This is because AMG's efficiency is affected by the slow convergence of just a few exceptional error components, while CG typically eliminates these particular frequencies very efficiently.

2. AMG Algorithm

AMG algorithms are solvers of linear systems of equations which are based on multigrid principles but do not explicitly use the geometry of grids.

Consider the system of linear equations

$$AU = F \tag{2.1}$$

where $A = (a_{ij})_{n\times n}, U = (u_1, \cdots, u_n)^T, F = (f_1, \cdots, f_n)^T$. A sequence of systems of equations is generated as

$$A^m U^m = F^m, \tag{2.2}$$

where $A^m = (a_{ij}^m)_{n_m \times n_m}, U^m = (u_1^m, \cdots, u_{n_m}^m)^T$, $F^m = (f_1^m, \cdots, f_{n_m}^m)^T$, $m = 1, \cdots, M, n = n_1 > \cdots > n_M$, $A^1 = A, U^1 = U, F^1 = F$. These equations formally play the same role as the coarse grid equations defined in the geometric multigrid (GMG) method. A grid Ω^m can be regarded as a set of unknowns $u_j^m (1 \leq j \leq n_m)$.

In general, there are two phases required in a AMG method: (1) the preparation phase or the setup phase, in which the five components: coarse grids Ω^m, transfer operators I_{m+1}^m and I_m^{m+1}, coarse operator A^{m+1} and smoothing operator G^m are constructed; (2) the solver phase, i.e. the general multigrid cycling procedure, in which the system of equations is solved.

In this paper, we use the AMG algorithm whose components are constructed as follows:

(CO1) Coarse grids $\Omega^m (m = 1, \cdots, n)$, where the finest grid-$\Omega^1$ is chosen fine enough to provide the desired accuracy, and the coarsest is chosen so that the exact solution of the problems on that grids is negligible compared to that of one relaxation sweep on the finest grid. The coarse grid-Ω^{m+1} is chosen as a subset in Ω^m, which is denoted by C^m. The remainder subset $\Omega^m - C^m$ is denoted by F^m. A point i is said to be strongly connected to j, if

$$|a_{ij}^m| \geq \theta \cdot \max_{k \neq i} |a_{ik}^m|, 0 < \theta \leq 1. \tag{2.3}$$

Let S_i^m denote the set of all strongly connection points of the point i and let $C_i^m = C^m \bigcap S_i^m$. In general, we require C_i^m for $\forall i \in F^m$ to be satisfy the following:
(CR1) If $i \in F^m$, and $j \in S_i^m$, either $j \in C_i^m$ or j must strongly depend on C_i^m. This is our primary criterion in the choice of C^m and F^m.
(CR2) The connection between variables in C^m and F^m should be as small as possible.

In practice, it is impossible to strictly satisfy both criteria (CR1) and (CR2) for all systems of equations. However, (CR2) is generally used as a guideline to construct C^m such that (CR1) is held. Now define the set of points which are strongly connected to i by $S_i^T = \{j : i \in S_j^m\}$, and for a set P, let $|P|$ denote the number of elements in P. The following two-part process is suggested by Ruge and Stüben. First, a basic choice for the C-point is performed as follows:

(1) Set $C^m = \emptyset, F^m = \emptyset, U = \Omega^m$, and $\lambda = |S_i^T|$ for all i,

(2) Pick an $i \in U$ with maximal λ_i, and set $C^m = C^m \bigcup \{i\}, U = U - \{i\}$,

(3) For all $j \in S_i^T \bigcap U$, perform (4) and (5),

(4) Set $F^m = F^m \bigcup \{j\}$ and $U = U - \{j\}$,

(5) For all $l \in S_j^m \bigcap U$, set $\lambda_l = \lambda_l + 1$,

(6) For all $j \in S_i^m \bigcap U$, set $\lambda_j = \lambda_j - 1$,

(7) If $U = \emptyset$, stop. Otherwise, goto (2).

The first part attempts to enforce the criterion (CR2) by distributing the C-point uniformly over the grid. The second part is combined with the computation of interpolation weights, in which the tentative F-point resulting from the first part are tested to ensure that the criterion (CR1) holds. The new C-points will be added as necessary. It should be noted that the steps (1)-(7) need only $O(n)$ operations when an efficient implementation is used.

(CO2) Interpolation operators I_{m+1}^m, that is, each variable in C^m interpolates directly from the corresponding variable in Ω^{m+1} with a weighting of 1, and each variable $i \in F^m$ interpolates from the smaller set C_i^m.

In [9], based on two geometric assumptions:
(G1) In the neighborhood N_i^m of a point $i \in \Omega^m$, the larger the quantity $|a_{ij}^m|$

is, the closer point j is to the point i,
(G2) An algebraically smooth error is also geometrically smooth between points i and j if $a_{ij}^m < 0$ or $|a_{ij}^m|$ is small, and it is geometrically oscillations if $a_{ij}^m > 0$ is large,

Chang gave the following interpolation formula for the variable $i \in F^m$

$$e_i^m = \sum_{j \in C_i^m} w_{ij}^m e_j^{m+1}, \forall i \in F^m, \tag{2.4}$$

and

$$\begin{aligned} w_{ik}^m &= -\frac{\bar{a}_{ik}^m}{\bar{a}_{ii}^m}, k \in C_i^m \\ \bar{a}_{ii}^m &= a_{ii}^m - \sum_{j \in D_i^{(1)}} |a_{ij}^m| - \sum_{j \in D_i^{(3)}} a_{ij}^m + 0.5 \sum_{j \in D_i^{(4)}} a_{ij}^m \\ \bar{a}_{ik}^m &= a_{ik}^m + \sum_{j \in D_i^{(2)}} a_{ij}^m g_{jk}^m + 2 \sum_{j \in D_i^{(3)}} a_{ij}^m g_{jk}^m + 0.5 \sum_{j \in D_i^{(4)}} a_{ij}^m g_{jk}^m. \end{aligned} \tag{2.5}$$

where

$$\begin{aligned} g_{jk}^m &= \frac{|a_{jk}^m|}{\sum_{k \in C_i^m} |a_{jk}^m|}, j \in D_i^m, k \in C_i^m, \\ D_i^{(1)} &= \{j : j \in D_i^w, l_{ij}^m = 0, a_{ij}^m \neq 0\}, \\ D_i^{(3)} &= \begin{array}{l} \{j : j \in D_i^w, l_{ij}^m > 0, \xi_{ij}^m \geq 0.5, a_{ij}^m < 0\} \bigcup \\ \{j : j \in D_i^s, \eta_{ij}^m < 0.75, \xi_{ij}^m \geq 0.5, a_{ij}^m < 0\}, \end{array} \\ D_i^{(4)} &= \{j : j \in D_i^s, \eta_{ij}^m > 2, \xi_{ij}^m \geq 0.5, a_{ij}^m < 0\}, \\ D_i^{(2)} &= \{j : j \in D_i^m \backslash (D_i^{(1)} \bigcup D_i^{(3)} \bigcup D_i^{(4)})\}; \end{aligned}$$

and

$$\begin{aligned} \xi_{ij}^m &= \frac{-\sum_{k \in C_i^m} a_{jk}^m}{\sum_{k \in C_i^m} |a_{jk}^m|}, \quad \eta_{ij}^m = \frac{|a_{ji}^m| l_{ij}^m}{\sum_{k \in C_i^m} |a_{jk}^m|}, \\ l_{ij}^m &= |S_{ij}^m|, \quad S_{ij} = \{k : k \in C_i^m, a_{jk}^m \neq 0\}, \\ D_i^m &= N_i^m - C_i^m, \quad D_i^s = D_i^m \bigcap S_i^m, \\ D_i^w &= D_i^m - D_i^s, N_i^m = \{j : j \in \Omega^m, j \neq i, a_{ij}^m \neq 0\}. \end{aligned}$$

Now that the coarse grid and interpolation operator are defined, the restriction operator I_m^{m+1} and the coarse grid operators A^{m+1} can be defined by the Galerkin type algorithm:

(CO3) The restriction operator $I_m^{m+1} : G(\Omega^m) \to G(\Omega^{m+1})$. In AMG, after the coarse grid and interpolation operator are defined, I_m^{m+1} and the following coarse-grid operator A^{m+1} can be defined by the Galerkin type algorithm: $I_m^{m+1} = (I_{m+1}^m)^T$ and $A^{m+1} = I_m^{m+1} A^m I_{m+1}^m$. This ensures that the correction from the exact solution of the coarse-grid problem is the best approximation in the range of interpolation, where "best" is meant in the energy norm.

(CO4) The coarse-grid operator $A^{m+1} : G(\Omega^{m+1}) \to G(\Omega^{m+1})$.

(CO5) A relaxation method for each grid $G^m : G(\Omega^m) \to G(\Omega^m)$, which is usually chosen as a fixed iterative procedure, for example, Gauss-Seidel or Jacobi relaxations with some parameter ω^m.

Thus, we have defined all the components necessary for the AMG solution process.

3. Conjugate Gradient Acceleration

First, we discuss the general Krylov subspace acceleration.

The acceleration of multigid by iterant recombination starts from successive approximations $u_h^1, u_h^2, \ldots, u_h^m$ from previous multigrid cycles. In order to find an improved approximation $u_{h,A}$, we consider a linear combination of the $\tilde{m}+1$ lastest approximations $u_h^{m-i}, i = 0, \ldots, \tilde{m}$,

$$u_{h,A} = u_h^m + \sum_{i=1}^{\tilde{m}} \alpha_i (u_h^{m-i} - u_h^m), \tag{3.1}$$

(assuming as $m \geq \tilde{m}$) with $\sum \alpha_i = 1$.

For linear equations, the corresponding defect, $r_{h,A} = f_h - A_h u_{h,A}$ is given by

$$r_{h,A} = r_h^m + \sum_{i=1}^{\tilde{m}} \alpha_i (r_h^{m-i} - r_h^m), \tag{3.2}$$

where $r_h^{m-i} = f_h - A_h u_h^{m-i}$. In order to obtain an improved approximation $u_{h,A}$, the parameters α_i are determined in such a way that the defect $r_{h.A}$ is minimized.

In general, we will minimize $r_{h,A}$, i.e.

$$\left\| r_h^m + \sum_{i=1}^{\tilde{m}} \alpha_i (r_h^{m-i} - r_h^m) \right\|$$

with respect to the l_2-norm $\|.\|_2$.

Especially, when we choose a corresponding minimization is performed in the A_h^{-1}-norm ($\|x\|_{A_h^{-1}} = (x, A_h^{-1}x)$), where A_h is the matrix corresponding to the discrete problem, the following CG acceleration algorithm is obtained:

(1) Choose some initial value u_h^0, and after performing a complete AMG cycle, we obtain u_h^1;

(2) Compute $r_h^0 = f_h - A_h u_h^0, p_0 = u_h^1 - u_h^0, q_0 = (r_k^0)^T r_h^0, k = 1$;

(3) Compute $\alpha_k = p_{k-1}/p_{k-1}^T A_h p_{k-1}, \tilde{u}_h^k = u_h^{k-1} + \alpha_k p_{k-1}$, thus the current approximation u_h^{k-1} is replaced by $\tilde{u}_h^k$;

(4) Compute $r_h^k = r_h^{k-1} - \alpha_k A_h p_{k-1}, q_k = (r_h^k)^T r_h^k, \beta_k = q_k/q_{k-1}, p_k = r_h^k + \beta_k p_{k-1}, k = k + 1$;

(5) With this replaced approximation $\tilde{u}_h^k$, the next AMG cycle is performed leading to a new iterant u_h^k;

(6) If $q_k < q_0\varepsilon$ with ε given parameter of convergence criterion, stop; otherwise, goto (3).

4. Analysis of CG Acceleration

In this section, we only give a simple picture of convergence analysis from the point of multigrid as a preconditioner.

First, in the framework of preconditioned conjugate gradient (PCG) methods, the rate of convergence of CG methods can be estimated in the following way.

Let the linear system to be solved be denoted by $Au = f$, and the CG method be applied to the preconditioned system

$$PAP^T(P^{-T}u) = Pf, \tag{4.1}$$

where $P^T P \approx A^{-1}$.

The CG algorithm has the fundamental property

$$Pr_h^n = \phi_n^0(PAP^T)Pr_h^0 \tag{4.2}$$

with ϕ_n^0 satisfying

$$\|\phi_n^0(PAP^T)Pr_h^0\|_{A^{-1}} = min_{\phi_n}\{\|\phi_n(PAP^T)Pr_h^0\|_{A^{-1}}, \phi_n \in \Theta_n^0\}$$

and the set Θ_n^0 by

$$\Theta_n^0 = \{\theta_n : \theta_n \text{ is a polynomial of degree} \leq n \text{ and } \theta_n(0) = 1\}.$$

It follows that

$$\|Pr_h^n\|_{A^{-1}}^2 = \|Pr_h^0\|_{A^{-1}}^2 min\{max[\psi(\lambda)^2, \lambda \in S(K)], \psi \in \Theta_n^0\} \quad (4.3)$$

where $S(K)$ is the set of eigenvalues of $K = PAP^T$.

From this, it may be shown that

$$\|Pr_h^n\|_{A^{-1}} / \|Pr_h^0\|_{A^{-1}} \leq 2e^{\frac{-2n}{\sqrt{Cond_2(K)}}} \quad (4.4)$$

with $Cond_2(K)$ the condition number measured in the spectral norm.

Now if the AMG iteration matrix M is symmetric, choosing $P = M^{-1}$ in the PCG algorithm gives us CG acceleration of AMG methods.

In the case of deterioration of AMG convergence, quite often only a few eigenmodes are slow to converge. That is, certain error components may remain large since they can not be reduced by smoothing procedures combined with coarse grid approximations. This means that $S(K)$ will be highly clustered around just a few values. So that $\psi(\lambda)$ will be small on $S(K)$ for $n \approx n_0$ with n_0 the number of clusters, indicating that n_0 iterations will suffice.

This is the reason for the satisfactory acceleration of AMG methods by conjugate gradient.

5. Numerical Experiments

In this section, we point out that if the AMG algorithm is well designed and fits the problem it will converge very fast, making conjugate gradient acceleration superflous or even wasteful.

The AMG algorithm we have proposed and used in the paper is shown efficient and robust for many types of problems including Poisson equation, Toeplitz matrix in signal processing problems, queuing network problems (singular problems), and elasticity problems.

However, the above AMG does not converge fast when it is applied to the very difficult biharmonic equation and the thorny problem whose error between any two horizontal gridlines is strongly related.

Here we do not try to remedy this by improving the algorithm, but employ CG acceleration to make some comparisons with the performance of AMG as a solver and a preconditioner.

Particular attentions are focused on V-cycle convergence factors, CPU-time consumed and memory required.

The following notations are used for the results reported in all tables:
ρ: asymptotic convergence factor,
t_p: computing time for the setup phase,
t_S: computing time for the solution phase,
N: number of iterations for convergence defined by $\|r^N\|/\|r^0\| \leq 10^{-6}$, where r^N is the residual vector at the Nth iteration,
EQ: total number of matrix equations, σ^A: ratio of the space occupied by all operators to the space at the finest grid,
σ^Ω: ratio of the total number of points on all grids to that on the finest grid,
Method I: The AMG Method,
Method II: the AMG acceleration by iterant recombination.

In all computations, the initial iteration u^0 is taken to be random numbers uniformly distributed in [0,1], and the Gauss-Seidel relaxation is used as the smoothing operator and $\theta = 0.25$.

Only one smoothing step is applied before and after coarse-grid correction steps, and smoothing is in C-F order, that is, first the C-variables are relaxed by plain Gauss-Seidel and then the F-variables.

Problem 1 Biharmonic problem on a unit square.

Let

$$\Delta^2 u = 0, \text{ in } \Omega,$$
$$u = 0, \text{ on } \partial\Omega,$$
$$\frac{\partial u}{\partial n} = 0, \text{ on } \partial\Omega,$$

with the following 13-point finite difference stencil

$$\begin{bmatrix} & & 1 & & \\ & 2 & -8 & 2 & \\ 1 & -8 & 20 & -8 & 1 \\ & 2 & -8 & 2 & \\ & & 1 & & \end{bmatrix}.$$

The resulting matrix equation is symmetric and not diagonally dominant. Furthermore, the matrix is very ill-conditioned with a condition number of $O(h^{-4})$. Therefore, this problem provides a good test case of the robustness and efficiency for various algorithms.

Table 1 compares the performance of AMG and its acceleration of CG technique.

Table 1 Computation results for Biharmonic problem

method	EQ	ρ	N	t_p	t_S	σ^A	σ^Ω
I	16×16	0.454	18	0.05	0.11	2.02	1.62
	32×32	0.788	58	0.11	0.77	2.13	1.65
	48×48	0.820	70	0.25	1.40	2.18	1.66
	64×64	0.814	67	0.49	2.92	2.20	1.66
II	16×16	0.242	10	0.05	0.11	2.02	1.62
	32×32	0.499	20	0.05	0.50	2.13	1.65
	48×48	0.598	27	0.21	1.10	2.18	1.66
	64×64	0.656	33	0.44	1.97	2.20	1.66

Problem 2 We consider a 5-point difference stenci

$$L_h^{hs} = \frac{1}{h^2}\begin{bmatrix} & 1 & \\ -1 & 4 & -1 \\ & 1 & \end{bmatrix}_h ,$$

whose corresponding matrix has the very special property that algebraically smooth error is geometrically smooth only in x-direction but strong oscillatory in y-direction.

The AMG algorithm in [9] and Gauss-Seidel-type MG algorithms in [13] can not solve the problem efficiently, while CG acceleration gives satisfactory results in Table 2.

Table 2 Computation results for L_h^{hs}

method	EQ	ρ	N	t_p	t_S	σ^A	σ^Ω
I	16×16	0.543	23	0.05	0.05	2.12	1.67
	32×32	0.797	61	0.06	0.27	2.16	1.70
	48×48	0.893	122	0.11	1.70	2.18	1.70
	64×64	0.917	159	0.17	3.51	2.20	1.72
II	16×16	0.204	9	0.06	0.05	2.12	1.67
	32×32	0.442	17	0.05	0.17	2.16	1.70
	48×48	0.574	25	0.06	0.50	2.18	1.70
	64×64	0.657	33	0.11	1.27	2.20	1.72

Remark. For the above difficult test problems, error reduction is significantly less efficient for some very specific error components, which causes a few eigenvalues of the AMG iteration matrix to be considerably closer to 1 than all the rest. Since the largest eigenvalue determines the spectral radius, these specific error components are then responsible for the poor AMG convergence.

From numerical results in the tables, we observe that such a situation is well suited for CG acceleration: the eigenvectors belonging to isolated eigenvalues can be expected to be captured by CG acceleration.

6. Conclusions

AMG is originally designed to be used as a solver, but the above theoretical analysis and numerical results demonstrate the efficacy of AMGCG. Here we do not intend to answer the question whether AMG should be used as a solver or as a preconditioner, however we point out that if we treat some more complex problems or if when we can not identify the cause of trouble, conjugate gradient acceleration is an easy and very efficient way out.

Acknowledgments

This work is partly supported by Morningside Center of Mathematics (MCM), Chinese Academy of Sciences (CAS). We wish to thank Prof. J. Xu for various discussions related to this work.

References

[1] A. Brandt, Multi-level adaptive solutions to boundary-value problems, Math. Comput. 31(1977), pp.333-390.

[2] A. Brandt, S. McCormick, and J. Ruge, Algebraic Multigrid (AMG) for sparse matrix equations, in Sparsity and Its Applications, D. Evans, Ed., Cambridge University Press, Cambridge, 1985, pp.257-284.

[3] R. Chan, Q. Chang and H. Sun, Multigrid Method for Ill-conditioned Symmetric Toeplitz Systems, SIAM J. Sci. Comput., 19(1998), pp.516-529.

[4] T. Chan, S. Go and L. Zikatanov, Lecture Notes on Multilevel Methods for Elliptic Problems on Unstructured Grids, LC 28th CFD, 1997.

[5] T. Chan, J. Xu and L. Zikatanov, An Agglomeration Multigrid Method for Unstructured Grids, TRCAM 98-8, UCLA, CA, 1998.

[6] Q. Chang and Z. Huang, Efficient Algebraic Multigrid Algorithms and their Convergence, SIAM J. Sci. Comput., to appear.

[7] Q. Chang, S. Ma and G. Lei, Algebraic Multigrid for Queuing Networks, Intern. J. Computer Math., 70(1999), pp.539-552.

[8] Q. Chang and Y. Wong, Recent Developments in Algebraic Multigrid Methods, Proceedings of Copper Mountain Conference on Iterative Methods, Colorado, 1 (1992).

[9] Q. Chang, Y. Wong and H. Fu, On the Algebraic Multigrid Methods, Journal of Computational Physics, 125(1996), pp.279-292.

[10] Q. Chang, Y. Wong and H. Fu, Algebraic Multigrid and Its Application to Euler Equations, Proceedings of Second International Conference on Computational Physics, World Scientific, Singapore, 1993.

[11] A. Cleary, R. Falgout, V. Henson, J. Jones, T. Manteuffel, S. McCormick, G. Miranda, and J. Ruge, Robustness and Scalability of Algebraic Multigrid, SIAM J. Sci. Comput., 21(2000), pp.1886-1908.

[12] W. Hackbusch, Multigrid Methods and Applications, Springer-lerlag, 1985.

[13] Z. Huang and Q. Chang, Gauss-Seidel-type Multigrid Methods, J. Comput. Math., to appear.

[14] C. Oosterlee and T. washio, Krylov Subspace acceleration of Nonlinear Multigrid with Application to Recirculating Flows, SIAM J. Sci. Comput. 21(2000), pp.1670-1690.

[15] J. Ruge and K. Stuben, Algebraic Multigrid, in Multigrid Methods, Frontiers in Applied Mathematics 3, S. McCormick, Ed., SIAM, Philadelphia, 1987, pp.73-130.

[16] J. Xu, Theory of Multilevel Methods , Ph.D. thesis, Cornell University, Ithaca, NY, 1989.

[17] J. Xu, Iterative Methods by Space Decomposition and Subspace Correction, SIAM Rev., 34(1992).

BASIC STRUCTURES OF SUPERCONVERGENCE IN FINITE ELEMENT ANALYSIS

Chuanmiao Chen
Institute of Computation, Hunan Normal University,
Changsha, Hunan, 410081, PRC
cmchen@sparc2.hunnu.edu.cn

Abstract Superconvergence for second order elliptic finite elements on uniform meshes is discussed. The element orthogonality analysis method and the orthogonality correction technique are especially emphasized. There are two basic structures of superconvergence, i.e. Gauss-Lobatto points and symmetric points. Their accuracy and global property are also analysed. Four main principles in using superconvergence are proposed.

Keywords: Finite Element, Superconvergence, Two Classes of Structures.

1. Introduction

We consider second order elliptic problems in a convex polygon Ω : find $u \in H_0^1(\Omega) = \{v \in H^1(\Omega), v = 0 \text{ on } \partial\Omega\}$ such that $A(u, v) = (f, v), v \in H_0^1(\Omega)$ and its nth-degree finite element approximation $u_h \in S_0^h \subset H_0^1(\Omega)$ satisfying

$$A(u - u_h, v) = 0, \quad v \in S_0^h, \tag{1}$$

where the bilinear form

$$A(u, v) = \int_\Omega (a_{ij} D_i u D_j v + a_i D_i u v + a_0 u v) dx$$

is $H_0^1(\Omega)$-coercive. Denote by $W^{m,p}(\Omega), ||u||_{m,p,\Omega}$ the Sobolev space and its corresponding norm, respectively. If $p = 2$ simply $H^m = W^{m,2}$ and $||u||_{m,\Omega} = ||u||_{m,2,\Omega}$. It is well known that, under some assumptions, the following estimates of the error $e = u - u_h$ hold:

$$||u - u_h||_{s,\Omega} \leq C h^{n+1-s} ||u||_{n+1,\Omega}, \quad s = 0, 1, \tag{2}$$

Recent Progress in Computational and Applied PDES
Edited by Tony Chan *et al.*, Kluwer Academic/Plenum Publishers, 2002

which, in general, cannot be improved. However, u_h or its gradient Du_h at some specific points x^* possibly has higher order accuracy (superconvergence):

$$(u - u_h)(x^*) = O(h^{n+1+\alpha}) \text{ or } D(u - u_h)(x^*) = O(h^{n+\alpha}),\ \alpha > 0. \quad (3)$$

With the use of the above property, we can get the numerical results with high accuracy. In other words, to get the desired accuracy, it is enough to implement finite element computation in coarser meshes.

In 1973, J.Douglas-T.Dupont [19,20], de Boor-B.Swartz [18] and V.Thomee [30] analyzed superconvergence (for one-dimensional problem. Later, superconvergence was studied in many countries with several methods. For example, 1). The tensor product method (Douglas-Dupont-Wheeler [21]); 2). The local average method (Bramble-Schatz [3], Thomee [31]); and 3). The element analysis method (Oganesyan-Rukhovetz [28], Zlamal-Lesaint [25,37,38], Chen [4,5], Lin et al. [26]).

In recent years, three most promising methods are developed, namely i.e.

1). The local symmetric theory (Schatz-Sloan-Wahlbin [29]); 2). The computer-based method (Babuška-Strouboulis [1,2]); and 3). The element orthogonality analysis (an up to date treatment of element analysis method, Chen [12,13,14,16,17]).

The purpose of this paper is to present an overview of basic structures of the superconvergence properties for second order elliptic finite elements. In particular, we shall introduce Element Orthogonality Analysis (EOA), which is a unified method to study superconvergence. About the review papers, see also [23,24].

2. The Element Orthogonality Analysis

The basic idea is to construct a superclose function $u_I \in S_0^h$ to u_h by some orthogonal expansion of u directly, which is equivalent to requiring the weak estimate

$$A(u_h - u_I, v) = A(u - u_I, v) = O(h^{n+s+\alpha})||v||_{1+s,1,\Omega},\ s = 0, 1. \quad (4)$$

If it holds, taking $v = g_h$ (discrete Green's Function) or $v = G_h$ (discrete Gradient-type Green's function) we have

$$u_h - u_I = O(h^{n+1+\alpha} \ln h),\ D(u_h - u_I) = O(h^{n+\alpha} \ln h),\ x \in \Omega. \quad (5)$$

It follows from the equality $D^s(u-u_h) = D^s(u-u_I)+O(h^{n+1-s+\alpha} \ln h)$, $s = 0, 1$, that the roots of $D^s R := D^s(u - u_I)$ are superconvergence points of $D^s e := D^s(u - u_h)$. Therefore, the key is to construct the desired superclose function $u_I \in S_0^h$. This is an important art in the finite element analysis.

2.1 The Orthogonal Expansion in an Element

As a simple example, we first discuss one-dimensional problems. Subdivide the interval $\Omega = (0, 1)$ by nodes

$$x_0 = 0 < x_1 < x_2 < \dots < x_N = 1.$$

Denote by $\tau_j = (x_{j-1}, x_j), \bar{x}_j = (x_{j-1}+x_j)/2, h_j = (x_j - x_{j-1})/2, 1 \leq j \leq N$ the element, its midpoint and half-step length, respectively. Assume that the subdivision is quasiuniform. Take a transformation $x = \bar{x}_j + h_j t, t \in E = (-1, 1)$ and denote $u(t) = u(\bar{x}_j + h_j t)$. Obviously, $\partial_t^k u(t) = h_j^k D_x^k u(x)$. We introduce Legendre polynomials

$$\begin{aligned} &l_0 = 1, l_1 = t, l_2 = \frac{1}{2}(3t^2 - 1), \dots \\ &l_n(t) = \gamma_n \partial_t^n (t^2 - 1)^n, \ \gamma_n = 1/(2^n n!), \end{aligned} \tag{6}$$

and M-type polynomials

$$\begin{aligned} M_0 = 1, M_1 = t, M_2 = \tfrac{1}{2}(t^2 - 1), M_3 = \tfrac{1}{2}(t^3 - t), \dots, \\ M_{n+1}(t) = \gamma_n \partial_t^{n-1} (t^2 - 1)^n. \end{aligned} \tag{7}$$

Denote by $t'_j, 1 \leq j \leq n$ the roots of $l_n(t)$ (i.e. nth -order Gauss points) and by $t_j^0, 0 \leq j \leq n$ the roots of $M_n(t)$ (i.e. nth-order Lobatto points). They will play an important role in the studying of superconvergence later.

In a standard element $\tau = (-h, h)$, we expand $\partial u(t)$ in E as an orthogonal series

$$\partial u(t) = \sum_{j=0}^{\infty} b_{j+1} l_j(t), \ b_{j+1} = (j + 1/2)(\partial u, l_j) = O(h^{j+1}).$$

Integrating in t, it leads to an M-type polynomial series

$$u(t) = \sum_{j=0}^{\infty} b_j M_j(t), \ b_0 = (u(-1) + u(1))/2.$$

Denote its part sum and remainder by

$$u_n(t) = \sum_{j=0}^{n} b_j M_j(t), \ \ R^*(t) = u - u_n = \sum_{j=n+1}^{\infty} b_j M_j(t). \tag{8}$$

Obviously $u_n = u$ at $t = \pm 1$ for $n \geqslant 1$. It guarantees that $u_n(x)$ constructed in each element forms a continuous function in the whole domain Ω, i.e. $u_n(x) \in$

S_0^h. We consider the $(n-1)$th simplest case $A(u,v) = \int_0^1 u_x v_x dx$. Because the remainder $R_t^* \perp P_{n-1}$ ($(n-1)th$-degree polynomial), it leads to

$$A(u_h - u_n, v) = \sum_{j=1}^{N} \int_{x_{j-1}}^{x_j} R_x^* v_x dx = \sum_{j=1}^{N} h_j^{-1} \int_{-1}^{1} R_t^* v_t dt = 0, \; v \in S_0^h.$$

Thus in this case $u_h = u_n$ is just the finite element solution, i.e. $u - u_h = R_n^*(t)$. It follows that $(n+1)$th-order Lobatto points and nth-order Gauss points in each element are superconvergence points of u_h and $D_x u_h$, respectively. In general case with variable coefficients, similar results of superconvergence hold. This method is also successful in multi-dimensional case, for example, rectangular elements, linear and quadratic triangular elements. A series of superconvergence results are obtained. To treat general equations with variable coefficients, we proposed an element cancellation technique, which is a useful and necessary tool to prove the weak estimates (4).

2.2 The Orthogonality Correction

However, when we simply use the orthogonal expansion in an element, it is found that its applicable fields are still limited. To overcome this defect, in recent years, we have suggested the Orthogonality Correction Technique, i.e. add some lower degree terms into the remainder R_n^* such that the new remainder

$$R = u - u_I = u_n^* + R_n^*, \; u_n^* = \sum_{j=2}^{n} b_j^* M_j(t), \; u_I = u_n - u_n^* \tag{9}$$

satisfies more orthogonal conditions in the element, where $b_j^*, 2 \le j \le n$, are the constants to be determined. Denote by $v = \sum_{j=0}^{n} \beta_i M_i(t)$ the test function. Consider an element integration in $\tau = (-h, h)$

$$J(\tau) = A_\tau(R, v) = h^{-1} \int_{-1}^{1} (a_{11} R_t v_t + a_0 h^2 R v) dt$$

$$= h^{-1} \sum_{i=0}^{n} \beta_i \Big(\sum_{j=2}^{n} b_j^* c_{ij} + \sum_{j=n+1}^{\infty} b_j c_{ij}\Big) \tag{10}$$

where the constants

$$c_{0j} = \int_{-1}^{1} a_0 h^2 M_j(t) dt = O(h^j), \; j \ge 2,$$

$$c_{ij} = \int_{-1}^{1} (a_{11} l_{i-1}(t) l_{j-1}(t) + a_0 M_i(t) M_j(t)) dt = O(h^{|i-j|}),\ i \geq 1.$$

Now, we require that all coefficients b_j^* satisfy the following orthogonal conditions

$$\sum_{j=2}^{n} b_j^* c_{ij} + r_i = 0,\ \ r_i = \sum_{j=n+1}^{\infty} b_j c_{ij} = O(h^{2n+2-i}),\ 2 \leq i \leq n. \tag{11}$$

This is a linear algebraic system of equations, which is absolutely diagonally dominant and then uniquely solvable. Thus we have the following estimates

$$b_j^* = O(h^{2n+2-i}),\ 2 \leq i \leq n. \tag{12}$$

With this choice, the element integration is simplified to

$$J(\tau) = h^{-1}(O(h^{2n+2})|\beta_0| + O(h^{2n+1})|\beta_1|)$$

$$= O(h^{2n})||u||_{n+1,p,\tau}||v||_{1,p',\tau}.$$

With the use of this technique, we can get a superclose function $u_I = u_n - u_n^* \in S_0^h$ such that

$$A(u_h - u_I, v) = A(R, v) = O(h^{2n})||u||_{n+1,p,\Omega}||v||_{1,p',\Omega}, 1 \leq p \leq \infty. \tag{13}$$

In particular, taking $v = u_h - u_I$, $p = 2$ and using the imbedded theorem, we obtain an optimal superconvergence estimate

$$\max_{x \in \Omega} |(u_h - u_I)(x)| + ||u_h - u_I||_{1,\Omega} \leq Ch^{2n}||u||_{n+1,\Omega}. \tag{14}$$

If taking $p = \infty$ and $v = G_h$ (discrete gradient-type Green's function), we further have

$$\max_{x \in \Omega} |D_x(u_h - u_I)(x)| \leq Ch^{2n}||u||_{n+1,\infty,\Omega}. \tag{15}$$

From these inequalities, we have a new error expression in an element

$$e = u - u_h = \sum_{j=2}^{n} b_j^* M_j(t) + \sum_{j=n+1}^{\infty} b_j M_j(t) + r_h, \tag{16}$$

where

$$r_h = O(h^{2n})||u||_{n+1,\Omega},\ \ D_x r_h = O(h^{2n})||u||_{n+1,\infty,\Omega}.$$

In particular, there is an optimal superconvergence $e(x_j) = O(h^{2n})||u||_{n+1,\Omega}$ at node x_j. This result was proven by Douglas-Dupont [19,20], here, however, is derived by EOA. Based on high degree interpolation of u_h or Du_h in two adjacent elements, superconvergence with two order higher (ultraconvergence) is also derived by Chen et al. [17]. It is very important that this method can be also applied to multi-dimensional case (of course, more complicated), and many new results can be derived. In practice, six types of elements are often used, i.e. interval, rectangle, triangle, hexahedron, tetrahedron and triangular prism elements. We found that they have two classes of basic structures for superconvergence on uniform meshes.

3. Structure 1: Gauss-Lobatto Points

In one-dimensional case with variable coefficients, the following three results of superconvergence are known.

1.1) At all nth-order Gauss points x' the derivative Du_h satisfies (Chen [6])

$$D(u - u_h)(x') = O(h^{n+1})||u||_{n+2,\infty,\Omega},\ n \geq 1. \tag{17}$$

If n is odd, the result holds for the averaging derivative $\bar{D}u_h$ at all inner node x_j.

1.2) At all $(n+1)$th-order Lobatto points x^0 (Chen [6])

$$(u - u_h)(x^0) = O(h^{n+2})||u||_{n+2,\infty,\Omega},\ n \geq 2. \tag{18}$$

1.3) At each node x_j, u_h has optimal superconvergence (Douglas-Dupont)

$$(u - u_h)(x_j) = O(h^{2n})||u||_{n+1,\Omega},\ n \geq 2. \tag{19}$$

Note that the result 1.3) can be proved directly by using of the one-dimensional Green's function. But, in multi-dimensional case, Green's function method fails, whereas the EOA is still valid. We make the following classification.

1 The nth-degree polynomial $P_n = \{x^i y^j \mid 0 \leq i, j \leq n, i + j \leq n\}$ in a triangle.

2 The regular serendipity family in a rectangle

$$Q_\lambda(n) = \{x^i y^j \mid (i,j) \in I_{n,\lambda}\},\ 1 \leq \lambda \leq n,$$

where the index set

$$I_{n,\lambda} \subset \{(i,j) | 0 \leq i, j \leq n, i + j \leq n + \lambda\}.$$

3 The defective family (or intermediate family) in a rectangle

$$Q^*(n) = P_n \bigoplus \{x^n y, xy^n\},\ n \geq 3.$$

The superconvergence results for regular rectangular family $Q_1(n)$ are proved as follows.

The result 1.1) and 1.2) (but with factor $\ln h$) are still valid and their superconvergence points for function u_h and gradient Du_h are the product of one-dimensional case, respectively (Zlamal [37,38] and Chen [5,8]).

The result 1.3) is changed to the form

1.3') At each angular node (x_i, y_j), $u_h \in Q_2(n), n \geq 3$,

$$u - u_h = O(h^{n+3} \ln h), \tag{20}$$

if the coefficients a_{11} and a_{22} are constant and $a_{12} = a_1 = a_2 = a_0 = 0$ (Douglas-Dupont-Wheeler used the tensor product method [21]).

When a_{11} and a_{22} are variable and $a_{12} = 0$, we have to use the orthogonality correction technique, the result (20) for odd $n \geq 3$ is proved by Chen [14]. In the case of even $n \geq 2$, ultraconvergence (two order higher) for n-th degree interpolation of the gradient $I_n(Du_h)$ at some specific points (x^*, y^*) is also derived by Chen [14],

$$Du - I_n(Du_h) = O(h^{n+2} \ln h)||u||_{n+2,\infty,\Omega}, \text{ for even } n \geq 2. \tag{21}$$

4. Structure 2: Symmetric Points T_h

We consider arbitrary degree triangular elements P_n and the defective rectangular elements $Q^*(n), n \geq 3$ on uniform mashes. The vertices, center and side middle points on rectangular meshes, and the vertices and side middle points on triangular meshes are called the symmetric points T_h. The following results on interior symmetric points set T_h are proved.

2.1) The averaging gradient has

$$\bar{D}(u - u_h) = O(h^{n+1} \ln h)||u||_{n+2,\infty,\Omega}, \text{ for odd } n \geq 1. \tag{22}$$

2.2) The solution has

$$u - u_h = O(h^{n+2} \ln h)||u||_{n+2,\infty,\Omega}, \text{ for even } n \geq 2. \tag{23}$$

For $n = 1$ or 2 degree triangular elements, these results in whole T_h are first proved by Chen [4,7] and Zhu [36], respectively. As the defective family $Q^*(n) = Q_1(n), n = 1, 2$, the corresponding results hold in whole T_h, see 1.1) and 1.2).

In general case, the results for nth-degree triangular element $u_h \in P_n$ are proved by Babuška et al [1,2], Schatz-Sloan-Wahlbin [29] and Chen [13,16]. The results for defective family $u_h \in Q^*(n) = P_n \bigoplus \{x^n y, xy^n\}$, $n \geq 3$, are studied by Zhang [34,35] and Chen [14] (for general case).

However, these results for $n \geq 3$, in general, hold only inside the domain. Besides these symmetric points, there are no longer other superconvergence points (Chen [13,16]). When second or third boundary value conditions are prescribed and $n \geq 3$ is odd, the result 2.2) may hold up to the boundary $\partial\Omega$, but which has not been proved till now. Note that the conclusions above also hold for the defective hexahedron and tetrahedron elements. Superconvergence points for triangular prism elements are the product of one-dimensional and triangular structures above.

5. Four Main Principles

We conclude by pointing out that based on EOA, a unified theory of superconvergence is established in this work. Most of the results are summarized into four main principles as follows.

1 Two basic structures of superconvergence on uniform meshes;

2 The domain with curved boundary can be subdivided by piecewise almost uniform meshes and the global superconvergence for lower degree elements is obtained;

3 The singular solution can be approximated by the finite elements on λ-graded meshes;

4 The three principles above hold for linear elliptic, parabolic and hyperbolic equations, system of equations and nonlinear problems, general domain and singular solution and so on.

Therefore, many problems can be solved by the above four principles. For details, see the author's recent book [14].

Acknowledgments

This work is supported by the Special Funds for Major State Basic Research Projects (Grant No.G1999032804) and National Naturel Science Foundation of China.

References

[1] Babuška I. and Strouboulis T., The Finite Element Method and its Reliability, to appear.

[2] Babuška I, Strouboulis T., Upadhyay C. and Gangaraj S, Computer-based proof of the existence of superconvergence points in the finite element method; superconvergence

of the derivatives in finite element solutions of Laplace's, Poisson's and the Elasicity equations, Numer. Methods for PDE, **12**(1996), pp. 347-392.

[3] Bramble J.H. and Schatz A.H., High order local accuracy by averaging in the finite element method, Math. Comp.**31**(1977), pp. 94-111.

[4] Chen C.M., Optimal points of the stresses approximated by triangular linear element in FEM, J. Xiangtan Univ. **1**(1978), pp. 77-90.

[5] Chen C.M., A new estimate for finite element method and optimal point theorem for stresses, Comm. Natural Science, Xiangtan Univ.**1**(1978), pp. 10-20 (with Zhu Q.D.).

[6] Chen C.M., Optimal points of approximation solution for Galerkin method for two-point boundary value problem, Numer. Math. J. Chinese Univ.**1**:1(1979), pp. 73-79, MR. 82e: 65086.

[7] Chen C.M., Optimal points of the stresses for triangular linear element, Numer. Math. J. Chinese Univ.**2**:2(1980), pp. 12-20, MR.83d:65279.

[8] Chen C.M., Superconvergence of finite element solution and its derivatives, Numer. Math. J. Univ. **3**:2(1981),118-125,MR. 82m:65100.

[9] Chen C.M., Finite Element Method and Its Analysis in Raising Accuracy, Hunan Science and Technique Press, Changsha,1982.

[10] Chen C.M., Superconvergence of finite element approximations to nonlinear elliptic problems, Proc.The China-France Symposium on FEM, ed.by Feng Kang and J. Lions, Science Press, Beijing, 1983, pp. 622-640 (in English), MR.85h:65235.

[11] Chen C.M., Element Analysis method and superconvergence, in: "Finite Element Methods", Lecture Notes in Pure and Applied Mathematics, Vol.196, ed. by M. Krizek, P. Neittaanmaki and R.Stenberg, Marcel Dekker, Inc. New York, 1998, pp.71-84 (in English).

[12] Chen C.M., Superconvergence for L^2-projection in finite elements, Chinese Science Bulletin, **43:1** (1998), pp. 22-24 (in English).

[13] Chen C.M., Superconvergence for triangular finite elements, Science in China, Series A, **42** (1999), pp. 917-924 (in English).

[14] Chen C.M., Structure Theory of Superconvergence for Finite Elements, Hunan Science and Technique Press, Changsha, 2001.

[15] Chen C.M. and Huang Y.Q., High Accuracy Theory of Finite Elements, Hunan Science and Technique Press, Changsha, 1995.

[16] Chen C.M., Jin J.C. and Shu S., Superconvergence for triangular cubic elements, Chinese Science Bulletin, **44** (1999), pp. 17-19 (in English).

[17] Chen C.M., Xie Z.Q. and Liu J.H., New error expansion for one-dimensional finite elements and ultraconvergence, J. Comput. Math., to appear

[18] de Boor C. and Swartz B., Collocation at Gaussian points, SIAM J. Numer. Anal., **10** (1973), pp. 582-606

[19] Douglas J. and Dupont T., Some superconvergence results for Galerkin methods for the approximate solution of two-point boundary value problems, Topics in Numerical Analysis, Academic Press, 1973, pp. 89-92.

[20] Douglas J. and Dupont T., Galerkin approximations for the two-point boundary problem using continuous, piecewise polynomial spaces, Numer. Math. **22**(974), pp. 99-109.

[21] Douglas J., Dupont T. and Wheeler M.F., An L^∞ estimate and a superconvergence result for a Galerkin method for elliptic equations based on tensor products of piecewise polynomials, RAIRO Model. Math. Anal. Numer. **8**(1974), pp. 61-66.

[22] Frehse J. and Rannacher R., Eine L^1-Fehlerabschatzung diskreter Grundlosungen in der Methods der finiten Elemente, Tagungsband "Finite Elemente", Bonn. Math. Schrift. **89** (1975), pp. 92-114.

[23] Krizek M. and Neittaamaki P., Superconvergence phenomenon in the finite element method arising from averiging gradients, Numer. Math.**45**(1984), pp. 105-116.

[24] Krizek M. and Neittaanmaki P., Bibliograph on Superconvergence, in: "Finite Element Methods", Lecture Notes in Pure and Applied Math., Vol.196, 1998, pp. 315-348.

[25] Lesaint P. and Zlamal M., Superconvergence of the gradient of finite element solutions, RAIRO Model. Math. Anal. Numer.,**13**(1979), pp. 139-166.

[26] Lin Q. and Yan N.N., Constructure and analysis of efficient finite elements, Hebei Univ. Press, Baoding, 1996, pp. 1-304.

[27] Rannacher R. and Scott R., Some optimal error estimates for piecewise linear finite element approximations, Math. Comp., **38** (1984), pp. 437-445.

[28] Oganesyan L.A. and Rukhovetz L.A., Study of the rate of convergence of variational difference schemces for second order elliptic equations in a two-dimensional field with a smooth boundary, USSS Comput. Math. Math. Phys.,**9**(1969), pp. 158-183.

[29] Schatz A., Sloan I. and Wahlbin L., Superconvergence in finite element methods and meshes which are locally symmetric with respect to a point, SIAM J. Numer. Anal. **33**(1996), pp. 505-521.

[30] Thomee V., Spline approximation and difference schemes for the heat equation, in: Mathmatical Foundations of the Finite Element method with Applications to PDE, Academic Press, 1972, pp. 711-746

[31] Thomee V., High order local approximation to derivatives in the finite element method, Math. Comp. **31**(1977), pp. 652-660.

[32] Wahlbin L., Superconvergence in Galerkin finite element methods, Springer-Verlag, Berlin Heidelberg, 1995, pp. 1-164.

[33] Wahlbin L., General principles of superconvergence in Galerkin finite element methods, in: "Finite Element Methods", Lecture Notes in Pure and Applied Math., Vol. 196, 1998, pp. 269-286.

[34] Zhang Z. M., Ultraconvergence of patch recovery technique, Math. Comp., **65** (1996), pp. 1431-1437.

[35] Zhang Z.M., Derivative superconvergence points in finite element solutions of Poisson's equation for the serendipity and intermediate families; a theoretical justfication, Math. Comp., **67** (1998), pp. 541-552.

[36] Zhu Q.D., Optimal points for quadratic triangular finite elements, J. Xiangtan University, **3** (1981), pp. 36-45.

[37] Zlamal M., Some superconvergence results in the finite element method, Lecture Notes in Math. 606, Springer, 1977, pp. 353-362.

[38] Zlamal M., Superconvergence and reduced integration in the finite element method, Math. Comp. **32** (1978), pp. 663-685.

A POSTERIORI ERROR ESTIMATES FOR MIXED FINITE ELEMENTS OF A QUADRATIC CONTROL PROBLEM *

Yanping Chen
ICAM, Xiangtan University, China
ypchen@xtu.edu.cn

Wenbin Liu
ICAM, Xiangtan University, China
CBS & IMS, Kent University, Canterbury, CT2 7NF England
W.B.Liu@ukc.ac.uk

Abstract In this paper, we present an a posteriori error analysis for the mixed finite element approximation of a linear quadratic control problem. We derive a posteriori error estimates for the coupled state and control approximations under some assumptions which hold in many applications. Such estimates, which are apparently not available in the literature, can be used to construct reliable adaptive mixed finite element methods for the control problem.

Keywords: adaptive refinement schemes, a posteriori error estimator, mixed finite element, linear quadratic control problem

1. Introduction

Efficient numerical methods are vital to any successful applications of optimal control in practical problems. Finite element approximation of optimal control problems plays a central role in numerical methods for these problems, see, e.g., [12, 27-29] and the references quoted therein.

In recent years, adaptive algorithms for the finite element approximation have been extensively investigated, beginning with the pioneering work in [3-4]. They ensure a higher density of nodes in certain area of the given domain,

*Supported by National Science Foundation of China, the Backbone Teachers Foundation of Chinese State Education Commission and the Special Funds for Major State Basic Research Projects.

where the solution is more difficult to approximate. At the heart of any adaptive finite element method is an a posteriori error estimator. The decision of whether further refinement of meshes is necessary is based on the estimate of the discretization error. If the further refinement is to be performed then the error estimator is used as a guide to show how the refinement might be accomplished most efficiently. The literature in this area is huge. Some of techniques directly relevant to our work can be found in [1, 3-4, 6, 8, 15-19, 24, 30].

Although adaptive finite element approximation is widely used in numerical simulations, it has not yet been fully utilized in optimal control. Initial attempts in this aspect have only been reported recently for some design problems, see, e.g., [2, 5]. However a posteriori error indicators of a heuristic nature are widely used in most applications. For instance, in some existing work on adaptive finite element approximation of optimal design, the mesh refinement is guided by a posteriori error estimators based on a posteriori error estimates solely from the state equation for a fixed control. Thus error information from approximation of the control design is not utilized. Although these methods may work well in some particular applications, they cannot be applied confidently in general. It is unlikely that the potential power of adaptive finite element approximation has been fully utilized due to the lack of more sophisticated a posteriori error indicators. Very recently, some error estimators of residual type were drived for some constrained control problems governed by elliptic, parabolic equations, and Stokes equations, see [20-23, 25-26]. The initial numerical results seem to be promising, see [13]. These error indicators are based on a posteriori error estimation of the discretization error for the state and the control (design). However to our best knowledge, there has been a lack of a posteriori error indicators for the important case where the objective functional of control problems includes flux of the state. Clearly one should consider mixed finite element discretisation for such control problems.

In this work, we derive a posteriori error estimates for the mixed finite element approximation of a linear quadratic control governed by the elliptic equation. We consider the following quadratic control problem:

$$\min_{u\in K\subset L^2(\Omega_U)} \left\{\frac{1}{2}\int_\Omega (\boldsymbol{p}-\boldsymbol{p}_0)^2dx + \frac{1}{2}\int_\Omega (y-y_0)^2dx + \frac{1}{2}\int_{\Omega_U} u^2dx\right\} \quad (1.1)$$

subject to

$$\begin{aligned} \operatorname{div}\boldsymbol{p} &= f + Bu \quad && \text{in } \Omega, && (1.2)\\ \boldsymbol{p} &= -A\nabla y, && \text{in } \Omega, && (1.3)\\ y &= 0, && \text{on } \partial\Omega, && (1.4) \end{aligned}$$

where the bounded open set $\Omega \subset \mathbb{R}^2$, is a convex polygon or has smooth boundary $\partial\Omega$, Ω_U is a bounded open set in $\mathbb{R}^2$ with Lipschitz boundary $\partial\Omega_U$, and

K is a closed convex set in $L^2(\Omega_U)$. Here, p_0 and y_0 are proper given functions, $f \in L^2(\Omega)$ and B is a continuous linear operator from $L^2(\Omega_U)$ to $L^2(\Omega)$. The coefficient matrix $A \in L^\infty(\Omega; \mathbb{R}^{2\times 2})$ is symmetric and uniformly ellpitic, i.e., $A(x)$ is a symmetric and positive definite 2×2-matrix, with eigenvalues $\lambda_j(x) \in \mathbb{R}$ satisfying

$$0 < c_A \leq \lambda_1(x),\ \lambda_2(x) \leq C_A \tag{1.5}$$

for almost all $x \in \Omega$.

We shall use the standard notation $W^{m,p}(\Omega)$ for Sobolev spaces on Ω with a norm $||\cdot||_{m,p}$ given by $||\phi||^p_{m,p} = \sum_{|\alpha|\leq m} ||D^\alpha \phi||^p_{L^p(\Omega)}$. We set $W_0^{m,p}(\Omega) = \{\phi \in W^{m,p}(\Omega) : \phi|_{\partial\Omega} = 0\}$. For $p = 2$, we denote $H^m(\Omega) = W^{m,2}(\Omega)$, $H_0^m(\Omega) = W_0^{m,2}(\Omega)$, $||\cdot||_m = ||\cdot||_{m,2}$ and $||\cdot|| = ||\cdot||_{0,2}$. In addition C denotes a general positive constant independent of h.

The outline of this paper is as follows. In Section 2, we shall give a brief review on the mixed finite element method, and construct the mixed finite element approximation for the convex optimal control problem (1.1)-(1.4). Then, we state preliminaries and the key properties of some interpolation operators in Section 3. In Section 4, we propose a posteriori error estimators for some intermediate errors for the RT, the BDM and the BDFM mixed method. Our analysis relies on a decomposition of the flux functions in the spirit of a generalized Helmholtz decomposition. Helmholtz decomposition was first used in [8] to prove efficiency and reliability of error estimators for the elliptic problem with the same mixed finite elements. Finally, a posteriori error bounds are derived for the control problem by applying the results in Section 3.

2. Mixed finite element approximation of the control problem

Let

$$\boldsymbol{V} = H(\mathrm{div};\Omega) = \{\boldsymbol{v} \in (L^2(\Omega))^2,\ \ \mathrm{div}\boldsymbol{v} \in L^2(\Omega)\}, \tag{2.1}$$

endowed with the norm:

$$||\boldsymbol{v}||_{H(\mathrm{div};\Omega)} = \left(||\boldsymbol{v}||^2_{0,\Omega} + ||\mathrm{div}\boldsymbol{v}||^2_{0,\Omega}\right)^{1/2},$$

and let

$$W = L^2(\Omega). \tag{2.2}$$

We also denote

$$U = L^2(\Omega_U). \tag{2.3}$$

To consider the mixed finite element approximation of the quadratic optimal control problem (1.1)-(1.4), we need a weak formula for the state equation (1.2)-(1.4). We recast (1.1)-(1.4) as the following weak form: find $(\boldsymbol{p}, y, u) \in \boldsymbol{V} \times W \times U$ such that (QCP)

$$\min_{u \in K \subset U} \frac{1}{2} \left\{ \|\boldsymbol{p} - \boldsymbol{p}_0\|^2_{(L^2(\Omega))^2} + \|y - y_0\|^2_W + \|u\|^2_U \right\} \tag{2.4}$$

$$(A^{-1}\boldsymbol{p}, \boldsymbol{v}) - (y, \text{div}\boldsymbol{v}) = 0, \quad \forall\ \boldsymbol{v} \in \boldsymbol{V}, \tag{2.5}$$

$$(\text{div}\boldsymbol{p}, w) = (f + Bu, w), \quad \forall\ w \in W, \tag{2.6}$$

where the inner product in $L^2(\Omega)$ or $L^2(\Omega)^2$ is indicated by $(\cdot, \cdot)$, K is a closed convex set in U, and B is a continuous linear operator from U to $L^2(\Omega)$. It is well known (see, e.g., [14]) that the quadratic control problem (QCP) (2.4)-(2.6) has a unique solution $(\boldsymbol{p}, y, u)$, and that a triplet $(\boldsymbol{p}, y, u)$ is the solution of (QCP) (2.4)-(2.6) if and only if there is a co-state $(\boldsymbol{q}, z) \in \boldsymbol{V} \times W$ such that $(\boldsymbol{p}, y, \boldsymbol{q}, z, u)$ satisfies the following optimal conditions: (QCP-OPT)

$$\begin{aligned}
(A^{-1}\boldsymbol{p}, \boldsymbol{v}) - (y, \text{div}\boldsymbol{v}) &= 0, & \forall\ \boldsymbol{v} \in \boldsymbol{V}, \quad &(2.7)\\
(\text{div}\boldsymbol{p}, w) &= (f + Bu, w), & \forall\ w \in W, \quad &(2.8)\\
(A^{-1}\boldsymbol{q}, \boldsymbol{v}) - (z, \text{div}\boldsymbol{v}) &= -(\boldsymbol{p} - \boldsymbol{p}_0, \boldsymbol{v}), & \forall\ \boldsymbol{v} \in \boldsymbol{V}, \quad &(2.9)\\
(\text{div}\boldsymbol{q}, w) &= (y - y_0, w), & \forall\ w \in W, \quad &(2.10)\\
(u + B^* z, \tilde{u} - u)_U &\geq 0, & \forall\ \tilde{u} \in K, \quad &(2.11)
\end{aligned}$$

where B^* is the adjoint operator of B, and $(\cdot, \cdot)_U$ is the inner product of U. In the rest of the paper, we shall simply write the product as $(\cdot, \cdot)$ whenever no confusion should be caused.

For ease of exposition we will assume that Ω and Ω_U are both polygons. Let $\mathcal{T}_h$ and $\mathcal{T}_h(\Omega_U)$ be regular (in the sense of [9]) triangulation or rectangulation of Ω and Ω_U respectively. They are assumed to satisfy the angle condition that there is a positive constant C such that for all $T \in \mathcal{T}_h$ ($T_U \in \mathcal{T}_h(\Omega_U)$)

$$C^{-1}h_T^2 \leq |T| \leq C h_T^2, \quad C^{-1}h_{T_U}^2 \leq |T_U| \leq C h_{T_U}^2 \tag{2.12}$$

where $|T|$ ($|T_U|$) is the area of T (T_U) and h_T (h_{T_U}) is the diameter of T (T_U). Let $h = \max h_T$ ($h_U = \max h_{T_U}$).

Let $\boldsymbol{V}_h \times W_h \subset \boldsymbol{V} \times W$ denote the RT, BDM, or BDFM space of index k associated with the triangulation or rectangulation $\mathcal{T}^h$ of Ω ([7]), where $k \geq 0$. Here, RT indicates entries for the Raviart-Thomas elements, BDM for the Brezzi-Douglas-Marini elements, and BDFM for the Brezzi-Douglas-Fortin-Marini elements.

Associated with $\mathcal{T}_h(\Omega_U)$ is another finite dimensional subspace U_h of U:

$$U_h := \{\tilde{u}_h \in U : \ \forall T \in \mathcal{T}_h(\Omega_U),\ \tilde{u}_h|_T \in P_k(T)\}. \tag{2.13}$$

The mixed finite element approximation of (QCP) (2.4)-(2.6) is as follows $(\mathrm{QCP})_h$: find $(\boldsymbol{p}_h, y_h, u_h) \in \boldsymbol{V}_h \times W_h \times U_h$ such that

$$\min_{u_h \in K_h \subset U_h} \frac{1}{2} \left\{ ||\boldsymbol{p}_h - \boldsymbol{p}_0||^2_{(L^2(\Omega))^2} + ||y_h - y_0||^2_W + ||u_h||^2_U \right\} \tag{2.14}$$

$$(A^{-1}\boldsymbol{p}_h, \boldsymbol{v}_h) - (y_h, \mathrm{div}\boldsymbol{v}_h) = 0, \quad \forall\ \boldsymbol{v}_h \in \boldsymbol{V}_h, \tag{2.15}$$

$$(\mathrm{div}\boldsymbol{p}_h, w_h) = (f + Bu_h, w_h), \quad \forall\ w_h \in W_h, \tag{2.16}$$

where K_h is a closed convex set in U_h. This control problem $(\mathrm{QCP})_h$ (2.16)-(2.18) again has a unique solution $(\boldsymbol{p}_h, y_h, u_h)$, and that a triplet $(\boldsymbol{p}_h, y_h, u_h) \in \boldsymbol{V}_h \times W_h \times U_h$ is the solution of $(\mathrm{QCP})_h$ (2.14)-(2.16) if and only if there is a co-state $(\boldsymbol{q}_h, z_h) \in \boldsymbol{V}_h \times W_h$ such that $(\boldsymbol{p}_h, y_h, \boldsymbol{q}_h, z_h, u_h)$ satisfies the following optimal conditions: $(\mathrm{QCP\text{-}OPT})_h$

$$\begin{aligned}
(A^{-1}\boldsymbol{p}_h, \boldsymbol{v}_h) - (y_h, \mathrm{div}\boldsymbol{v}_h) &= 0, & \forall\ \boldsymbol{v}_h \in \boldsymbol{V}_h, \quad (2.17)\\
(\mathrm{div}\boldsymbol{p}_h, w_h) &= (f + Bu_h, w_h), & \forall\ w_h \in W_h, \quad (2.18)\\
(A^{-1}\boldsymbol{q}_h, \boldsymbol{v}_h) - (z_h, \mathrm{div}\boldsymbol{v}_h) &= -(\boldsymbol{p}_h - \boldsymbol{p}_0, \boldsymbol{v}_h), & \forall\ \boldsymbol{v}_h \in \boldsymbol{V}_h, \quad (2.19)\\
(\mathrm{div}\boldsymbol{q}_h, w_h) &= (y_h - y_0, w_h), & \forall\ w_h \in W_h, \quad (2.20)\\
(u_h + B^* z_h, \tilde{u}_h - u_h) &\geq 0, & \forall\ \tilde{u}_h \in K_h, \quad (2.21)
\end{aligned}$$

where B^* is the adjoint operator of B.

3. Preliminaries and some interpolation operators

Since the domain Ω has a smooth boundary or is convex with a polygonal boundary, the Poincare's inequality holds

$$||\phi - \bar{\phi}||_{0,\Omega} \leq C||\nabla\phi||_{0,\Omega}, \quad \forall\ \phi \in H^1(\Omega), \tag{3.1}$$

where

$$\bar{\phi} = \int_\Omega \phi/|\Omega|.$$

Let $\bigcup \mathcal{T}_h$ denote the set of triangular or rectangle elements (open) in $\mathcal{T}_h$. We define.

$$W^{m,p}\left(\bigcup \mathcal{T}_h\right) := \{\phi \in L^p(\Omega) :\ \forall\ T \in \mathcal{T}_h,\ \phi|_T \in W^{m,p}(T)\}$$

and consider local versions of these differential operators (understood in the distributional sense), namely, div_h, curl_h: $H^1\left(\bigcup \mathcal{T}_h\right)^2 \to L^2(\Omega)$ and ∇_h, Curl_h: $H^1\left(\bigcup \mathcal{T}_h\right) \to L^2(\Omega)^2$ defined such that, e.g.,

$$\mathrm{div}_h \boldsymbol{v}|_T := \mathrm{div}(\boldsymbol{v}|_T) \quad (T \in \mathcal{T}_h).$$

Since, we assumed that $A \in L^\infty(\Omega; \mathbb{R}^{2\times 2}_{\text{sym}})$ is uniformly elliptic on Ω. Then, by the Lax-Milgram lemma, the operator

$$- \operatorname{div}(A\nabla\cdot) : \; H_0^1(\Omega) \to H^{-1}(\Omega) \\ \text{is invertible and the norm of the inverse is bounded.} \tag{3.2}$$

Moreover, since $\Omega \in \mathbb{R}^2$ is convex polygon or has a smooth boundary, the condition $A \in C^{1,0}(\overline{\Omega})$ implies that

$$-\operatorname{div}(A\nabla\cdot) : \; H_0^1(\Omega) \cup H^2(\Omega) \to L^2(\Omega) \text{ is invertible} \tag{3.3}$$

and there exists a constant $C > 0$ such that

$$||\phi||_{2,\bigcup \mathcal{T}_h} \leq C||\operatorname{div}(A\nabla\phi)||_{0,\Omega}, \tag{3.4}$$

for all $\phi \in H_0^1(\Omega)$ such that $\operatorname{div}(A\nabla\phi) \in L^2(\Omega)$.

Let $\mathcal{E}_h$ denote the set of element sides in $\mathcal{T}_h$. The local mesh size h is defined on both Ω and $\bigcup \mathcal{E}_h$ by $h|_T := h_T$ for $T \in \mathcal{T}_h$ and $h_E := h_E$ for $E \in \mathcal{E}_h$, respectively.

For all $E \in \mathcal{E}_h$ we fix one direction of a unit normal on E pointing outside of Ω in case that $E \subset \partial\Omega$. We define an operator J: $H^1(\bigcup \mathcal{T}_h) \to L^2(\bigcup \mathcal{E}_h)$, for $\phi \in H^1(\bigcup \mathcal{T}_h)$ by

$$J(\phi)|_E := (\phi_{T_+})|_E - (\phi_{T_-})|_E \text{ if } E = \overline{T_+} \bigcap \overline{T_-}, \tag{3.5}$$

$E \in \mathcal{E}_h$; T_+, $T_- \in \mathcal{T}_h$ and ν_E points from T_+ into its neighbor element T_-; while

$$J(\phi)|_E := (\phi_T)|_E \text{ if } E = \overline{T} \bigcap \partial\Omega \;\; (E \in \mathcal{E}_h; \; T \in \mathcal{T}_h). \tag{3.6}$$

It is clear that $J(\phi)|_E$ represents the jump of the function ϕ across the edge E.

We define $S^0(\mathcal{T}_h) \subset L^2(\Omega)$ as the piecewise constant space, and $S^1(\mathcal{T}_h) \subset H^1(\Omega)$ or $S_0^1(\mathcal{T}_h) \subset H_0^1(\Omega)$ as continuous and piecewise linear functions; piecewise is understood with respect to $\mathcal{T}_h$. We consider the Clement's interpolation operator I_h: $H^1(\Omega) \to S^1(\mathcal{T}_h)$ which satisfies

$$||\phi - I_h\phi||_{0,T} \leq C||h\phi||_{1,\omega_T} \quad \forall\, \phi \in H_0^1(\Omega), \tag{3.7}$$

$$||\phi - I_h\phi||_{0,E} \leq C||h^{1/2}\phi||_{1,\omega_E} \quad \forall\, \phi \in H_0^1(\Omega), \tag{3.8}$$

for each $T \in \mathcal{T}_h$ and $E \in \mathcal{E}_h$, with which we associate neighborhoods

$$\omega_T := \bigcup \left\{T' \in \mathcal{T}_h : \; \overline{T} \bigcap \overline{T'} \neq \emptyset\right\} \text{ and } \omega_E := \bigcup \{T \in \mathcal{T}_h : \; E \subset \overline{T}\}.$$

Moreover, the maximal number of elements in ω_T is h-independently bounded by the angle condition.

Now, we define the standard $L^2(\Omega)$-orthogonal projection P_h: $W \rightarrow W_h$, which satisfies [7]: for any $\phi \in W$

$$\int_T (\phi - P_h\phi) w_h dxdy = 0 \quad \forall\, w_h \in W_h,\ T \in \mathcal{T}_h. \tag{3.9}$$

Since $S^0(\mathcal{T}_h) \subset W_h \subset H^1(\bigcup \mathcal{T}_h)$, we have the approximate property

$$||h^{-1} \cdot (\phi - P_h\phi)||_{0,\Omega} \leq C||\nabla_h \phi||_{0,\Omega} \quad (\phi \in H^1(\bigcup \mathcal{T}_h). \tag{3.10}$$

Next, let us define the projection operator $\Pi_h : \boldsymbol{V} \rightarrow \boldsymbol{V}_h$, which satisfies: for any $\boldsymbol{q} \in \boldsymbol{V}$

$$\int_E w_h(\boldsymbol{q} - \Pi_h \boldsymbol{q}) \cdot \boldsymbol{\nu_E} ds = 0, \ \forall\, w_h \in W_h,\ E \in \mathcal{E}_h, \tag{3.11}$$

$$\int_T (\boldsymbol{q} - \Pi_h \boldsymbol{q}) \cdot \boldsymbol{v}_h dxdy = 0, \ \forall\, \boldsymbol{v}_h \in \boldsymbol{V}_h,\ T \in \mathcal{T}_h. \tag{3.12}$$

Further, the interpolant Π_h satisfies a local error estimate:

$$||h^{-1} \cdot (\boldsymbol{q} - \Pi_h \boldsymbol{q})||_{0,\Omega} \leq C|\boldsymbol{q}|_{1,\bigcup \mathcal{T}_h} \quad (\boldsymbol{q} \in H^1(\bigcup \mathcal{T}_h) \bigcap \boldsymbol{V}). \tag{3.13}$$

4. A posteriori error estimates for optimal control problems

For any $\tilde{u}_h \in U_h$, let $(\boldsymbol{p}(\tilde{u}_h), y(\tilde{u}_h)) \in \boldsymbol{V} \times W$ be the solution of the following equations:

$$(A^{-1}\boldsymbol{p}(\tilde{u}_h), \boldsymbol{v}) - (y(\tilde{u}_h), \mathrm{div}\boldsymbol{v}) = 0, \quad \forall\ \boldsymbol{v} \in \boldsymbol{V}, \tag{4.1}$$

$$(\mathrm{div}\boldsymbol{p}(\tilde{u}_h), w) = (f + B\tilde{u}_h, w), \quad \forall\ w \in W, \tag{4.2}$$

Let $(\boldsymbol{p}, y, u) \in \boldsymbol{V} \times W \times U$ and $(\boldsymbol{p}_h, y_h, u_h) \in \boldsymbol{V}_h \times W_h \times U_h$ be the solutions of (QCP)(2.4)-(2.6) and (QCP)$_h$ (2.14)-(2.16) respectively.

Set some intermediate errors:

$$\boldsymbol{\varepsilon}_1 := \boldsymbol{p}(u_h) - \boldsymbol{p}_h \quad \text{and} \quad e_1 := y(u_h) - y_h. \tag{4.3}$$

Let us first note the following error equations from (2.15)-(2.16) and (4.1)-(4.2):

$$(A^{-1}\boldsymbol{\varepsilon}_1, \boldsymbol{v}_h) - (e_1, \mathrm{div}\boldsymbol{v}_h) = 0, \quad \forall\ \boldsymbol{v}_h \in \boldsymbol{V}_h, \tag{4.4}$$

$$(\mathrm{div}\boldsymbol{\varepsilon}_1, w_h) = 0, \quad \forall\ w_h \in W_h, \tag{4.5}$$

By (3.2) and the uniqueness of the solutions for (4.1)-(4.2), we can get that $y(u_h) \in H_0^1(\Omega)$ and

$$p(u_h) = -A\nabla y(u_h) \quad \text{and} \quad \text{div}p(u_h) = f + Bu_h \quad \text{in } \Omega. \tag{4.6}$$

We first establish some intermediate a posteriori error estimates:

Lemma 4.1. *For the RT, the BDM, or the BDFM elements there is a positive constant C, which only depends on A, Ω, and on the shape of the elements and their polynomial degree k, such that*

$$||p(u_h) - p_h||_{H(\text{div};\Omega)} + ||y(u_h) - y_h||_{L^2(\Omega)} \leq C\eta_1, \tag{4.7}$$

where

$$\eta_1 := \left(\sum_{T \in \mathcal{T}_h} \eta_{1T}^2 \right)^{1/2}, \tag{4.8}$$

and for any element $T \in \mathcal{T}_h$

$$\begin{aligned} \eta_{1T}^2 := \; & ||f + Bu_h - \text{div}p_h||_{0,T}^2 + h_T^2 \cdot ||\text{curl}(A^{-1}p_h)||_{0,T}^2 \\ & + h_T^2 \cdot \min_{w_h \in W_h} ||A^{-1}p_h - \nabla_h w_h||_{0,T}^2 \\ & + ||h_E^{1/2} J(A^{-1}p_h \cdot \tau)||_{0,\partial T}^2. \end{aligned} \tag{4.9}$$

Moreover, the reverse inequality of (4.7) holds as well provided that

$$\text{on each } T \in \mathcal{T}_h, \quad A^{-1}p_h|_T \in P_l \quad \text{and} \quad \nabla_h y_h|_T \in P_l. \tag{4.10}$$

Lemma 4.2. *For the RT, the BDM, or the BDFM elements there is a positive constant C, which only depends on A, Ω, and on the shape of the elements and their polynomial degrees k and l, such that*

$$C\eta_1 \leq ||p(u_h) - p_h||_{H(\text{div};\Omega)} + ||y(u_h) - y_h||_{L^2(\Omega)}. \tag{4.11}$$

Let $(p, y, q, z, u) \in (V \times W)^2 \times U$ and $(p_h, y_h, q_h, z_h, u_h) \in (V_h \times W_h)^2 \times U_h$ be the solutions of (QCP-OPT) (2.7)-(2.11) and (QCP-OPT)$_h$ (2.17)-(2.21) respectively.

For any $u_h \in U_h$, let $(q(u_h), z(u_h)) \in V \times W$ be the solution of the following equations:

$$\begin{aligned} (A^{-1}p(u_h), v) - (y(u_h), \text{div}v) &= 0, & (4.12)\\ (\text{div}p(u_h), w) &= (f + Bu_h, w), & (4.13)\\ (A^{-1}q(u_h), v) - (z(u_h), \text{div}v) &= -(p(u_h) - p_0), v), & (4.14)\\ (\text{div}q(u_h), w) &= (y(u_h) - y_0, w), & (4.15) \end{aligned}$$

for all $\boldsymbol{v} \in \boldsymbol{V}$ and $w \in W$.

Define some further intermediate errors:

$$\varepsilon_2 := \boldsymbol{q}(u_h) - \boldsymbol{q}_h \quad \text{and} \quad e_2 := z(u_h) - z_h. \tag{4.16}$$

We can also derive the following result:

Lemma 4.3. *For the RT, the BDM, or the BDFM elements there is a positive constant C which only depends on A, Ω, and on the shape of the elements and their polynomial degree k, such that*

$$\begin{aligned}&||\boldsymbol{q}(u_h) - \boldsymbol{q}_h||_{H(\text{div};\Omega)} + ||z(u_h) - z_h||_{L^2(\Omega)} \\ &\quad \leq C\eta_2 + C||\boldsymbol{p}_h - \boldsymbol{p}(u_h)|| + C||y_h - y(u_h)|| \leq C(\eta_1 + \eta_2),\end{aligned} \tag{4.17}$$

where η_1 is defined in Lemma 4.1 and

$$\eta_2 := \left(\sum_{T \in \mathcal{T}_h} \eta_{2T}^2\right)^{1/2}, \tag{4.18}$$

and for any element $T \in \mathcal{T}_h$

$$\begin{aligned}\eta_{2T}^2 := \; & ||y_h - y_0 - \text{div}\boldsymbol{q}_h||_{0,T}^2 + h_T^2 \cdot ||\text{curl}(A^{-1}\boldsymbol{q}_h)||_{0,T}^2 \\ & + h_T^2 \cdot \min_{w_h \in W_h} ||A^{-1}\boldsymbol{q}_h + \boldsymbol{p}_h - \boldsymbol{p}_0 - \nabla_h w_h||_{0,T}^2 \\ & + ||h_E^{1/2} J(A^{-1}\boldsymbol{q}_h \cdot \boldsymbol{\tau})||_{0,\partial T}^2.\end{aligned} \tag{4.19}$$

In this paper, we do not assume that the discretized constraint set K_h is contained by K.

Let $(\boldsymbol{p}, y, \boldsymbol{q}, z, u) \in (\boldsymbol{V} \times W)^2 \times U$ and $(\boldsymbol{p}_h, y_h, \boldsymbol{q}_h, z_h, u_h) \in (\boldsymbol{V}_h \times W_h)^2 \times U_h$ be the solutions of (QCP-OPT) (2.7)-(2.11) and (QCP-OPT)$_h$ (2.17)-(2.21) respectively.

With these intermediate errors, we can decompose the errors as following

$$\begin{aligned} \boldsymbol{p} - \boldsymbol{p}_h &= \boldsymbol{p} - \boldsymbol{p}(u_h) + \boldsymbol{p}(u_h) - \boldsymbol{p}_h &&:= \epsilon_1 + \varepsilon_1, \\ y - y_h &= y - y(u_h) + y(u_h) - y_h &&:= r_1 + e_1, \\ \boldsymbol{q} - \boldsymbol{q}_h &= \boldsymbol{q} - \boldsymbol{q}(u_h) + \boldsymbol{q}(u_h) - \boldsymbol{q}_h &&:= \epsilon_2 + \varepsilon_2, \\ z - z_h &= z - z(u_h) + z(u_h) - z_h &&:= r_2 + e_2. \end{aligned} \tag{4.20}$$

From (2.7)-(2.8) and (4.12)-(4.13), (2.9)-(2.10) and (4.14)-(4.15), we derive the following error equations:

$$(A^{-1}\epsilon_1, \boldsymbol{v}) - (r_1, \text{div}\boldsymbol{v}) = 0, \tag{4.21}$$

$$(\text{div}\epsilon_1, w) = (B(u - u_h), w), \tag{4.22}$$

$$(A^{-1}\epsilon_2, \boldsymbol{v}) - (r_2, \text{div}\boldsymbol{v}) = (\boldsymbol{p}(u_h) - \boldsymbol{p}, \boldsymbol{v}), \tag{4.23}$$

$$(\text{div}\epsilon_2, w) = (y - y(u_h), w), \tag{4.24}$$

for all $v \in V$ and $w \in W$. The assumption that $A \in L^\infty(\Omega; \mathbb{R}^{2\times 2})$ implies

$$||\epsilon_1||_{H(\text{div};\Omega)} + ||r_1||_{L^2(\Omega)} \leq C||u - u_h||_U, \tag{4.25}$$

Similarly, we have

$$||\epsilon_2||_{H(\text{div};\Omega)} + ||r_2||_{L^2(\Omega)} \leq C||u - u_h||_U. \tag{4.26}$$

In the following we estimate $||u - u_h||_U$ and then obtain the main results:

Theorem 4.1. *Let $(\boldsymbol{p}, y, \boldsymbol{q}, z, u) \in (\boldsymbol{V} \times W)^2 \times U$ and $(\boldsymbol{p}_h, y_h, \boldsymbol{q}_h, z_h, u_h) \in (\boldsymbol{V}_h \times W_h)^2 \times U_h$ be the solutions of (QCP) (2.7)-(2.11) and $(QCP)_h$ (2.17)-(2.21) respectively. Assume that*

$$(u_h + B^* z_h)|_{T_U} \in H^s(T_U) \ \ (s = 0 \text{ or } 1) \quad \text{for any } T_U \in \mathcal{T}_h(\Omega_U)), \tag{4.27}$$

and that there is a $\tilde{u}_h \in K_h$ such that

$$|(u_h + B^* z_h, \tilde{u}_h - u)| \leq C \sum_{T_U} h_{T_U} ||u_h + B^* z_h||_{H^s(T_U)} ||u - u_h||^s_{L^2(T_U)}. \tag{4.28}$$

Then for all $\tilde{u} \in K$ we have

$$\begin{aligned} ||u - u_h||_U^2 \ \leq \ & C\Big\{ \textstyle\sum_{T_U} h_{T_U}^{1+s} ||u_h + B^* z_h||^{1+s}_{H^s(T_U)} + ||z_h - z(u_h)||^2_{L^2(\Omega)} \\ & + |(u_h + B^* z_h, u_h - \tilde{u})| + |(B^*(z_h - z(u_h)), u_h - \tilde{u})| \\ & + |(B^*(z(u_h) - z), u_h - \tilde{u})| + |(u_h - u, u_h - \tilde{u})| \Big\}, \end{aligned} \tag{4.29}$$

where $z(u_h)$ is defined by (4.12)-(4.15). Moreover,

$$||\boldsymbol{p} - \boldsymbol{p}_h||_{H(\text{div};\Omega)} + ||y - y_h||_{L^2(\Omega)} \leq C(\eta_1 + \eta_2 + ||u - u_h||_U), \tag{4.30}$$

$$||\boldsymbol{q} - \boldsymbol{q}_h||_{H(\text{div};\Omega)} + ||z - z_h||_{L^2(\Omega)} \leq C(\eta_1 + \eta_2 + ||u - u_h||_U), \tag{4.31}$$

where η_1 is defined in Lemma 4.1 and η_2 is defined in Lemma 4.3.

References

[1] M.Ainsworth, J.T.Oden, A Posteriori Error Estimation in Finite Element Analysis, John Wiley & Sons 2000.

[2] P.Alotto, etc, Mesh adaption and optimisation techniques in magnet design, IEEE Transactions on Magnetics, Vol.32, 1996.

[3] I.Babuska, W.C.Rheinboldt, Error estimates for adaptive finite element computations, SIAM, J. Numer. Anal., Vol.5, 1978, 736-754.

[4] I.Babuska, T.Strouboulis, The Finite Element Method and its Reliability, Oxford University Press, 2001.

[5] N.V.Banichuk, Mesh refinement for shape optimisation, Structural optimisation, Vol.9, 1995, 45-51

[6] D.Braess, R.Verfurth, A posteriori error estimators for the Raviart-Thomas element, Preprint 175/1994 Fakultat fur Mathematik der Ruhr-Universitat Bochum.

[7] F.Brezzi, M.Fortin, Mixed and Hybrid Finite Element Methods, Springer-Verlag 1991. MR 92d:65187

[8] C.Carstensen, A posteriori error estimate for the mixed finite element method, Math. Comp., Vol.66, No.218, 1997, 465-476.

[9] P.G.Ciarlet, The Finite Element Method For Elliptic Problems, North-Holland, Amsterdam 1978. MR 58:25001

[10] P.Clement, Approximation by finite element functions using local regularization, RAIRO Ser. Rougr Anal. Numer. R-2 77-84(1975). Mr 53:4569

[11] P.Grisvard, Elliptic Problems in Nonsmooth Domains. Pitman 1985. MR 86m:35044

[12] J.Haslinger, P.Neittaanmaki, Finite Element Approximation for Optimal Shape Design, John Wiley and Sons, Chichester, 1989

[13] R.Li, W.B.Liu, H.P.Ma, and T.Tang, Adaptive finite element approximation for elliptic optimal control, accepted in SIAM J. Control.and.Optim., 2002

[14] J.L.Lions, Optimal Control of Systems Governed by Partial Differential Equations, Springer-Verlag, Berlin, 1971.

[15] W.B.Liu, J.W.Barrett, Quasi-norm error bounds for the finite element approximation of degenerate quasilinear elliptic variational inequalities, RAIRO Numer. Anal. Vol.28, 1994, 725-744.

[16] W.B.Liu, J.W.Barrett, Quasi-norm error bounds for the finite element approximation of degenerate quasilinear parabolic variational inequalities, Numerical Functional Analysis and Optimisation, Vol.16, 1995, 1309-1321.

[17] W.B.Liu, J.Rubio, Optimal conditions for elliptic variational inequalities of the second kind, IAM J. Control, Vol.8, 1991, 211-230.

[18] W.B.Liu, J.Rubio, Optimal conditions for strongly monotone variational inequalities, Applied Math. Optim., Vol.27, 1993, 291-312.

[19] W.B.Liu, D.Tiba, Error estimates for the finite element approximation of a class of nonlinear optimal control problems, J. Numer. Func. Optim., Vol.22, 2001, 953-972.

[20] W.B.Liu, N.N.Yan, A posteriori error estimates for a model boundary optimal control problem, J. Comp. Appl. Math., Vol.120, 2000, 159-173.

[21] W.B.Liu, N.N.Yan, A posteriori error analysis for convex boundary control problems, SIAM J. Numer. Anal., Vol.39, 2000, 73-99.

[22] W.B.Liu, N.N.Yan, A posteriori error analysis for convex distributed optimal control problems, Advan.Comp.Math., Vol.15, 2001, 285-309.

[23] W.B.Liu, N.N.Yan, A posteriori error estimates for a nonlinear control problems, in EUNMA'99 proceedings, Science Press, 2000.

[24] W.B.Liu, N.N.Yan, A posteriori error estimators for a class of variational inequalities, JSC, Vol.35, 2000, 361-393.

[25] W.B.Liu, N.N.Yan, A posteriori error estimates for parabolic optimal control problems, accepted in Numer. Math., 2002

[26] W.B.Liu, N.N.Yan, A posteriori error estimates for control problems governed by Stokes equations, submitted.

[27] P.Neittaanmaki, D.Tiba, Optimal Control of Nonlinear Parabolic Systems: Theory, algorithms and applications, M.Dekker, New York, 1994.

[28] O.Pironneau, Optimal Shape Design for Elliptic Systems, Springer-Verlag, Berlin, 1984.

[29] D.Tiba, Lectures on The Optimal Control of Elliptic Equations, University of Jyvaskyla Press, Finland, 1995.

[30] R.Verfurth, A Review of A Posteriori Error Estimation and Adaptive Mesh-refinment Techniques, Wiley-Teubner 1996.

SUPERCONVERGENCE OF LEAST-SQUARES MIXED FINITE ELEMENT APPROXIMATIONS OVER QUADRILATERALS*

Yanping Chen, Manping Zhang
ICAM, Xiangtan University, Xiangtan, 411105, China
ypchen@xtu.edu.cn
zhangmanping@sina.com.cn

Abstract A least-squares mixed finite element method over quadrilaterals is formulated and applied for a class of second order elliptic problems. A superconvergence result is established for approximate solutions. The superconvergence indicates an accuracy of $O(h^{r+2})$ for the least-squares mixed finite element approximation if Raviart-Thomas elements of order r are employed with optimal error estimate of $O(h^{r+1})$.

Keywords: superconvergence; elliptic problem; interpolation projection; least-squares mixed finite element; quadrilateral

1. Introduction

It is proved that least-squares mixed finite element methods lead to symmetric algebraic systems and are not subject to the Ladyzhenskaya Babuška Brezzi (LBB) consistency requirement. The objective of this article is to investigate superconvergence phenomena for second-order elliptic problems by using this method.

We consider the following second-order elliptic boundary-value problem

$$\begin{aligned} -\mathrm{div}(A\mathrm{grad}u) &= f \quad \text{in } \Omega \\ u &= 0 \quad \text{on } \Gamma_D \\ (-A\mathrm{grad}u)\cdot \boldsymbol{n} &= 0 \quad \text{on } \Gamma_N \end{aligned} \tag{1.1}$$

where $\Omega \subset R^2$ is an open bounded domain with Lipschitz boundary Γ such that $\Gamma = \Gamma_D \cup \Gamma_N$, A is a symmetric, positive definite matrix of coefficients and $\boldsymbol{n}$

*Supported by National Science Foundation of China, the Ministry of Education and the Special Funds for Major State Basic Research Projects.

is the unit outward vector normal to Γ. Introducing the flux $\sigma = -A\text{grad}u$, we get the first order system

$$\begin{aligned} \sigma + A\text{grad}u &= 0 \quad \text{in } \Omega \\ \text{div}\sigma - f &= 0 \quad \text{in } \Omega \\ u &= 0 \quad \text{on } \Gamma_D \\ \sigma \cdot n &= 0 \quad \text{on } \Gamma_N \end{aligned} \tag{1.2}$$

Our attention is focused on finite element partitions of Ω into convex quadrilaterals. The quadrilateral elements were constructed by using local mapping techniques for any stable rectangular space such as the Raviart-Thomas [13] or Brezzi-Douglas-Fortin-Marini [2] spaces. We shall investigate superconvergence result by using the L^2-projection and some mixed finite element projections and the integral identities technique developed by Q.Lin and his collaborates [11-12].

There have been many superconvergence results [1, 3, 10-12, 14-16] for the finite element methods and, related to the mixed finite element for elliptic problems [8-9]. Besides, Chen investigated in [5] superconvergence phenomema for second-order elliptic problems with diagonal coefficient matrix by least-squares mixed based rectangulation. This article shows that similar superconvergence holds true when quadrilateral elements are employed in the least-squares mixed finite elment method.

We shall formulate in section 2 the problem and prove the existence and uniqueness of the least-squares mixed weak form. In Section 3, we shall establish the construction of the least-squares mixed elements over quadrilaterals. In section 4, the interpolation operators are defined and some preliminary lemmas are presented, and in section 5, the main result of this paper is stated.

2. Problem formulation

Assume that the matrix of coefficients $A = (a_{ij}(x))_{2\times 2}, x \in \overline{\Omega}$, is symmetric positive definite and the coefficients $a_{ij}(x)$ are bounded, i.e., there exist positive constants α_1 and α_2 such that

$$\alpha_1 \xi^T \xi \leq \xi^T A \xi \leq \alpha_2 \xi^T \xi \tag{2.1}$$

for any vectors $\xi \in R^2$ and $x \in \overline{\Omega}$.

We shall use the standard notation for Sobolev spaces $H^m(\Omega)$ with norm $\|\cdot\|_{m,\Omega}$ and seminorms $|\cdot|_{i,\Omega}$, $0 \leq i \leq m$; as usual, $L^2(\Omega) = H^0(\Omega)$. Let $(H^m(\Omega))^2$ be the corresponding product space and

$$H(\text{div};\Omega) = \{\tau \in (L^2(\Omega))^2,\ \text{div}\tau \in L^2(\Omega)\}$$

$$V = \{v \in H^1(\Omega) : \ v = 0 \ \text{ on } \Gamma_D\}, \tag{2.2}$$

with norm

$$||v||_{1,\Omega} = \left(||v||^2_{0,\Omega} + ||\text{grad}v||^2_{0,\Omega}\right)^{1/2}.$$

By the Poincaré-Friedrichs inequality, there exists a constant C_F such that

$$||v||_0^2 \leq C_F^2 ||\text{grad}v||_0^2. \tag{2.3}$$

Let $\boldsymbol{\tau} = (\tau_1, \tau_2)$ be a smooth vector function and $v \in H^1(\Omega)$, and denote

$$\text{div}\boldsymbol{\tau} = \partial_1\tau_1 + \partial_2\tau_2, \qquad \text{grad}v = (\partial_1 v, \partial_2 v).$$

Introduce the space

$$\boldsymbol{W} = \{\tau \in H(\text{div};\Omega): \ \boldsymbol{\tau}\cdot\boldsymbol{n} = 0 \quad \text{on } \Gamma_N\} \tag{2.4}$$

with norm

$$||\tau||_{H(\text{div};\Omega)} = \left(||\tau||^2_{0,\Omega} + ||\text{div}\tau||^2_{0,\Omega}\right)^{1/2}.$$

The least-squares minimization problem then is: find $u \in V$, $\boldsymbol{\sigma} \in \boldsymbol{W}$ such that

$$J(u,\boldsymbol{\sigma}) = \inf_{v\in V, \boldsymbol{\tau}\in \boldsymbol{W}} J(v,\boldsymbol{\tau}),$$

where

$$J(v,\boldsymbol{\tau}) = (\text{div}\boldsymbol{\tau} - f, \text{div}\boldsymbol{\tau} - f) + (\boldsymbol{\tau} + A\text{grad}v, A^{-1}\boldsymbol{\tau} + \text{grad}v). \tag{2.5}$$

Note that we have applied a weight A^{-1} to the square of $\boldsymbol{\tau} + A\text{grad}v$. $J(v,\boldsymbol{\tau})$ is equivalent to the follwing functional

$$J_1(v,\boldsymbol{\tau}) = (\text{div}\boldsymbol{\tau} - f, \text{div}\boldsymbol{\tau} - f) + (\boldsymbol{\tau} + A\text{grad}v, \boldsymbol{\tau} + A\text{grad}v),$$

but it is a more balanced and better scaled form (see [3]).

The corresponding variational problem is: find $u \in V$, $\boldsymbol{\sigma} \in W$ such that

$$a(u,\boldsymbol{\sigma};v,\boldsymbol{\tau}) = (f, \text{div}\boldsymbol{\tau}) \ \ \forall\ v \in V,\ \ \boldsymbol{\tau} \in \boldsymbol{W}, \tag{2.6}$$

where

$$a(u,\boldsymbol{\sigma};v,\boldsymbol{\tau}) = (\text{div}\boldsymbol{\sigma}, \text{div}\boldsymbol{\tau}) + (\boldsymbol{\sigma} + A\text{grad}u, A^{-1}\boldsymbol{\tau} + \text{grad}v). \tag{2.7}$$

Theorem 2.1. There exists a constant $C > 0$ such that for all $v \in V$, $\boldsymbol{\tau} \in \boldsymbol{W}$,

$$C\left(||v||^2_{1,\Omega} + ||\boldsymbol{\tau}||^2_{H(\text{div};\Omega)}\right) \leq a(v,\boldsymbol{\tau};v,\boldsymbol{\tau}). \tag{2.8}$$

Proof: Expanding $a(\cdot;\cdot)$, adding and subtracting a term involving β (>0), which will be specified later, and regrouping terms, yield

$$\begin{aligned} a(v,\boldsymbol{\tau};v,\boldsymbol{\tau}) &= (\mathrm{div}\boldsymbol{\tau},\mathrm{div}\boldsymbol{\tau}) + (\boldsymbol{\tau}+A\mathrm{grad}v, A^{-1}\boldsymbol{\tau}+\mathrm{grad}v) \\ &= (\mathrm{div}\boldsymbol{\tau},\mathrm{div}\boldsymbol{\tau}) + (\boldsymbol{\tau},A^{-1}\boldsymbol{\tau}) + 2(\boldsymbol{\tau},\mathrm{grad}v) + (A\mathrm{grad}v,\mathrm{grad}v) \\ &= (\mathrm{div}\boldsymbol{\tau},\mathrm{div}\boldsymbol{\tau}) - 2(\mathrm{div}\boldsymbol{\tau},\beta v) + (\beta^2 v, v) - (\beta^2 v, v) \\ &\quad +2(\mathrm{div}\boldsymbol{\tau},\beta v) + (\boldsymbol{\tau},A^{-1}\boldsymbol{\tau}) + 2(\boldsymbol{\tau},\mathrm{grad}v) + (A\mathrm{grad}v,\mathrm{grad}v). \end{aligned}$$

Integrating by parts in the fifth term, and setting $v=0$ on Γ_D, $\boldsymbol{\tau}\cdot\boldsymbol{n}=0$ on Γ_N, give

$$\begin{aligned} a(v,\boldsymbol{\tau};v,\boldsymbol{\tau}) &= (\mathrm{div}\boldsymbol{\tau}-\beta v, \mathrm{div}\boldsymbol{\tau}-\beta v) - (\beta^2 v, v) + (\boldsymbol{\tau},A^{-1}\boldsymbol{\tau}) \\ &\quad +2(\boldsymbol{\tau},\mathrm{grad}v) - 2(\boldsymbol{\tau},\beta\mathrm{grad}v) - ((1-\beta)^2 A\mathrm{grad}v,\mathrm{grad}v) \\ &\quad -((1-\beta)^2 A\mathrm{grad}v,\mathrm{grad}v) + (A\mathrm{grad}v,\mathrm{grad}v) \\ &\geq +(\boldsymbol{\tau}+(1-\beta)A\mathrm{grad}v, A^{-1}(\boldsymbol{\tau}+(1-\beta)A\mathrm{grad}v)) \\ &\quad -(\beta^2 v, v) + ((2\beta-\beta^2)A\mathrm{grad}v,\mathrm{grad}v) \\ &\geq ((-\beta^2 C_F^2 + (2\beta-\beta^2)\alpha_1)\mathrm{grad}v,\mathrm{grad}v), \end{aligned} \tag{2.9}$$

where we have used (2.1) and (2.2). Let $\beta=\alpha_1/(\alpha_2+C_F^2)$. We have

$$\beta(2\alpha_1-\beta(\alpha_1+C_F^2)) = \frac{\alpha_1^2}{\alpha_2+C_F^2} > 0.$$

Hence

$$a(v,\boldsymbol{\tau};v,\boldsymbol{\tau}) \geq C\|\mathrm{grad}v\|_0^2 \geq C\|v\|_1^2. \tag{2.10}$$

Obviously, from (2.7)

$$a(v,\boldsymbol{\tau};v,\boldsymbol{\tau}) \geq (\boldsymbol{\tau}+A\mathrm{grad}v, A^{-1}(\boldsymbol{\tau}+A\mathrm{grad}v)). \tag{2.11}$$

$$a(v,\boldsymbol{\tau};v,\boldsymbol{\tau}) \geq (\mathrm{div}\boldsymbol{\tau}+cv, \mathrm{div}\boldsymbol{\tau}+cv). \tag{2.12}$$

Then, it follows from (2.31) in [12] that

$$\begin{aligned} \|\boldsymbol{\tau}\|_0^2 &\leq \alpha_2^{-1}(\boldsymbol{\tau},A^{-1}\boldsymbol{\tau}) \\ &\leq 2a_2^{-1}(\boldsymbol{\tau}+A\mathrm{grad}v, A^{-1}\boldsymbol{\tau}+\mathrm{grad}v) \\ &\quad +2a_2^{-1}(A\mathrm{grad}v,\mathrm{grad}v) \\ &\leq Ca(v,\boldsymbol{\tau};v,\boldsymbol{\tau}), \end{aligned} \tag{2.13}$$

$$\begin{aligned} \|\mathrm{div}\boldsymbol{\tau}\|_0^2 &\leq 2\|\mathrm{div}\boldsymbol{\tau}+cv\|_0^2 + 2\|cv\|_0^2 \\ &\leq Ca(v,\boldsymbol{\tau};v,\boldsymbol{\tau}). \end{aligned} \tag{2.14}$$

Combining (2.11)-(2.14), we get (2.8). □

Now we apply the Lax-Milgram lemma to establish the existence and uniqueness of the solutions to problem (2.6).

Theorem 2.2. Let f $\in L^2(\Omega)$. Then the problem (2.6) has a unique solution $u\in V$, $\boldsymbol{\sigma}\in \boldsymbol{W}$.

3. The least-squares mixed finite element approach

The mixed method over quadrilaterals used in this paper is defined by using mapping techniques to the reference element $\widehat{K}$=[-1,1]×[-1,1]. Consider any quadrilateral K, where p_i in a counterclockwise direction stands for the coordinates of the corresponding vertex. There exists an affine invertible mapping (see [9]) $\widehat{F}_K : \widehat{K} \to K$, such that $K = \widehat{F}_K(\widehat{K})$. Let G be the Jacobi matrix (derivative) of $\widehat{F}_K$ and M=$|\det(G)|^{-1} G$.

On an arbitrary convex quadrilateral K, the Raviart-Thomas space is defined by

$$\begin{aligned} V_k(K) &= \{v = \widehat{v} \circ \widehat{F}_K^{-1} : \widehat{v} \in Q_{k,k}(\widehat{K})\}, \\ \boldsymbol{W}_r(K) &= \{\boldsymbol{\tau} = M\, \widehat{\boldsymbol{\tau}} \circ \widehat{F}_K^{-1} : \widehat{\boldsymbol{\tau}} \in Q_{r+1,r}(\widehat{K}) \times Q_{r,r+1}(\widehat{K})\}. \end{aligned} \tag{3.1}$$

where $Q_{m,n}(\widehat{K})$ indicates the space of polynomials of degree no more than m and n in $\widehat{x}$ and $\widehat{y}$, respectively.

Let $\mathcal{T}_h$ be a finite element partition of Ω into quadrilaterals, where h is the mesh parameter, generally denoting the biggest one of diameters of elements in partitions $\mathcal{T}_h$. The global finite element space over $\mathcal{T}_h$ is defined as follows

$$V_h = \{v_h \in C^0(\Omega) : \; v_h|_K \in V_k(K), \; \forall K \in \mathcal{T}_h, \; v_h = 0 \text{ on } \Gamma_D\}, \tag{3.2}$$

$$\boldsymbol{W}_h = \{\boldsymbol{\tau}_h \in H(\text{div};\Omega) : \; \boldsymbol{\tau}_h|_K \in \boldsymbol{W}_r(K) \quad \forall K \in \mathcal{T}_h, \; \boldsymbol{\tau}_h \cdot \boldsymbol{n} = 0 \text{ on } \Gamma_N\}. \tag{3.3}$$

where V_h is the pressure space and $\boldsymbol{W}_h$ is that for the velocity.

These spaces possess the following approximation properties:

$$\inf_{v_h \in V_h} \{||v - v_h||_{0,\Omega} + h||\text{grad}(v - v_h)||_{0,\Omega}\} \leq Ch^{k+1}||v||_{k+1,\Omega}, \tag{3.4}$$

$$\inf_{\boldsymbol{\tau}_h \in \boldsymbol{W}_h} ||\boldsymbol{\tau} - \boldsymbol{\tau}_h||_{0,\Omega} \leq Ch^{r+1}||\boldsymbol{\tau}||_{r+1,\Omega}, \tag{3.5}$$

$$\inf_{\boldsymbol{\tau}_h \in \boldsymbol{W}_h} ||\text{div}(\boldsymbol{\tau} - \boldsymbol{\tau}_h)||_{0,\Omega} \leq Ch^{r+1}||\boldsymbol{\tau}||_{r+2,\Omega}, \tag{3.6}$$

for any $v \in H^{k+1}(\Omega) \cap V$ and $\boldsymbol{\tau} \in (H^{r_1+1}(\Omega))^2 \cap \boldsymbol{W}$.

The corresponding finite element approximation to (2.6) is: find $u_h \in V_h$, $\boldsymbol{\sigma}_h \in \boldsymbol{W}_h$ such that

$$a(u_h, \boldsymbol{\sigma}_h; v_h, \boldsymbol{\tau}_h) = (f, \text{div}\boldsymbol{\tau}_h), \quad \forall v_h \in V_h, \; \boldsymbol{\tau}_h \in \boldsymbol{W}_h. \tag{3.7}$$

From Theorem 2.2 we conclude that problem (3.7) has a unique solution since $V_h \subset V$, $\boldsymbol{W}_h \subset \boldsymbol{W}$. Moreover,using (2.6) for the exact solution of (1.2) we get the orthogonality property:

$$a(u - u_h, \boldsymbol{\sigma} - \boldsymbol{\sigma}_h; v_h, \boldsymbol{\tau}_h) = 0, \quad \forall v_h \in V_h, \; \boldsymbol{\tau}_h \in \boldsymbol{W}_h. \tag{3.8}$$

4. Interpolant projection operators

Let $\widehat{P}_k\widehat{u} \in \widehat{V}_k(\widehat{K})$ and $\widehat{\sigma} \in \widehat{\mathbf{W}}_r(\widehat{K})$ be the standard finite element interpolants of $\widehat{u}$ and $\widehat{\sigma}$, respectively; then for all $u \in V, \sigma \in W$, the projections $P_h u$ and $\pi_h \sigma$ are separately defined as follows

$$P_h u = \widehat{P}_k \widehat{u} \circ \widehat{F}_K^{-1}, \tag{4.1}$$

$$\pi_h \sigma = M(\widehat{\pi}_h(M^{-1}\widehat{\sigma})). \tag{4.2}$$

From approximation theory [2, 6-7, 13], these interpolation functions have the following approximate properties:

$$||u - P_h u||_{0,\Omega} + h||\text{grad}(u - P_h u)||_{0,\Omega} \leq Ch^{k+1}||u||_{k+1,\Omega}, \tag{4.3}$$

$$||\sigma - \pi_h \sigma||_{0,\Omega} \leq Ch^{r+1}||\sigma||_{r+1,\Omega}, \tag{4.4}$$

$$||\text{div}(\sigma - \pi_h \sigma)||_{0,\Omega} \leq Ch^{r+1}||\sigma||_{r+2,\Omega}. \tag{4.5}$$

Lemma 4.1. Assume that the finite element partition $\mathcal{T}_h$ is h^2*-uniform* (see [9]) and $\pi_h\sigma$ is the projection of σ defined above, then there exists a constant C such that

$$(\sigma - \pi_h\sigma, \tau_h)_{0,K} \leq Ch^{r+2}||\sigma||_{r+2,K}||\tau_h||_{0,K}, \quad \forall \tau_h \in \mathbf{W}_h, \; K \in \mathcal{T}_h. \tag{4.6}$$

Proof. Let

$$\widetilde{\sigma} = M^{-1}\widehat{\sigma}, \; \widetilde{v} = M^{-1}\widehat{v}.$$

Then $\widehat{\sigma} = M\widetilde{\sigma}, \widehat{v} = M\widetilde{v}$, and from (4.2)

$$\widehat{\pi_h\sigma} = M(\widehat{\pi}_h\widetilde{\sigma}).$$

With $B_K = M^t M \det(G_K)$ and by changing variables, we have that

$$\begin{aligned}(\sigma - \pi_h\sigma, \tau_h)_{0,K} &= \int_K (\sigma - \pi_h\sigma)\cdot \tau_h dxdy \\ &= \int_{\widehat{K}} (\widehat{\sigma} - \widehat{\pi_h\sigma})\cdot \widehat{\tau}_h \det(G) d\widehat{x}d\widehat{y} \\ &= \int_{\widehat{K}} (B_K - \overline{B}_K)(\widetilde{\sigma} - \widehat{\pi}_h\widetilde{\sigma})\cdot \widetilde{\tau}_h d\widehat{x}d\widehat{y} \\ &\quad + \int_{\widehat{K}} B_K(\widetilde{\sigma} - \widehat{\pi}_h\widetilde{\sigma})\cdot\widetilde{\tau}_h d\widehat{x}d\widehat{y},\end{aligned} \tag{4.7}$$

where $\overline{B}$ is the average of B_K on the reference element $\widehat{K}$. Since the qudrilateral K is an h^2*-parallelogram,* we have

$$| \, B_K - \overline{B}_K \, | \leq Ch,$$

for some constant C.

By Lemmas 5.2 and 5.5 in [9], there exists a constant C such that

$$\begin{aligned} \|\tilde{\sigma} - \hat{\pi}_h\tilde{\sigma}\|_{0,\hat{K}}\|\tilde{\tau}_h\|_{0,\hat{K}} &\leq C \mid \tilde{\sigma} \mid_{r+2,\hat{K}} \|\tilde{\tau}_h\|_{0,\hat{K}} \\ &\leq Ch^{r+2}\|\sigma\|_{r+2,K}\|\tau_h\|_{0,K}. \end{aligned} \tag{4.8}$$

Thus, substituting above into (4.7) yields the following estimates

$$\begin{aligned} &(\sigma - \pi_h\sigma, \tau_h)_{0,K} \\ &\quad\leq Ch^{r+3}\|\sigma\|_{r+2,K}\|\tau_h\|_{0,K} + Ch^{r+2}\|\sigma\|_{r+2,K}\|\tau_h\|_{0,K} \\ &\quad\leq Ch^{r+2}\|\sigma\|_{r+2,K}\|\tau_h\|_{0,K}. \end{aligned} \tag{4.9}$$

which is the desired result. □

Lemma 4.2. Assume that the finite element partition $\mathcal{T}_h$ is h^2-*uniform* and $\pi_h\sigma$ is the projection of σ defined above, let k=r+1, then

$$(\sigma - \pi_h\sigma, \text{grad}v_h)_{0,K} \leq Ch^{r+2}\|\sigma\|_{r+2,K}\|v_h\|_{1,K}, \ \forall\, v_h \in V_h,\ K \in \mathcal{T}_h. \tag{4.10}$$

Proof. It follows from substitution of variables that

$$\int_K (\sigma - \pi_h\sigma)\cdot \text{grad}v_h dxdy = \int_{\hat{K}} (\hat{\sigma} - \hat{\pi}_h\hat{\sigma})\cdot \text{grad}\hat{v}_h D\hat{F}_K^{-1} \det(G) d\hat{x}d\hat{y}, \tag{4.11}$$

By Lemma 3.3 in [5], withe $x_K = y_K = 0$, and $h_K = h'_K = 1$, we can have

$$(\hat{\sigma} - \hat{\pi}_h\hat{\sigma}, \text{grad}\hat{v}_h)_{0,\hat{K}} \leq C \mid \hat{\sigma} \mid_{r+2,\hat{K}} \mid \hat{v}_h \mid_{r+1,\hat{K}} . \tag{4.12}$$

Substituting (4.12) into (4.11) gives

$$\begin{aligned} &\textstyle\int_K (\sigma - \pi_h\sigma)\cdot \text{grad}v_h dxdy \\ &\quad\leq \mid \hat{F}_K^{-1} \mid_{1,\infty,K} \mid \det(G) \mid_{0,\infty,K} \mid \textstyle\int_{\hat{K}} (\hat{\sigma} - \hat{\pi}_h\hat{\sigma})\cdot \text{grad}\hat{v}_h d\hat{x}d\hat{y} \mid \\ &\quad\leq Ch^{-1}h^2 \mid \hat{\sigma} \mid_{r+2,\hat{K}} \mid \hat{v}_h \mid_{r+1,\hat{K}} \\ &\quad\leq Ch^{-1}h^2h^{r+1}\|\sigma\|_{r+2,K}\|\hat{v}_h\|_{1,\hat{K}} \\ &\quad\leq Ch^{r+2}\|\sigma\|_{r+2,K}\|v_h\|_{1,K}. \end{aligned} \tag{4.13}$$

where we have used the standard inverse inequality for finite element functions. Hence, the proof of Lemma 4.2 is completed. □

Now we introduce the following space

$$\widetilde{V}_h = \{v_h \in L^2(\Omega) : v_h|_K \in Q_r(K).\ \forall\, K \in \mathcal{T}_h\}$$

It is obvious that

$$\text{div}\sigma \in \widetilde{V}_h,\ \forall\, \sigma \in \boldsymbol{W}_h. \tag{4.14}$$

The definitions of the interpolation functions $P_h u$ and $\pi_h\sigma$ lead to the following lemma.

Lemma 4.3. Assume that the finite element partition $\mathcal{T}_h$ is h^2-*uniform*, $\pi_h\sigma$ and $P_h u$ are separately defined as in (4.1) and (4.2). Then

$$(u - P_h u, \mathrm{div}\boldsymbol{\tau}_h)_{0,K} = 0, \qquad \forall\, \boldsymbol{\tau}_h \in \boldsymbol{W}_h, K \in \mathcal{T}_h, \tag{4.15}$$

$$(\mathrm{div}(\sigma - \pi_h\sigma), v_h)_{0,K} = 0, \qquad \forall\, v_h \in \widetilde{V}_h, K \in \mathcal{T}_h. \tag{4.16}$$

5. Main superconvergence result

For any $v \in V$, we define a projection $S_h v \in V_h$ such that

$$(A\mathrm{grad}(v - S_h v), \mathrm{grad} v_h)_{0,\Omega} = 0 \tag{5.1}$$

for all $v_h \in V_h$. Then, from standard finite element theory [6], we have the estimate

$$||v - S_h v||_{0,\Omega} + h||v - S_h v||_{1,\Omega} \leq Ch^{k+1}||v||_{k+1,\Omega}. \tag{5.2}$$

By the high-accuracy theory of the finite element method [4, 11-12, 15], we also have

$$||S_h u - P_h u||_{1,\Omega} \leq Ch^{k+1}||u||_{k+2}. \tag{5.3}$$

Theorem 5.1. Assume that the finite element partition $\mathcal{T}_h$ is h^2-*uniform* and $(u_h, \boldsymbol{\sigma}_h)$ is the solution of (3.7) by using quadrilateral elements of Raviart-Thomas of order r. If the exact solution u and $\boldsymbol{\sigma}$ of (2.6) satisfies

$$u \in H^{k+2}(\Omega), \qquad \boldsymbol{\sigma} \in [H^{r+2}(\Omega)]^2.$$

If $k = r + 1$, then

$$||u_h - P_h u||_{1,\Omega} + ||\sigma_h - \pi_h\sigma||_{H(\mathrm{div};\Omega)} \leq Ch^{r+2}\left(||u||_{r+3,\Omega} + ||\sigma||_{r+2,\Omega}\right). \tag{5.4}$$

Proof. From the coercivity of the bilinear form in Theorem 2.1, and (3.8),

$$\begin{aligned}
&||u_h - P_h u||_{1,\Omega}^2 + ||\sigma_h - \pi_h\sigma||_{H(\mathrm{div};\Omega)}^2 \\
&\leq Ca(u_h - P_h u, \boldsymbol{\sigma}_h - \pi_h\boldsymbol{\sigma}; u_h - P_h u, \boldsymbol{\sigma}_h - \pi_h\boldsymbol{\sigma}) \\
&= Ca(u - P_h u, \boldsymbol{\sigma} - \pi_h\boldsymbol{\sigma}; u_h - P_h u, \boldsymbol{\sigma}_h - \pi_h\boldsymbol{\sigma}) \\
&= C\Big((\mathrm{div}(\boldsymbol{\sigma} - \pi_h\boldsymbol{\sigma}), \mathrm{div}(\boldsymbol{\sigma}_h - \pi_h\boldsymbol{\sigma}))_{0,\Omega} \\
&+ (\boldsymbol{\sigma} - \pi_h\boldsymbol{\sigma}, A^{-1}(\boldsymbol{\sigma}_h - \pi_h\boldsymbol{\sigma}))_{0,\Omega} + (\boldsymbol{\sigma} - \pi_h\boldsymbol{\sigma}, \mathrm{grad}(u_h - P_h u))_{0,\Omega} \\
&+ (\mathrm{grad}(u - P_h u), \boldsymbol{\sigma}_h - \pi_h\boldsymbol{\sigma})_{0,\Omega} + (A\mathrm{grad}(u - P_h u), \mathrm{grad}(u_h - P_h u))_{0,\Omega}\Big) \\
&= \sum_{i=1}^{5} I_i.
\end{aligned} \tag{5.5}$$

Now, let us bound the terms $\{I_i\}$.

It follows from (4.14) that

$$\mathrm{div}(\sigma_h - \pi_h\sigma) \in \widetilde{V}_h.$$

By (4.16) in Lemma 4.3, we have

$$I_1 = (\mathrm{div}(\sigma - \pi_h\sigma), \mathrm{div}(\sigma_h - \pi_h\sigma))_{0,\Omega} = 0. \tag{5.6}$$

For any $K \in \mathcal{T}_h$, let $\overline{A}_K$ be the average of A^{-1} on the element K. Since the coefficients are smooth and $\mathcal{T}_h$ is h^2-*uniform*

$$|A^{-1} - \overline{A}_K| \leq Ch, \qquad \text{on each } K \in \mathcal{T}_h,$$

where the constant C is independent of the element K, and now from the approximation property (4.4) and Lemma 4.1 we see that

$$\begin{aligned} I_2 &= \Big| \sum_{K \in \mathcal{T}_h} \Big[(\sigma - \pi_h\sigma, (A^{-1} - \overline{A}_K)(\sigma_h - \pi_h\sigma))_{0,K} \\ &\quad + (\sigma - \pi_h\sigma, \overline{A}_K(\sigma_h - \pi_h\sigma))_{0,K}\Big]\Big| \\ &\leq \sum_{K \in \mathcal{T}_h} \Big[Ch||\sigma - \pi_h\sigma||_{0,K}||\sigma_h - \pi_h\sigma||_{0,K} \\ &\quad + |\overline{A}_K| \cdot |(\sigma - \pi_h\sigma, \sigma_h - \pi_h\sigma)_{0,K}|\Big] \\ &\leq Ch^{r+2}||\sigma||_{r+2,\Omega}||\sigma_h - \pi_h\sigma||_{0,\Omega}. \end{aligned} \tag{5.7}$$

Using Lemma 4.2, we have

$$I_3 \leq Ch^{r+2}||\sigma||_{r+2,\Omega}||u_h - P_hu||_{1,\Omega}. \tag{5.8}$$

Next, we know from the approximate property of P_h that

$$\begin{aligned} I_4 &= -(u - P_hu, \mathrm{div}(\sigma_h - \pi_h\sigma))_{0,\Omega} \\ &\leq C||u - P_hu||_{0,\Omega}||\sigma_h - \pi_h\sigma||_{H(\mathrm{div};\Omega)} \\ &\leq Ch^{k+1}||u||_{k+1,\Omega}||\sigma_h - \pi_h\sigma||_{H(\mathrm{div};\Omega)}. \end{aligned} \tag{5.9}$$

We can estimate the last term I_5 by using (5.1)-(5.3)

$$\begin{aligned} I_5 &= (A\mathrm{grad}(S_hu - P_hu), \mathrm{grad}(u_h - P_hu))_{0,\Omega} \\ &\leq ||A||_{0,\infty}||S_hu - P_hu||_{1,\Omega}||u_h - P_hu||_{1,\Omega} \\ &\leq Ch^{k+1}||u||_{k+2,\Omega}||u_h - P_hu||_{1;\Omega}. \end{aligned} \tag{5.10}$$

Therefore, for $k = r + 1$, applying (5.6)-(5.10), we obtain the theorem. □

References

[1] I. Babuška, T.Strouboulis. The Finite Element Method and its Reliability. Oxford University Press, 1999.

[2] F. Brezzi, J. Douglas, M. Fortin, and L. Marini. Efficient rectangular mixed finite elements in two and three spaces variables. RAIRO Model. Math. Anal. Numer., 21(1987), pp. 581-604.

[3] Z. Cai, R. Lazarov, T. A. Manteuffel, S. F. Mccormick. First-order system least squares for second-order partial differential equations: Part I. SIAM J. Numer. Anal., 31(1994), no.6, pp. 1785-1799.

[4] C. M. Chen, Y. Q. Huang. High Accuracy Theory of Finite Elements. Hunan science Press, 1994.

[5] Y. P. Chen. Superconvergence for elliptic problems by least-squares mixed finite element, Internat. J. Numer. Methods Engrg. (To appear)

[6] P. G. Ciarlet. The Finite Element Method for Elliptic Problems. North Holland, Amsterdam, New York, Oxford, 1978.

[7] J. Douglas, Jr., J. E. Roberts. Global estimates for mixed finite element methods for second order elliptic equations. Math. Comp., 44(1985), no. 169, pp. 39-52.

[8] R. E. Ewing, R. D. Lazarov, J. Wang. Superconvergence of the velocity along the Gauss lines in mixed finite element methods. SIAM J. Numer. Anal., 28(1991), no. 4, pp. 1015-1029.

[9] R. E. Ewing, M. M. Liu, J. Wang. Superconvergence of mixed finite element approximations over quadrilaterals. SIAM J. Numer. Anal., 36(1999), no. 3, pp. 772-787.

[10] M. Křížk, P. Neittaanmäki, R. Stenberg. Finite Element Methods: Superconvergence, Post-Processing, and A Posteriori Estimates. Marcel Dekker, Inc. New York. Basel. Hong Kong, 1998.

[11] Q. Lin, N. N. Yan. High Efficiency FEM Construction and Analysis. Hebei University Press, 1996.

[12] Q. Lin, Q. D. Zhu. The Preprocessing and Postprocessing for The Finite Element Method. Shanghai Sci & Tech. Press, 1994.

[13] P. A. Raviart, J. M. Thomas. A mixed finite element method for 2nd order elliptic problems. Mathematics Aspects of the Finite Element Methods, Lecture Notes in Math. 606, 1977, pp. 292-315.

[14] A. H. Schatz, L. B. Wahlbin. Interior maximum-norm estimates for finite element methods. Math. Comp., 64(1995), pp. 907-928.

[15] L. B. Wahlbin. Superconvergence in Galerkinn Finite Element Methods. Springer Lecture Notes in Math. 1605, Springer-Verlag, New York, 1995.

[16] O. C. Zienkiewicz, J. Z. Zhu. The superconvergenct patch recovery and a posteriori error estimates, I: The recovery technique. Internat. J. Numer. Methods Engrg., 33(1992), pp. 1331-1364.

A POSTERIORI ERROR ANALYSIS AND ADAPTIVE METHODS FOR PARABOLIC PROBLEMS*

Zhiming Chen
LSEC, Institute of Computational Mathematics, Academy of Mathematics and System Sciences, Chinese Academy of Sciences, P.O. Box 2719, Beijing 100080, China
zmchen@lsec.cc.ac.cn

Abstract We report the recent progress in deriving sharp a posteriori error estimates for linear and nonlinear parabolic problems. We show how to use special properties of a linear dual problem in non-divergence form with vanishing diffusion and strong advection to derive $L^1 L^1$ norm estimate for the continuous casting problem. This estimate exhibits a mild explicit dependence on velocity. We next use a direct energy estimate method to develop an efficient and reliable a posteriori error estimator for linear parabolic equations which does not depend on any regularity assumption on the underlying elliptic operator. A convergent adaptive algorithm with variable time-step sizes and space meshes is proposed and studied which, at each time step, delays the mesh coarsening until the final iteration of the adaptive procedure, allowing only mesh and time-step size refinements before. The key ingredient in the convergence analysis is a new coarsening strategy.

Keywords: A posteriroi error estimates, parabolic, finite element

1. Introduction

A posteriori error estimates are computable quantities in terms of the discrete solution and data that measure the actual discrete errors without the knowledge of limit solutions. They are essential in designing algorithms for mesh and time-step size modification which equi-distribute the computational effort and optimize the computation. Ever since the pioneering work of Babuška and Rheinboldt [6], the adaptive finite element methods based on a posteriori error estimates have become a central theme in scientific and engineering computations. There are considerable efforts in the literature devoted to the development

*Partially supported by China NSF Grant 10025102 and China National Key Project "Large Scale Scientific Computation Research" (G1999032802).

of efficient adaptive algorithms for various linear and nonlinear parabolic partial differential equations. In particular, a posteriori error estimates are derived in Bieterman and Babuška [1], [2] and Moore [14] for one dimensional and in Eriksson and Johnson [12], [13], Verfürth [20] for multi-dimensional linear and mildly nonlinear parabolic problems. Efficient adaptive procedures based on a posteriori error estimates are also developed in Chen and Dai [4], Chen, Nochetto and Schmidt [9], Nochetto, Schmidt and Verdi [16] for solving nonlinear partial differential equations arising from physical and industrial processes.

The continuous casting problem is a convection-dominated nonlinearly degenerate diffusion problem. It is discretized implicitly in time via the method of characteristics, and in space via continuous piecewise linear finite elements. The first objective of this paper is to show how to derive an a posteriori error estimate for the L^1L^1 norm of temperature which exhibits a mild explicit dependence on velocity. The analysis is based on special properties of a linear dual problem in non-divergence form with vanishing diffusion and strong advection.

The main tool in deriving a posteriori error estimates in [4], [9], [12], [13], [16] is the analysis of linear dual problems of the corresponding *error* equations. The derived a posteriori error estimates, however, depend on the H^2 regularity assumption on the underlying elliptic operator. The next objective of the paper is to remove such a rather restrictive assumption and derive an a posteriori error estimate which bounds the full energy error of the approximate solution for a linear parabolic problem with discontinuous coefficients. Moreover, a local lower bound is proved for the associated local a posteriori error estimator. The method to prove the upper bound is a direct energy estimate argument which has been used in Chen and Nochetto [8] for elliptic obstacle problem and in Chen, Nochetto and Schmidt [10] for the phase relaxation model, a system of one parabolic equation coupled with one variational inequality. Based on this error estimator, we develop a convergent adaptive algorithm with variable time-step sizes and space meshes which, at each time step, delays the mesh coarsening until the final iteration of the adaptive procedure, allowing only mesh and time-step size refinements before.

2. Sharp a Posteriori Error Estimate for the Continuous Casting Problem

2.1 The Setting

Let the ingot occupy a cylindrical domain Ω with large aspect ratio. Let $0 < L < +\infty$ be the length of the ingot and $\Gamma \subset \mathbf{R}^d$ for $d = 1$ or 2 be its (polygonal) cross section. We show $\Omega = \Gamma \times (0, L)$ in Figure 1, and hereafter write $x = (y, z) \in \Omega$ with $y \in \Gamma$ and $0 < z < L$.

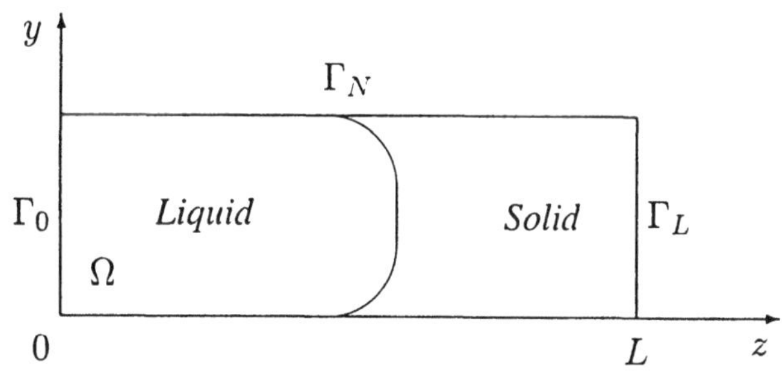

Figure 1: The domain Ω

We study the following *convection-dominated nonlinearly degenerate diffusion* problem

$$\partial_t u + v(t)\partial_z u - \Delta\theta = 0 \quad \text{in} \quad Q_T, \tag{2.1}$$

$$\theta = \beta(u) \quad \text{in} \quad Q_T, \tag{2.2}$$

$$\theta = g_D \quad \text{on} \quad \Gamma_0 \times (0,T), \tag{2.3}$$

$$\partial_\nu\theta + p(\theta - \theta_{\text{ext}}) = 0 \quad \text{on} \quad \Gamma_N \times (0,T), \tag{2.4}$$

$$u(x,0) = u_0(x) \quad \text{in} \quad \Omega, \tag{2.5}$$

where $Q_T = \Omega \times (0,T)$,

$$\Gamma_0 = \Gamma \times \{0\}, \quad \Gamma_L = \Gamma \times \{L\}, \quad \Gamma_N = \partial\Gamma \times (0,L),$$

and $\theta + \theta_c$ is the absolute temperature, θ_c is the melting temperature, u is the enthalpy, $v(t) > 0$ is the extraction velocity of the ingot, ν is the unit outer normal to $\partial\Omega$, and θ_{ext} is the external temperature. The mapping $\beta : \mathbf{R} \to \mathbf{R}$ is Lipschitz continuous and monotone increasing; since β is not strictly increasing, (2.1) is *degenerate* parabolic. The missing *outflow* boundary condition on Γ_L is unclear because the ingot moves at the casting speed and is cut shorter from time to time. It is thus evident that any standard boundary condition could only be an approximation. For simplicity, we impose a Dirichlet outflow condition

$$\theta = g_D < 0 \quad \text{on} \quad \Gamma_L \times (0,T). \tag{2.6}$$

A Robin type boundary condition on Γ_L is also possible (see the discussion in [9]). It is convenient to denote by Γ_D the Dirichlet part of $\partial\Omega$, that is $\Gamma_0 \cup \Gamma_L$.

The importance of simulating and controlling the continuous casting process in the production of steel, copper, and other metals is recognized in industry. The extraction velocity $v(t)$ as well as the cooling conditions on the mold and water spray region are known to be decisive in determining material properties of the ingot. Avoiding excessive thermal stresses and material defects is an essential, and rather empirical, aspect of the continuous casting process.

The system (2.1)-(2.5) is a special case of general Stefan problems with prescribed convection Rulla [18]. An outflow Dirichlet condition together with an inflow Neumann condition is assumed in [18] to guarantee uniqueness of weak solutions; our more realistic boundary data (2.3) and (2.6) violate this restriction. Under the additional assumption that the free boundary does not touch the inflow boundary Γ_0, uniqueness of weak solutions to (2.1)-(2.5) and (2.6) is shown in Rodrigues and Yi [17].

We now introduce the fully discrete problem, which combines continuous piecewise linear finite elements in space with characteristic finite differences in time. In fact, we use the method of characteristics to discretize the convection. We denote by τ_n the n-th time step and set

$$t^n := \sum_{i=1}^{n} \tau_i, \quad \varphi^n(\cdot) := \varphi(\cdot, t^n)$$

for any function φ continuous in $(t^{n-1}, t^n]$. Let N be the total number of time steps, that is $t^N \geq T$.

Let $\mathbf{V}_0 = \{v \in H^1(\Omega) : v = 0 \text{ on } \Gamma_D\}$ and $\mathbf{V}^*$ the dual space of $\mathbf{V}_0$. We denote by $\mathcal{M}^n$ a uniformly regular partition of Ω into simplexes. The mesh $\mathcal{M}^n$ is obtained by refinement/coarsening of $\mathcal{M}^{n-1}$, and thus $\mathcal{M}^n$ and $\mathcal{M}^{n-1}$ are *compatible*. Let $\mathbf{V}^n$ indicate the usual space of C^0 piecewise linear finite elements over $\mathcal{M}^n$ and $\mathbf{V}_0^n = \mathbf{V}^n \cap \mathbf{V}_0$. Let $\{x_k^n\}_{k=1}^{K^n}$ denote the interior nodes of $\mathcal{M}^n$. Let $I^n : C(\bar{\Omega}) \to \mathbf{V}^n$ be the usual Lagrange interpolation operator, namely $(I^n\varphi)(x_k^n) = \varphi(x_k^n)$ for all $1 \leq k \leq K^n$. Finally, let the discrete inner products $\langle \cdot, \cdot \rangle^n$ and $\langle\langle \cdot, \cdot \rangle\rangle_E^n$ be the sum over $S \in \mathcal{M}^n$ of the element scalar products

$$\langle \varphi, \chi \rangle_S^n = \int_S I^n(\varphi\chi)dx, \quad \langle\langle \varphi, \chi \rangle\rangle_S^n = \int_{S\cap E} I^n(\varphi\chi)d\sigma,$$

for any piecewise uniformly continuous functions φ, χ. The discrete initial enthalpy $U^0 \in \mathbf{V}^0$ is defined at nodes x_k^0 of $\mathcal{M}^0 = \mathcal{M}^1$ to be

$$U^0(x_k^0) := u_0(x_k^0) \quad \forall\, x_k^0 \in \Omega \backslash F_0, \quad U^0(x_k^0) := 0 \quad \forall\, x_k^0 \in F_0.$$

Hence, U^0 is easy to evaluate in practice.

Discrete Problem. *Given $U^{n-1}, \Theta^{n-1} \in \mathbf{V}^{n-1}$, then $\mathcal{M}^{n-1}$ and τ^{n-1} are modified as described below to get $\mathcal{M}^n$ and τ_n and thereafter $U^n, \Theta^n \in \mathbf{V}^n$ computed according to the following prescription*, for any $\varphi \in \mathbf{V}_0^n$,

$$\Theta^n = I^n\beta(U^n), \quad \Theta^n - I^n g_D^n \in \mathbf{V}_0^n, \quad \bar{U}^{n-1} := U^{n-1}(\bar{x}^{n-1}),$$

$$\frac{1}{\tau_n}\langle U^n - I^n\bar{U}^{n-1}, \varphi\rangle^n + \langle \nabla\Theta^n, \nabla\varphi\rangle + \langle\langle p^n(\Theta^n - \theta_{\text{ext}}^n), \varphi\rangle\rangle_{\Gamma_N}^n = 0,$$

where $\bar{x}^{n-1} = x - v^{n-1}\tau_n e_z$.

In view of the constitutive relation $\Theta^n = I^n\beta(U^n)$ being enforced only at the nodes, and the use of mass lumping, the discrete problem yields a monotone operator in $\mathbf{R}^{K^n}$ which is easy to implement and solve by nonlinear SOR [16], for example.

2.2 A Posteriori Error Analysis

The a posteriori error analysis depends on the parabolic duality argument which we now discuss. We consider a linear backward parabolic problem in non-divergence form, which can be viewed as the adjoint formal derivative of (2.1). For any $U \in BV(0,T;L^2(\Omega))$, we define

$$b(x,t) = \begin{cases} \dfrac{\beta(u)-\beta(U)}{u-U} & \text{if } u \neq U, \\ \beta_1 & \text{otherwise}. \end{cases} \tag{2.7}$$

We assume that $\beta(s) = 0$ for $s \in [0,\lambda]$ and $0 < \beta_1 \le \beta'(s) \le \beta_2$ for a.e. $s \in \mathbf{R}\backslash[0,\lambda]$; $\lambda > 0$ is the latent heat. Thus $0 \le b(x,t) \le \beta_2$, for a.e. $(x,t) \in Q_T$. Let $b_\delta \in C^2(\bar{Q}_T)$ be a regularization of b satisfying

$$b_\delta \ge \delta > 0, \quad 0 \le b_\delta - b \le \delta \quad \text{a.e. in } Q_T, \tag{2.8}$$

where $0 < \delta \le 1$ is a parameter to be chosen later. For arbitrary $t_* \in (0,T]$ and $\chi \in L^\infty(Q_T)$, let ψ be the solution of the following linear backward parabolic problem

$$\partial_t\psi + v(t)\partial_z\psi + b_\delta\Delta\psi = -b^{1/2}\chi \quad \text{in} \quad \Omega\times(0,t_*), \tag{2.9}$$

$$\psi = 0 \quad \text{on} \quad \Gamma_D\times(0,t_*), \tag{2.10}$$

$$\partial_\nu\psi + p\psi = 0 \quad \text{on} \quad \Gamma_N\times(0,t_*), \tag{2.11}$$

$$\psi(x,t_*) = 0 \quad \text{in} \quad \Omega. \tag{2.12}$$

We set $Q_* = \Omega\times(0,t_*)$. Existence of a unique solution $\psi \in W_q^{2,1}(Q_*)$ for any $q \ge 2$ of (2.9)-(2.12) follows from the theory of nonlinear strictly parabolic problems. Note that we impose a Dirichlet outflow boundary condition on Γ_0, which yields a boundary layer for ψ.

The key point in deriving sharp a posteriori error estimate now is to derive sharp stability estimates for this dual problem. We start with a simple, but essential, non-degeneracy property first proved in [7, Lemma 3.2].

Lemma 2.1. *Let $\xi, \rho \in \mathbf{R}$ satisfy $|\beta(\xi)| \ge \rho > 0$. Then we have*

$$|\xi - \eta| \le \Big(\frac{1}{\beta_1} + \frac{\lambda}{\rho}\Big)|\beta(\xi) - \beta(\eta)|, \quad \forall\, \eta \in \mathbf{R}.$$

Now we introduce the uniquess condition: $\exists\, \varepsilon_0, \rho_0 > 0$ such that $\theta \geq \rho_0$ a.e. in $\Gamma \times [0, \varepsilon_0] \times [0, T]$. Under this condition we know from Lemma 2.1 that there exists $r > 0$ depending on $= \rho_0$ such that

$$b(x,t) \geq r \ \text{ in } \Gamma \times \big([0,\varepsilon_0] \cup [L-\varepsilon_1, L]\big) \times [0,T].$$

We set $A = \|\chi\|_{L^\infty(Q_*)}$ and assume $0 < v_0 V \leq v(t) \leq V$ for $t \in [0,T]$ and $|v'(t)| \leq v_1 V$ a.e. $t \in [0,T]$, with $v_0, v_1 > 0$ constants. We also let $V \geq 1$.

Lemma 2.2. *The following a priori bound is valid for all $x \in \bar{\Omega}$ and $0 \leq t \leq t_* \leq T$*

$$|\psi(x,t)| \leq \frac{\beta_2^{1/2} L}{v_0 V} \|\chi\|_{L^\infty(Q_*)}.$$

Proof. We consider the barrier function $\Lambda(z) = (\beta_2^{1/2} A / v_0 V)(L - z)$ for $0 \leq z \leq L$. ◇

Lemma 2.3. *There exists $C > 0$ independent of V and t_* such that for all $0 \leq t_* \leq T$*

$$|\partial_\nu \psi| \leq C \|\chi\|_{L^\infty(Q_*)} \quad on\ \Gamma_0 \times (0, t_*).$$

Proof. We consider the barrier function

$$\sigma(z) = k\Big(1 - e^{-Vz/r}\Big) - \Big(\frac{\beta_2^{1/2} A}{v_0 V}\Big) z, \quad 0 \leq z \leq \varepsilon_0,$$

with $k = \beta_2^{1/2} A(L + \varepsilon_0)/v_0 V(1 - e^{-V\varepsilon_0/r})$. ◇

Based on above two estimates, we can prove the following stability estimates required in our a posteriori error analysis.

Lemma 2.4. *There exists $C > 0$ independent of V and t_* such that for all $0 \leq t_* \leq T$*

$$\max_{0 \leq t \leq t_*} \|\nabla\psi(\cdot,t)\|^2_{L^2(\Omega)} + \int_0^{t_*}\int_\Omega b_\delta |\Delta\psi|^2 dx dt \leq C V t_* \|\chi\|^2_{L^\infty(Q_*)},$$

$$\int_0^{t_*}\int_\Omega \Big(|\partial_t\psi + v(t)\partial_z\psi|^2 + \delta|D^2\psi|\Big) dx dt \leq C V t_* \|\chi\|^2_{L^\infty(Q_*)}.$$

With these stability estimates, we can derive the following a posteriori error estimate which depends mildly on the casting velocity V.

Theorem 2.1. *There exists a constant $C > 0$ independent of V and t^m such that the following a posteriori error estimate holds for all $t^m \in [0,T]$.*

$$\int_0^{t^m} \|\beta(u) - \beta(U)\|_{L^1(\Omega)} dt \leq C (V t^m)^{1/2} \Big(\mathcal{E}_0 + \sum_{i=5}^{10} \mathcal{E}_i + \Big(\Lambda_m \sum_{i=1}^{4} \mathcal{E}_i\Big)^{1/2}\Big),$$

where

$$\Lambda_m = \Big(\sum_{n=1}^{m} \tau_n\big(1 + \lambda|\Omega| + \| \Theta^n \|^2_{L^2(\Omega)}\big)\Big)^{1/2}$$

and the error indicators $\mathcal{E}_i$ depending only the discrete solutions U^n, Θ^n and the known data.

The detailed proof of this theorem as well as extensive numerical experiments can be found in [9].

3. Reliable and Efficient a Posteriori Error Estimate for a Linear Parabolic Problem

Let Ω be a polyhedron domain in $\mathbf{R}^d (d = 1, 2, 3)$, $\Gamma = \partial\Omega$ and $T > 0$, In this section we consider the following linear parabolic equation:

$$\begin{aligned} &\frac{\partial u}{\partial t} - (a(x)\nabla u) = f \quad \text{in } \Omega \times (0, T), \\ &u = 0 \quad \text{on } \Gamma \times (0, T), \quad u(x, 0) = u_0(x) \quad \text{in } \Omega, \end{aligned} \tag{3.1}$$

where $f \in L^2(0, T; L^2(\Omega))$, $u_0 \in L^2(\Omega)$ and $a(x)$ is a piecewise constant function. Let τ_n be the n-th time-step size and $t^n = \sum_{i=1}^n \tau_i$. Denote by $\mathcal{M}^n$ a uniformly regular partition of Ω into simplices such that $a(x)$ is a constant on each element $K \in \mathcal{M}^n$. We use refinement/coarsening procedures based on the bisection algorithm, which lead to compatible consecutive meshes whose minimum angles are bounded uniformly away from zero. Let V^n indicate the usual space of conforming linear finite elements over $\mathcal{M}^n$ and $V_0^n = V^n \cap H_0^1(\Omega)$. In this section, we consider the following simple fully discrete finite element scheme for solving (3.1): For $n = 1, 2, \cdots$, find $U_h^n \in V_0^n$ such that

$$\Big\langle \frac{U_h^n - U_h^{n-1}}{\tau_n}, v \Big\rangle + \langle a\nabla U_h^n, \nabla v\rangle = \langle \bar{f}^n, v\rangle \quad \forall v \in V_0^n, \tag{3.2}$$

where $\langle\cdot,\cdot\rangle$ stands for the inner product on $L^2(\Omega)$, and

$$\bar{f}^n = \frac{1}{\tau_n}\int_{t^{n-1}}^{t^n} f(x, t)\, dt.$$

Let $U^n \in H_0^1(\Omega)$ be the solution of the following semi-discrete problem

$$\Big\langle \frac{U^n - U^{n-1}}{\tau_n}, \varphi \Big\rangle + \langle a\nabla U^n, \nabla\varphi\rangle = \langle \bar{f}^n, \varphi\rangle \quad \forall\varphi \in H_0^1(\Omega). \tag{3.3}$$

We observe that by modifying the time-step size τ_n, we are essentially controlling the error between $u^n = u(x, t^n)$ and U^n, not between u^n and U_h^n.

Moreover, let $U_*^n \in H_0^1(\Omega)$ be the solution of the following auxiliary problem

$$\left\langle \frac{U_*^n - U_h^{n-1}}{\tau_n}, \varphi \right\rangle + \langle a\nabla U_*^n, \nabla\varphi\rangle = \langle \bar{f}^n, \varphi\rangle \quad \forall \varphi \in H_0^1(\Omega), \tag{3.4}$$

then we also observe that for fixed time-step size τ_n, by adapting the mesh $\mathcal{M}^n$ we are essentially controlling the error between U_h^n and U_*^n, not between U_h^n and U^n (or the real solution u). These two observations play an important role in the subsequent analysis to prove the local lower bound for the space a posteriori error estimator and in the development of a convergent refinement/coarsening strategy.

We define the interior residual

$$R^n := \bar{f}^n - \frac{U_h^n - U_h^{n-1}}{\tau_n},$$

along with the jump residual across $e \in \mathcal{B}^n$

$$J_e^n := [\![a\nabla U_h^n]\!]_e \cdot \nu_e = (a\nabla U_h^n|_{K_1} - a\nabla U_h^n|_{K_2}) \cdot \nu_e, \quad e = \partial K_1 \cap \partial K_2,$$

using the convention that the unit normal vector ν_e to e points from K_2 to K_1. We observe that integration by parts implies

$$\langle a\nabla U_h^n, \nabla\varphi\rangle = -\sum_{e\in\mathcal{B}^n} \int_e J_e^n \varphi ds \quad \forall \varphi \in H_0^1(\Omega).$$

Theorem 3.1 (Chen and Jia [6]). *For any integer $1 \le m \le N$, there exists a constant $C > 0$ depending only on the minimum angle of meshes $\mathcal{M}^n, n = 1, 2, \cdots, m$, and the coefficient $a(x)$ such that the following a posteriori error estimate holds*

$$\begin{aligned} &\frac{1}{2}\| u^m - U_h^m \|_{L^2(\Omega)}^2 + \sum_{n=1}^{m} \int_{t^{n-1}}^{t^n} |\!|\!| u - U_h^n |\!|\!|_\Omega^2 dt \\ \le\ & \| u_0 - U_h^0 \|_{L^2(\Omega)}^2 + \sum_{n=1}^{m} \tau_n \eta_{\text{time}}^n + C \sum_{n=1}^{m} \tau_n \eta_{\text{space}}^n \\ +\ & 2\Big(\sum_{n=1}^{m} \int_{t^{n-1}}^{t^n} \| f - \bar{f}^n \|_{L^2(\Omega)} dt\Big)^2, \end{aligned}$$

where the time error estimator η_{time}^n and space error estimator η_{space}^n are given by

$$\eta_{\text{time}}^n = \frac{1}{3} |\!|\!| U_h^n - U_h^{n-1} |\!|\!|_\Omega^2, \quad \eta_{\text{space}}^n = \sum_{e\in\mathcal{B}^n} \eta_e^n$$

with the local error estimator η_e^n defined as

$$\eta_e^n = \frac{1}{2} \sum_{K\in\Omega_e} h_K^2 \| R^n \|_{L^2(K)}^2 + h_e \| J_e^n \|_{L^2(e)}^2.$$

Here Ω_e is the collection of two elements sharing the common side $e \in \mathcal{B}^n$.

Let $P^n : L^2(\Omega) \to V_0^n$ be the L^2 projection operator. We remark that the coarsening errors involving $U_h^{n-1} - P^n U_h^{n-1}$ which appeared in previous studies [9], [12], [16] are not present in our a posteriori error estimates. They are hidden in the space error term $\| h_n R^n \|_{L^2(\Omega)}^2$ and time error term $|||U_h^n - U_h^{n-1}|||_\Omega^2$. Here h_n is the piecewise constant function which equals to h_K on each $K \in \mathcal{M}^n$.

Now we consider the following local lower bound of the space error estimator which ensures over-refinement will not occur for the refinement strategy based on our space error estimator. For any $K \in \mathcal{M}^n$ and $\varphi \in L^2(\Omega)$, we define $P_K\varphi = \frac{1}{|K|}\int_K \varphi dx$, the average of φ over K.

Theorem 3.2 (Chen and Jia [6]). *Suppose that there exists a constant $\hat{C} > 0$ such that $h_K^2 \le \hat{C}\tau_n$ for any $K \in \mathcal{M}^n$. Then there exist constants $C_2, C_3 > 0$ depending only on $\hat{C}$, the minimum angle of $\mathcal{M}^n$ and the coefficient $a(x)$ such that for any $e \in \mathcal{B}^n$, the following estimate holds*

$$\begin{aligned} \eta_e^n &\le C_2 \sum_{K\in\Omega_e} \left(\frac{1}{\tau_n} \| U_*^n - U_h^n \|_{L^2(K)}^2 + |||U_*^n - U_h^n|||_K^2 \right) \\ &+ C_3 \sum_{K\in\Omega_e} h_K^2 \| R^n - P_K R^n \|_{L^2(K)}^2. \end{aligned}$$

Based on the local error indicators, the usual adaptive algorithm solving the parabolic problem (3.1) at the n-th time step reads as follows

Solve $\to$ Estimate $\to$ Refine/Coarsen.

Here the refinement/coarsening procedure includes both the mesh and time-step size modifications. There are several possibilities proposed in the literature on the strategies of the adaptive control of meshes and time-step sizes. We refer to Schmidt and Siebert [19] for a nice review on various adaptive algorithms and their implementation details. In this paper, at each time step n, we propose the following algorithm to modify the time-step size τ_n and mesh $\mathcal{M}^n$ starting from the initial time-step size $\tau_{n,0} = \tau_{n-1}$ and initial mesh $\mathcal{M}^{n,0} = \mathcal{M}^{n-1}$:

1 Refine the time-step size $\tau_{n,0}$ and the mesh $\mathcal{M}^{n,0}$ so that for the solution $U_h^n \in V_0^n$ of (3.2) on the final mesh $\mathcal{M}^n$ with final time-step

size τ_n, the associated space and time error estimators are less than the prescribed tolerances respectively;

2 Coarsen the mesh $\mathcal{M}^n$ so that for the solution U_H^n on the coarsened mesh $\mathcal{M}_H^n$, the associated coarsening error estimator less than some prescribed tolerance;

3 Enlarge the initial time-step size $\tau_{n+1,0}$ for next time step if the current time error estimator is much less than the tolerance.

In [6] we extend the convergence analysis of adaptive finite element methods developed for linear elliptic problems in Dörfler [11] and Morin, Nochetto and Siebert [15] to prove the iteration in Step 1 of above algorithm terminates in finite number of steps. The main difficulty now is the treatment of the so-called data oscillation of the residual $\bar{f}^n - (U_h^n - U_h^{n-1})/\tau_n$ which changes at each refinement procedure.

The choice of the coarsening error estimator and coarsening strategy in Step 2 is a subtle issue. The error incurred due to the over-coarsening can only be reduced by re-refining the coarsened mesh. Thus over-coarsening leads to unnecessary solution of discrete problems, that is usually expensive and undesirable. Our coarsening error indicator is based on a direct control of the error between the coarsened discrete solution and the limit solution U_*^n of (3.4). More precisely, let $\mathcal{M}_H^n$ be a coarsening of $\mathcal{M}^n$, and U_H^n, U_h^n be the corresponding solutions of (3.2) with fixed time-step size τ_n. Since $V_0^{n,H} \subset V_0^n$, we deduce by Galerkin orthogonality

$$\begin{aligned} &\frac{1}{\tau_n}\| U_*^n - U_H^n \|_{L^2(\Omega)}^2 + |\!|\!| U_*^n - U_H^n |\!|\!|_\Omega^2 \\ \le\ &\frac{1}{\tau_n}\| U_*^n - U_h^n \|_{L^2(\Omega)}^2 + |\!|\!| U_*^n - U_h^n |\!|\!|_\Omega^2 \\ +\ &\frac{1}{\tau_n}\| U_h^n - I_H^n U_h^n \|_{L^2(\Omega)}^2 + |\!|\!| U_h^n - I_H^n U_h^n |\!|\!|_\Omega^2, \end{aligned}$$

where $|\!|\!|\varphi|\!|\!|_\Omega = \langle a\nabla\varphi, \nabla\varphi\rangle^{1/2}$ is the energy norm of $\varphi \in H^1(\Omega)$, and $I_H^n : C(\bar{\Omega}) \to V^{n,H}$ is the usual linear finite element interpolant over $\mathcal{M}_H^n$. Our coarsening error indicator is then is given by

$$\eta_{\text{coarse}}^n := \frac{1}{\tau_n}\| U_h^n - I_H^n U_h^n \|_{L^2(\Omega)}^2 + |\!|\!| U_h^n - I_H^n U_h^n |\!|\!|_\Omega^2.$$

which does not depend on the coarsened solution U_H^n. It is shown in [6] that this direct error control leads to only one coarsening step. The implementation issue of the coarsening startegy is discussed in detail in [6].

Acknowledgments

This paper is based on joint works with Ricardo H. Nochetto, Alfred Schmidt and Feng Jia. The author would like to thank them for their contribution and efforts.

References

[1] M. Bieterman and I. Babuška, *The finite element method for parabolic equations: (I) a posteriori estimation,* Numer. Math. 40 (1982), 339-371.

[2] M. Bieterman and I. Babuška, *The finite element method for parabolic equations: (II) a posteriori estimation and adaptive approach,* Numer. Math. 40 (1982), 373-406.

[3] I. Babuška and C. Rheinboldt, *Error estimates for adaptive finite element computations,* SIAM J. Numer. Anal. 15 (1978), 736-754.

[4] Z. Chen and S. Dai, *Adaptive Galerkin methods with error control for a dynamical Ginzburg-Landau model in superconductivity,* SIAM J. Numer. Anal. 38 (2001), 1961-1985.

[5] Z. Chen and S. Dai, *On the efficiency of adaptive finite element methods for elliptic problems with discontinuous coefficients,* (2001), submitted.

[6] Z. Chen and F. Jia, *A convergent adaptive finite element algorithm with reliable and efficient error control for linear parabolic problems,* (2001), submitted.

[7] Z. Chen and L. Jiang, *Approximation of a two-phase continuous casting Stefan problem,* J. Partial Differential Equations 11 (1998), 59-72.

[8] Z. Chen and R.H. Nochetto, *Residule type a posteriori error estimates for elliptic obstacle problems,* Numer. Math. 84 (2000), 527-548.

[9] Z. Chen, R.H. Nochetto and A. Schmidt, *A characteristic Galerkin method with adaptive error control for continuous casting problem,* Comput. Methods Appl. Mech. Engrg. 189 (2000), 249-276.

[10] Z. Chen, R.H. Nochetto and A. Schmidt, *Error control and adaptivity for a phase relaxation model,* Math. Model. Numer. Anal. 34 (2000), 775-797.

[11] W. Dörfler, *A convergent adaptive algorithm for Possion's equations,* SIAM J. Numer. Anal. 33 (1996), 1106-1124.

[12] K. Eriksson and C. Johnson, *Adaptive finite element methods for parabolic problems I: A linear model problem,* SIAM J. Numer. Anal. 28 (1991), 43-77.

[13] K. Eriksson and C. Johnson, *Adaptive finite element methods for parabolic problems IV: Nonlinear problems,* SIAM J. Numer. Anal. 32 (1995), pp. 1729- 1749.

[14] P.K. Moore, *A posteriori error estimation with finite element semi- and fully discrete methods for nonlinear parabolic equations in one space dimension,* SIAM J. Numer. Anal. 31 (1994), 149-169.

[15] P. Morin, R.H. Nochetto and K.G. Siebert, *Data oscillation and convergence of adaptive FEM,* SIAM J. Numer. Anal. 38 (2000), 466-488.

[16] R.H. Nochetto, A. Schmidt and C. Verdi, *A posteriori error estimation and adaptivity for degenerate parabolic problems,* Math. Comp. 69 (2000), 1-24.

[17] J.F. Rodrigues and F. Yi, *On a two-phase continuous casting Stefan problem with nonlinear flux,* Euro. J. Appl. Math. 1 (1990), pp. 259-278.

[18] J. Rulla, *Weak solutions to Stefan problems with prescribed convection*, SIAM J. Math. Anal. 18 (1987), pp. 1784-1800.

[19] A. Schmidt and K.G. Siebert, *ALBERT: An adaptive hierarchical finite element toolbox*, IAM, University of Freiburg, 2000. http://www.mathematik.uni-freiburg.de/IAM/Research/projectsdz/albert.

[20] R. Verfürth, *A posteriori error estimates for nonlinear problems:* $L^{p,r}(0,T;W^{1,\rho}(\Omega))$ *Finite element discretization of parabolic equations*, Numerical Methods in Partial Differential Equations 14 (1998), 487-518.

NUMERICAL COMPUTATION OF QUANTIZED VORTICES IN THE BOSE-EINSTEIN CONDENSATE

Qiang Du*
Lab for Scientific and Engineering Computation, Academia Sinica, Beijing, China
Hong Kong University of Science & Technology, Hong Kong, China
qdu@lsec.cc.ac.cn

Abstract The theoretical analysis of many recent experimental work on a single component Bose-Einstein condensate has been based on the mean-field Gross-Pitaevskii equations. We discuss a few algorithms for solving the Gross-Pitaevskii equations and we use them to compute the quantized vortices in Bose-Einstein condensate.

Keywords: vortices, Bose-Einstein condensate, Gross-Pitaevskii equation, numerical schemes, finite element, finite difference, operator-splitting, symplectic and multi-symplectic integration

1. Introduction

Bose-Einstein condensation was first predicted in 1924 by Bose and Einstein. However, the experimental confirmation of Bose-Einstein condensation in atomic gases was only achieved recently in 1995. It was considered as such an important development that the centennial Nobel Prize for Physics has been awarded to the researchers who created this so-called fifth state of matter in the laboratory, Eric Cornell, Wolfgang Ketterle, and Carl Wieman. These researchers were also cited for their early fundamental studies of the properties of the condensates.

One of the early questions asked of BECs were whether they were superfluids. A particularly interesting signature of superfluids is the ability to support quantized circulation. The existence of quantized vortices in the Bose-Einstein condensate has been well documented, see for instance [7, 8, 12, 22, 25, 26, 34, 31, 32, 38]. The vortex cores in the Bose-Einstein condensate are about

*Rearch supported in part by the state major basic research project G19990128 and a grant from FR-HK joint research scheme.

a thousand times larger than those in the superfluid ^{4}He so that they can be examined more closely: "*two of the most interesting things in a vortex's life are its birth and death. Now we can look at both*", claimed by E. Cornell. In the last few years, there has been a great surge in the studies of quantized vortices in the Bose-Einstein condensate both experimentally and theoretically.

Theoretical studies of vortices in the BEC experiments have often been made in the framework of the nonlinear Gross-Pitaevskii equation, well known for superfluids, but which provides a very good description of Bose-Einstein condensates: it is assumed that the N particles of the gas are condensed in the same state for which the wave function ϕ minimizes the Gross-Pitaevskii energy.

The quantized vortices may be observed in a Bose-Einstein condensate with either optical traps or magnetic traps. For simplicity, we focus on the case of a condensate being placed in a rotating magnetic trap, though much of our discussion can be applied to the case of an optical trap as well. By introducing a rotating frame at the angular velocity $\tilde{\Omega} = \tilde{\Omega}\mathbf{e}_z$, the trapping potential becomes time independent, and the wave function ϕ minimizes the energy

$$\mathcal{E}_{3D}(\phi) = \int \frac{\hbar^2}{2m}|\nabla\phi|^2 + \frac{m}{2}\sum_{\alpha}\omega_\alpha^2 r_\alpha^2|\phi|^2 + \frac{N}{2}g_{3D}|\phi|^4 - \hbar\tilde{\Omega}\cdot(i\phi, \nabla\phi\times\mathbf{x}), \qquad (1.1)$$

under the constraint

$$\int |\phi|^2 = 1.$$

Here, for any complex quantities u, v and their complex conjugates $\bar{u}$, $\bar{v}$, $(u, v) = (u\bar{v} + \bar{u}v)/2$. The terms in the energy correspond to the kinetic energy, the trapping potential energy, the interaction energy and the inertial due to the change of frame.

In the recent works of [3, 4], a rigorous mathematical framework for the study of the energy $\mathcal{E}_{3D}$ and its two dimensional version has been established in the Thomas Fermi limit. The asymptotic energy expansion and limiting behavior of its minimizers are characterized. The study was based on the observation that the Gross-Pitaevskii energy has a striking similarity with the high-kappa, high-field limit of the Ginzburg-Landau free energy used in the modeling of superconductors [10, 16, 19].

Another consequence of this close resemblance is the fact that we were able to construct numerical integration codes for the Gross-Pitaevskii equations by modifying an extensive battery of codes developed over the years for the numerical simulation of vortex dynamics in Ginzburg-Landau models [17, 14, 18, 19].

Similar to [3], we introduce the characteristic length $d = (\hbar/m\omega_x)^{1/2}$ and re-scale the distance by $R = d/\sqrt{\varepsilon}$ where $\varepsilon^2 = \hbar^2/(2Ngm)$. Define the new variable $u(\mathbf{r}) = R\psi(\mathbf{x})$ where $\mathbf{x} = R\mathbf{r}$. Also let $\omega = \omega_x, \omega_y = \lambda_y\omega, \omega_z = \lambda_z\omega$ with $0 \leq \lambda_y, \lambda_z \leq 1$ and set $\varnothing = \tilde{\Omega}/\varepsilon\omega$. The nondimensionalized energy can then be rewritten as:

$$E_0(u) = \int \frac{1}{2}|\nabla u|^2 + \frac{1}{2\varepsilon^2}(x^2 + \lambda_y^2 y^2 + \lambda_z^2 z^2)|u|^2 + \frac{1}{4\varepsilon^2}|u|^4 + \Omega\cdot(iu, \nabla u \times \mathbf{r})\,.$$

Due to the constraint that the integral of $|u|^2$ must equal to 1, it is equivalent to minimize

$$E(u) = \int |\nabla u|^2 + 2\Omega \cdot (iu, \nabla u \times \mathbf{r}) + \frac{1}{2\varepsilon^2}|u|^4 - \frac{1}{\varepsilon^2}a(\mathbf{r})|u|^2 \qquad (1.2)$$

where $a(\mathbf{r}) = \alpha - (x^2 + \lambda_y^2 y^2 + \lambda_z^2 z^2)$ for some constant α which is chosen so that the integral of $a(\mathbf{r})$ on the ellipsoid $\mathcal{D} = \{a > 0\} = \{x^2 + \lambda_y^2 y^2 + \lambda_z^2 z^2 < \alpha\}$ is equal to 1.

Corresponding to the physical experiments conducted recently, we may take $\Omega = (0, 0, \varnothing)^T$. The form of the energy (1.0) is close to the Ginzburg-Landau free energy functional for superconductors [17] in the high-kappa, high field limit [10, 16]. Indeed, the energy in (1.2) can be rewritten as

$$E(u) = \int_{\mathcal{D}} \left\{ |(\nabla - i\mathbf{A})u|^2 + \frac{1}{2\varepsilon^2}(a_\varepsilon(\mathbf{r}) - |u|^2)^2 \right\} + c_\varepsilon \qquad (1.3)$$

where $a_\varepsilon(\mathbf{r}) = a(\mathbf{r}) - \varepsilon^2\varnothing^2 r^2$, $\mathbf{A}$ is a vector potential defined by

$$\mathbf{A} = \begin{pmatrix} y \\ -x \\ 0 \end{pmatrix} \varnothing\,,$$

and the constant c_ε is given by

$$c_\varepsilon = \int_{\mathcal{D}} \{\frac{1}{2\varepsilon^2}(a^2(\mathbf{r}) - a_\varepsilon^2(\mathbf{r}))\}.$$

The vector $\mathbf{A}$ may be viewed as a given magnetic vector potential so that the magnetic field plays the role of the angular velocity of the rotation since curl $\mathbf{A} = 2\varnothing$. The trapping potential in the BEC may be linked to the inhomogeneities modelled in the Ginzburg-Landau model to study the pinning mechanism [19].

The unique feature of the Gross-Pitaevskii model is the addition of the constraint on the L^2 norm of u, which would eliminate the trivial state $u = 0$ from consideration. This complication has been worked around in the theoretical analysis developed by [3, 4] in the Thomas-Fermi limit, largely due to the fact that

$$\int_{\mathcal{D}} a(\mathbf{r})d\mathbf{r} = 1.$$

An interesting signature of the BEC that has attracted a lot of attentions is the nucleation of vortices at high angular velocity. The energy minimizers and critical velocities of vortex nucleation have been studied analytically in [3, 4] in the Thomas-Fermi limit $\varepsilon \longrightarrow 0$. The main ingredient of the analysis lies in the decoupling of the energy into three sources: a part coming from the state without vortices, another part from contribution of individual vortices and an additional part produced due to the rotation. In many of the experiments, ε ranges in $10^{-3} \sim 10^{-2}$. We have carried out numerical simulations that produced detailed bifurcation diagrams which confirmed the theoretical analysis for ε in these ranges.

We now present a number of numerical algorithms that are useful in the study of energy minimizers of the Gross-Pitaevskii energy as well as their dynamical properties.

2. The model equations

We begin by describe the differential equations associated with the minimizers of the Gross-Pitaevskii energy as well as the time-dependent G-P equations.

2.1 The steady state equation

The minimizers of the G-P energy satisfies the equation:

$$-(\nabla - i\mathbf{A})^2 u + \frac{1}{\varepsilon^2}|u|^2 u - \frac{a_\varepsilon(\mathbf{r})}{\varepsilon^2} u = \mu_\varepsilon(u) u \ \text{ in } \mathcal{D} \tag{2.1}$$

where $\mu_\varepsilon(u)$ is a constant multiplier corresponding to the constraint

$$\int_{\mathcal{D}} |u|^2 = 1\ .$$

One may impose the boundary condition $u = 0$ since no particle is allowed to go outside the trap. One may also elect to apply a natural variational boundary condition that will alter very little the behavior of the solution if the computation domain is chosen to be slightly larger than $\mathcal{D}$.

2.2 The time-dependent Gross-Pitaevskii equation

The dynamics of the BEC may be modelled by the time-dependent G-P equation:

$$i\frac{\partial u}{\partial t} - (\nabla - i\mathbf{A})^2 u + \frac{1}{\varepsilon^2}|u|^2 u - \frac{a_\varepsilon(\mathbf{r})}{\varepsilon^2} u = 0 \tag{2.2}$$

in $\mathcal{D}$ with initial condition $u(\mathbf{r}, 0) = u_0(\mathbf{r})$ in $\mathcal{D}$ and boundary condition $u = 0$ or $(\nabla - i\mathbf{A})\, u \cdot n = 0$ on $\partial\mathcal{D}$. The constraint $\int_{\mathcal{D}} |u|^2 = 1$ is automatically preserved at all time and the energy remains constant as well.

2.3 Evolution in the imaginary time

To numerically compute the minimizers of (1.2), we consider the time-dependent equation in the imaginary time:

$$\frac{\partial u}{\partial t} - (\nabla - i\mathbf{A})^2 u + \frac{1}{\varepsilon^2}|u|^2 u - \frac{a_\varepsilon(\mathbf{r})}{\varepsilon^2} u = \mu_\varepsilon(u) u \qquad (2.3)$$

in $\mathcal{D}$ with initial condition $u(\mathbf{r}, 0) = u_0(\mathbf{r})$ in $\mathcal{D}$ and boundary condition $u = 0$ on $\partial\mathcal{D}$. $\mu_\varepsilon(u)$ again denotes the Lagrange multiplier. If u_0 satisfies the constraint $\|u_0\| = 1$, then we may determine $\mu_\varepsilon(u)$ by

$$\mu_\varepsilon(u) = \int_{\mathcal{D}} \left\{ |(\nabla - i\mathbf{A})\, u|^2 + \frac{1}{\varepsilon^2}|u|^4 - \frac{a_\varepsilon(\mathbf{r})}{\varepsilon^2}|u|^2 \right\} d\mathcal{D}\,,$$

In [3], discussion on the existence and uniqueness of the weak solution has been given. The long time asymptotic behavior may also be examined using techniques similar to that in [30].

2.4 The Thomas-Fermi regime

In the experimental setting, ε often takes small value, thus, analytical studies of solutions of the Gross-Pitaeskii equation may be made in the $\varepsilon \to 0$ limit, that is, in the so-called Thomas-Fermi regime.

As an illustration, for the two dimensional version of the energy, a renormalized energy has been derived in [3] based on the asymptotic behavior as $\varepsilon \to 0$. It was shown that the energetically favorable locations of the vortices $\{(x_i, y_i)\}$ of a n-vortex minimizer of the energy is determined by minimizing

$$\sum_{i \neq j} \ln\left(|x_i - x_j|^2 + \frac{|y_i - y_j|^2}{\lambda^2} \right) - \alpha \sum_i (x_i^2 + y_i^2)\,.$$

where α is a given positive constant and λ is the aspect ratio of two dimensional harmonic magnetic trap.

Away from the Thomas-Fermi regime, most of the existing studies rely on some results of numerical simulations. It is thus of great interests to discuss various types of numerical schemes applicable to the solution of the Gross-Pitaevskii equations.

3. Numerical schemes

There are various ways to solve the time-dependent Gross-Pitaevskii equations, see for example [8] or [22]. Here, we outline several possible numerical schemes.

To solve the steady state equation and to integrate the time dependent equation in imaginary time, we take the advantage of the similarity with the solution of the high-kappa high-field time-dependent Ginzburg-Landau equations [16], and adapt a code developed in [16, 17, 18, 19]. For integration in real time, the Hamiltonian structure of the G-P equation can be utilized.

3.1 Spatial discretization

For spatial discretization, there are several possibilities, including finite element approximation [17, 19], gauge invariant difference approximation [15] and finite volume approximations [20].

For instance, for the gauge invariant difference approximation, let h be the spatial mesh size, j, k be the grid indices in the $x - y$ plane, one may introduce the **link variables** $\Phi_{j+1/2,k} = exp(-i\mathbf{A}^1_{j+1/2,k}h)$, $\Phi_{j,k+1/2} = exp(-i\mathbf{A}^2_{j+1/2,k}h)$. Then, we may use the approximation

$$
\begin{aligned}
(\nabla - i\mathbf{A})^2 u_{jkl} \approx\; & \bar{\Phi}_{j-1/2,j}\frac{u_{(j-1)kl} - u_{jkl}}{h^2} + \Phi_{j+1/2,k}\frac{u_{(j+1)kl} - u_{jkl}}{h^2} \\
& + \bar{\Phi}_{j,k-1/2}\frac{u_{j(k-1)l} - u_{jkl}}{h^2} + \Phi_{j+1/2,k}\frac{u_{j(k+1)l} - u_{jkl}}{h^2} \\
& + \frac{u_{jk(l+1)} + u_{jk(l-1)} - 2u_{jkl}}{h^2} .
\end{aligned}
$$

Such a spatial difference scheme conveniently preserves the symmetry of the resulting discrete operator. Based on this, one may easily get a semi-discrete in space scheme for both the steady state and the time-dependent G-P equation.

For unstructured triangular grids, a finite volume method can be obtained as a generalization of the above technique [20].

3.2 Solving the equation in imaginary time

For time-discretization, it is important to get asymptotically stable schemes for large time which in general requiring the use of implicit schemes with no limitations on the time step size.

Let $\{u_n\}$ be approximate solutions of $\{u(t_n)\}$ at discrete time $\{t_n\}$ with time-step $\Delta t_n = t_n - t_{n-1}$. In [3], two time-discretization schemes have been introduced:

A first order backward-Euler in time discretization:. Given u_{n-1}, we solve for u^*:

$$
\begin{aligned}
& \frac{u^* - u_{n-1}}{\Delta t_n} - (\nabla - i\mathbf{A})^2 u^* - \mu(u_{n-1})u^* \\
& \quad + \frac{1}{\varepsilon^2}|u^*|^2 u^* - \frac{1}{\varepsilon^2} a_\varepsilon u^* = 0 \qquad (3.1)
\end{aligned}
$$

Then, we apply the projection $u_n = u^*/\|u^*\|$. Both the backward Euler step and the projection step give only first order in time accuracy.

A norm-preserving, energy-decreasing second order scheme. For any u, v, let $f(u,v) = (|u|^2 + |v|^2)(u+v)/2$. Given u_{n-1}, we first solve for u^*:

$$\frac{2(u^* - u_{n-1})}{\Delta t_n} - (\nabla - i\mathbf{A})^2 u^* - \nu(u^*)u^* + \frac{1}{\varepsilon^2} f(2u^* - u_{n-1}, u_{n-1}) - \frac{1}{\varepsilon^2} a_\varepsilon u^* = 0 \tag{3.2}$$

where $\nu(u^*)$ is given by

$$\nu(u^*) \int_{\mathcal{D}} |u^*|^2 = \int_{\mathcal{D}} \left\{ |(\nabla - i\mathbf{A}) u^*|^2 \right\} + \int \left\{ \frac{1}{\varepsilon^2} f(2u^* - u_{n-1}, u_{n-1})\bar{u}^* - \frac{a_\varepsilon}{\varepsilon^2}|u^*|^2 \right\} .$$

Then, $u_n = 2u^* - u_{n-1}$.

During the discrete time evolution, the energy decreases while the norm is preserved. This discrete scheme is the second order in time and unconditionally stable. It captures some essential features of the continuous dynamic system, making it suitable for long time integration and for studies of meta-stabilities of the solutions.

3.3 Solving the equation in real time

When one is interested in the dynamics of vortices in real time, the time-dependent Gross-Pitaevskii equation needs to be solved efficiently and accurately in time. We now present two general approaches: one that uses time-splitting (operator splitting) techniques and one that preserves certain intrinsic properties of the equation.

3.4 Time-splitting scheme

Similar to the study on many evolutionary equations that include the time dependent Schrodinger equations, one can adopt a time or operator-splitting scheme that has been discussed by many authors [29]:

Given u_{n-1}, one may proceed by alternating solve the following subproblems:

1. For each position $\mathbf{r}$, solve the ODE:

$$iu_t = (a_\varepsilon(\mathbf{r}) - |u|^2)u.$$

2. For each (x,y), solve the linear equation in time and z :

$$iu_t = \partial_{zz} u \qquad \text{in} \qquad [t, t+\lambda_2] .$$

3. For each z, solve the linear equation in time and x, y :

$$iu_t = (\nabla - i\mathbf{A})^2 u \,, \quad \text{in} \quad [t, t+\lambda_3] .$$

Here, λ_i's may be viewed as fractional time-steps. The three subproblems may be solved alternatingly and repeatedly.

Note that an explicit solution of the ODE systems in the step 1 is given by

$$u(\mathbf{r}, t+\lambda_1) = u(\mathbf{r}, t) \exp\left(i\lambda_1(|u(\mathbf{r}, t)|^2 - a_\varepsilon(\mathbf{r}))/\varepsilon^2\right) .$$

Steps 2 and 3 can be further discretized using an Euler or other time integration schemes to preserve some properties of the original equation and to ensure better stability. For instance, a Crank-Nicolson scheme for steps 2 and 3 would preserve the L^2 norm of the solution which is a property enjoyed by the time-dependent G-P equation.

The differential operators involved in those two steps commute with each other, a key property to allow construction of high order approximation schemes. Due to the linearity of the problems involved in those two steps, fast solvers may be applied. We refer to [5, 28, 36] for further discussions on the operator splitting strategies as well as applications to time-dependent Schrodinger equations.

3.5 Symplectic and multi-symplectic scheme

The time-dependent Gross-Pitaevskii equation is a Hamiltonian system which enjoys both the symplectic and multi-symplectic properties. Thus, it is desirable to use discrete schemes that preserve the symplectic [11] and multi-symplectic structures [6, 11, 33].

Symplectic integrator. : one may use a standard practice to rewrite the time-dependent G-P equation as a Hamiltonian system. Let $u = p + iq$ where p, q are the real and imaginary part of u, and let $E(p, q)$ denote $E(u)$ as in the equation (1.2), then we have

$$\frac{d}{dt}\begin{pmatrix} p \\ q \end{pmatrix} = J \begin{pmatrix} \partial/\partial p \\ \partial/\partial q \end{pmatrix} E(p,q) \quad \text{where} \quad J = \begin{pmatrix} 0 & -1 \\ 1 & 0 \end{pmatrix} .$$

For $Z_t = J\nabla_Z E(Z)$, the following second order symplectic schemes may be applied [11]:

$$Z_n - Z_{n-1} = \Delta t J \nabla_Z E\left(\frac{Z_n + Z_{n-1}}{2}\right),$$

and a fourth order R-K symplectic integrator is given by

$$\begin{aligned} K_1 &= Z_{n-1} + \frac{\Delta t}{12} J\left(3\nabla_Z E(K_1) + (3 - 2\sqrt{3})E(K_2)\right) . \\ K_2 &= Z_{n-1} + \frac{\Delta t}{12} J\left((3 + 2\sqrt{3})\nabla_Z E(K_1) + 3E(K_2)\right) . \end{aligned}$$

with

$$Z_n - Z_{n-1} = \frac{\Delta t}{2} J\left(\nabla_Z E(K_1) + \nabla_Z E(K_2)\right) .$$

Multi-symplectic integrator. : More recently, multi-symplectic schemes for integrating Hamiltonian systems of nonlinear PDEs have received much attention [6, 11, 33]. The time-dependent G-P equation also possesses a multi-symplectic structure, namely, let $Z = (p, q, \mathbf{v}, \mathbf{w})$ with $\mathbf{v} = \nabla p - \mathbf{A}q$, $\mathbf{w} = \nabla q + \mathbf{A}p$), and

$$S(Z) = \frac{1}{2}(|\mathbf{v}|^2 + |\mathbf{w}|^2) + \mathbf{A}\cdot\mathbf{v}q - \mathbf{A}\cdot\mathbf{w}p + \frac{1}{4}(a(\mathbf{r}) - |p|^2 - |q|^2)^2) \, .$$

then, the G-P equation can be rewritten as a multi-symplectic Hamiltonian system:

$$M\frac{\partial}{\partial t}Z + K_1\frac{\partial}{\partial x}Z + K_2\frac{\partial}{\partial y}Z + K_3\frac{\partial}{\partial z}Z = \nabla_Z S(Z) \, , \tag{3.3}$$

where $\nabla_Z S(Z)$ denotes the gradient of the function $S = S(Z)$ with respect to the variable Z and

$$M = \begin{pmatrix} J & \mathbf{O} \\ \mathbf{O} & \mathbf{O}_{6\times 6} \end{pmatrix} .$$

$$K_1 = \begin{pmatrix} 0 & 0 & 1 & 0 & 0 & 0 & 0 & 0 \\ 0 & 0 & 0 & 0 & 0 & 1 & 0 & 0 \\ -1 & 0 & 0 & 0 & 1 & 0 & 0 & 0 \\ 0 & 0 & 0 & 0 & 0 & 0 & 0 & 0 \\ 0 & 0 & 0 & 0 & 0 & 0 & 0 & 0 \\ 0 & -1 & 0 & 0 & 0 & 0 & 0 & 0 \\ 0 & 0 & 0 & 0 & 0 & 0 & 0 & 0 \\ 0 & 0 & 0 & 0 & 0 & 0 & 0 & 0 \end{pmatrix},$$

$$K_2 = \begin{pmatrix} 0 & 0 & 0 & 1 & 0 & 0 & 0 & 0 \\ 0 & 0 & 0 & 0 & 0 & 0 & 1 & 0 \\ 0 & 0 & 0 & 0 & 0 & 0 & 0 & 0 \\ -1 & 0 & 0 & 0 & 0 & 0 & 0 & 0 \\ 0 & 0 & 0 & 0 & 0 & 0 & 0 & 0 \\ 0 & 0 & 0 & 0 & 0 & 0 & 0 & 0 \\ 0 & -1 & 0 & 0 & 0 & 0 & 0 & 0 \\ 0 & 0 & 0 & 0 & 0 & 0 & 0 & 0 \end{pmatrix},$$

and

$$K_3 = \begin{pmatrix} 0 & 0 & 0 & 0 & 1 & 0 & 0 & 0 \\ 0 & 0 & 0 & 0 & 0 & 0 & 0 & 1 \\ 0 & 0 & 0 & 0 & 0 & 0 & 0 & 0 \\ 0 & 0 & 0 & 0 & 0 & 0 & 0 & 0 \\ -1 & 0 & 0 & 0 & 0 & 0 & 0 & 0 \\ 0 & 0 & 0 & 0 & 0 & 0 & 0 & 0 \\ 0 & 0 & 0 & 0 & 0 & 0 & 0 & 0 \\ 0 & -1 & 0 & 0 & 0 & 0 & 0 & 0 \end{pmatrix}.$$

The system (3.3) has a multi-symplectic conservation law:

$$\frac{\partial}{\partial t}(-dp \wedge dq) + \nabla \cdot (dp \wedge d\mathrm{v} + dq \wedge d\mathrm{w}) = 0 \ ,$$

as well as the local energy conservation law:

$$\frac{\partial}{\partial t}\left[S - \frac{1}{2}\left(ZK_1 Z_x + ZK_2 Z_y + ZK_3 Z_z\right)\right] + \nabla \cdot Z \left(\sum_{j=1}^{3} K_j\right) Z_t = 0 \ .$$

Giving a uniform Cartesian grid, let Z_c denote the center average of the Z on the eight vertices, Z_{fx+} and Z_{fx-} be the averages of Z on the four vertices in each face in the x-direction. $Z_{fy\pm}$ and $Z_{fz\pm}$ follow similar convention. Let $Z^{n+1/2}$ be the averages of Z^n and Z^{n+1} at two consecutive time steps. Then based on the approach used in [27], the following difference scheme preserves the multi-symplectic property in the discrete sense:

$$M\frac{Z_c^{n+1} - Z_c^n}{\Delta t} + K_1\frac{Z_{fx+}^{n+1/2} - Z_{fx-}^{n+1/2}}{\Delta x} + K_2\frac{Z_{fy+}^{n+1/2} - Z_{fy-}^{n+1/2}}{\Delta y}$$
$$+K_3\frac{Z_{fz+}^{n+1/2} - Z_{fz-}^{n+1/2}}{\Delta y} = \nabla_Z S(Z_c^{n+1/2}) \ .$$

Developing efficient iterative schemes for the solution of the above difference approximation will be an immediate need to make the multi-symplectic integrator effective for multi-dimensional PDEs. For instance, it has been suggested in [11, 27] that if the nonlinear terms on the right hand side is completely lagged behind, the resulting linear systems at each iteration would share the same coefficient matrix for all iteration and all time steps. The coefficient matrix is sparse, but not symmetric. Direct calculation of the inverse has been used in problems with only one spatial dimension, though in higher dimensional case, memory requirement will constrain such a strategy. Still, an efficient fast solver may be feasible in higher space dimensions.

One may similarly develop spectral or pseudospectral spatial discretization in settings where a periodic boundary condition is applicable.

3.6 Some numerical examples

The recent experimental works have produced vortices ranging from a few to a few hundred. In figure 1, a two-dimensional simulation of a minimizer of the G-P energy was given along with an experimental picture produced by the MIT group (http://cua.mit.edu/ketterle_group/home.htm). Here, $\varepsilon = 10^{-2}$ and we have used a symmetric trap, i.e., a trap with $\lambda_y = 1$, the angular velocity was taken to be $\Omega = 150$. The initial condition is given to be a vortex free state which exhibits a parabolic profile.

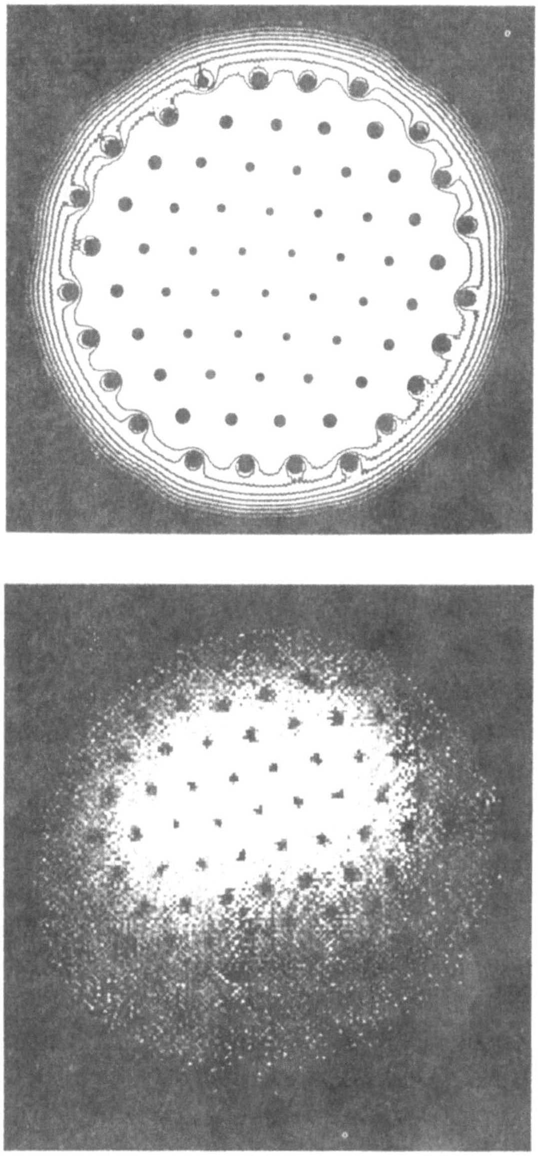

Figure 1. Vortex solutions. Top: contour plots of $|u|$ at $\Omega = 155$. Bottom: MIT experments

Other numerical solutions including those for anisotropic traps have been reported in [3]. One of the important issues remain to be studied further is related to spontaneous nucleation of the vortices and the dynamic and thermodynamic stability of the vortex solution. This can be facilitated by rigorous mathematical analysis as well as extensive numerical simulations.

4. Conclusion

We have presented some numerical integration schemes for the solution of the Gross-Pitaevskii equations. They are applicable to various forms of the equations, thus allow us to numerical investigate many model equations closely related to physical experiments that are currently underway. The numerical simulations will be conducted in cooperation with physicists so to gain further insight on the properties of the Bose-Einstein condensate. We expect to report more of our findings in future publications.

One naturally looks beyond the recent exciting scientific understanding of the Bose-Einstein to seek for technological application. We end the paper by quoting the press release of the Royal Swedish Academy of Sciences: "*It is interesting to speculate on areas for the application of BEC. The new* control *of matter which this technology involves is going to bring revolutionary applications in such fields as precision measurement and nanotechnology*". No doubt that numerical simulation will be becoming a useful tool in the future development.

Acknowledgment

Much of the theory and numerical schemes presented in the first few sections was developed in collaboration with A. Aftalion of CNRS and the University of Paris, VI., see [3].

References

[1] J. Abo-Shaeer, C. Raman, J. Vogels, and W. Ketterle, **Science, 292** 476 (2001).

[2] A.Aftalion, E.Sandier and S.Serfaty, to appear in J. Math. Pures et Appl. (2000).

[3] A.Aftalion and Q. Du, Phy. Rev. A, December 2001.

[4] A.Aftalion and T.Riviere, cond-mat/0105208.

[5] W. Bao, S. Jin and P. Markowich, J. Comp. Phys., to appear.

[6] T. Bridges and S. Reich, Physics Letters A, **284** 184 (2001).

[7] D.Butts and D.Rokhsar, **Nature 397**, 327 (1999).

[8] Y.Castin and R.Dum, Eur. Phys. J. D, **7**, 399 (1999).

[9] T. Chan and L. Shen, SIAM J. Numer. Anal., **24** 336 (1987).

[10] S. Chapman, Q. Du, M. Gunzburger and J. Peterson, Adv. Math.Sci. Appl. **5**, 193 (1995).

[11] J. Chen, M. Qin and Y. Tang, CCAST-WL reading **6** 125 (2001)

[12] F.Dalfovo, S.Giorgini, L.Pitaevskii and S.Stringari, Rev. Mod. Phys. **71**,463 (1999).

[13] F.Dalfovo, L.Pitaevskii and S.Stringari, Phys. Rev. A **54**, 4213 (1996).

[14] J. Deang, Q. Du and M. Gunzburger, Phy. Rev. B, **64**, 50256, 2001

[15] Q. Du, Math Comp, **67** 965 (1997).

[16] Q. Du, P. Gray, SIAM Appl Math, **56**, 1060 (1996).

[17] Q. Du, M.D. Gunzburger and J.S. Peterson, SIAM Review, **34**, 54 (1992).

[18] Q. Du, M.D. Gunzburger and J.S. Peterson, Phys. Rev. B, **46**, 9027 (1992);

[19] Q. Du, M.D. Gunzburger and J.S. Peterson, Phys. Rev. B, **51**, 16194 (1995).

[20] Q. Du, R. Nicolaides and X. Wu, SIAM Numer Anal, **35**, 1049 (1998).

[21] D.L.Feder, C.W.Clark and B.I.Schneider, Phys. Rev. A, **61** 011601(R) (1999).

[22] D.L.Feder, C.W.Clark and B.I.Schneider, Phys. Rev. Lett., **82**, 4956 (1999).

[23] K. Feng, J. Comput. Math, **4**, 279 (1986).

[24] A.L.Fetter and D.L.Feder, Phys. Rev. A, **58**, 3185 (1998).

[25] A.L.Fetter, Phys. Rev A, **148**, 429 (1965).

[26] A.L.Fetter and A.A.Svidzinsky, cond-mat/0102003.

[27] J. Hong and M. Qin, CCAST-WL reading **6** 1 (2001)

[28] R. Kellog, Math Comp, **23**, 23 (1969)

[29] R. Klein and A.J. Majda, Physica D., **53** 267 (1991).

[30] F. Lin, Q. Du, SIAM Math Anal, **28**, 1265 (1997).

[31] K.W. Madison, F. Chevy, W. Wohlleben and J. Dalibard, Phys. Rev. Lett., **84**, 806 (2000).

[32] K.W. Madison, F. Chevy, W. Wohlleben and J. Dalibard, J.Mod.Opt., **47**, 2715 (2000).

[33] J. Marsden G. Patrick and S. Shkoller, Comm. Math Phys **199** 351 (1999).

[34] M.R.Matthews et al. Phys. Rev. Lett., **83**, 2498 (1999).

[35] R. Onofrio et al. Phys. Rev. Lett., **85**, 2228 (2000).

[36] D. Peaceman and H. Rachford, J. Soc. Ind. Appl. Math., **3**, 28 (1955)

[37] J.S.Stiessberger and W.Zwerger, PRA **62** 061601 (2000).

[38] A.A.Svidzinsky and A.L.Fetter, Phys. Rev. Lett., **84**, 5919 (2000).

AN OPTIMIZATION ALGORITHM FOR THE METEOROLOGICAL DATA ASSIMILATION PROBLEM

Eva Eggeling
Fraunhofer-Institute for Algorithms and Scientific Computing (SCAI), Schloss Birlinghoven, Sankt Augustin, Germany
eva.eggeling@scai.fraunhofer.de

Shlomo Ta'asan
Department of Mathematical Sciences, Carnegie Mellon University, Pittsburgh, PA 15213, USA
shlomo@andrew.cmu.edu

Abstract The aim of this paper is to present an efficient optimization algorithm for solving the meteorological data assimilation problem. In our study this inverse problem is formulated as a constrained optimization problem. An accurate model for the problem involves the Navier-Stokes equations and in certain cases the Euler equations. In order to understand diverse intricate aspects of the problem we have focused on the Euler case and formulated three model problems which aim at tackling some of the basic difficulties faced in the real problem. We study the effect of dissipation in finite difference schemes on the identifiability in the problem. Basically, dissipation will result in bad estimation far from the measurement locations due to loss of information as the waves propagate. We demonstrate results for several cases, including the advection equation and a wave equation. For this purpose we use different measurement types in terms of the location of the measurements, the amount and noise level of the data.

Keywords: Constrained optimization, Data assimilation, Adjoint method, Condition number, Dissipation, Fourier analysis

1. The Data Assimilation Problem

The meteorological data assimilation problem can be defined as the problem to represent the current state of the atmosphere as accurate as possible, based on incomplete and noisy measurements in the past. For details we refer to the literature, e.g. [1]. We formulate this inverse problem as a constrained optimization problem:

In q measuring points $\mathbf{x}_j \in I\!R^2$ measurements $d(\mathbf{x}_j, t)$ are given. The time-dependent function $U(\mathbf{x}, t)$ is assumed to satisfy the initial value problem

$$\begin{aligned} U_t(\mathbf{x}, t) &= LU(\mathbf{x}, t) \\ U(\mathbf{x}, 0) &= \alpha(\mathbf{x}), \end{aligned} \tag{1}$$

where L is a differential operator. In real applications, L is the Navier-Stokes or the Euler operator. The initial value α is not known and has to be determined from measurements of the solution $U(x, t)$, at different locations during some time interval. Let the measuring point be $\mathbf{x}_j \in I\!R^2, j = 1, \ldots, q$, where the value of U is given by the measurements $d(x_j, t)$, which are noisy in practice. This has to be addressed by any algorithm for the problem.

The unknown α is determined by solving the following minimization problem

$$\min_{\alpha} \int_0^T \sum_{j=1}^{q} \| U(\mathbf{x}_j, t) - d(\mathbf{x}_j, t) \|^2 \, dt, \tag{2}$$

where U is the solution of the initial value problem (1). We may refer to α as the *design variable*.

2. The Model problems

We restrict our study to issues related to the Euler equation. The main feature as far as the optimization concerns with respect to the PDE is the *wave propagation*. The Euler equations involve the propagation of waves in the direction of the streamlines (Fig. 1) as well as pressure waves propagating in all directions (in subsonic flow)(Fig. 2).

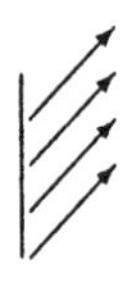

Figure 1.

Figure 2.

To understand the effect of each type of wave propagation on the identifiability of α we have defined a sequence of model problems such that each type of wave can be analyzed independently: model problem A for wave types shown in Fig. 1, model problem B for those of Fig. 2, and model problem C, being the linearized shallow water equations. Problem C combines the difficulties of the other two.

Let $\Omega = [0,1]^2$, and consider the following model problems:

A **Advection equation:**

$$\begin{aligned} u_t + au_x(\mathbf{x},t)x + bu_y(\mathbf{x},t) &= 0 \\ u(\mathbf{x},0) &= \alpha(\mathbf{x}) \end{aligned}$$

B **System of wave equations:**

$$\begin{pmatrix} u \\ v \end{pmatrix}_t + \begin{pmatrix} 1 & 0 \\ 0 & -1 \end{pmatrix} \begin{pmatrix} u \\ v \end{pmatrix}_x + \begin{pmatrix} 0 & 1 \\ 1 & 0 \end{pmatrix} \begin{pmatrix} u \\ v \end{pmatrix}_y = 0$$

with initial conditions
$$\begin{aligned} u(\mathbf{x},0) &= \alpha_1(\mathbf{x}) \\ v(\mathbf{x},0) &= \alpha_2(\mathbf{x}), \end{aligned}$$

C **Linearized Shallow Water Equations** see [4].

We assume periodic boundary conditions in all cases. Each of these models focuses on an aspect of the full meteorological data assimilation problem. In this paper all the analysis and numerical simulations were performed for problem A, due to space limitation.

2.1 Data distribution

A crucial issue in the data assimilation problem is the distribution of measurement data. In practice, the measurements are given only in some regions, with noise, and maybe not for the complete time interval. This affects the ability to identify the unknown function α. We have experimented with different spatial arrangements for the measurements data. See Fig. 3(a)- 3(e).

3. Analysis of the continuous problem

3.1 Problem A, the advection equation

The quality of the optimization result is closely related to the given measurements. So, it is important to analyze the relation between the initial value of the PDE and the given measurements. Especially, it is interesting to understand how much data are necessary to reconstruct the initial value within a given error.

The solution of the advection equation is $g(x - at)$, for initial condition $\alpha(x) = g(x)$. Let $f(t) = g(x_0 - at)$, where x_0 is a point where the measurements are taken. From the relation of f and g it is clear that for given f we can find g. That is, the inverse problem subject to problem A is well posed.

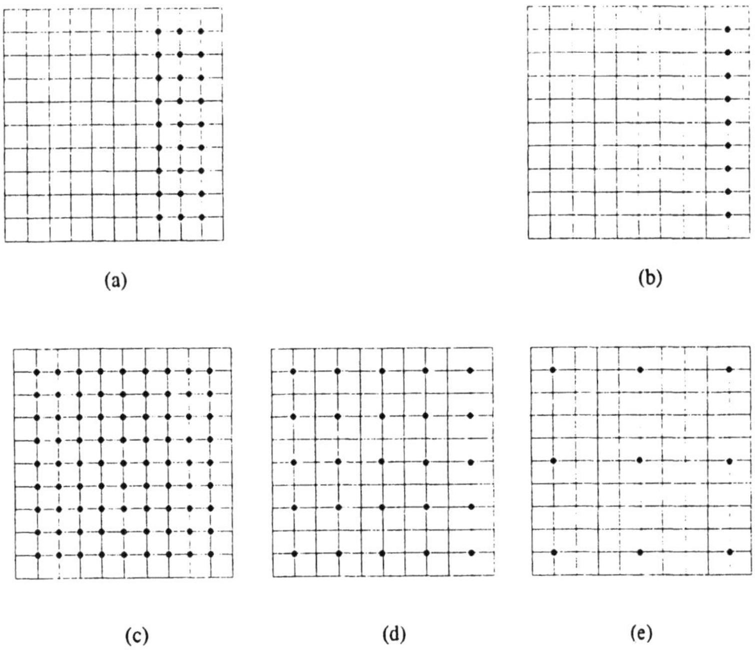

Figure 3. Different data distributions: 3(a) and 3(b)) localized data in parts of the domain, 3(c)- 3(e) decaying amount of data

However, since most schemes involve dissipation it is useful to modify problem A slightly to understand the effect of dissipation. Thus we consider,

$$\begin{aligned} u_t + a u_x &= \varepsilon u_{xx}, \\ u(x,0) &= \alpha(x), \end{aligned} \tag{3}$$

and study the identifiability of α as a function of ε. Considering the specific optimization problem (2) subject to equation (3) leads to the study of the mapping,

$$\mathcal{T} : \alpha(x) \mapsto u(0,t), \tag{4}$$

which contains information about the relation between the initial condition and the given measurements, here localized at $x = 0$. This mapping is of importance, because it is closely related to the Hessian $\mathcal{H}$ of the optimization problem through the relation $\mathcal{H} = \mathcal{T}^*\mathcal{T}$. For more details, we refer to [2].

3.2 Wave-Packet Analysis of $\mathcal{T}$

A proper analysis of the operator $\mathcal{T}$ requires microlocal considerations. That is, simultaneously localizing in the real space and in the frequency space. The usual Fourier analysis is insufficient here.

We consider the initial value to be in the shape of a wave packet and look at problem (3). This means that $u(x,t)$ is localized in space, say around $-x_0$, and $\hat{u}(\kappa,0)$ is localized around $\kappa=\kappa_0$, where we have used $\hat{u}$ to denote the Fourier transform of u,

$$u(x,t)=\int_{-\infty}^{\infty}\hat{u}(\kappa,t)e^{i\kappa x}\,d\kappa. \tag{5}$$

Then we obtain

$$\begin{aligned} u(x,0) &= \int_{-\infty}^{\infty}\hat{u}(\kappa,0)e^{i\kappa x}d\kappa = e^{i\kappa_0 x}\int_{-\infty}^{\infty}\hat{u}(\kappa,0)e^{i(\kappa-\kappa_0)x}\,d\kappa, \\ &= e^{i\kappa_0 x}\,g(x+x_0), \end{aligned}$$

where g is centered at 0 with $supp(g)\subset(-\delta,\delta)$. In addition we assume that the measurements are taken at $x=0$, hence x_0 measures the distance to the measurements.

We solve (3) using the Fourier transform,

$$\hat{u}_t+a(i\kappa)\hat{u}=-\varepsilon\kappa^2\hat{u} \quad\Longrightarrow\quad \hat{u}(\kappa,t)=\hat{u}(\kappa,0)e^{-ia\kappa t}e^{-\varepsilon\kappa^2 t} \tag{6}$$

Then, equation (5) and (6) lead to

$$\begin{aligned} u(x,t) &= \int_{-\infty}^{\infty}\hat{u}(\kappa,0)e^{-\varepsilon\kappa^2 t}e^{-ia\kappa t}e^{i\kappa x}d\kappa \\ &\approx e^{-\varepsilon\kappa_0^2 t}e^{i\kappa_0 x}e^{-i\kappa_0 at}\int_{-\infty}^{\infty}\hat{u}(\kappa,0)e^{i(\kappa-\kappa_0)(x-at)}\,d\kappa \\ &= e^{-\varepsilon\kappa_0^2 t}e^{i\kappa_0(x-at)}g(x+x_0-at). \end{aligned}$$

For $x=0$, it follows that

$$\begin{aligned} u(0,t) &\approx e^{-\varepsilon\kappa_0^2 t}e^{-i\kappa_0 at}\int_{-\infty}^{\infty}\hat{u}(\kappa,0)e^{i(\kappa-\kappa_0)(-at)}\,d\kappa \\ &\approx e^{-\varepsilon\kappa_0^2 x_0/a}e^{-i\kappa_0 at}g(x_0-at). \end{aligned} \tag{7}$$

In analogy to the previous example, we use $x_0/a=t$, because g is localized at 0. The norm of $u(0,t)$ depends on κ_0 and x_0/a. We can have a $O(1)$ signal in $\alpha(x)=u(x,0)$ but from this last formula we see that for large x_0, which means for far away signals, the norm of $u(0,t)$ is exponentially small. The result of this analysis is that $\mathcal{T} : u(x,0) \mapsto u(0,t)$ has the following properties:

$$\begin{array}{lrcl} \text{frequency:} & \kappa_0 & \mapsto & -a\kappa_0 \\ \text{center} & :-x_0 & \mapsto & \frac{x_0}{a} \\ \text{amplitude:} & A & \mapsto & e^{-\varepsilon\kappa_0^2\frac{x_0}{a}}A, \\ \text{dilation} & :\; 2\delta & \mapsto & \frac{2\delta}{a}. \end{array}$$

The consequence is, that the Hessian matrix of the related optimization problem (3) has a very small eigenvalue, which means that (3) gets ill-posed. △

4. The Discrete Problem

We know from the previous section that the PDE behaves well for $\varepsilon = 0$. But even if the continuous problem is well-posed we may get into troubles after discretizing the equations. Therefore it is important to analyze and understand the effect of a discretization scheme on the optimization process, because the performance of the optimization algorithm will depend on it.

4.1 The Discretization schemes

We use the following discretization schemes for the state and costate equations of the optimization problem:

Lax-Friedrichs (2D)

$$\frac{U^{n+1} - \bar{U}^n}{dt} = L_h U^n \qquad (8)$$

$$\bar{U}^n = \frac{1}{4}(U^n_{i,j+1} + U^n_{i-1,j} + U^n_{i,j-1} + U^n_{i+1,j})$$

Lax-Wendroff

$$\frac{U^{n+1} - U^n}{dt} = L_h U^n + \frac{dt}{2} L_h^2 U^n \qquad (9)$$

Runge-Kutta

$$U^{n+1} = \left[I + dt\, L_h + \frac{1}{2}(dt\, L_h)^2 + \frac{1}{6}(dt\, L_h)^3 + \frac{1}{24}(dt\, L_h)^4 \right] U^n \qquad (10)$$

L is the differential operator. We use a central discretization scheme in space for the corresponding discrete differential operator L_h, see [4].

4.2 Wave-Packet Analysis of $\mathcal{T}_h$

Now, we analyze what happens to the mapping $\mathcal{T} : \alpha(x) \mapsto u(0,t)$ in different numerical schemes to understand the effect of the discretization on the optimization process. We especially pay attention to properties like speed of the waves, frequency of the oscillations and decay of the waves in time. We consider the one-dimensional version of problem A. The Fourier transform of a discrete function $u(x,t)$ can be written as

$$u(x,t) = u(jh,t) = \int_{-\pi}^{\pi} \hat{u}(\theta,t)e^{i\theta j}d\theta = \int_{-\pi}^{\pi} \hat{u}(\theta,t)e^{i\theta x/h}d\theta. \tag{11}$$

Spatial discretization is given by

$$(L_h u)(x,t) = u_t + a\frac{u(x+h,t) - u(x-h,t)}{2h} = 0,$$

where h is the grid size in space. The symbol of this discrete operator is $\hat{L}_h(\theta) = ia\sin(\theta)/h$. We now analyze the different schemes:

Lax-Friedrichs scheme. Let k denote the step size in time. The Fourier representation of (8) is:

$$\begin{aligned} \hat{u}(\theta,t+k) &= [\cos\theta - i(\frac{a\sin\theta}{h})k]\,\hat{u}(\theta,t) \\ \hat{u}(\theta,t) &= [\cos\theta - i(\frac{a\sin\theta}{h})k]^{t/k}\,\hat{u}(\theta,0). \end{aligned} \tag{12}$$

We denote the norm of $(\cos\theta - \frac{i(a\sin\theta)k}{h})$ by $\rho(\theta)$ and its complex angle by $\Theta(\theta)$. For simplicity, we write $\rho(\theta_0)$ and $\Theta(\theta_0)$ respectively as ρ_0 and Θ_0. It is easy to see that $\Theta'(\theta) = -\frac{a}{\rho^2}(k/h)$. Therefore, $\Theta(\theta) \approx \Theta_0 - \frac{a}{\rho_0^2}(k/h)(\theta-\theta_0)$. This leads to the following representation,

$$\begin{aligned} u(x,t) &= \int_{-\pi}^{\pi} \hat{u}(\theta,t)e^{i\theta x/h}d\theta = \int_{-\pi}^{\pi} \hat{u}(\theta,0)\rho^{t/k}e^{i\Theta(t/k)}e^{i\theta x/h}d\theta \\ &\approx \rho_0^{t/k}e^{i\frac{\Theta_0}{k}t}e^{i\theta_0(x/h)}\int \hat{u}(\theta,0)\,e^{i(x-\frac{a}{\rho_0^2}t)/h(\theta-\theta_0)}d\theta \\ u(0,t) &= \rho_0^{t/k}e^{i\frac{\Theta_0}{k}t}\int \hat{u}(\theta,0)\,e^{-i(\frac{a}{\rho_0^2})\frac{t}{h}(\theta-\theta_0)}d\theta. \end{aligned} \tag{13}$$

We can observe from (13) that:

- The frequency in the measurements is given by Θ_0/k.
- $(\rho_0^{t/k})$ measures the *dissipation* of the discretization scheme. We can see already from (12) that $\rho^{t/k} = \| \cos\theta - i\frac{a\sin\theta k}{h} \|^{t/k}$ represents the decay of the wave after t time steps.
- $\rho_0^{1/k}$ corresponds to $e^{-\varepsilon\kappa_0^2}$ in equation (7) of section 3.1.

In analogy with the continuous case, we can observe also here from (13), that for far away signals the related optimization problem becomes ill-posed.

Lax-Wendroff. As the Fourier representation of (9) is

$$\hat{u}(\theta,t) = [1 + (ak/h)^2(\cos\theta - 1) - i(ak/h)\sin\theta]^{t/k}\hat{u}(0,\theta).$$

we define analogously to the previous discretization, $\rho(\theta)$ and $\Theta(\theta)$ respectively as the norm and the angle of $(1 + (\frac{ak}{h})^2(\cos\theta - 1) - i(\frac{ak}{h})\sin\theta)$, and use ρ_0 and Θ_0 to denote $\rho(\theta_0)$ and $\Theta(\theta_0)$. We find, that

$$\Theta(\theta) = \Theta_0 - \frac{\cos\theta_0 - (\frac{ak}{h})^2(\cos\theta_0 - 1)}{\rho_0^2}(\frac{ak}{h})(\theta - \theta_0).$$

With $\beta_0 := \cos\theta_0 - (\frac{ak}{h})^2(\cos\theta_0 - 1)$, we obtain

$$\begin{aligned} u(x,t) &= \int_{-\pi}^{\pi} \hat{u}(\theta,t)e^{i\theta x/h}d\theta \\ &\approx \rho_0^{t/k}e^{i\Theta_0(t/k)}e^{i\theta_0(x/h)}\int \hat{u}(0,\theta)e^{i(x-\frac{\beta_0 a}{\rho_0^2}t)(\theta-\theta_0)/h}d\theta., \end{aligned}$$

and

$$u(0,t) = \rho_0^{t/k}e^{i\frac{\Theta_0}{k}t}\int \hat{u}(0,\theta)e^{-i(a\beta_0/\rho_0^2)\frac{t}{h}(\theta-\theta_0)}d\theta. \tag{14}$$

From representation (14), we can draw the following conclusions.

- The speed of the wave is $a\beta_0/\rho_0^2$
- The frequency of the oscillation in measurements is Θ_0/k.
- The decay of the wave is given by $\rho^{t/k}$

Also here it is clear that signals that arrive from a long distance, are damped so much that their identification in α becomes ill-posed. As expected, the discretizations have different effects on the optimization, because of their different properties. With respect to the dissipation, the Lax-Friedrichs scheme is worse than the Lax-Wendroff discretization.The latter appears to be a good candidate for the optimization. Looking at the speed of the wave, we observe that problems with high frequencies may occur using the Lax-Wendroff scheme. Low frequencies can be handled very well.

5. The algorithm

The algorithm for solving the constrained optimization problem is based on a Quasi-Newton method, [3, 5] . For the calculation of the gradient we use the adjoint algorithm, and for the update of the Hessian, denoted by $\mathcal{H}$, the BFGS (Broyden, Fletcher, Goldfarb, Shanno) updating formula [5]. This leads to the following **Algorithm:**

- given initial value
- calculate the gradient
- search direction $\xi = -\text{grad}$
- Initial Hessian $\mathcal{H} = I$
- DO
  - line search
    * perturb initial condition α in (1)
    * calculate perturbed gradient
    * calculate step size δ
    * update alpha
  - calculate new gradient
  - stopping criterion
  - update the Hessian
  - update the search direction $\xi = \xi - \mathcal{H}\cdot$ grad

  ENDO

6. Condition number of the numerical Hessian matrix

The *condition number* of a Hessian matrix $\mathcal{H}_h$, defined as the quotient of the largest and smallest eigenvalue of the matrix

$$Cond(\mathcal{H}_h) = \lambda_{\max}/\lambda_{\min}$$

can be used for the classification of the mathematical model: If $Cond(\mathcal{H})_h$ is small, we will deal with an easy problem, if it is large the problem is difficult and requires many optimization iterations.
If the quotient $\lambda_{max}/\lambda_{min}$ grows further, the problem gets ill-posed. We will perform some numerical tests, using the Lax-Wendroff discretization and calculate $Cond(\mathcal{H}_h)$ of the numerical Hessian matrix. We analyze $Cond(\mathcal{H}_h)$ as a function of the amount of measurements, and as a function of the number of the design variables for a fixed number of given data. For problem A, we observe that:

- for a fixed number of design variables, $Cond(\mathcal{H}_h)$ grows for fewer and fewer measurements and the problem gets ill-posed (Table 1),
- for a fixed grid size and a fixed amount of measurements, $Cond(\mathcal{H}_h)$ grows if we have more and more design variables. The problem also gets ill-posed in this case (Table 2).

Table 1. Cond($\mathcal{H}_h$) for a fixed number of design variables ($\frac{n-1}{2}$) on a $n \times n$ grid and a decaying amount of given data

		$Cond(\mathcal{H}_h)$	
data distribution	*n= 17*	*n=33*	*n=65*
total domain	2	2	2
half domain	7	14	125
quarter domain	33	315	∞
4 lines	33	1578	∞
3 lines	93	∞	∞

Table 2. Cond($\mathcal{H}_h$) for measurements on three vertical lines of grid points on a $n \times n$ grid and a growing number of design variables

		$Cond(\mathcal{H}_h)$	
♯ *design variables*	*n=17*	$n = 33$	$n = 65$
3	12	15	20
4	13	39	22
5	16	17	23
6	23	41	22
7	44	44	64
8	-	50	30
16	-	-	77

Comparing the results for $Cond(\mathcal{H}_h)$ with the number of optimization cycles needed to solve the optimization problem, we see that they show the same behaviour. The problem is said to be solved, if the error in the design variable $\alpha(\mathrm{x})$ is less than 10^{-6}. For the cases where we could not solve the optimization problem, we set $Cond(\mathcal{H}_h) = \infty$ in the Tables. If $Cond(\mathcal{H}_h)$ is reasonably small, we are able to find the correct solution within an acceptable number of optimization cycles. Table 3 shows the number of optimization iterations needed. It can be compared to Table 1.

7. Summary of the results

On the PDE level we can solve the advection equation (problem A) exactly. On the differential level the optimization based on the advection equation seems to be a well-posed problem therefore. However, the discrete problem becomes ill-posed for far away signals. Based on Fourier analysis we prefer the Lax-Wendroff discretization in the numerical experiments presented here. We will analyze other numerical schemes in the future. The condition number of the

Table 3. Number of optimization cycles on a $n \times n$ grid for $(\frac{n-1}{2})$ design variables and a decaying amount of given data

data distribution		♯ *optimization cycles*	
	n=17	*n=33*	*n=65*
total domain	15	17	19
half domain	28	39	88
quarter domain	53	-	-
3 lines	93	-	-

numerical Hessian shows that for a fixed grid size the problem gets ill-posed for fewer and fewer measurements and for the number of design variables increasing. This fits to the numerical results. The results for only a few measurements may be also influenced by the distance of the signal and not just by their amount. For a very small amount of measurements (3 lines close to each other, near the boundary), we can solve the optimization problem related to the advection equation up to $n/4$ design variables, where $n \times n$ is the size of the grid,

- on a 17×17 grid up to 7 design variables in 44 optimization cycles,
- on a 33×33 grid up to 8 design variables in 50 cycles
- on a 65×65 grid up to 16 design variables in 77 cycles.

With measurements at every grid point we can solve the optimization problem with respect to the advection equation for the maximum number of design variables in 17 cycles for small, medium-sized and large grids. The condition number of the numerical Hessian $Cond(\mathcal{H}_h) = 2$ in these cases. The quality (noise) of the measurements may influence the degree of difficulty of the problem (well- or ill-posed).

All observations lead to the conclusion, that far away from given measurements we do not have a good identifiability and we should use there a coarse grid representation of the design variable α. This has not been tested.

References

[1] Daley R.: *Atmospheric data analysis*, Cambridge University Press, 1991.

[2] Ta'asan, Shlomo, *Theoretical Tools for Problem Setup*, in *Inverse Design and Optimization Methods*, VKI Lecture Notes 1977-05, April 1997.

[3] Gill P.E., Murray W., Wright, *Practical Optimization*, Academic Press, 1981.

[4] Hirsch, C., *Numerical Computation of Internal and External Flows*, Vol. 1, Wiley, Chichester, 1989.

[5] Kelley, C.T., *Iterative Methods for Optimization*, SIAM series on Frontiers in applied mathematics, 1999.

ANALYTIC ASPECTS OF YANG-MILLS FIELDS

Gang Tian*
Department of Mathematics, Massachusetts Institute of Technology
Cambridge, MA 02139

Abstract The Yang-Mills equation have played a fundamental role in our study of physics and geometry and topology in last few decades. Its regularity theory is crucial to our understanding and mathematical applications of its solutions. In this note, we briefly discuss some analytic aspects and recent progress on the Yang-Mills equation in an Euclidean space.

In the following, unless specified, we assume for simplicity that M is an open subset in $\mathbb{R}^n$ with the euclidean metric. Let G be a compact subgroup in $\mathbf{SO}(r)$ and g be its Lie algebra. Then g is a collection of $r \times r$ matrices closed under the standard Lie bracket. But we should emphasis that all our discussions here are valid for any differential manifold with a Riemannian metric and any compact Lie group G.

Our discussions in this note are for elliptic Yang-Mill equation. Many results here can be extended to the Yang-Mills-Higgs equation. One can also study the theory of the Yang-Mills equation on Lorentzian manifolds. The resulting equation is of weakly hyperbolic type and is very hard to study.

1. Yang-Mills Fields

First we recall that a connection on M with values in g is of the form

$$A = A_i dx_i, \quad A_i \in \mathbf{g} \tag{1.1}$$

where $x_1, \cdots, x_n$ are euclidean coordinates of $\mathbb{R}^n$ and A_i are matrices in g. Its curvature can be computed as follows:

$$F_A = dA + A \wedge A = F_{ij} dx_i \wedge dx_j \tag{1.2}$$

and

$$F_{ij} = \frac{1}{2}(\partial_i A_j - \partial_j A_i + [A_i, A_j]). \tag{1.3}$$

*Supported partially by NSF grants and a Simons fund

where ∂_i denotes the i-th partial derivative and $[A, B] = AB - BA$ is the Lie bracket of g.

The Yang-Mills functional is defined on the space of fields and given by

$$\mathcal{Y}(A) = \frac{1}{4\pi^2} \int_M |F_A|^2 dV_g, \tag{1.4}$$

where $|F_A|^2 = -\sum_{i,j} \mathrm{tr}(F_{ij}F_{ij})$. The Yang-Mills equation is simply its Euler-Lagrange equation

$$\sum_{i=1}^{n} (\partial_i F_{ij} - [F_{ij}, A_i]) = 0, \quad \forall j. \tag{1.5}$$

If we denote by D_A the differential operator $dB - [B, A]$ and D_A^* be its adjoint, then (1.5) can be written simply as $D_A^* F_A = 0$. On the other hand, being the curvature of a field, A automatically satisfies the second Bianchi identity $D_A F_A = 0$, that is

$$\partial_k F_{ij} + \partial_i F_{jk} + \partial_j F_{ki} = [A_k, F_{ij}] + [A_i, F_{jk}] + [A_j, F_{ki}], \quad \forall i, j, k. \tag{1.6}$$

We will call A a Yang-Mills field [1]if it satisfies (1.5).

The gauge group $\mathcal{G}$ consists of all smooth maps form M into $\mathbf{G} \subset \mathbf{SO}(r)$. It acts on the space of fields by assigning A to $\sigma(A) = \sigma A \sigma^{-1} - \sigma d\sigma^{-1}$ for each $\sigma \in \mathcal{G}$. Clearly, the Yang-Mills functional is invariant under the action of $\mathcal{G}$, so does the Yang-Mills equation. In particular, it implies that the Yang-Mills equation is not elliptic.

The simplest Yang-Mills fields are provided by harmonic one forms: If $G = U(1)$, then $\mathrm{g} = i\mathbb{R}$ and A is simply an one-form and the Yang-Mills equation is $d^*dA = 0$, the gauge transformation is given by $\sigma = e^{ia} \mapsto A + ida$. It follows that modulo gauge transformations, Yang-Mills $U(1)$-fields are in one-to-one correspondence with harmonic one forms.

2. Anti-Self-Dual Solutions

We will briefly describe a few special Yang-Mills fields.

Let Ω be a closed differential form on M of degree $n - 4$. Let us introduce Ω-anti-self-dual instantons, or simply, asd fields if no possible confusion may occur. Recall that the Hodge operator $*$ on differential forms is defined by

$$*(dx_{\sigma(1)} \wedge \cdots \wedge dx_{\sigma(l)}) = \mathrm{sign}(\sigma) dx_{\sigma(l+1)} \wedge \cdots \wedge dx_{\sigma(n)}, \quad l = 1, \cdots, n, \tag{2.1}$$

where σ is any permutation of $\{1, \cdots, n\}$. We say that a field A is an Ω-anti-self-dual instanton if its curvature $F_A = \sum F_{ij} dx_i \wedge dx_j$ satisfies

$$*(F_A \wedge \Omega) = -F_A, \tag{2.2}$$

or equivalently

$$\sum_{i,j} F_{ij} dx_i \wedge dx_j \wedge \Omega = -\sum_{i,j} F_{ij} * (dx_i \wedge dx_j).$$

If A is Ω-anti-self-dual and D_A denotes its associated covariant derivative, then $D_A F_A = -D_A(\Omega \wedge F_A) = 0$, so A is a Yang-Mills field. Clearly, the anti-self-duality is invariant under gauge transformations. So anti-self-dual instantons provide a special class of Yang-Mills solutions. One can show that these solutions are minima of the Yang-Mills functional among fields with certain topological constraints.

If $n = 4$, we may simply take $\Omega = 1$ and then the anti-self-dual equation becomes

$$F_{12} = -F_{34},\ F_{13} = -F_{42},\ F_{14} = -F_{23}. \tag{2.3}$$

If $M \subset \mathbb{R}^{2m}$ and $\Omega = \omega^{m-2}$, where ω is the standard symplectic form

$$\sum_{i=1}^{m} dx_i \wedge dx_{i+m}.$$

Then Ω-anti-self-dual instantons are simply Hermitian-Yang-Mills connections (cf. [Ti]), which were studied extensively in the geometry of holomorphic vector bundles. Other examples of Ω-anti-self-dual instantons include complex anti-self-dual instantons of Donaldson-Thomas (cf. [Ti]).

3. Coulomb Gauge

A *gauge transformation* is a measurable map $\sigma : M \to G$. This group acts on fields by the formula

$$\sigma(A) := \sigma \cdot A \cdot \sigma^{-1} - d\sigma \cdot \sigma^{-1}. \tag{3.1}$$

We call A and $\sigma(A)$ *gauge equivalent*. Observe that

$$F(\sigma(A)) = \sigma \cdot F(A) \cdot \sigma^{-1}. \tag{3.2}$$

Definition 3.1. *A field A is said to be in a* Coulomb gauge on M *if it satisfies the condition $d_* A = 0$ on the interior of M, and $A \cdot \nu = 0$ on the boundary ∂M, where ν is a unit normal of ∂M.*

The Yang-Mills equation is not elliptic because of gauge transformations. However, it is elliptic modulo gauge transformations. To see it, we assume that $A = \sum_i A_i dx_i$ is in a Coulomb gauge, then $\sum_i \partial_i A_i = 0$, and the Yang-Mills equation reads

$$\sum_{i=1}^{n} \left(\partial_i^2 A_j - [2\partial_i A_j - \partial_j A_i, A_i] - [[A_i, A_j], A_i]\right) = 0, \quad \forall j.$$

Given any smooth field A, locally, there is always a Coulomb gauge σ, that is

$$\sum_{i=1}^{n} \partial_i(\sigma A_i \sigma^{-1}) + (\partial_i \sigma)\sigma^{-1} = 0. \tag{3.3}$$

Its local solvability is obvious. The reason for constructing Coulomb gauges is: if A is in a Coulomb gauge, we expect A to have one more derivative of regularity than $F(A)$. The question then arises: *given an arbitrary field A, under what conditions can we find a gauge equivalent Coulomb gauge $A_{coulomb}$ which has a one more derivative of regularity than $F(A)$?*

The following was proved in [Uh2].

Theorem 3.1. *Let $A = A_i dx_i$ be any field with $A_i \in L^p(B_1(p), g)$ for some $p \geq n/2$, where $B_1(p)$ is a unit ball in $\mathbb{R}^n$. Then there exists $\epsilon(n) > 0$ and $c(n) > 0$ such that if $||F_A||_{n/2} \leq \epsilon(n)$, where $|| \cdot ||_q$ denotes the L^q-norm in $B_1(p)$, then there is a Coulomb gauge σ satisfying (3.3) and $||\sigma(A)||_p \leq c(n)||F_A||_p$.*

In four dimensions this assumption is perfect for applications to equations such as the Yang-Mills equations, however it is a bit too strong in higher dimensions to study such questions as the dimension of singularities of Yang-Mills fields. In [TT], Tao and I used the Morrey space $M_2^{n/2}$, which is larger than $L^{n/2}$. Let us recall some results from [TT]:

Definition 3.2. *If M is a domain and $1 \leq q \leq p$, we define the Morrey spaces $M_q^p(\Omega)$ to be those locally L^q functions (possibly vector-valued) whose norm*

$$||f||_{M_q^p} := \sup_{x_0 \in \mathfrak{R}^n; 0 < r \leq 1} r^{n(\frac{1}{p}-\frac{1}{q})} \left(\int_{B(x_0,r)\cap\Omega} |f|^q \right)^{1/q}$$

is finite. Here and in the sequel $B(x,r)$ denotes the open ball in $\mathfrak{R}^n$ with center x and radius r.

We also define Morrey-Sobolev spaces $M_{q,k}^p$ for integers $k \geq 0$ by the formula

$$||f||_{M_{q,k}^p(\Omega)} := \sum_{j=0}^{k} ||\nabla^j f||_{M_q^p(\Omega)}.$$

The norm $M_2^{n/2}$ arises very naturally in Price's monotonicity formula for the curvature of Yang-Mills fields (cf. Section 4). The norm M_q^p has the scaling of L^p, but the functions are only L^q integrable. >From Hölder's inequality we see that all L^p functions are in M_q^p, but not conversely. Note that the M_q^p norm depends only on the magnitude of f. Also, we have

$$||F(\sigma(A))||_{M_q^p} = ||F(A)||_{M_q^p}. \tag{3.4}$$

The space $M_2^{n/2}$ is the space which we shall be placing our curvatures. Using the heuristic that a field requires one more derivative than the curvature, and a gauge transform requires two more derivatives, we thus hope to place fields and gauge transforms in $M_{2,1}^{n/2}$ and $M_{2,2}^{n/2}$ respectively.

For any $\epsilon > 0$, let $\mathfrak{U}_\epsilon(M)$ denote the set of all smooth fields on M which satisfy the bound

$$\|F(A)\|_{M_2^{n/2}(\Omega)} \leq \epsilon. \tag{3.5}$$

We observe that this space is invariant under gauge transformations.

We now show that Coulomb gauges can be prescribed on the unit cube $[0, 1]^n$ as long as the field is smooth and $M_2^{n/2}$ norm of the curvature is sufficiently small. Here is a result of [TT].

Theorem 3.2. *If $0 < \epsilon \ll 1$ is sufficiently small, then every field A in $\mathfrak{U}_\epsilon([0, 1]^n)$ is gauge equivalent to a Coulomb gauge $A_{coulomb}$ which obeys the bound*

$$\|A_{coulomb}\|_{M_{2,1}^{n/2}([0,1]^n)} \leq C\|F(A)\|_{M_2^{n/2}([0,1]^n)}; \tag{3.6}$$

A similar result was proved by Meyer and Rivirie. In [TT], we apply this theorem to constructing Coulomb gauges even for nonsmooth fields. For example, let $M = [0, 1]^n$ and S be a closed subset stratified by submanifolds of dimension no more than $n - 4$, if A is any field on $M\backslash S$ with $\|F(A)\|_{M_2^{n/2}(\Omega)}$ sufficiently small, then there is a Coulomb gauge $A_{coulomb}$ of A on $M\backslash S$. It is also true for any closed subset S of Hausdorff dimension no more than $n - 4$ and with local simply-connectedness property in a suitable sense (cf. [TT]).

4. Monotonicity and Curvature Estimates

Given any vector field X on M with compact support, we can integrate it to get an one-parameter group of diffeomorphisms $\phi_t : M \mapsto M$. Put $A_t = \phi_t^*(A)$. Then $A_0 = A$ and A_t coincides with A near the boundary of M. If A is a smooth Yang-Mills field, differentiating $\mathcal{Y}(A_t)$ on t at $t = 0$, one can derive as Price did in [Pr]

$$\int_M (|F_A|^2 \text{div} X - 4 \sum_{i,j=1}^{n} F_{ij} F_{kj} \partial_i X^k) dV = 0, \tag{4.1}$$

where $X = X^k \partial_k$. This is very important even though it is nothing but the first variation of $\mathcal{Y}$ along X. Let us derive some of its consequences. Let $p \in M$ such that the ball $B_{\rho_0}(p)$ with radius ρ_0 and center p is contained inside M. Then taking X to be $\xi(r) r \partial_r$, where r is the distance from p and ξ is a cut-off function in $B_{\rho_0}(p)$, we can get the monotonicity formula of Price.

Theorem 4.1. *([Pr]) Let A be any Yang-Mills field on M. Then for any $0 \leq \sigma \leq \tau \leq \rho_0$, we have*

$$\tau^{4-n} \int_{B_\tau(p)} |F_A|^2 dV - \sigma^{4-n} \int_{B_\sigma(p)} |F_A|^2 dV$$

$$= 4 \int_{B_\tau(p)\setminus B_\sigma(p)} r^{4-n} \sum_i |F_A(\partial_r, \partial_i)|^2 dV. \tag{4.2}$$

In particular, $\rho^{4-n} \int_{B_\rho(p)} |F_A|^2 dV$ is nondecreasing with ρ.

An application of this monotonicity is the following curvature estimate which was proved by K. Uhlenbeck and Nakajima ([Na]).

Theorem 4.2. *Let A be any Yang-Mills field on U. Then there are $\epsilon = \epsilon(n) > 0$ and $C = C(n) > 0$, such that for any $B_\rho(p) \subset M$, we have*

$$|F_A|(p) \leq \frac{C}{\rho^2} \left(\rho^{4-n} \int_{B_\rho(p)} |F_A|^2 dV \right)^{\frac{1}{2}}, \tag{4.3}$$

whenever $\rho^{4-n} \int_{B_\rho(p)} |F_A|^2 dV \leq \varepsilon$.

We can associate a measure μ_A to each field A as follows: For any continuous function f with compact support, we define

$$\int_M f \mu_A = \int_M f |F_A|^2 dV. \tag{4.4}$$

We can simply write $\mu_A = |F_A|^2 dV$. By the monotonicity, we have is a nondecreasing function $\rho^{4-n} \mu_A(B_\rho(p))$.

Now we let $\{A_i\}$ be a sequence of Yang-Mills fields such that for each compact subset $K \subset M$, $\mu_i(K)$ are uniformly bounded, where μ_i is the measure associated to A_i. Then a subsequence $\{\mu_a\}$ of $\{\mu_i\}$ converges weakly to a measure μ. Because of the monotonicity for μ_i, one can easily show that $\rho^{4-n}\mu(B_\rho(p))$ is an nondecreasing function for each $p \in M$. Define the density function of μ by

$$\Theta_\mu(p) = \lim_{\rho \to 0} \rho^{4-n} \mu(B_\rho(p)). \tag{4.5}$$

Because of the monotonicity for μ, this density Θ_μ is well-defined, nonnegative and upper-semi-continuous. It follows that the support S of Θ_μ is a locally closed subset of M such that the Hausdorff measure $\mathcal{H}^{n-4}(S \cap K)$ is finite for any compact subset K. Furthermore, it follows from Theorem 4.2 that $\Theta_\mu(p) \geq \epsilon$ for any $p \in S$ and the curvature of A_a is uniformly bounded on any compact subset in $M \backslash S$. Then, using Theorem 4.2, one can show the following theorem which is due to Uhlenbeck.

Theorem 4.3. *Let A_a, μ_a, μ and S be as above. Then there are gauge transformations σ_a such that by taking a subsequence if necessary, $\sigma_a(A_a)$ converges smoothly to a Yang-Mills field A defined on $M \backslash S$. Moreover, $\mu_A \leq \mu$.*

By an admissible Yang-Mills field, we mean a smooth Yang-Mills field A defined outside a locally closed subset $S(A)$ in M, such that $\mathcal{H}^{n-4}(S(A) \cap K) < \infty$ and $\mu_A(K) < \infty$ for any compact subset $K \subset M$. Clearly, the limiting field in Theorem 4.3 is admissible. In fact, Following Uhlenbeck, one can easily extend Theorem 4.3 to any sequence of admissible Yang-Mills fields. We will assume that $S(A)$ is the singular set of an admissible Yang-Mills field A. If $S(A) = \emptyset$, then A is smooth.

5. Structure of Blow-up Loci

Let $\{A_i\}$ be a sequence of smooth Yang-Mills fields such that its associated measures μ_i converge weakly to a measure μ. As before, we denote by Θ_μ the density and by S the support of μ. By Theorem 4.3 and taking a subsequence if necessary, we may assume that there are gauge transformations σ_i such that $\sigma_i(A_i)$ converge to an admissible Yang-Mills field A outside S.

Now we will examine structure of S. Let μ_A be the measure associated to A. Define

$$S_b(\{A_i\}) = \overline{\{p \in M \mid \Theta_\mu(p) > 0, \ \lim_{r \to 0} r^{4-n} \int_{B_r(p)} |F_A|^2 dV = 0\}}. \quad (5.1)$$

This set is called the blow-up locus of $\{A_i\}$. If no confusion occurs, we will simply write S_b for this blow-up locus. It is easy to see that $\mathcal{H}^{n-4}(\overline{S \backslash S_b}) = 0$. The following proposition was proved in [Ti]. It gives the first regularity on the blow-up locus.

Theorem 5.1. *Let $\{A_i\}$ be the above sequence of Yang-Mills fields which converge to A. Then its blow-up locus S_b is $\mathcal{H}^{n-4}$-rectifiable, that is for $\mathcal{H}^{n-4}$-a.e. p in S_b, there is a unique tangent space T_pS_b of S_b at p. Moreover, for any smooth function f with compact support, we have*

$$\int_M f d\mu = \int_M f d\mu_A + \int_{S_b} f \Theta_\mu d\mathcal{H}^{n-4}. \quad (5.2)$$

Furthermore, there are constraints on the geometry of the blow-up loci.

Theorem 5.2. *([Ti]) For any vector field X with compact support in M, we have*

$$-\int_{S_b} \mathrm{div}_{S_b} X \Theta_\mu \, d\mathcal{H}^{n-4} = \int_M (|F_A|^2 \mathrm{div} X - 4 \sum_{i,j=1}^{n} F_{ij} F_{kj} \partial_i X^k) dV, \tag{5.3}$$

where $\mathrm{div}_{S_b} X$ *denotes the divergence of* X *along* S_b *and* F_{ij} *are the components of* F_A.

We say that an admissible field A is stationary if (4.1) holds for any vector field with compact support.

Corollary 5.1. *Let A be in the above theorem. If A is stationary, then S_b is stationary, that is S has no boundary in M and its generalized mean curvature vanishes.*

I doubt that A is stationary in general, but it is stationary when A is anti-self-dual (see the following). If $A = 0$, then S_b is stationary and the curvature of A_i concentrates near a minimal variety of codimension 4. It leads to the question: *Let S be a minimal submanifold of dimension $n-4$ in general position, is S the limit of a sequence of Yang-Mills fields (possibly with respect to different metrics)?*

We will call the above (A, S_b, Θ_μ) a generalized Yang-Mills field. Two generalized Yang-Mills (A, S_b, Θ) and (A', S_b', Θ') if A and A' are gauge equivalent on an open dense subset. The set of all generalized Yang-Mills fields modulo gauge transformations is precompact.

Theorem 4.2 can be also used to prove the existence of tangent cones for generalized Yang-Mills fields. Let A be a stationary admissible Yang-Mills field with singular set $S(A)$. For any $\lambda > 0$ and $p \in S(A)$, we can define

$$A_\lambda(q) = \lambda \sum_i A_i(p + \lambda(q-p)) dx_i,$$

where $A = \sum_i A_i dx_i$. Then there are sequences $\{\lambda(i)\}$ such that $\lim_{i \to \infty} \lambda(i) = 0$ and $A_{\lambda(i)}$ converge to a field A^c outside S_c with $\mathcal{H}^{n-4}(S^c \cap B_R(0)) < \infty$ for any $R > 0$. Further, measures $|F_{A_i}|^2 dV$ converge weakly to a measure μ_c with density Θ_c. It follows from Theorem 4.2

Corollary 5.2. *Let $A_{\lambda(i)}$, A_c, S_c, Θ_c be as above. Then we have that $\partial_r \Theta_c = 0$, $a \cdot S_c = S_c$ and $F_{A_c}(\partial_r, \cdot) = 0$.*

6. Compactifying Spaces of Anti-Self-Dual Instantons

If A_i are Ω-anti-self-dual instantons, then its weak limit A is Ω-anti-self-dual wherever it is well-defined. As before, we will call such an A admissible Ω-anti-self-dual instanton or simply asd solution if there is no confusion.

Theorem 6.1. *([Ti]) Assume that Ω is a parallel form of degree $n-4$. Let A be an admissible Ω-anti-self-dual instanton on M. Then A is stationary.*

Now we assume that $\{A_i\}$ be a sequence of Ω-anti-self-dual instantons which converge to an admissible Ω-anti-self-dual instanton A, where Ω is a form on M of degree $n-4$. Let $S_b \subset M$ be the blow-up locus of $\{A_i\}$ with the density Θ_μ. Note that μ_i is the measure associated to A_i and $\lim_{i\to\infty} \mu_i = \mu$.

For any admissible field A', we can associate a current $C_2(A')$ as follows: For any smooth form φ with compact support in M, we define

$$C_2(A) = \frac{1}{8\pi^2} \int_M \operatorname{tr}(F_{A'} \wedge F_{A'}) \wedge \varphi. \tag{6.1}$$

Clearly, if A' is smooth, it is nothing else but the current represented by the Chern-Weil form defining the second Chern class, so it is closed. In general, it was proved in [Ti] that $C_2(A')$ is closed in M.

Since S_b is rectifiable (Theorem 5.1), we can also define a current $C_2(S_b, \Theta_\mu)$ by

$$C_2(S_b, \Theta_\mu) = \frac{1}{8\pi^2} \int_M (\varphi, \Omega) \Theta_\mu d\mathcal{H}^{n-4}. \tag{6.2}$$

Theorem 6.2. *([Ti]) Let A_i, A et al be as above. Then $\frac{1}{8\pi^2}\Theta_\mu$ is integer-valued and S_b is calibrated by Ω, that is for $\mathcal{H}^{n-4}$-a.e. $p \in S_b$ where T_pS_b exists, the restriction of Ω to T_pS_b coincides with the induced volume form. Moreover, we have*

$$\lim_{i\to\infty} C_2(A_i) = C_2(A) + C_2(S_b, \Theta_\mu). \tag{6.3}$$

In particular, for any compact K, we have

$$\lim_{i\to\infty} \mu_i(K) = \mu_A(K) + \int_{S_b \cap K} \Theta_\mu d\mathcal{H}^{n-4}.$$

A simplified situation of Theorem 6.2 can be described as follows: Let $\pi : \mathbb{R}^n \mapsto \mathbb{R}^4$ be an orthogonal projection and B be an asd instanton on $\mathbb{R}^4$, then the pull-back $A = \pi^* B$ is Ω-asd if and only if $L = \pi^{-1}(0)$ is an Ω-calibrated subspace. This can be checked directly. As before, we ask if *an Ω-calibrated submanifold is the limit of a sequence of Ω-asd instantons.*

It is well known (cf. [HL]) that if $|\Omega| \leq 1$, then any integral current calibrated by Ω is minimizing in its homology class. It follows

Corollary 6.1. *Assume that $|\Omega| \leq 1$. Let S_b be the blow-up locus of a sequence of asd instantons A_i converging to A and Θ_μ be its associated density. Then $C_2(S_b, \Theta_\mu)$ is an area-minimizing integral current.*

The support S_b of $C_2(S_b, \Theta_\mu)$ may not be smooth. However, one can show that a dense open subset of S_b is smooth. Further, we do expect

Conjecture 6.1. *Let Ω be any closed differential form with $|\Omega| \leq 1$, then Ω-calibrated integral current is supported on the closure N of a smooth manifold N_0 such that $N \backslash N_0$ is of codimension at least two.*

Let us give an example. Assume that $n = 2m$. Fix an identification $\mathbb{R}^n = \mathbb{C}^m$. Let ω be given in complex coordinates $z_1, \cdots, z_m$ by

$$\omega = \frac{\sqrt{-1}}{2} \sum_{i=1}^{m} dz_i \wedge d\bar{z}_i.$$

Put $\Omega = \omega^{m-2}/(m-2)!$. Then an Ω-asd instanton A is simply a Hermitian-Yang-Mills connection, that is $F_A^{0,2} = 0$ and $F_A^{1,1} \cdot \omega = 0$, where $F_A^{k,l}$ is the (k,l)-part of F_A. Moreover, a subspace $L \subset \mathbb{R}^n$ of codimension 4 is Ω-calibrated if and only if L is a complex subspace in $\mathbb{C}^m$. Let S be the blow-up locus of a sequence of Hermitian-Yang-Mills connections and Θ be its associated density. Then $C_2(S, \Theta)$ is a closed integral current whose tangent spaces are complex subspaces. It follows from a result of J. King [Ki] that there are positive integers m_a and irreducible complex subvarieties V_a such that for any smooth φ with compact support in M,

$$C_2(S, \Theta)(\varphi) = \sum_a m_a \int_{V_a} \varphi.$$

It can be also proved that if A is an admissible asd instanton with respect to $\omega^{m-2}/(m-2)!$, then there is a gauge transformation τ such that $\tau(A)$ extends to be a smooth field outside a complex subvariety of codimension greater than 2 (cf. [TY]).

Now we give a geometric application. Let M be a compact n-manifold with a Riemannian metric g and Ω be a closed form of degree $n-4$. Let E be a vector bundle over M. Denote by $\mathfrak{M}_{\Omega,E}$ the collection of all gauge equivalence classes of Ω-asd instantons of E over M. This is usually refered as the moduli space of Ω-asd instantons. In general, $\mathfrak{M}_{\Omega,E}$ may not be compact. The problem is how to compactify it.

A generalized Ω-asd instanton is made of an admissible Ω-asd instanton A of E, which extends to a smooth field over $M \backslash S(A)$ for a closed subset $S(A)$ with $\mathcal{H}^{n-4}(S(A)) = 0$, and a closed integral current $C = C_2(S, \Theta)$ calibrated by Ω, such that cohomologically,

$$[C_2(A)] + [C_2(S, \Theta)] = C_2(E). \tag{6.4}$$

where $C_2(E)$ denotes the second Chern class of E. Two generalized Ω-asd instantons (A, C), (A', C') are equivalent if and only if $C = C'$ and there is a gauge transformation σ on $M \backslash S(A) \cup S(A')$, such that $\sigma(A) = A'$ on $M \backslash S(A) \cup S(A')$. We denote by $[A, C]$ the gauge equivalence class of (A, C). We identify $[A, 0]$ with $[A]$ in $\mathcal{M}_{\Omega,E}$ if A extends to a smooth field of E over M modulo a gauge transformation. We define $\overline{\mathfrak{M}}_{\Omega,E}$ to be set of all gauge equivalence classes of generalized Ω-asd instantons of E over M.

The topology of $\overline{\mathfrak{M}}_{\Omega,E}$ can be defined as follows: a sequence $[A_i, C_i]$ converges to $[A, C]$ in $\overline{\mathfrak{M}}_{\Omega,E}$ if and only if there are representatives (A_i, C_i) such that their associated currents $C_2(A_i, C_i)$ converge weakly to $C_2(A, C)$ as currents, where

$$C_2(A', C') = C_2(A') + C_2(S', \Theta'), \quad C' = (S', \Theta').$$

It is not hard to show that by taking a subsequence if necessary, $\tau_i(A_i)$ converges to A outside $S(A)$ and the support of C for some gauge transformations τ_i.

Theorem 6.3. *([Ti]) For any M, g, Ω and E as above, $\overline{\mathfrak{M}}_{\Omega,E}$ is compact with respect to this topology.*

7. Removable Singularity Theorem

Let A be a stationary admissible Yang-Mills field. The basic regularity problem is whether there is a gauge transformation τ such that $\tau(A)$ can be extended across S or part of S.

We say that a locally closed subset S is stratified by submanifolds if it is a finite union of locally closed submanifolds. Clearly, the Hausdorff dimension S is equal to the maximal dimension of these submanifolds. The following was proved in [TT].

Theorem 7.1. *Let A be a stationary admissible Yang-Mills field. Assume that its singular set $S(A)$ is stratified by submanifolds of dimension $n-4$. Then there is an $\epsilon > 0$, which depends only on n, such that for any $B_\rho(p) \subset\in S(A)$, if*

$$\rho^{4-n} \int_{B_\rho(p)} |F_A|^2 \, dV < \epsilon. \tag{7.1}$$

then there is a gauge transformation σ near p such that $\sigma(A)$ extends to be a smooth field near p.

When $n \leq 3$, $S(A)$ is empty. When $n = 4$, $S(A)$ consists of finitely many points and (4.1) holds for any admissible Yang-Mills fields. Hence, this theorem is simply the removable singularity theorem of K. Uhlenbeck for Yang-Mills fields on 4-manifolds [Uh1]. When $n > 4$, this theorem was proved in [Ti] under extra gauge conditions. With some extra efforts, one can weaken the assumption on the singular set $S(A)$. We will discuss it in another paper.

Corollary 7.1. *Let A be a stationary admissible Yang-Mills field. Assume that its singular set $S(A)$ is stratified by submanifolds of dimension $n-4$. Then there is a gauge transformation σ such that $\sigma(A)$ is smooth outside a locally closed subset S' consisting of submanifolds of dimension less than $n-4$. If $n=4$, then $\sigma(A)$ is actually smooth.*

In general, we propose

Conjecture 7.1. *Let A be a stationary admissible Yang-Mills field, then there is a gauge transformation σ such that $\sigma(A)$ extends to be a smooth field outside a locally closed subset with locally finite Hausdorff measure of dimension $n-5$.*

Further, we have

Conjecture 7.2. *If A is an admissible asd field, then there is a gauge transformation τ such that $\mathcal{H}^{n-6}(S(\tau(A))\cap K) < \infty$ for any compact $K \subset M$, where $S(\tau(A))$ denotes the singular set of $\tau(A)$.*

Notes

1. In mathematical literatures, Yang-Mills fields are often called as Yang-Mills connections

References

[Fe] H. Federer, Geometric Measure Theory. Springer. Berlin-Heidelberg-New York, (1969).

[HL] R. Harvey and H.B. Lawson, "Calibrated geometries," *Acta. Math.,.* **148** (1982), pp. 47–157.

[Ki] J. King, "The currents defined by analytic varieties," *Acta. Math.*, **127** (1971), pp. 185-220.

[Na] H. Nakajima, "Compactness of the moduli space of Yang-Mills connections in higher dimensions" *J. Math. Soc. Japan*, **40** (1988).

[Pr] P. Price, "A monotonicity formula for Yang-Mills fields," *Manuscripta Math.*, **43** (1983), pp. 131-166.

[Ti] G. Tian, "Gauge Theory and Calibrated Geometry, I", Annals of Mathematics, **151** (2000), no. 1.

[TT] T. Tao and G. Tian, "'A Singularity Removal Theorem For Yang-Mills Fields in Higher Dimension", Preprint, 2001.

[TY] G. Tian and B.Z. Yang, "Compactifying Moduli of Hermitian-Yang-Mills Connections", Preprint, 2001.

[Uh1] K.K. Uhlenbeck, "Removable Singularities in Yang-Mills Fields," *Comm. in Math. Phys.*, **83** (Springer-Verlag 1982), pp. 11-29.

[Uh2] K.K. Uhlenbeck, "Connections with L^p Bounds on Curvature," *Comm. in Math. Phys.*, **83** (Springer-Verlag 1982), pp. 31-42.

THE SHAPE OPTIMIZATION OF AXISYMMETRIC STRUCTURES BASED ON FICTITIOUS LOADS VARIABLE*

Shuguang Gong, Yunqing Huang
Department of Mathematics, Xiangtan University, Hunan 411105, China
gongsg@xtu.edu.cn
huangyq@xtu.edu.cn

Guilan Xie, Bo Hong
Institute of Mechanical Engineering, Xiangtan University, Hunan 411105, China

Abstract A numerical method is developed for shape optimization of axisymmetric structures based on choosing the sets of fictitious loads acting on a structure's 'control point' as the optimizing design variable. The axisymmetric structure is decomposed by using an isoparametric parabolic-type element that has high accuracy for design sensitivity analysis. An efficient analytic method, based on a fictitious load variable, is developed for sensitivity analysis in shape optimization of the axisymmetric structure. Two optimization examples have been tested successfully.

Keywords: Fictitious Loads, FEM, Shape Optimization, Axisymmetric Structures

1. Introduction

The parts of an axisymmetric structure are often used in many kinds of machine, such as axial or discal parts in the rotating facility, pressure vessel subject to pressure loads etc. The failure of these parts is, in the majority of cases, due to local stress concentration in the transitions. The maximal interest of shape optimization is to improve the structural stress-distributing status of transitions, while achieving the optimum weight or area, by changing from initial boundary to optimized shape. Numerical methods for shape optimization

*This work is supported by the special funds for major state basic research projects.

have been developed since the 1960s [1] - [4]. The design variable in the shape optimization may be finite element nodal coordinates [5], coefficients of splines [6, 7] or polynomials [8, 9], or other geometric parameters [10]. A shape optimization method based on a natural design variable was first developed by Belegundu and Rajan [11, 12]. The fictitious loads acting on an auxiliary structure was chosen as the design variable of shape optimization.

The main objective of this paper is to develop an efficient method of shape optimization for axisymmetric structure, which is to use isoparametric parabolic-type element that has high design sensitivity analysis accuracy [13]. By means of a linear relationship between nodal displacement and fictitious loads, an efficient sensitivity analysis method is developed in shape optimization of axisymmetric structures. The cost function is to minimize the maximum Von Mises stress at a boundary of transitions. Finally, two numerical examples will be tested.

2. The Optimization Method

The shape optimal design problem under investigation is to minimize an objective function, $f(b)$, subject to inequality constraints, it can be stated as:

$$\begin{cases} Minimize\ f(b) \\ subject\ to \\ \quad b_i^l \leq b_i \leq b_i^u, \quad i = 1, 2, \cdots, m \\ \quad g_j(b) \leq 0, \quad\quad j = 1, 2, \cdots, q_c \end{cases} \tag{1}$$

where $f(b)$ is the Von Mises stress at a point in the structure, or the structural weight, b_i^l, b_i^u the lower and upper constraints of the i-th design variables, respectively, $g_j(b)$ structural response, m the number of design variable.

Using a finite element displacement method in the linear elastic body, the finite element equation for the static analysis can be written as:

$$K\Delta = F \tag{2}$$

where K is the structural stiffness matrix, F is the load vector, and Δ is a vector of nodal displacements.

In this paper, fictitious loads at control nodes are selected as the design variable. If we apply a unit fictitious load at the control nodes each time, and solve for the nodal displacement δ^i under a unit load from (2), then the nodal displacement due to these fictitious loads are added onto the current shape C^0 to obtain a new shape [11] as follows:

$$C(b) = C^0 + \sum_{i=1}^{m} b_i \delta^i \tag{3}$$

where C is a vector consisting of the X- and Y-coordinates of all the grid points, and b is a vector of the design variable. Each vector δ^i is obtained from the equations

$$K_b \delta^i = q^i \tag{4}$$

where q^i is a load vector with all zeros except for a unit value at the i-th location, K_b is usually different from the structural stiffness matrix K in (2) since the boundary conditions differ from those used in the original problems. The details may be found in [11, 12]

It is clear from (3) that the movement of nodes is only relavant to the magnitude of the fictitious loads. If we can control the maximum of fictitious loads in the optimizing iteration, then the movement of nodes location and the distortion of mesh will be controlled. Thus we can make sure in advance that the maximal nodal displacement and the maximal value of fictitious load can be obtained in each optimizing iteration. At the same time, to increase the rate of descent of the objection function, the valid value of design variable is degressive from the maximum b_{jmax} to satisfy some given accuracy in the iteration:

$$b_{jmax} = \frac{\delta^i_{jmax}}{\| C(b_j) - C^0 \|_\infty}. \tag{5}$$

Combining the above ideas, the basic iteration procedure of shape optimization is given as follows:

Step 1: Let C^0 be a vector of the current shape nodal location. Given the maximal displacement δ_{max}, set $b_j = 0, j = 1, 2, \cdots, m$.

Step 2: Apply a unit load, q^i, at the control point, one at a time, and solve for δ^i from (4) and, solve for b_{jmax} from (5).

Step 3: Using a discrete approach to calculate the sensitivity analysis coefficient. Use the sensitivity coefficients in a nonlinear programming to obtain the fictitious load b_j at the control point.

Step 4: Define the new shape by (3).

Step 5: Repeat above procedures until some given conditions are satisfied.

3. The sensitivity analysis

The derivatives of the objective function and constraint functions with respect to the design variable provide the variational trends of the structures for optimization. Calculation of these derivatives is known as sensitivity analysis [14]. The response of a continuum structure to the change of design variables is an implicit function of the variables. Therefore, it is difficult to calculate accurate gradients of the function for sensitivity analysis.

In this paper, we will calculate the gradients of the function by using discrete approaches. Taking derivatives of equation (2) with respect to the design variable b gives:

$$\frac{\partial K}{\partial b}\Delta + K\frac{\partial \Delta}{\partial b} = \frac{\partial F}{\partial b} \tag{6}$$

Since the load F is a constant in the present problem, we have:

$$K\frac{\partial \Delta}{\partial b} = -\frac{\partial K}{\partial b}\Delta \tag{7}$$

The derivatives of stress components σ in an element and of Von Mises stress σ_e at a point can be calculated from

$$\frac{\partial \sigma}{\partial b} = D(\frac{\partial B}{\partial b}\Delta^e + B\frac{\partial \Delta^e}{\partial b}) \tag{8}$$

$$\begin{aligned}\frac{\partial \sigma_e}{\partial b} &= \frac{1}{2\sigma_e}[(\sigma_1 - \sigma_2)\frac{\partial(\sigma_1 - \sigma_2)}{\partial b} + (\sigma_2 - \sigma_3)\frac{\partial(\sigma_2 - \sigma_3)}{\partial b} \\ &\quad + (\sigma_3 - \sigma_1)\frac{\partial(\sigma_3 - \sigma_1)}{\partial b}]\end{aligned} \tag{9}$$

The derivatives of weight W and area A for an axisymmetric structure element can be written as:

$$\frac{\partial W^e}{\partial b} = 2\pi\int_{-1}^{1}\int_{-1}^{1}\rho[\frac{\partial r}{\partial b}|J| + r\frac{\partial |J|}{\partial b}]d\xi d\eta \tag{10}$$

$$\frac{\partial A^e}{\partial b} = \int_{-1}^{1}\int_{-1}^{1}\frac{\partial |J|}{\partial b}d\xi d\eta \tag{11}$$

In the above equations, it is important to calculate the derivatives of stiffness matrix K. The calculation of $\partial K/\partial b$ is done by many methods such as the divided difference scheme [11], semi-analytic method [15] or finite difference approximation [16]. In this paper, the calculation of $\partial K/\partial b$ can be done by a new method that is called as analytic methods. The procedure will be obtained below:

With the known methodology of 2D 8-nodes isoparametric element [17], the derivatives of an element stiffness matrix K^e with reslect to the design variable b for axisymmetric structures can be formulated as :

$$\frac{\partial K^e}{\partial b} = 2\pi\int_{-1}^{1}\int_{-1}^{1}[\frac{\partial B_i^T}{\partial b}DB_j r|J| + B_i^T D\frac{\partial B_j}{\partial b}r|J|$$

$$+B_i^T DB_j \frac{\partial r}{\partial b}|J| + B_i^T DB_j r \frac{\partial |J|}{\partial b}] d\xi d\eta \tag{12}$$

where

$$\frac{\partial B_i}{\partial b} = \begin{pmatrix} \frac{\partial}{\partial b}\frac{\partial N_i}{\partial r} & 0 \\ 0 & \frac{\partial}{\partial b}\frac{\partial N_i}{\partial z} \\ -\frac{N_i}{r^2}\frac{\partial r}{\partial b} & 0 \\ \frac{\partial}{\partial b}\frac{\partial N_i}{\partial z} & \frac{\partial}{\partial b}\frac{\partial N_i}{\partial r} \end{pmatrix} \quad i = 1, 2, \cdots, 8 \tag{13}$$

and

$$\frac{\partial}{\partial b}\begin{pmatrix} \frac{\partial N_i}{\partial r} \\ \frac{\partial N_i}{\partial z} \end{pmatrix} = \frac{\partial J^{-1}}{\partial b}\begin{pmatrix} \frac{\partial N_i}{\partial \xi} \\ \frac{\partial N_i}{\partial \eta} \end{pmatrix} \tag{14}$$

where J is the Jacobian matrix. Taking derivatives of J with respect to design variable b, we have:

$$\frac{\partial J}{\partial b} = \frac{\partial}{\partial b}\begin{pmatrix} \frac{\partial r}{\partial \xi} & \frac{\partial z}{\partial \xi} \\ \frac{\partial r}{\partial \eta} & \frac{\partial z}{\partial \eta} \end{pmatrix} = \frac{\partial}{\partial b}\sum_{i=1}^{8}\begin{pmatrix} \frac{\partial N_i}{\partial \xi} r_i & \frac{\partial N_i}{\partial \xi} z_i \\ \frac{\partial N_i}{\partial \eta} r_i & \frac{\partial N_i}{\partial \eta} z_i \end{pmatrix}$$

$$= \begin{pmatrix} \frac{\partial N_1}{\partial \xi} & \frac{\partial N_2}{\partial \xi} & \cdots & \frac{\partial N_8}{\partial \xi} \\ \frac{\partial N_1}{\partial \eta} & \frac{\partial N_2}{\partial \eta} & \cdots & \frac{\partial N_8}{\partial \eta} \end{pmatrix} \frac{\partial}{\partial b}\begin{pmatrix} r_1 & z_1 \\ r_2 & z_2 \\ \vdots & \vdots \\ r_8 & z_8 \end{pmatrix}. \tag{15}$$

Similar to the definition of the partial derivatives, we have

$$\frac{\partial r_i}{\partial b_j} \approx \frac{r_i(b_1, \ldots, b_j + \alpha, \cdots, b_m) - r_i(b_1, \cdots, b_m)}{\alpha} \tag{16}$$

$$\frac{\partial z_i}{\partial b_j} \approx \frac{z_i(b_1, \ldots, b_j + \alpha, \cdots, b_m) - z_i(b_1, \cdots, b_m)}{\alpha} \tag{17}$$

where α can be seen as a load applied on the structure, no other effect is considered here. Let $b_j = 0, (j = 1, 2, \cdots, m)$ we have

$$\frac{\partial r_i}{\partial b_j} \approx \frac{r_i(0, \ldots, \alpha, \cdots, 0) - r_i(0, \cdots, 0)}{\alpha} = \frac{u_i^j}{\alpha} \tag{18}$$

$$\frac{\partial z_i}{\partial b_j} \approx \frac{z_i(0, \ldots, \alpha, \cdots, 0) - z_i(0, \cdots, 0)}{\alpha} = \frac{v_i^j}{\alpha} \tag{19}$$

where u_i^j, v_i^j are the i-th nodal r and z displacements under the unit load q^j respectively, determined by the expression (4).

Combining the above ideas, we summarize the calculating procedure of the sensitivity analysis as follows:

Step 1: Set $\alpha = 1, j = 1, 2, \cdots, m$.

Step 2: Apply the j-th unit load on the control point, and solve for δ^j from (4). Thus from (18) and (19), one can form the vector of $\partial r_i/\partial b_j$ and $\partial z_i/\partial b_j$.

Step 3: Calculate $\partial B_i/\partial b_j$ and $\partial J/\partial b_j$ from (13) - (15), and form the stiffness matrix of element $\partial K^e/\partial b_j$ by using (12).

Step 4: Assemble the global matrix $\partial K/\partial b_j$ similar to the assembly of the globale stiffness matrix K.

Step 5: Solve the equation (7) to obtain the vector $\partial\Delta/\partial b_j$. Thus one can obtain $\partial\sigma/\partial b_j$ from (8), $\partial\sigma_e/\partial b_j$ from (9), $\partial W^e/\partial b_j$ from (10) and $\partial A^e/\partial b_j$ from (11), respectively.

4. Numerical examples

To verify the validity of the porposed numerical procedures, a computer program written in FORTRAN was developed for the sensitivity analysis and cyclic variable methods. Two test problems will be examined in this section.

4.1 The transition problem

The section of original shape subjected to uniform tensile load in radius direction is shown in Fig.1, which was obtained by simplifying a turbine axis [18]. Due to symmetry, only its upper half section is analyzed. The finite element mesh, shown in Fig.1, consists of 65 quadriateral isoparametric elements and 232 nodes.

The objective of optimization is to minimize the maximum Vom Mises stress at a point along the A-B profile, by changing the boundary shape of transition region A-B. The transition profile shape after optimization is shown in Fig.2 and the Von Mises stress distribution along the A-B-C boundary is shown in Fig.3. The gradient of the objective function in the optimization iteration is shown in Fig.4.

The number of design variables is 11. The initial and final areas are 100 and 93.067, respectively, yielding about 7% reduction. The maximum Von Mises stress was reduced form 123.28 to 98.8 and the extent of reduction is approximate 20%.

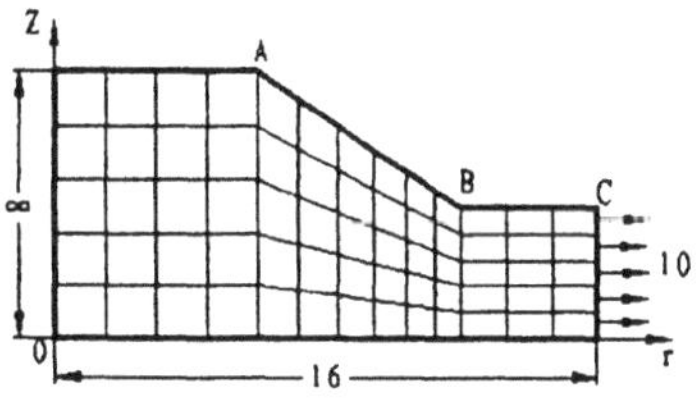

Figure 1. Initial shape model

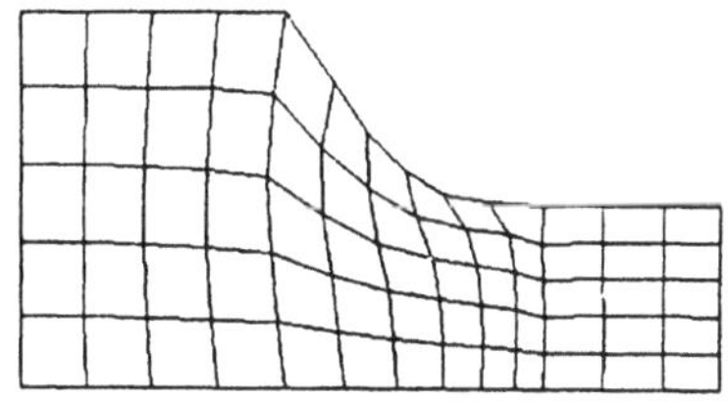

Figure 2. Optimized profile

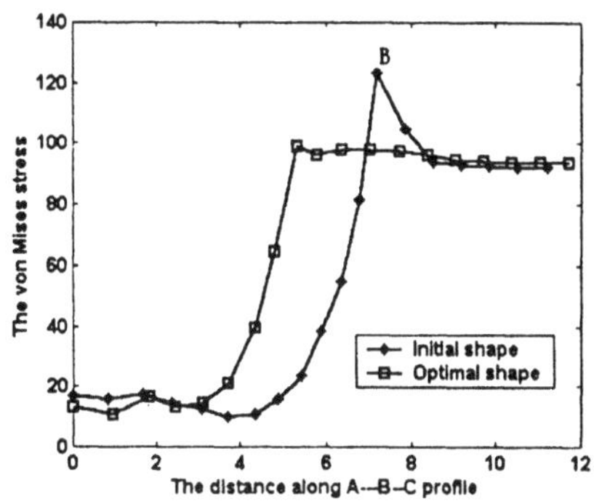

Figure 3. The Stress distribution along A-B-C profile

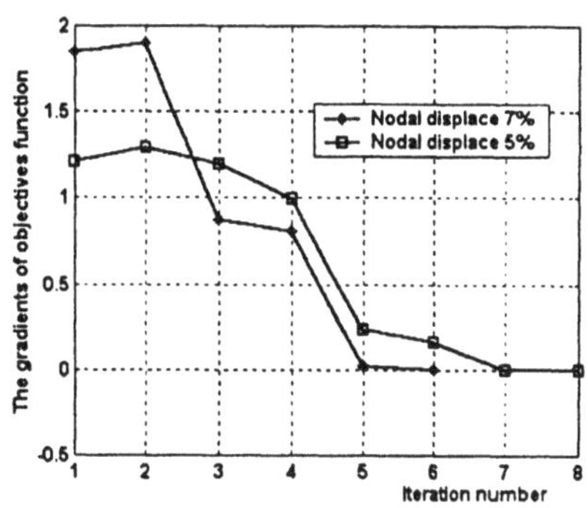

Figure 4. The gradient of objective function

4.2 The flange design problem

A flange structure model, subject to inner pressure, is shown in Fig.5. It contains 47 elements and 176 nodes. The nodal Von Mises stress distributed status of the flange is shown in Fig.6.

The goal of the optimization is to minimize the Von Mises stress of transition. The number of design variables is 12. The optimized shape is shown in Fig.7, and the stress distribution of transition is shown in Fig.8. The gradient of the objective function is shown in Fig.9.

By comparing the stress in Fig. 6 and Fig. 8, we see that the structure after optimization has more uniform stress distribution. The maximum Von Mises stress is reduced from 20.896 to 13.741, yielding a 34% reduction. The augmentation of area is reduced from 4068.5 to 4151.845, yielding a 2% augmentation.

5. Conclusion

The numerical results obtained by using the proposed method show that the shape optimization of axisymmetric structure has been achieved successfully.

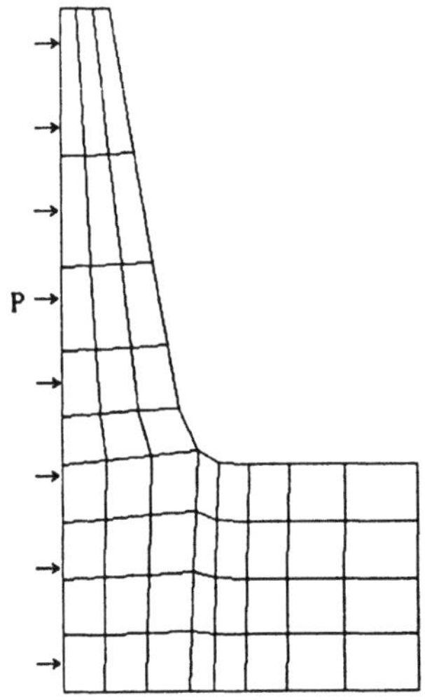

Figure 5. Initial flange model

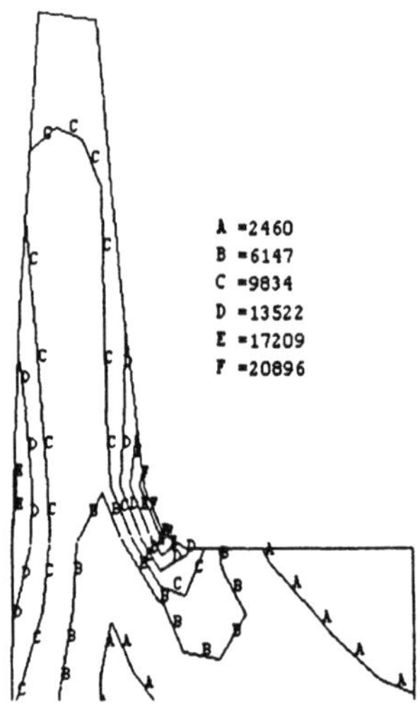

Figure 6. The Stress distribution of transition

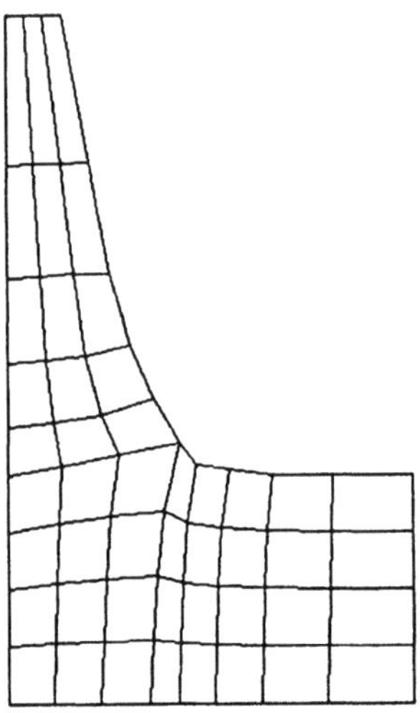

Figure 7. Optimized flange shape

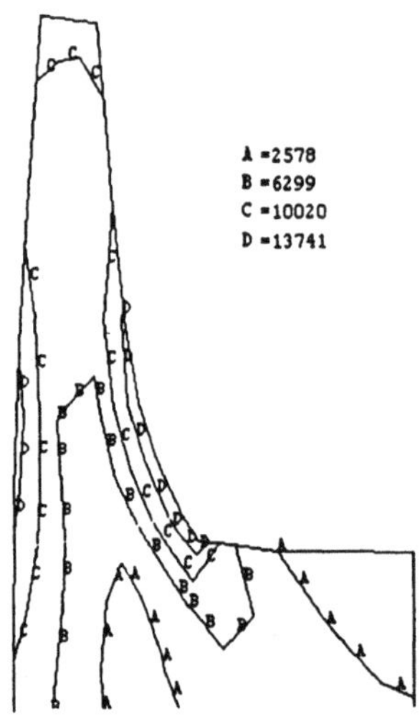

Figure 8. The stress distribution after optimizing

While the extent of the change of area is relatively small, by controlling the maximal displacement of the grid points, the proposed method is seen to produce low element distortions. If augmentation of area is relatively large, then the adaptive mesh or refined finite element mesh may be used.

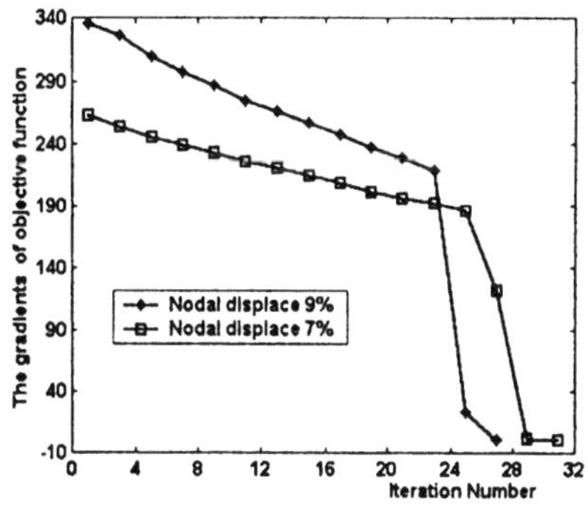

Figure 9. The gradient of objective function

The analytic method of sensitivity analysis based on the fictitious loads was discussed. The value of the gradient function is convergent, as shown in Fig. 4 and Fig. 9. Another advantage of this method is that, in the optimization iteration, the gradient value can be calculated by using some standard finite element subroutines. If the quadrilateral isoparametric element is used, then smoother and more optimal shapes may be obtained. Similar observation has been made in [12].

References

[1] G. N. Vanderplaats, *Structural optimization: Past, Present, and Future*, AIAA Journal, 20 (1982), pp. 992-1000

[2] T. H. Raphael and V. G. Ramana, *Structural shape optimization: a survey*, Computer Methods in Applied Mechanics and Engineering, 57 (1986), pp. 91-106

[3] Ding Yunliang, *Shape optimization of structures: a literature survey*, Computer & Structures, 24 (1986), pp. 985-1004

[4] Hsu Yehliang, *Review of structure shape optimization*, Computers in Industry, 25 Nov.(1994), pp. 3-13

[5] O. C. Zienkiewicz and J. S. Compbell, *Shape optimization structural and sequential linear programming*, in: R. H. Gallagher and O. C. Zienkiewicz, eds., Optimum Structural Design (Wiley, New York, 1973) pp. 109-126.

[6] V. Braibant and C. Fleury, *Shape optimal design using B-splines*, Computer Methods in Applied Mechanics and Engineering, 44 (1984), pp. 247-267

[7] M. H. Imam, *Three-dimensional shape optimization*, International Journal of Numerical Methods and Engineering, 18 (1982), pp. 661-673

[8] W. J. Stroud, C. B. Dexter and M. Stein, *Automated preliminary design of simplified wing structure to satisfy strength and flutter requirements*, NASA TN D-6534, (1971)

[9] S.S. Bhavikatti and C.V. Ramakrishnan, *Optimum shape design of rotating risks*, Computers & Structure, 11 (1980), pp. 397-401

[10] Gutkowski Witold and Dems Krzysztof, *Shape optimization of a 2D body subjected to several loading conditions*, Engineering Optimization, 29 (1997), pp. 293-311

[11] A. D. Belegundu and S. D. Rajan, *A shape optimization approach based on natural design variable and shape function*, Computer Methods in Applied Mechanics and Engineering, 66 (1988), pp. 87-106

[12] S. D. Rajan and A. D. Belegundu, *Shape optimization design using fictitious loads*, AIAA Journal, 27 (1989), pp. 102-107.

[13] Yang Renjye and Chol K. Kyung, *Accuracy of finite element shape design sensitivity analysis*, Journal of Struct. Mech. 13 (1985), pp. 223-239

[14] Cheu Tsu-Chien, *Sensitivity analysis and shape optimization of axisymmetric structures*, International Journal for Numerical Methods in Engineering, 28 (1989), pp. 95-108

[15] Gu Yuanxian and Cheng Gengdong, *Research and application of numerical methods of structural shape optimization*, Computational Structural Methanics and Applications, 10 (1993), pp. 321-335

[16] M. Adelman Howard and T. Haftka Raphael, *Sensitivity analysis of discrete structural system*, AAIA Journal, 24 (1986), pp. 823-832

[17] Gan Shunxian, *The finite element technology and programming*, Beijing Technology University Press, 1988

[18] Zhang Xiaodong, Zhang Qiang and Luo Yabo, *Optimization of shape design for axisymmetric structures*, Journal of Jiang Han Petroleum Institute, 20 (1998), pp. 73-76.

A NEW KIND OF PRECONDITIONER FOR INTERFACE EQUATIONS OF MORTAR MULTIPLIERS ON SUBSPACES*

Qiya Hu
Institute of Computational Mathematics,
Chinese Academy of Sciences, Beijing 100080, China
hqy@lsec.cc.ac.cn

Abstract In this paper we discuss non-overlapping domain decomposition methods with nonmatching grids for three-dimensional elliptic problems, in which interface unknown is chosen as mortar multiplier. We develope a class of preconditioners for the interface equation derived by projection on a suitable subspace. For our preconditioner, each local solver is defined on the common face between two neighbouring subdomains unlike the existing preconditioners, and it can be implemented in a more efficient way. It will be shown that the condition number of the preconditioned system grows only as the logarithm of the dimension of the local problem associated with an individual substructure.

Keywords: domain decomposition, nonmathing grids, Lagrange multiplier, interface equation, preconditioner, condition number

1. Introduction

In recent years, there is a growing interest in the domain decomposition methods (DDMs) with Lagrange multipliers, early studied in [7], [8], [9] and [18]. This kind of DDM has many advantages over the traditional ones for the case of nonmatching grids (refer to [1], [13], [16] and [25]).

FETI (see [9]) is one of the DDMs with Lagrange multipliers, in which the floating subdomains are handled in a simple way. FETI includes two main ingredients: (a) introducing a pseudoinverse and a projection operator to derive an interface equation of the multiplier; (b) applying the projected PCG method to solve this interface equation with a Dirichlet preconditioner. Each local solver in the Dirichlet preconditioner is defined on the boundary of a sub-

*The work is supported by Special Funds for Major State Basic Research Projects of China G1999032804
The proofs of the results given here will be provided in another paper

domain, which can be implemented by solving a local Dirichlet problem. The condition number of the preconditioned system has been estimated in [20] under some assumptions. The original FETI seems to be less effective to the case of nonmatching grids or the discontinuous coefficient problems, and so more complicated interface preconditioners have been constructed in [14], [17] and [22].

In [13] (see also [12]), the authors discussed the DDMs with polynomial Lagrange multipliers, and constructed a cheap interface preconditioner for completely positive definite problems (without any floating subdomain). For this preconditioner, each local solver is defined on the common face between two neighbouring subdomains. Since the local polynomial multiplier space has low dimensions, the local problem can be solved exactly. This preconditioner is also very efficient to the cases of nonmatching grids and discontinuous coefficient problems.

In the present paper, we discuss the DDMs with mortar multipliers, and extend the idea of [13] to the case with floating subdomains. To this end, we will use the interface equation in FETI. The main contributions of this paper are: (1) estimate the condition number of the preconditioned system based on a new kind of dual norm; (2) develope a kinds of cheap local solvers by using an extension theorem of $H^{\frac{1}{2}}$-norm.

The outline of the remainder of the paper is as follows. In §2, we build the interface equation of the Lagrange multiplier. In §3, we construct the interface preconditioner and give the convergence result. In §4, we study the local solvers.

2. Domain decomposition

2.1 Saddle-point formulation

Consider the model problem

$$\begin{cases} -div(a\nabla u) = f, & in\ \Omega, \\ u = 0, & on\ \partial\Omega, \end{cases} \tag{2.1}$$

where Ω is a bounded, connected Lipschitz domain in $\mathcal{R}^3$ and $a \in L^\infty(\Omega)$ is a positive function.

Let $H_0^1(\Omega)$ denote the standard Sobolev space, and define the bilinear form

$$\mathcal{A}(v,w) = \int_\Omega a\nabla v \cdot \nabla w dx, \quad v,w \in H_0^1(\Omega).$$

Let $(\cdot,\cdot)$ denote the $L^2(\Omega)$-inner product. The weak formulation of (2.1) in $H_0^1(\Omega)$ reads: Find $u \in H_0^1(\Omega)$ such that

$$\mathcal{A}(u,v) = (f,v), \quad \forall v \in H_0^1(\Omega). \tag{2.2}$$

In the following, we define a discrete problem of (2.2). Let the domain Ω be decomposed into $\bar{\Omega} = \bigcup_{k=1}^{N} \bar{\Omega}_k$. Assume that

(i) $\Omega_i \cap \Omega_j = \emptyset$ when $i \neq j$. If $\bar{\Omega}_i \cap \bar{\Omega}_j \neq \emptyset$ for some $i \neq j$, we denote it by Γ_{ij}. Define $\Gamma = \cup \Gamma_{ij}$;

(ii) each subdomain Ω_i has the same "siz" d in the usual sense (see [6] and [28]).

As usual, we assume that each Ω_k is a polyhedron. With each subdomain Ω_k we associate a regular triangulation $\mathcal{T}_k$ made of elements that are either hexahedra or tetrahedra. We denote by h_k the mesh size of $\mathcal{T}_k$, i.e., h_k denotes the maximum diameter of any hexahedra or tetrahedra of the mesh $\mathcal{T}_k$. The triangulations of the subdomains are independent of each other and generally do not match at the interfaces between subdomains. Hence, each interface Γ_{ij} is provided with two different ($2D$) meshes. Define $V(\Omega_k)$ as the space consisting of continuous piecewise linear functions associated with $\mathcal{T}_k$. When $\partial\Omega_k \cap \partial\Omega \neq \emptyset$, we require any function in $V(\Omega_k)$ vanishes on $\partial\Omega_k \cap \partial\Omega$. Define $V(\partial\Omega_k) = V(\Omega_k)|_{\partial\Omega_k}$.

To describe an approximate space on global Ω, we need a Lagrange multiplier space. This multiplier space can be defined in various ways, for example, the polynomial multiplier space (see [8] and [18]), the mortar multiplier space (see [3] and [24]), and the dual multiplier space (see [15] and [26]). In this paper, we consider the mortar multiplier space.

Let $\mathcal{T}_{ij}$ and $\mathcal{T}_{ji}$ denote the triangulation on Γ_{ij} associated with $\mathcal{T}_i$ and $\mathcal{T}_j$ respectively. For each $\Gamma_{ij} \subset \Gamma$, we choose $\mathcal{T}_{ij}$ or $\mathcal{T}_{ji}$ to define the *local* multiplier space, for example, choose $\mathcal{T}_{ij}$.

For simplicity of exposition, we will use the spaces (on $\Gamma_{ij} \subset \Gamma$) defined by

$$V_i(\Gamma_{ij}) = V(\partial\Omega_i)|_{\Gamma_{ij}}, \quad V_j(\Gamma_{ij}) = V(\partial\Omega_j)|_{\Gamma_{ij}}$$
$$V_i^0(\Gamma_{ij}) = V_i(\Gamma_{ij}) \cap H_0^1(\Gamma_{ij}), \quad V_j^0(\Gamma_{ij}) = V_j(\Gamma_{ij}) \cap H_0^1(\Gamma_{ij}).$$

Now, we define the local multiplier space $W(\Gamma_{ij}) \subset V_i(\Gamma_{ij})$ (see [4]).

Since we will not involve concrete structure of $W(\Gamma_{ij})$, we consider only the case where the face Γ_{ij} is meshed with triangular elements. We denote by x_m, $1 \leq m \leq n(i,j)$, the set of all vertices of the triangles and distinguish the internal nodes that belong to Γ_{ij} (numbered from 1 to $n_0(i,j)$) from those that belong to the boundary of Γ_{ij} (numbered from $n_0(i,j)+1$ to $n(i,j)$). With all these nodes are associated the shape functions ϕ_m so that any element φ of $W(\Gamma_{ij})$ can be written as

$$\varphi = \sum_{m=1}^{n(i,j)} \varphi(x_m)\phi_m,$$

and those elements that belong to $V_i^0(\Gamma_{ij})$ can be written as

$$\varphi = \sum_{m=1}^{n_0(i,j)} \varphi(x_m)\phi_m.$$

The vertices x_m, $n_0(i,j)+1 \leq m \leq n(i,j)$ belong to the same triangles as internal nodes within Γ_{ij}. We denote by x_m^l, $1 \leq l \leq Q(m)$ those vertices inside Γ_{ij} that belong to a side of a triangle with end point x_m. For each m such that $n_0(i,j)+1 \leq m \leq n(i,j)$, we choose $Q(m)$ positive numbers a_m^l with the property that $\sum_{l=1}^{Q(m)} a_m^l = 1$. The definition of the space $W(\Gamma_{ij})$ is then

$$W(\Gamma_{ij}) = \{\varphi \in V_i(\Gamma_{ij}) : \forall m, n_0(i,j)+1 \leq m \leq n(i,j), \varphi(x_m) = \sum_{l=1}^{Q(m)} a_m^l \varphi(x_m^l)\},$$

which can also be written as

$$W(\Gamma_{ij}) = \{\varphi \in V_i(\Gamma_{ij}) : \varphi = \sum_{m=1}^{n_0(i,j)} \varphi(x_m)\phi_m + \sum_{m=n_0(i,j)+1}^{n(i,j)} [\sum_{l=1}^{Q(m)} a_m^l \varphi(x_m^l)]\phi_m\}.$$

Note that $dim(W(\Gamma_{ij})) = dim(V_0^i(\Gamma_{ij}))$. Define $W(\Gamma) = \prod_{\Gamma_{ij} \subset \Gamma} W(\Gamma_{ij})$. Let $P_{ij} : L^2(\Gamma_{ij}) \to W(\Gamma_{ij})$ be the L^2 projection operator. For $v \in V(\Omega)$, set $v|_{\Omega_k} = v_k$. Define $V(\Omega) = \prod_{k=1}^{N} V(\Omega_k)$ and

$$\tilde{V}(\Omega) = \{v \in V(\Omega) : \; P_{ij}(v_i|_{\Gamma_{ij}} - v_j|_{\Gamma_{ij}}) = 0 \; for \; each \; \Gamma_{ij} \subset \Gamma\},$$

where $v = (v_1, v_2, \cdots, v_N)$. Note that we do not require $\tilde{V}(\Omega) \subset H_0^1(\Omega)$.

Define the *local* bilinear form

$$\mathcal{A}_k(v,w) = \int_{\Omega_k} a\nabla v \cdot \nabla w dx, \quad v, w \in H^1(\Omega_k).$$

The discrete problem of (2.2) is: *Find* $u_h \in \tilde{V}(\Omega)$ *such that*

$$\sum_{k=1}^{N} \mathcal{A}_k(u_k, v_k) = (f, v) \quad \forall v \in \tilde{V}(\Omega). \tag{2.3}$$

It has been shown in [4] that the unique solution u_h of the system (2.3) has the optimal error estimate. By introducing the *sign* function

$$\sigma_{ij} = \begin{cases} 1, & i < j \\ -1, & i > j, \end{cases}$$

we define the *weak* trace operator $B_k : V(\Omega_k) \to W(\Gamma)$ as follows:

$$(B_k u)|_{\Gamma_{ij}} = \begin{cases} \sigma_{ij} P_{ij}(u_k|_{\Gamma_{ij}}), & \Gamma_{ij} \subset \partial\Omega_k, \\ 0, & \Gamma_{ij} \not\subset \partial\Omega_k. \end{cases}$$

Define the operators $A : V(\Omega) \to V(\Omega)$ and $B : V(\Omega) \to W(\Gamma)$ respectively by

$$(Av, w) = \sum_{k=1}^{N} \mathcal{A}_k(v_k, w_k), \quad \forall w \in V(\Omega)$$

and

$$Bv = \sum_{k=1}^{N} B_k v_k, \quad v \in V(\Omega).$$

It is clear that A is a symmetric and positive semi-definite operator. Moreover, A has the block structure

$$A = diag(A_1 \; A_2 \cdots A_N),$$

where A_k is the operator defined by the bilinear form $\mathcal{A}_k(\cdot, \cdot)$.

It is easy to see that the space $\tilde{V}(\Omega)$ can be written as

$$\tilde{V}(\Omega) = \{v \in V(\Omega) : Bv = 0\}.$$

Then (2.3) is equivalent to the saddle-point problem: *Find* $(\bar{u}, \lambda) \in V(\Omega) \times W(\Gamma)$ *such that*

$$\begin{cases} A\bar{u} + B^t \lambda = f, \\ B\bar{u} = 0. \end{cases} \tag{2.4}$$

Hereafter, $B^t : W(\Gamma) \to V(\Omega)$ denotes the dual of B, which satisfies

$$(B^t \mu, v) = \langle \mu, Bv \rangle, \quad \forall \mu \in W(\Gamma), \; v \in V(\Omega)$$

with $\langle \cdot, \cdot \rangle$ denoting the $L^2(\Gamma)$ inner product.

It is known that the interface unknown λ is the Lagrange multiplier for the constraint $B\bar{u} = 0$. Moreover, the solution of (2.4) is also unique, which is equivalent to the condition that

$$ker(A) \cap ker(B) = \{0\}. \tag{2.5}$$

Although the operator A is block diagonal, the system (2.4) can not be solved in the standard way (refer to [18] and [13]). The main difficulty is that the *local* operators A_k associated with the internal subdomains are singular.

2.2 Interface equation

For convenience's sake, we derive the interface equation of λ in operator language (compare [10] and [20]).

Let A^+ be the pseudoinverse of the operator A. A solution u_h of the first equation in (2.4) exists if and only if $f - B^t\lambda \in range(A)$. Hence,

$$u_h = A^+(f - B^t\lambda) + u_0 \ \ if \ \ f - B^t\lambda \perp ker(A), \tag{2.6}$$

where $u_0 \in ker(A)$ remains to be determined. Substituting u_h from (2.6) into the second equation of (2.4), yields

$$BA^+(f - B^t\lambda) + Bu_0 = 0. \tag{2.7}$$

Moreover, by the condition $f - B^t\lambda \perp ker(A)$, we deduce

$$\langle \lambda, Bv_0 \rangle = (f, v_0), \quad \forall v_0 \in ker(A). \tag{2.8}$$

Define

$$\bar{W} = \{\mu \in W(\Gamma) : \mu \perp Bv_0, \ \forall v_0 \in ker(A)\},$$

and let $P : W(\Gamma) \to \bar{W}$ denote the L^2 projection with respect to the L^2 inner product on Γ. Since $PBu_0 = 0$, it follows by (2.7) that

$$PS\lambda = Pg, \tag{2.9}$$

where

$$S = BA^+B^t, \ \ g = BA^+f.$$

Any solution λ of (2.8) and (2.9) yields the same solution u_h of (2.4) by using (2.6) if u_0 is determined by (2.7)

$$\langle Bu_0, Bv_0 \rangle = -\langle BA^+(f - B^t\lambda), Bv_0 \rangle, \quad \forall v_0 \in ker(A).$$

Let $\lambda_0 \in W(\Gamma)$ be a particular solution of (2.8). Then, any solution of (2.9) may be written in the form

$$\lambda = \lambda_0 + \bar{\lambda}, \quad \bar{\lambda} \in \bar{W}.$$

Substituting it into (2.9), yields

$$PS\bar{\lambda} = \bar{g}, \quad \bar{\lambda} \in \bar{W} \tag{2.10}$$

with $\bar{g} = P(g - S\lambda_0)$.

The operator $PS : \bar{W} \to \bar{W}'$ is symmetric and positive definite on the subspace $\bar{W}$ (refer to [21]), so we can solve (2.9) by PCG method on this subspace (refer to [10]). The effectiveness of this method depends on preconditioner of PS. Note that the action of the projection P is very cheap to implement, because $ker(A)$ consists of piecewise constant functions with respect to the initial division of Ω.

3. Preconditioner

In this section, we construct a substructuring preconditioner for the interface operator PS. To make the preconditioner to be more precise, we introduce positive constants α_k and β_k, which are defined by

$$\alpha_k \leq a(x) \leq \beta_k, \quad \forall x \in \Omega_k \ (k = 1, \cdots, N). \tag{3.1}$$

3.1 Motivation

Since the operators A and B have natural block structures, the operator S can be written as

$$S = \sum_{k=1}^{N} B_k A_k^+ B_k^t,$$

where A_k^+ is the pseudoinverse of the local operator $A_k : V(\Omega_k) \rightarrow V(\Omega_k)$. Let I_k denote the identity operator on $V(\Omega_k)$. Define $\bar{A}_k = A_k + d^{-2}\alpha_k I_k$ and

$$\bar{S} = \sum_{k=1}^{N} B_k \bar{A}_k^{-1} B_k^t.$$

Let n_k denote the dimension number of the space $V(\Omega_k)$. Throughout this paper, let C denote the generic constant independent of n_k and d. For convenience, following [28], the symbols $\lesssim$, $\gtrsim$ and $\eqsim$ will be used in the rest of this paper. $x_1 \lesssim y_1$, $x_2 \gtrsim y_2$ and $x_3 \eqsim y_3$, mean that $x_1 \leq C_1 y_1$, $x_2 \geq C_2 y_2$ and $C_3 x_3 \leq y_3 \leq C_3 x_3$ for some constants C_1, C_2, C_3 and C_3 and are independent of the dimensional number of the approximate spaces.

To explain our idea, we need an equivalence result.

Lemma 3.1. *The operator S is spectrally equivalent to the operator $\bar{S}$ on the subspace $\bar{W}$.*

The above lemma tells us that PS and $P\bar{S}$ are also spectrally equivalent on $\bar{W}$, and so we only need to construct an efficient preconditioner for $P\bar{S}$. The preconditioner would possess the form of PM^{-1}, where $M : W(\Gamma) \rightarrow W(\Gamma)$ is an efficient preconditioner for $\bar{S}$. For a face $\Gamma_{ij} \subset \Gamma$, let $I_{ij}^t : W(\Gamma_{ij}) \rightarrow W(\Gamma)$ denote the zero extension operator. It is clear that $W(\Gamma)$ has the direct sum decomposition

$$W(\Gamma) = \sum_{\Gamma_{ij}} I_{ij}^t W(\Gamma_{ij}).$$

Thus, we may define M as a block-diagonal operator, such that each block of M is spectrally equivalent to the restriction operator of $\bar{S}$ on a local space $W(\Gamma_{ij})$.

3.2 Main result

For each face $\Gamma_{ij} \subset \Gamma$, let $I_{ij} : W(\Gamma) \to W(\Gamma_{ij})$ denote the natural restriction operator, which is the dual of I_{ij}^t with respect to the $L^2(\Gamma)$ inner product. For each k, define $R_k : W(\Gamma) \to V(\Omega_k)$ by $R_k = \bar{A}_k^{-1} B_k^t$. Let $\langle \cdot, \cdot \rangle_{\Gamma_{ij}}$ denote the L^2 inner product on the local interface Γ_{ij}, and let the operators S_{ij}^i, $S_{ij}^j : W(\Gamma_{ij}) \to W(\Gamma_{ij})$ be defined by

$$\langle S_{ij}^i \lambda_{ij}, \mu_{ij} \rangle_{\Gamma_{ij}} = (\bar{A}_i R_i I_{ij}^t \lambda_{ij}, R_i I_{ij}^t \mu_{ij})_{\Omega_i}, \quad \forall \mu_{ij} \in W(\Gamma_{ij})$$
$$\langle S_{ij}^j \lambda_{ij}, \mu_{ij} \rangle_{\Gamma_{ij}} = (\bar{A}_j R_j I_{ij}^t \lambda_{ij}, R_j I_{ij}^t \mu_{ij})_{\Omega_j}, \quad \forall \mu_{ij} \in W(\Gamma_{ij}).$$

Define $S_{ij} = S_{ij}^i + S_{ij}^j$. It is easy to see that S_{ij} is just the restriction of the operator $\bar{S}$ on $W(\Gamma_{ij})$.

Let $\Lambda_{ij} : W(\Gamma_{ij}) \to W(\Gamma_{ij})$ be a symmetric and positive definite operator. Assume that the operator Λ_{ij} is spectrally equivalent to the operator S_{ij}^{-1}. We define the preconditioner M by

$$M^{-1} = \sum_{\Gamma_{ij}} I_{ij}^t \Lambda_{ij} I_{ij}. \tag{3.2}$$

Throughout this paper, we assume that the local multiplier space $W(\Gamma_{ij})$ is associated with the local trace space $V_i(\Gamma_{ij})$ (to distinguish it from $V_j(\Gamma_{ij})$). When the coefficient $a(x)$ has a large jump across the local interface Γ_{ij}, we need a particular choice of the index i. Note that in applications there are a few such local interfaces at most. In this case, choose the index i such that **one** of the following conditions is satisfied.

H_1: $\alpha_i \leq \alpha_j$;

H_2: $V_i^0(\Gamma_{ij}) \subset V_j^0(\Gamma_{ij})$ or $h_j \ll h_i$.

Remark 3.1. *The relation $V_i(\Gamma_{ij}) \subset V_j(\Gamma_{ij})$ means that the grids on Γ_{ij} are matching or nested. It is clear that we do not need to choose a particular index i for the case of matching grids. If the condition H_2 can not be satisfied, we then choose the index i in according to the condition H_1. Note that this particular choice of the index i will not influence applications of our method.*

Theorem 3.1. *Let the operator M^{-1} be defined by (3.2). Then,*

$$cond(PM^{-1}PS) \leq C[1 + \log(d/h)]^2, \tag{3.3}$$

where $h = \min_{1 \leq k \leq N} h_k$. If the condition H_1 **or** *H_2 is satisfied, the constant C in (3.3) is bounded by $\max_{1 \leq k \leq N} (\beta_k / \alpha_k)$, which is independent of the jump of the coefficient $a(x)$ across the local interface Γ_{ij}.*

The proof of this theorem depends on a squence of Lemmas. Here, we omit it.

4. On the local solvers Λ_{ij}

When the local multiplier space $W(\Gamma_{ij})$ has low dimensions, the action of the inverse S_{ij}^{-1} can be implemented exactly. Otherwise, we have to develope a cheaper solver which would be spectrally equivalent to S_{ij}^{-1}. To this end, we need two additional results.

4.1 Extension theorem of $H^{\frac{1}{2}}$-norm

The following result is of interest itself, the proof of which is technical.

Theorem 4.1. *(extension theorem) Let Ω_i be a tetrahedron or hexahedron, and Γ_{ij} be a face of Ω_i. Then, there is a linear extension operator $E : V_i(\Gamma_{ij}) \rightarrow V(\partial\Omega_i)$, such that*

$$(Ev)|_{\Gamma_{ij}} = v \;\; and \;\; \|Ev\|_{\frac{1}{2},\partial\Omega_i} \lesssim \|v\|_{\frac{1}{2},\Gamma_{ij}}, \quad v \in V_i(\Gamma_{ij}). \tag{4.1}$$

4.2 Spectrally equivalent operator to S_{ij}^{-1}

The following result is the basis to design a cheap Λ_{ij}. The proof will use Theorem 4.1.

Theorem 4.2. *Let $\tilde{\Lambda}_{ij} : W(\Gamma_{ij}) \rightarrow W(\Gamma_{ij})$ be a symmetric and positive definite operator satisfying $\langle\tilde{\Lambda}_{ij}\cdot,\cdot\rangle_{\Gamma_{ij}} \eqsim \|\cdot\|^2_{\frac{1}{2},\Gamma_{ij}}$. Then, for any $\mu_{ij} \in W(\Gamma_{ij})$, we have (when H_1 is satisfied)*

$$\alpha_i\langle\tilde{\Lambda}_{ij}\mu_{ij},\mu_{ij}\rangle_{\Gamma_{ij}} \lesssim \langle S_{ij}^{-1}\mu_{ij},\mu_{ij}\rangle_{\Gamma_{ij}} \lesssim \beta_i\langle\tilde{\Lambda}_{ij}\mu_{ij},\mu_{ij}\rangle_{\Gamma_{ij}}, \tag{4.2}$$

or (when H_2 is satisfied)

$$\frac{1}{\alpha_i^{-1}+\alpha_j^{-1}}\cdot\langle\tilde{\Lambda}_{ij}\mu_{ij},\mu_{ij}\rangle_{\Gamma_{ij}} \lesssim \langle S_{ij}^{-1}\mu_{ij},\mu_{ij}\rangle_{\Gamma_{ij}} \lesssim \frac{1}{\beta_i^{-1}+\beta_j^{-1}}\cdot\langle\tilde{\Lambda}_{ij}\mu_{ij},\mu_{ij}\rangle_{\Gamma_{ij}}. \tag{4.3}$$

4.3 Cheap local solvers

It is easy to see that

$$\frac{\alpha_i^{-1}+\alpha_j^{-1}}{\beta_i^{-1}+\beta_j^{-1}} \leq \max\{\frac{\beta_i}{\alpha_i},\frac{\beta_j}{\alpha_j}\}.$$

Thus, it follows by Theorem 4.2 that the desired operator Λ_{ij} would be defined as

$$\Lambda_{ij} = \alpha_i\tilde{\Lambda}_{ij} \; (when \; H_1 \; is \; satisfied)$$

or

$$\Lambda_{ij} = \frac{1}{\alpha_i^{-1} + \alpha_j^{-1}} \cdot \tilde{\Lambda}_{ij} \; (when\ H_2\ is\ satisfied),$$

so that the condition number of the preconditioned system is independent of the jump of the coefficient $a(x)$ across the local interface Γ_{ij}.

We first give a choice of the local operator $\tilde{\Lambda}_{ij}$ by using discrete norms of linear finite element function.

For each $\Gamma_{ij} \subset \Gamma$, define the operator $\tilde{\Lambda}_{ij} : W(\Gamma_{ij}) \to W(\Gamma_{ij})$ by

$$\langle \tilde{\Lambda}_{ij}\varphi, \psi\rangle_{\Gamma_{ij}} = h_i^4 \sum_{\substack{x_m, x_n \in \mathcal{T}_{ij} \\ x_m \neq x_n}} \frac{(\varphi(x_m) - \varphi(x_n))(\psi(x_m) - \psi(x_n))}{|x_m - x_n|^3} \tag{4.4}$$

$$+ \; d^{-1} h_i^2 \sum_{x_l \in \mathcal{T}_{ij}} \varphi(x_l)\psi(x_l), \quad \varphi, \psi \in W(\Gamma_{ij}). \tag{4.5}$$

In particular, for the nodal basis $\{\varphi_l\}$ of $W(\Gamma_{ij})$, we have

$$\langle \tilde{\Lambda}_{ij}\varphi_m, \varphi_n\rangle_{\Gamma_{ij}} = \begin{cases} \sum_{\substack{x_l \in \mathcal{T}_{ij} \\ x_l \neq x_m}} \frac{h_i^4}{|x_l - x_m|^3} + \frac{h_i^2}{d}, & if\ m = n \\ \frac{-h_i^4}{|x_m - x_n|^3}, & if\ m \neq n. \end{cases}$$

When $\psi = \varphi$, the right hand side of (4.5) is just the discrete form of the norm $\|\varphi\|^2_{\frac{1}{2},\Gamma_{ij}}$ (refer to [28]), so $\langle \tilde{\Lambda}_{ij}\cdot, \cdot\rangle_{\Gamma_{ij}}$ is spectrally equivalent to the norm $\|\cdot\|^2_{\frac{1}{2},\Gamma_{ij}}$ on $W(\Gamma_{ij})$.

Now, we describe the action of the local solver $\tilde{\Lambda}_{ij}$. We only need to explain how to get $\tilde{\Lambda}_{ij}\mu \in W(\Gamma_{ij})$ from $\mu \in W(\Gamma_{ij})$.

As in Section 2, set $dim(W(\Gamma_{ij})) = n_0(i,j)$. Let M_{ij} denote the (sparse) mass matrix with the entries $\langle \varphi_m, \varphi_n\rangle_{\Gamma_{ij}}$, and K_{ij} denote the matrix with the entries $\langle \tilde{\Lambda}_{ij}\varphi_m, \varphi_n\rangle_{\Gamma_{ij}}$ $(m, n = 1, \cdots, n_0(i,j))$. For $\mu \in W(\Gamma_{ij})$, let μ and $\tilde{\Lambda}_{ij}\mu$ be written as

$$\mu = \sum_l a_l \varphi_l \; and \; \tilde{\Lambda}_{ij}\mu = \sum_m z_m \varphi_m.$$

Define the vectors $b = (a_1\, a_2 \cdots a_{n_0(i,j)})^t$ and $\chi = (z_1\, z_2 \cdots z_{n_0(i,j)})^t$. By the equation

$$\sum_m z_m \varphi_m = \sum_l a_l \tilde{\Lambda}_{ij}\varphi_l,$$

we know that the unknown χ can be obtained by

$$M_{ij}\chi = K_{ij}b. \tag{4.6}$$

Remark 4.1. *We would like to compare the arithmetic complexity of the local solver* $\tilde{\Lambda}_{ij}$ *with that of the existing local solvers. Here, we consider only the number of multiplication of the direct algorithm. It is easy to see that solving* χ *by (4.6) needs only* $O(n_0^2(i,j))$ *multiplications. But, computation of* S_{ij}^{-1} *(see Section 3) or solving a Dirichlet problem on* Ω_i *(for FETI method) needs* $O(n_0^3(i,j))$ *multiplications, since* S_{ij} *results in a dense stiffness matrix, and a Dirichlet problem on* Ω_i *has* $O(n_0^{\frac{3}{2}}(i,j))$ *unknowns. We point out that we do not need to make a particular requirement for the meshes on* Γ_{ij} *here . This is a very important merit in DDMs for three- dimensional problems. If the meshes on* Γ_{ij} *have particular structure, we can decrease the number of the multiplications by defining a special operator* $\tilde{\Lambda}_{ij}$.

References

[1] Y. Achdou, Yu. Kuznetnov and O. Pironneau, *Substructuring Preconditioners for the* Q_1 *mortar element method*, Numer. Math., 1995, Vol.71, pp. 419-449

[2] Y. Achdou, Y. Maday, and O. Widlund, *Iterative substructuring preconditioners for mortar element methods in two dimensions*, SIAM J. Numer. Anal., **36**(1999), pp. 551-580

[3] F. Belgacem, *The mortar finite element method with Lagrange multipliers*, Numer. Math., **84**(1999), pp. 173-197

[4] F. Belgacem and Y. Maday, *The mortar element method for three dimensional finite elements*, M^2AN **31**(1997), pp. 289-302

[5] C. Bernardi, Y. Maday and A. Patera, *A new nonconforming approach to domain decomposition:the mortar element method*, In: Pitman, H.Brezis and J. Lions (eds) Nonlinear partial differential equations and their applications, 1989

[6] J. Bramble, J. Pasciak and A. Schatz, *The construction of preconditioners for elliptic problems by substructuring*, IV. Math. Comp., **53**(1989), pp. 1-24.

[7] Q. Dinh, R. Glowinski and J. Periaux, *Solving elliptic problems by domain decomposition methods with applications*, in Elliptic Problem Solver II, Academic Press, New York, 1982

[8] M. Dorr, *On the discretization of interdomain coupling in elliptic boundary-value problems*, in Domain Decomposition Methods, T. F. Chan...(eds), 1989, SIAM. Philadelphia, pp. 17-37.

[9] C. Farhat and F. Roux, *A method of finite element tearing and interconnecting and its parallel solution algorithm*, Internat. J. Numer. Methods Engrg, **32**(1991), pp. 1205-1227

[10] C. Farhat, J. Mandel, F. Roux, *Optimal convergence properties of the FETI Domain Decomposition Method*, Comput. Methods. Appl. Mech. Engrg.,**115** (1994), pp. 367-388.

[11] Q. Hu and G. Liang, *A general framework to construct interface preconditioners*, Chinese J. Num. Math.& Appl.,**21**(1999), pp. 83-95 (Published in New York)

[12] Q. Hu, *Study of Domain Decomposition Methods with Non-matching Grids*, Ph. D thises, Institute of Mathematics, Chinese Academy of Science, Beijing, 1998

[13] Q. Hu, G. Liang and J. Lui, *The construction of preconditioner for domain decomposition methods with Lagrangian multipliers*, J. Comp. Math., **19**(2001), pp. 213-224

[14] A. Klawonn and O. Widlund, *FETI and Neumann-Neumann iterative substructuring methods: connections and new results*, to appear in Comm. Pure Appl. Math.

[15] C. Kim, R. Lazarov, J. Pasciak and P. Vassilevski, *Multiplier spaces for the mortar finite element method in three dimensions*, Submitted

[16] Y. Kuznetsov, *Efficient iterative solvers for elliptic finite element problems on non-matching grids*, Russian J. Numer. Anal. and Math. Modeling, **10**(1995), 3, pp. 187-211

[17] C. Lacour, *Iterative substructuring preconditioners for the mortar finite element method*, In P. Bjorstad, M. Espedal, and D. Keyes, editors, Ninth International Conference of Domain Decomposition Methods, 1997.

[18] G. Liang and J. He, *The non-conforming domain decomposition method for elliptic problems with Lagrangian multipliers*, Chinese J. Num. Math. Appl.15:1(1993), pp. 8-19

[19] G. Liang and P. Liang, *Non-conforming domain decomposition with the hybrid finite element method*, Math. Numer. Sinica, 1989, Vol.11, No.3, pp. 323-332

[20] J. Mandel and R. Tezaur, *Convergence of a substructuring methods with Lagrangian multipliers*, Numer. Math., **73**(1996), pp. 473-487

[21] J. Mandel, R. Tezaur and C. Farhat, *A scalable substructuring method by Lagrange multipliers for plate bending problems*, SIAM J. Numer. Anal., **36**(1999), pp. 1370-1391

[22] D. Rixen and C. Farhat, *A simple and efficient extension of a class of substructure based preconditioners to heterogeousnstructural mechanics problems*, Int. J. Numer. Mech. Engng., **44**(1999), pp. 489-516

[23] B. Smith, P. Bjorstad and W. Gropp, *Domain Decomposition: Parallel multilevel methods for elliptic partial differential equations*, Cambridge University Press, 1996.

[24] P. Tallec, *Domain Decomposition Methods in Computational Mechanies*, Comput. Mech. Adv., 2: pp. 1321-220, 1994.

[25] P. Tallec, T. Sassi and M. Vidrascu, *Three-dimensional domain decomposition methods with nonmatching grids and unstructured coarse solvers*, Contemporary Mathematics, 1994, Vol.180, pp. 61-74.

[26] B. Wohlmuth, *A mortar finite element method using dual spaces for the Lagrange multiplier*, to appear

[27] J. Xu, *Iterative methods by space decomposition and subspace correction*, SIAM Review, **34**(1992), pp. 581-613.

[28] J. Xu and J. Zou, *Some non-overlapping domain decomposition methods*, SIAM Review, **24**(1998).

AN OPTIMAL ERROR ESTIMATE FOR AN H-P CLOUDS GALERKIN METHOD *

Jun Hu, Yunqing Huang
Department of Mathematics, Xiangtan University, Xiangtan, 411105, China
hujunlya@263.net
huangyq@xtu.edu.cn

Weimin Xue
Department of Mathematics, Hong Kong Baptist University
Kowloon Tong, Hong Kong, China
wmxue@hkbu.edu.hk

Abstract In this paper, we investigate the consistency and the approximation properties of h-p clouds methods. For this purpose, a special partition of unity function space in which inverse inequalities can be established is constructed. The optimal error estimate for the h-p clouds Galerkin methods is then established. The convergence rates are measured by the radius of influence domains of weight functions instead of the mesh size as usually used in the finite element analysis.

Keywords: h-p clouds, error estimate, partition of unity, meshless(mesh free) methods, moving least square, reproducing kernel particle

1. Introduction

There is a growing interest in the so-called "meshless" methods. It may be partly traced to high costs involved in meshing and remeshing procedures. Modelling of adapting domain geometry, fracture, fragment and similar phenomena requires considerable remeshing efforts, which can easily constitute the largest portion of analysis costs.

Recent literatures on the convergence of meshless methods can be found in [1, 7, 8, 10]. Some general results on the partition of unity finite element methods are provided in [1] by using technique of Taylor expansion, Liu et. al in [15] explored the convergence and error estimate of RKPM interpolation

*Subsidized by the Special Fund for Major State Basic Research Projects and State Educational Ministry.

under the assumption that the function to be approximated belongs to C^{m+1} for some $m \geq 1$. The analysis and proofs in [15] are not rigorous, though; e.g. no conditions are identified under which the method is defined, and some statements and proofs seem to be incorrect. A rigorous theoretical analysis of RKPM is given in [10], where in particular optimal order error estimates are derived with the use of the theory of averaged Taylor polynomials introduced in [4]. Duarte and Oden's technique in [7] of proving the local approximation properties for h-p clouds methods is similar to that of showing the convergence for finite element method. For global approximation property of h-p clouds methods, [7] just shows the case of k equal to zero, which can be regarded as a special case of the result of [1]. For arbitrary integer k, the essential difficulties are how to glue the local convergence rates into a global one and how to present the global approximation function to the given function. Since local spaces are not complete and partition of unity is not only a tool for gluing local spaces but also admits some order consistency, a direct application of the result of Babuška and Melenk in [1] will not give the satisfied results.

In this paper, we investigate the consistence and the approximation properties of h-p clouds methods (k is an arbitrary integer) at first, and then we discuss the convergence of h-p clouds Galerkin methods. For this purpose, a special partition of unity function space in which inverse inequalities can be established is constructed. By employing the arguments in [10], applying various techniques of Taylor expansion, theories of average polynomials interpolation, inverse estimate and triangular inequalities, we obtain the optimal error estimate of h-p clouds Galerkin methods. The convergence rates are measured by the radius of influence domains of weight functions instead of the mesh size as usually used in the finite element analysis.

This paper is organized as follows. In section 2, we introduce meshless function spaces. In section 3, we investigate the consistence and the approximation properties of h-p clouds methods. Section 4 is devoted to the convergence of h-p clouds Galerkin methods. The last section gives some concluding remarks.

2. Meshless Function Spaces

Often used meshless methods include RKPM [12,13,14], MLSM [2,12,13], PUM [1], h-p clouds method, etc. It is well-known (and easy to verify) that RKPM is equivalent to the moving least squares approximations with shifted monomials. Let$\{\Phi_i\}_{i=1}^{N}$ be shape functions of MLSM [2,12,13] or RKPM [12,13,14]. Denote the MLSM of RKPM function spaces by

$$V = \{f \| f = \sum_{i=1}^{N} \Phi_i b_i\}, \tag{2.1}$$

respectively. The consistency for order k of RKPM or MLSM can be found in [2,12,13,14].

Set $A = PWP^T$, where the definitions of matrixes P and W can be found in [2,12,13]. P is often called moment matrix. We need to introduce the concept of regularity for a family sets of nodes (presented first in [10] in the context of RKPM).

Definition 2.1 A family of sets of nodes is said to be regular if there exists a constant $L_0 > 0$ such that $\max_{x \in \bar{\Omega}} \|A(x)^{-1}\|_2 \leq L_0$ for any set of nodes in the family. Here $\|\cdot\|$ is the matrix spectrum norm.

Let $\{X_i\}_{i=1}^N$ be a set of nodes of interest. A necessary condition for regularity of nodes is there exist node X_i and a constant c_0 independent of point x for any point x such that $\|(x - x_i)\|/h_i < c_0 < 1$, where h_i are the diameters of the supports of the window functions.

We focus on the case of circle supports of window functions and construct a special kind of partition of unity space here. Let $\{X_i\}_{i=1}^N$ be a set of nodes including the nodes on the boundary $\partial\Omega$ and $\{\Omega_i\}_{i=1}^N$ be an associated open cover of Ω, which satisfies the regularity and the overlapping condition,i.e, $\forall X \in \Omega$, $\exists M_2$ such that $card\{i \mid X \in \Omega_i\} \leq M_2$. Partition of unity $\{\varphi_i\}_{i=1}^N$ are defined by

$$\varphi_i = \omega_i / \sum_{i=1}^{N} \omega_i \tag{2.2}$$

where

$$\omega_i = \begin{cases} (1 - (\frac{\|X - X_i\|}{h_i})^2)^{p+3} & \frac{\|X - X_i\|}{h_i} \leq 1 \\ 0 & \frac{\|X - X_i\|}{h_i} \geq 1 \end{cases}$$

A special partition of unity space is defined as following

$$V^0 = span\{\sum_{i=1}^{N} \varphi_i V_i\}, \tag{2.3}$$

where $V_i = span\{1, \frac{X - X_i}{h_i}, \cdots, (\frac{X - X_i}{h_i})^p\}$.

Proposition 2.1 Let V^0 defined by (2.3), then $V^0 \subset C^{p+2}(\Omega)$.

For a given partition of unity $\{\Phi_i\}_{i=1}^N$ which admits the consistency of order k we denote the so called h-p clouds function space by

$$F^{k,p}(\Omega) = \{f \| f = \sum_{i=1}^{N} \Phi_i b_i + \sum_{i=1}^{N} \Phi_i \sum_{j} c_{ji} p_j(X)\}$$

where $p_j(X)$ consist of all basis of polynomials of degree greater than k and less than or equal to p.

3. Convergence of General h-p Clouds Approximations

The convergence of h-p clouds approximation is first investigated by Duarte and Oden [7]. They analyze the local convergence rates of h-p clouds approximations and obtain the global convergence of h-p clouds methods for the case of $k = 0$. For arbitrary integer k, the essential difficulties are how to present the approximate function and how to glue the local convergence rates into a global one. In this section, we shall consider the convergence of general h-p clouds approximations.

First, let us quote some basic results for the approximation properties on RKPM and MLSM from [10].

Theorem 3.1. *Let $\{\Omega_i\}_{i=1}^N$ be influence domains of MLSM or RKPM shape functions $\{\Phi_i(x)\}_{i=1}^N$ which cover Ω. Suppose that the cover satisfied the overlapping condition and the set of nodes admits regularity condition, then*

$$\max_{1\leq i\leq N}\max_{\beta:\beta=l}\|D^\beta\Phi_i\|_{L^\infty(\Omega_i)}\leq\frac{C}{h^l}\qquad 0\leq l\leq k,\tag{3.1}$$

where h is the maximum diameter of influence domains of shape functions.

Based on the shape functions ϕ_i, define the interpolation u_I of a given function u by

$$u_I(X)=\sum_{i=1}^N\Phi_i(X)u(X_i)$$

The interpolation approximation error estimates are stated by the following theorem.

Theorem 3.2. *Assume the conditions of Theorem 3.1 hold and suppose $u\in H^k(\Omega)\bigcap C^0(\Omega)$. If the shape functions Φ_i admit consistency of order k, then there hold*

$$\|u-u_I\|_{L^2(\Omega)}\leq ch^{k+1}\|u\|_{H^{k+1}(\Omega)}\tag{3.2}$$

$$\|\nabla(u-u_I)\|_{L^2(\Omega)}\leq ch^k\|u\|_{H^{k+1}(\Omega)}.\tag{3.3}$$

3.1 Some Technical Lemmas

We provide some preliminary results in this subsection.

Lemma 3.1. *([15] and [10]) Let $\{X_i\}_{i=1}^N\subset\Omega$ be a set of nodes and $\{\Omega_i\}_{i=1}^N$ be an associated open cover of Ω. Suppose the cover satisfies the overlapping condition and the nodes are regular. Let $\{\Phi_i\}_{i=1}^N$ be a partition of unity corresponding to the cover $\{\Omega_i\}_{i=1}^N$ with consistency of order k. Then the following* *identities hold.*

$$\sum_{i=1}^{N} X_i^\alpha \Phi_i(X) = X^\alpha \ \forall \mid \alpha \mid \leq k, \tag{3.4}$$

$$\sum_{i=1}^{N} (X_i - X)^\alpha \Phi_i(X) = \delta_{\alpha 0} \ \forall \mid \alpha \mid \leq k, \tag{3.5}$$

$$\sum_{i=1}^{N} (X_i - X)^\alpha D^\beta \Phi_i(X) = \beta! \delta_{\alpha\beta} \ \forall \mid \alpha \mid, \mid \beta \mid \leq k. \tag{3.6}$$

Lemma 3.2. (inverse estimate) *Let $f(X)$ and $g(X)$ be polynomials of order p in Ω_i with diameter h_i. Assume there exist constants $c_1 > 0$ and $c_2 > 0$ such that*

$$c_1 \leq \mid f(x) \mid \leq c_2,$$

then the following inverse estimates,

$$\|D^\beta \left(\frac{f(X)}{g(X)}\right)\|_{q,\Omega_i} \leq c h^{-|\beta|+|l|} \|D^l \left(\frac{f(X)}{g(X)}\right)\|_{q,\Omega_i} \ |l| \leq |\beta| \leq p. \tag{3.9}$$

Proof First of all, by similar arguments in finite element method through a linear transformation, we can show the inverse estimate for any polynomial $F(X)$,

$$\|\nabla^\beta F\|_{q,\Omega_i} = C h^{|l|-|\beta|} \|\nabla^l F\|_{q,\Omega_i} \qquad \forall 1 \leq q \leq \infty.$$

To prove (3.9), denote $z(X) = \frac{f(X)}{g(X)}$. A direct calculation gives

$$\begin{aligned}
|\nabla z(X)| &= \left| \frac{\nabla g(X)}{f(X)} - \frac{\nabla f(X) g(X)}{f^2(X)} \right| \\
&\leq \left| \frac{\nabla g(X)}{f(X)} \right| + \left| \frac{-\nabla f(X) g(X)}{f^2(x)} \right| \\
&\leq C\left(\mid \nabla g(X) \mid + \left| \frac{-\nabla f(X)}{f(X)} \right| \mid z(X) \mid \right) \\
&\leq C(\mid \nabla g(X) \mid + h^{-1} \mid z(X)) \mid,
\end{aligned}$$

which leads to

$$\begin{aligned}
\|\nabla z\|_{q,\Omega_i} &\leq C(\|\nabla g\|_{q,\Omega_i} + h^{-1} \|z\|_{q,\Omega_i}) \\
&\leq C h^{-1}(\|g\|_{q,\Omega_i} + \|z\|_{q,\Omega_i}) \\
&\leq C h^{-1} \|z\|_{q,\Omega_i}
\end{aligned}$$

It is obvious that

$$\nabla^{\alpha}\left(\frac{g}{f}\right)=\frac{G}{F},$$

where F and G are polynomials and $F = f^{|\alpha|+1}$. By induction, we can conclude that (3.9) holds.◇

3.2 Approximation Properties of V^0

Now we investigate the properties of the partition of unity function space V^0.

Lemma 3.3. *Let V be the special partition of unity function space defined in section 2.3. Assume the nodes $\{X_i\}_{i=1}^{N}$ admit the regularity condition. Then any components $v_i(X) = \varphi_i V(X) \in V^0$ satisfy the following inverse estimate,*

$$| D^{\beta} v_i |_{q,\Omega_i} \leq c h^{l-\beta} | D^{l} v_i |_{q,\Omega_i} \quad | l | \leq | \beta | \leq p+1 \tag{3.10}$$

Proof By the definition of the partition of unity function space, it suffices to establish the result for the following mode functions

$$z(x)=\omega_i(X)(\frac{X-X_i}{h_i})^l/\sum_{i=1}^{N}\omega_i \quad l\leq p$$

Set $g(X) = \omega_i(X)\left(\frac{X-X_i}{h_i}\right)^l$, $f(X) = \sum_{i=1}^{N} \omega_i(X)$. By the regularity of nodes, $\forall X \in \Omega$, there exists X_i such that $\|X - X_i\|/h_i \leq c_0 < 1$, where c_0 is a positive constant, therefore there are two positive constants c_1 and c_2 such that $c_1 \leq| f(X) |\leq c_2$. The conclusion immediately follows from lemma 3.2. ◇
A direct result of Theorem 3.6 is the following corollary:

Corollary 3.1 Assume the assumptions of lemma 3.2 hold. Let $\{\varphi_i\}_{i=1}^{N}$ be the special partition of unity defined in section 2.3 , then the following estimates hold:

$$| D^{\alpha}\varphi_i |_{L^{\infty}(\Omega)} \leq c_{\alpha}/h^{|\alpha|} \quad i=1,\cdots,N \tag{3.11}$$

where $c_{\alpha} > 0$ is a constant.

Now, we discuss convergence of partition of unity approximations, We have

Theorem 3.3. *Let V^0 be the special partition of unity function space defined in section 2.3 and assume $u(X) \in H^{p+1}(\Omega)$. Further, assume the nodes are regular. Then there exists $u_I(X) \in V^0$ such that*

$$| u - u_I |_{H^l(\Omega)} \leq c h^{p+1-l} | u |_{H^{p+1}(\Omega)} \quad 0 \leq l \leq p+1 \tag{3.12}$$

Proof $\forall \Omega_i$, let $Q_i^{p+1}u(X)$ be the TAPI (Taylor Average Polynomial Interpolation) of degree p of $u(x)$ over Ω_i (see [13,10] for details). By the definition of local approximation spaces of V_i, there exist b_{ji} on Ω_i such that

$$u_{iap} = \sum_j b_{ji} p_j(X) = Q_i^{p+1} u(X)$$

By the approximation estimates of TAPI [13,10]

$$| u - u_{iap} |_{H^l(\Omega_i \bigcap \Omega)} \leq c h^{p+1-l} | u |_{H^{p+1}(\Omega_i \bigcap \Omega)} \quad 0 \leq l \leq p+1$$

Denote

$$u_I(X) = \sum_{i=1}^{N} \varphi_i(X) u_{iap}(X).$$

Then

$$| u - u_I |_{H^l(\Omega)} = | u - \sum_{i=1}^{N} \varphi_i u_{iap} |_{H^l(\Omega)}$$

For given $\alpha \leq l$, by Hŏlder's inequality, Corollary 3.1 and approximation of TAPI, we have

$$\begin{aligned}
& \sum_{|\beta|=\alpha} \quad \|D^\beta[\varphi_i(u - u_{iap})]\|^2_{L^2(\Omega_i)} \\
= & \quad \sum_{|\beta|=\alpha} \| \sum_{j=0}^{\beta} C_\beta^j D^j \varphi_i D^{\beta-j}(u - u_{iap}) \|^2_{L^2(\Omega_i)} \\
\leq & \quad c \sum_{|\beta|=\alpha} \sum_{j=0}^{\beta} \|D^j \varphi_i\|^2_{L^\infty(\Omega_i)} \|D^{\beta-j}(u - u_{iap})\|^2_{L^2(\Omega_i)} \\
\leq & \quad c \sum_{|\beta|=\alpha} \sum_{j=0}^{\beta} \frac{1}{(h_i)^{2|j|}} (h_i)^{2p+2-2|\beta|+2|j|} \, | u |^2_{H^{p+1}(\Omega_i)} \\
\leq & \quad c h_i^{2p+2-2\alpha} \, | u |^2_{H^{p+1}(\Omega_i)}
\end{aligned}$$

where h_i is the radius of Ω_i, the diameter of the influence domain. Using overlapping conditions we obtain

$$
\begin{aligned}
| u - u_I |_{H^l(\Omega)}^2 &= | \sum_{i=1}^{N} \varphi_i(u - u_{iap}) |_{H^l(\Omega)}^2 \\
&\leq c \sum_{i=1}^{N} | \varphi_i(u - u_{iap}) |_{H^l(\Omega_i)}^2 \\
&= c \sum_{i=1}^{N} (\sum_{|\beta|=l} \| D^\beta(\varphi_i(u - u_{iap})) \|_{L^2(\Omega_i)}^2) \\
&\leq c \sum_{i=1}^{N} h_i^{2p+2-2l} | u |_{H^{p+1}(\Omega_i)}^2
\end{aligned}
$$

Let $h = \max(h_i)$. We have

$$
| u - u_I |_{H^l(\Omega)} \leq c \sum_{i=1}^{N} h_i^{p+1-l} | u |_{H^{p+1}(\Omega_i)} \leq ch^{p+1-l} | u |_{H^{p+1}(\Omega)},
$$

which completes the proof.◇

3.3 Convergence of General h-p clouds Approximations

Below we establish an important preliminary result.

Theorem 3.4. *Let $u(X) \in V^0$ be the special partition of unity function space and $F^{k,p}(\Omega)$ be the h-p clouds function space defined in section 2. Assume the nodes $\{X_i\}_{i=1}^N$ are regular and overlapping condition satisfied. Then, there exists $u_I(X) \in F^{k,p}(\Omega)$ such that*

$$
| u - u_I |_{W^{l,q}(\Omega)} \leq ch^{p+1-l} | u |_{W^{p+1,q}(\Omega)} \quad l = 0, 1
$$

Proof Set

$$
u_I(X) = \sum_{i=1}^{N} u_i \Phi_i + \sum_{i=1}^{N} \Phi_i \sum_j b_{ji} p_j
$$

where b_{ji} are to be coefficients which are evaluated later. For $X \in \Omega_j$, because $u(X) \in C^{p+2}$, expanding u_i at the point X, we obtain

$$
\begin{aligned}
u_i = u(X_i) &= \sum_{|\alpha|\le k} \frac{1}{\alpha!}(X_i - X)^\alpha D_X^\alpha u(X) \\
&+ \sum_{k+1\le|\alpha|\le p} \frac{1}{\alpha!}(X_i - X)^\alpha D_X^\alpha u(X) \\
&+ \sum_{|\alpha| = p+1} \frac{1}{\alpha!}(X_i - X)^\alpha D_X^\alpha u(X + \theta(X_i - X)).
\end{aligned}
$$

For $k+1 \le |\alpha| \le p$, Taylor expansion of $D_X^\alpha u(X)$ at the point X_i yields

$$
\begin{aligned}
D_X^\alpha u(X) &= \sum_{|\beta|\le p-|\alpha|} (-1)^\beta \frac{D_X^{\alpha+\beta} u(X_i)}{\beta!}(X_i - X)^\beta \\
&+ \sum_{|\beta|=p-|\alpha|+1} (-1)^\beta \frac{D_X^{\alpha+\beta} u(X_i + \tilde{\theta}(X - X_i)}{\beta!}(X_i - X)^\beta,
\end{aligned}
$$

which leads to

$$
\begin{aligned}
u(X_i) &= \sum_{|\alpha|\le k} \frac{1}{\alpha!}(X_i - X)^\alpha D_X^\alpha u(X) \\
&+ \sum_{k+1\le|\alpha|\le p} \frac{1}{\alpha!}(X_i - X)^\alpha \{ \sum_{|\beta|\le p-|\alpha|} (-1)^\beta \frac{D_X^{\alpha+\beta} u(X_i)}{\beta!}(X_i - X)^\beta \\
&+ \sum_{|\beta|=p-|\alpha|+1} (-1)^\beta \frac{D_X^{\alpha+\beta} u(X_i + \tilde{\theta}(X - X_i))}{\beta!}(X_i - X)^\beta \} \\
&+ \sum_{|\alpha| = p+1} \frac{1}{\alpha!}(X_i - X)^\alpha D_X^\alpha u(X + \theta(X_i - X)).
\end{aligned}
$$

Thus we have

$$\begin{aligned} u_I(X) &= \sum_{i=1}^{N} \Phi_i \sum_{|\alpha|\le k} \frac{1}{\alpha!}(X_i - X)^\alpha D_X^\alpha u(X) \\ &+ \sum_{i=1}^{N} \Phi_i \sum_{k+1\le|\alpha|\le p} \frac{1}{\alpha!}(X_i - X)^\alpha \cdot \\ &\quad \sum_{|\beta|\le p-|\alpha|} (-1)^\beta \frac{D_X^{\alpha+\beta} u(X_i)}{\beta!}(X_i - X)^\beta \\ &+ \sum_{i=1}^{N} \Phi_i \sum_j b_{ji} p_j + r, \end{aligned}$$

where the remainder r is given by

$$\begin{aligned} r &= \sum_{i=1}^{N} \Phi_i \sum_{k+1\le|\alpha|\le p} \frac{1}{\alpha!}(X_i - X)^\alpha \\ &\quad \sum_{|\beta|=p-|\alpha|+1} (-1)^\beta \frac{D_X^{\alpha+\beta} u(X_i + \theta(X - X_i))}{\beta!}(X_i - X)^\beta \\ &+ \sum_{i=1}^{N} \Phi_i \sum_{|\alpha| = p+1} \frac{1}{\alpha!}(X_i - X)^\alpha D_X^\alpha u(X + \theta(X_i - X)). \end{aligned}$$

Hence we can choose $b_{ji}(u, h_i, \alpha, \beta, p, i)$ such that

$$u_I(X) = \sum_{i=1}^{N} \Phi_i \sum_{|\alpha|\le k} \frac{1}{\alpha!}(X_i - X)^\alpha D_X^\alpha u(X) + r(X).$$

Moreover,

$$\begin{aligned} &\sum_{i=1}^{N} \Phi_i \sum_{|\alpha|\le k} \frac{1}{\alpha!}(X_i - X)^\alpha D_X^\alpha u(X) \\ &= \sum_{|\alpha|\le k} \frac{1}{\alpha!} D_X^\alpha u(X) \sum_{i=1}^{N} \Phi_i (X_i - X)^\alpha \\ &= \sum_{|\alpha|\le k} \frac{1}{\alpha!} D_X^\alpha u(X) \alpha! \delta_{\alpha 0} = u(X). \end{aligned}$$

So we derived

$$u_I(X) - u(X) = r(X) \quad in\Omega_i \tag{3.13}$$

For every i, it is easy to see that

$$\|r\|_{L^q(\Omega\bigcap\Omega_j)} \leq ch^{p+1} \mid u \mid_{W^{P+1,q}(\Omega\bigcap\Omega_j)}$$

$$\begin{aligned}\|\nabla r\|_{L^q(\Omega\bigcap\Omega_j)} &\leq ch^p \mid u \mid_{W^{p+1,q}(\Omega\bigcap\Omega_j)} \\ &+ ch^{p+1} \mid u \mid_{W^{p+2,q}(\Omega\bigcap\Omega_j)}\end{aligned}$$

By the inverse estimate (Lemma 3.3), we have

$$\|\nabla^\alpha(u-u_I)\|_{L^q(\Omega\bigcap\Omega_j)} \leq ch^{p+1-|\alpha|} \mid u \mid_{W^{p+1,q}(\Omega\bigcap\Omega_j)} \quad |\alpha| \leq 1$$

From the overlapping condition, one can derive

$$\mid u - u_I \mid_{W^{l,q}(\Omega)} \leq ch^{p+1-l} \mid u \mid_{W^{p+1,q}(\Omega)} \quad l = 0, 1$$

The proof is completed. ◇

By Theorem 3.3 and Theorem 3.4, we can establish the convergence rates of h-p clouds approximations.

Theorem 3.5. *Let $F^{k,p}$ be h-p clouds function space. Suppose that $u(X) \in H^{p+1}(\Omega)$. Then there exists $u_I(X) \in F^{k,p}$ such that*

$$\mid u - u_I \mid_{H^l(\Omega)} \leq ch^{p+1-l} \mid u \mid_{H^{p+1}(\Omega)} \quad l = 0, 1 \tag{3.14}$$

Proof By Theorem 3.3, there exists $u_0(x) \in V^0$ such that

$$\mid u - u_1 \mid_{H^l(\Omega)} \leq ch^{p+1-l} \mid u \mid_{H^{p+l}(\Omega)} \quad l = 0, 1,$$

and

$$\mid u_1 \mid_{H^{p+1}(\Omega)} \leq c \mid u \mid_{H^{p+1}(\Omega)}$$

By Theorem 3.4, we can find $u_I(x) \in F^{k,p}$ such that

$$\mid u_0 - u_I \mid_{H^l(\Omega)} \leq ch^{p+1-l} \mid u_1 \mid_{H^{p+1}(\Omega)} \quad l = 0, 1.$$

Hence for $l = 0, 1$ we have

$$\begin{aligned}\mid u - u_I \mid_{H^l(\Omega)} &\leq \mid u - u_0 \mid_{H^l(\Omega)} + \mid u_0 - u_I \mid_{H^l(\Omega)} \\ &\leq ch^{p+1-l} \mid u \mid_{H^{p+1}(\Omega)} + ch^{p+1-l} \mid u_0(x) \mid_{H^{p+1}(\Omega)} \\ &\leq ch^{p+1-l} \mid u \mid_{H^{p+1}(\Omega)}\end{aligned}$$

This completes the proof of this theorem ◇

Remark If a proper partition of unity function space is chosen (smoother), one can show that there is $u_I(X) \in F^{k,p}$ such that

$$\mid u - u_I \mid_{W^{l,q}(\Omega)} \leq ch^{p+1-l} \mid u \mid_{W^{p+1,q}(\Omega)} \quad 0 \leq l \leq p+1$$

Remark Choose $k = p$ in Theorem 3.5, we obtain the error estimates of RKPM and MLSM interpolations.

4. Convergence for Neumann Problems

Since a natural boundary condition requires less restrictions on both trial functions and weight functions, it is convenient to consider the following model problem

$$a(u,v) = f(v) \ \forall\, v \in H^1(\Omega) \tag{4.1}$$

where $a(u,v)$ satisfies the coercive condition and continuous condition, i.e. there exist constants $\alpha > 0$ and $\beta > 0$ such that

$$\alpha\|v\|_1^2 \leq a(v,v) \ \ \forall v \in H^1(\Omega) \tag{4.2}$$

$$|\, a(u,v) \,| \leq \beta\|u\|_1\|v\|_1 \ \ \forall u,v \in H^1(\Omega) \tag{4.3}$$

Meshless Galerkin methods are defined by: Find $u^h \in F^{k,p}$ such that

$$a(u^h, v) = f(v) \ \forall\, v \in F^{k,p}. \tag{4.4}$$

Now we derive the error estimates for h-p clouds Galerkin methods.

Theorem 4.1. *Suppose $u(X) \in H^{p+1}(\Omega)$ is the solution of the problem of (4.1), and $u^h(X)$ is Galerkin approximate solution of (4.4). Then there hold*

$$\|u - u^h\|_{1,\Omega} \leq ch^p\|u\|_{p+1,\Omega}$$

$$\|u - u^h\|_{0,\Omega} \leq ch^{p+1}\|u\|_{p+1,\Omega}$$

provided Ω is smooth or convex.

Proof By Céa lemma

$$\|u - u^h\|_{1,\Omega} \leq c \inf_{v(x)\in F^{k,p}} \|u - v\|_{1,\Omega}$$

Theorem 3.5 leads to

$$\|u - u^h\|_{1,\Omega} \leq ch^p\|u\|_{p+1,\Omega}$$

By a standard duality argument for smooth or convex domain, we have

$$\|u - u^h\|_{0,\Omega} \leq ch^{p+1}\|u\|_{p+1,\Omega}$$

5. Concluding remarks

We provide an analysis for a general h-p clouds Galerkin methods. In particular, we proved the convergence and obtain the optimal error estimates. Generally speaking, meshless methods share the following potential advantages

- Modelling for two and three dimensional objects only requires nodes. No data or assumed structure on the interconnections of the nodes is needed.
- Adaptivity become relatively easy, since it is only necessary to add nodes.
- Problems such as progressive crack growth and moving interfaces can be easily handled without remeshing.
- Fast convergence rates can be obtained by properly choosing local approximation spaces.

However, there are also some disadvantages. For example,

- It is not easy to deal with the essential boundary value conditions though there is some progress on this topic, see, e.g. [5,6,8,11,16,19].
- How to evaluate the integrals in the meshless methods is still a major problem.

References

[1] Babuška. I and J. M. Melenk, The partition of unity finite element method, Int. J. Num. Meth. Eng. 40(1997), pp. 727-758.

[2] Belytschko, T., Y. Y. Lu and L. Gu. Element-free Galerkin methods, Int. J. Numer. Meth. Engr 37 (1994), pp. 229-256.

[3] T. Belytsho, Y. Krongauz, D. Organ, M. Fleming and P. Krysl, Meshless methods: An overview and recent developments.

[4] S. C. Brenner and L. R. Scott. The Mathematical Theory of Finite Element Methods. Springer-Verlag, New York, 1994.

[5] Chen J. S. and Wang H. P, New boundary condition treatments in meshfree computation problems, Comp. Meth. Appl. Nech. Engr, 187 2000.

[6] Chen, J. S., Wu, C. T., Yoon, S., and You, Y., "A Stabilized Conforming Nodal Integration for Galerkin Meshfree Methods," International Journal for Numerical Methods in Engineering, 50 (2001), pp. 435-466.

[7] Duarte, C. A. and J. T. Oden. H-p clouds-an h-p meshless method. Numer. Meth. Part. Diff. Equat, (1996), pp. 1-34.

[8] J. Gosz and W. K. Liu, Admissible approximations for essential boundary conditions in the reproducing kernel method, Comp. Mech 19 (1996), pp. 120-135.

[9] M. Griebel and M. A. Schweitzer. A particle-partition of unity method for the solution of elliptic, parabolic and hyperbolic pdes.

[10] W. Han and X. Meng. Error analysis of the Reproducing Kernel Particle Method, Computer Methods in Applied Mechanics and Engineering, 190 (2001), pp. 6157-6181.

[11] W. Han, G. J. Wagner and W. K. Liu, Convergence analysis of a hierarchical enrichment of Dirichlet boundary conditions in a meshfree method, to appear.

[12] Hu Jun. Analysis for a kind of h-p clouds Galerkin method and lower approximation of eigenvalues. MSc. Dissertationion, Xiangtan University, 2001.

[13] Hu Jun and Huang Yun Qing. Analysis for a kind of meshless Galerkin Method, The Fourth International Conference on Engineering and Computation, 2001, Beijing, China.

[14] W. K. Liu, S. Jun, and Y. F. Zhang. Reproducing kernel particle methods. Int. J. Numer. Meth. Engr, 20 (1995), pp. 1081-1106.

[15] W. K. Liu, S. Li, and T. Betytscho. Moving least-square reproducing kernel methods. part I: methodology and convergence, Computer Methods in Applied Mechanics and Engineering 143(1997), pp. 113-154.

[16] Krongauz Y., Belytschko T., Enforcement of Essential Boundary Conditions in Meshless Approximations Using Finite Elements, Computer Methods in Applied Mechanics and Engng, 131 (1996), pp. 133-145.

[17] P. Krysl and T.Belytschko. Element-free Galerkin method: Convergence of the continuous and discontinuous shape functions. Comp. Meth. Appl. Mech. Engr. 148 (1996), pp. 257-277

[18] B. Nayroles, G. Touzot, and P. Villon. Generalizing the finite element method: diffuse approximation and diffuse elements. Comp. Mech. 10 (1992), pp. 307-318.

[19] G. J. Wagner and W. K. Liu. Hierarchical enrichment for bridging scales and mesh-free boundary conditions. Int. J. Numer. Meth. Engr 50 (2001), pp. 507-524.

SOME PROBLEMS IN LARGE SCALE NON-HERMITIAN MATRIX COMPUTATIONS

Zhongxiao Jia*
Department of Mathematical Sciences, Tsinghua University, Beijing 100084, P.R. China

Abstract Large scale matrix eigenproblems arise in a lot of disciplines and scientific and engineering computations. Efficient and reliable numerical methods for solving them are extremely important as they play a vital role in innumerable scientific applications. In this paper, we attempt to review state of the art of theory and algorithms of commonly used numerical methods for large scale nonsymmetric (non-Hermitian) eigenproblems. We finally pose some challenging problems on the existing three kinds of popular solvers: orthogonal projection methods, oblique projection methods and refined projection methods.

Keywords: Large scale matrix, eigenproblem, projection method, Ritz value, Ritz vector, refined Ritz vector

1. Introduction

In many areas of applied sciences and engineering there invoke many important applications related to the eigenproblems of large unsymmetric matrice. For the large scale eigenproblems, all standard tools encounter obvious difficulties with speed and storage as well as flow of data within memory in the hierarchy such that they are very inefficient, and thus usually come at their wits' end. Even though in some cases they could be used for large problems by some technical treatments, noticing that many matrices coming from applications are large sparse, these tools would destroy the special structure of matrices involved, while this will lead to disasterous consequence. In a word, those efficient standard techniques suitable to small and medium sized matrices have to be abandoned, and we have to turn our attention into seeking efficient and reliable methods which can exploit sparsity and leave the matrix in question unchanged.

*Work supported by the Special Funds for the State Major Basic Research Projects (G1999032805) and the Foundation for Key Scholars in Chinese Universities.

Fortunately, recent evolution in computer hardware and software has stimulated a great deal of effects in attacking the eigenproblems of large unsymmetric matrices, and theory and algorithms have been intensively investigated since the early 1970's and especially 1980's. There have hitherto been principally three classes of basic projection methods available, as can be seen below, in which a common feature is that the action of the matrix A is only to form a subroutine of matrix by vector product as well as possibly its transpose by vector product and this just tallys what we appeal as regards storage and flow of data within memory in the hierarchy.

In Sections 2–4, we review three commonly used kinds of projection methods and state of the art of them. In Section 5, we pose some challenging problems.

2. Orthogonal projection methods

2.1 The methods

Given an m-dimensional subspace E, an orthogonal projection method on E seeks the Ritz pairs $(\tilde{\lambda}_i, \tilde{\varphi}_i)$ with $\|\tilde{\varphi}_i\| = 1$ that satisfy the following Rayleigh-Ritz approximation:

$$\begin{cases} \tilde{\varphi}_i \in E, \\ A\tilde{\varphi}_i - \tilde{\lambda}_i\tilde{\varphi}_i \perp E, \end{cases} \tag{1}$$

and use them to approximate some eigenpairs (λ_i, φ_i) of A.

The above relations can be written as

$$\begin{cases} By_i = \tilde{\lambda}_i y_i, \\ \tilde{\varphi}_i = Vy_i, \end{cases} \tag{2}$$

where $V = (v_1, v_2, ..., v_m)$ whose columns form an orthonormal basis of E, and $B = V^*AV$ is the projected matrix of A onto E in the basis $\{v_j\}_1^m$ or is called the matrix presentation of the restriction of A to E in the basis $\{v_j\}_1^m$.

A few well-known and most commonly used methods fall into this category: Arnoldi's method [1, 29] if $E = \mathcal{K}_m(v_1, A) = span\{v_1, Av_1, \ldots, A^{m-1}v_1\}$, a Krylov subspace; the block Arnoldi method [17, 30] if $E = \mathcal{K}_n(Q_1, A) = span\{Q_1, AQ_1, \ldots, A^{n-1}Q_1\}$, a block Krylov subspace; the subspace iteration method [29] if $E = span\{A^n X_0\}$, where X_0 is an $N \times m$ matrix whose columns are linearly independent; Davidson's method [8, 30]; the Jacobi–Davidson method [31].

2.2 State of the art

We first look at the subspace iteration. Parlett and Stewart have made the convergence analysis for the subspace iteration method [29]. To be more stable, Stewart suggests to compute Schur vectors and invariant subspaces instead of eigenvectors. This can avoid serious difficulties that the ill-conditioning of nonexistence of an eigenbasis causes and thus make the methods more robust; see [29]. A remarkable drawback of this method is that it may converge too slowly. In order to improve the speed of convergence, Rutishauser [29] suggests to use the Chebyshev polynomials to accelerate the methods when the matrix involved is symmetric. Saad [29] extends this technique to the unsymmetric case. Numerical experiments have shown that this acceleration can be quite effective.

Davidson's method is also an orthogonal projection method. This method was initially proposed to solve large symmetric eigenvalue problems arising from the quantum chemistry [8]. Later, Sadkane [30] generalized this method to the unsymmetric case. When the Davidson's method is run, the subspace E is expanded by solving an equation related to the last residual by some preconditioning techniques. A fatal drawback of this method is that it may perform very badly sometimes, independently of A.

The Jacobi-Davidson method [31] was proposed by Sleijpen and Van der Vorst in 1996. It expands the subspace step by step by solving a certain defect equation and computes Ritz vectors as approximate eigenvectors. However, the performance of this method depends strongly on efficient and accurate solution of a resulting large unsymmetric linear equations, which makes it very difficult to analyse its convergence and stability. Many aspects about this method are not clear nowdays.

The Arnoldi method [1, 29] is one of most important orthogonal projection methods. In it, the original matrix is partially reduced to an upper Hessenberg matrix by the Arnoldi process and one then computes the eigenvalues of the resulting Hessenberg matrix as approximations to the eigenvalues of the given matrix.

Saad [29] considers the convergence theory of the Arnoldi method for real simple eigenvalues and the eigenvalue with largest real part and the corresponding eigenvector, and proposed a restarting version of it and other variants. Later, the convergence theory of the Arnoldi method was investigated by Jia in [12, 13] for a general matrix that can be defective, in which a priori theoretical error bounds for eigenvectors are given and bounds for eigenvalues have been refined. The results there have shown that the approximate eigenvectors or Ritz vectors obtained by orthogonal projection methods may fail to converge. Jia and Stewart [21] removed the restriction that A is diagonalizable and analyzed the convergence of Ritz pairs by means of separation. They have shown that the Ritz values converge to the eigenvalues of A unconditionally while the convergence of Ritz vector requires that the corresponding Ritz value is well separated from the other Ritz values However, this condition can not always be satisfied. Thus, the Ritz vectors may fail to converge even if the corresponding Ritz values converge.

An important variant of the Arnoldi method is its block version, the block Arnoldi method [17, 30], which is an orthogonal projection method on a block Krylov subspace. In [12, 17], Jia establishes a convergence theory of the block Arnoldi method and its truncated version, the block incomplete orthogonalization method [14], when A is diagonalizable. The results show that the block method have two advantages: First, they are suitable and efficient for the case where the eigenvalues to be found are clustered; second, they are able to compute mutiple eigenvalues and clustered eigenvalues and determine the associated eigenspace. Analogous to the Arnoldi method, however, Ritz vectors may fail to converge.

Sorensen [32] suggests to use the Arnoldi algorithm with implicit restarting, which can make use of the information on the Krylov subspace generated previously by means of the shifted QR algorithm and reduce the computational cost at each restart. The key problem for this technique is how to choose the shifts involved. A popular choice [32] is to take the unwanted Ritz values as the shifts. Another shift scheme for the case of symmetric matrix is the Leja shifts [2]. The Leja shifts are often better than the exact shifts for computing a few smallest eigenvalues when the subspace size is very small.

The shift-and-invert Arnoldi method [29] is an important tool for the large sparse unsymmetric generalized eigenproblems. It is mathematically equivalent to the Arnoldi method for a spectral transformed eigenproblem. If the shift is suitably selected, the distribution of the spectrum of the shifted and inverted matrix may be favorable even if the eigenvalues close to the shift are clustered. Therefore, the Arnoldi method applied for the transformed eigenproblem may give a much faster convergence with eigenvalues close to the shift. Ruhe [28] proposes a rational Krylov sequence algorithm (RKS), which is different from the shift-and-invert Arnoldi in that the former selects a different shift for each step of Arnoldi process. Thus RKS can be viewed as a generalization of the shift-and-invert Arnoldi. Numerical experiments have shown that this scheme can be quite efficient.

3. Oblique projection methods

3.1 The methods

Given two subspaces L and K, an oblique projection method seeks the pairs $(\tilde{\lambda}_i, \tilde{\varphi}_i)$ with $\|\tilde{\varphi}_i\| = 1$ that satisfy the following condition

$$\begin{cases} \tilde{\varphi}_i \in K, \\ A\tilde{\varphi}_i - \tilde{\lambda}_i\tilde{\varphi}_i \perp L, \end{cases} \tag{3}$$

and use them to approximate some eigenpairs (λ_i, φ_i) of A. The subspace K is referred to as a right subspace and L a left subspace. Assume that W and V are both orthonormal matrices, whose columns form a basis of the right and left subspaces respectively. Then, the above relations can be written as

$$\left\{ \begin{array}{rcl} By_i & = & \tilde{\lambda}_i W^* V y_i, \\ \tilde{\varphi} & = & V y_i, \end{array} \right. \tag{4}$$

where $B = W^*AV$

A few commonly used methods fall into this category: The subspace iteration method [5, 11] if $K = \text{span}\{A^l X_0\}$ and $\text{span}\{(A^*)^l Y_0\}$; the biorthogonalization (or unsymmetric) Lanczos method [24, 29] if $K = \mathcal{K}_m(v_1, A)$ and $L = \mathcal{K}_m(w_1, A^*)$; the harmonic Arnoldi method [26] if $K = \mathcal{K}_m(v_1, A)$ and $L = A\mathcal{K}_m(v_1, A)$.

3.2 State of the art

Though oblique subspace iteration is more complex and less efficient in computer time and storage requirement than orthogonal subspace iteration, it can simultaneously compute the right and left eigenvectors or invariant subspaces associated with the dominant eigenvalues at the same time.

The harmonic Arnoldi method is used for finding interior eigenvalues of large unsymmetric matrix in question, whereas the Arnoldi method has trouble in computing them and to do so it must be combined with shift-and-invert technique. This is mainly because the Krylov subspace usually does not contain good approximations to the corresponding eigenvectors, but also due to properties of the orthogonal projection procedure.

A popular oblique projection method is the biorthogonalization Lanczos method proposed by Lanczos in 1950 [24], which tridiagonalizes successively a given general matrix to tridiagonal form starting with two biorthogonal vectors and then computes the eigenvalues of the resulting tridiagonal matrix as those of the original matrix. This method is an oblique projection method [29]. Unlike subspace iteration methods, conceptually speaking, the biorthogonalization Lanczos method itself is a direct procedure which reduces a given general matrix to tridiagonal form to completion if possible. In the very beginning since it came out, the method was soon replaced by more efficient Householder and Givens transformations and abandoned for a long time.

In the unsymmetric case, serious breakdowns may occur at any step of the reduction. Breakdowns have nothing to do with roundoffs and the ill-conditioning of the eigenproblem. Several authors have paid their attention to the subject of curing and avoiding serious breakdowns and some robust schemes havė been proposed [29], e.g., Parlett et al. [29] first suggest a look-ahead version. It utilizes 2×2 block pivots to improve the robustness and stability of the method. Later Cullum [7] extended their symmetric Lanczos algorithm without reorthogonalization to the unsymmetric case and suggested a new way for treating the resulting non-Hermitian tridiagonal matrices.

>From the formal orthogonal polynomials' point of view, Gutknecht [9, 10] gave a thorough analysis on the method and serious breakdowns. Under the assumption that no serious breakdown occurs, Ye [35] has made a convergence analysis for this method, showing that a few eigenvalues with largest (smallest) real parts of the original large matrix usually appear firstly as the eigenvalues of a small sized tridiagonal matrix. Bai [3] has shown that in finite precision convergence of a Ritz value implies the loss of biorthogonalization if this Ritz value is well conditioned and no breakdown occurs.

The biorthogonalization Lanczos method can be used to determine simultaneously the right and left eigenvectors associated with the wanted eigenvalues. However, one of the drawbacks lies in a fact that it requires the use of both matrix and its transpose. In some applications, for example, in studying the stability of a dynamic system governed by certain partial differential equation, the original matrix is not available explicitly but matrix by vector product is easy to form. In these cases, the transpose of these matrix is not available and cannot even be approximated by finite differences [29]. So the biorthogonalization Lanczos method is not usable at this time.

Compared to subspace iteration methods, it is recognized that in the symmetric case, the Lanczos algorithm performs much better than subspace iteration methods; in the unsymmetric case, some numerical experiments made by Saad indicate that Arnoldi's method is as well more effective than subspace iteration methods. In fact, it can be seen from their own convergence theory that subspace iteration methods without acceleration converge only linearly, while both the biorthogonalization Lanczos method and gneralized Lanczos methods converge exponentially, which is one of the crucial reasons why it is the case.

4. Refined projection methods

4.1 The methods

As mentioned above, orthogonal and oblique projection methods have the disadvantage of possible non-convergence of Ritz vectors. In order to correct this problem, a class of refined projection methods has been proposed by Jia [12, 15, 19, 20].

For each approximate eigenvalue μ_i (Ritz value, harmonic Ritz value, etc.), we seek a unit norm vector $u_i \in E$ that satisfies the condition

$$\|(A-\mu_i I)u_i\| = \min_{\substack{u \in E, \\ \|u\| = 1}} \|(A-\mu_i I)u\| \tag{5}$$

and use it to approximate φ_i. we refer to u_i as a *refined* Ritz vector (or refined approximate vector) with $\tilde{\lambda}_i$ in E. A few well-known and most commonly used methods fall into this category: The refined Arnoldi method [15] if $E = \mathcal{K}_m(v_1, A) = span\{v_1, Av_1, \ldots, A^{m-1}v_1\}$ and the μ_i are Ritz values; the refined block Arnoldi method [16] if $E = \mathcal{K}_n(Q_1, A) = span\{Q_1, AQ_1, \ldots, A^{n-1}Q_1\}$, a block Krylov subspace, and the μ_i are Ritz values; the refined subspace iteration method [20] if $E = span\{A^n X_0\}$, where X_0 is an $N \times m$ matrix whose columns are linearly independent and the μ are Ritz values; the refined harmonic Arnoldi method [22] if $K = \mathcal{K}_m(v_1, A)$ and $L = A\mathcal{K}_m(v_1, A)$ and the μ_i are harmonic Ritz values.

We look at the computation of u_i. Let V form an orthonormal basis of E. Then

$$\begin{aligned} \|(A - \mu_i I)u_i\| &= \min_{\|z\|=1} \|(A - \mu_i I)Vz\| && (6) \\ &= \|(A - \mu_i I)Vz_i\| && (7) \\ &= \sigma_{\min}((A - \mu_i I)V). && (8) \end{aligned}$$

Therefore, the solution of this problem is the right singular vector of $AV - \mu_i V$ associated with its smallest singular value. Thus we can compute refined Ritz vectors by computing the singular value decomposition of $AV - \mu_i V$.

Jia [20] compares the computational cost of this process with that of a Ritz vector; see also Stewart [33].

The above singular value decomposition is quite expensive for computing u_i. A much cheaper approach to computing u_i for a general E is proposed by Jia [20] as follows:

Form the cross-product matrix

$$\begin{aligned} C &= (AV - \mu_i V)^*(AV - \mu_i V) \\ &= (AV)^*(AV) - \bar{\mu}_i V^* AV - \mu V^* A^* V + |\mu|^2. \end{aligned}$$

Its computational cost is negligible since AV and V^*AV are already available when computing Ritz values. One then only solves the small eigenproblem of dimension m to get its eigenvector z_i associated with the smallest eigenvalue. Finally, form $u_i = Vz_i$ to get the refined Ritz vector.

The cross-product based algorithm is almost as accurate as the above conventional singular value decomposition algorithm if the ratio of the largest singular value and the second smallest one of $AV - \mu V$ is not very small [33]. Otherwise it may lose some accuracy.

If E is a (block) Krylov subspace $\mathcal{K}(A, v_1)$ or $\mathcal{K}_m(A, V_1)$, the computation cost of u_i can be considerably reduced and is only $O(m^3)$ [15, 16].

4.2 State of the art

Jia [15] first proves that refined Ritz vectors converge for a Krylov subspace, and he shows that the same conclusion holds for a block Krylov subspace [16].

A unified convergence theory for the refined projection methods has been established by Jia [19] and Jia and Stewart [21]. The former paper considers the convergence when A is diagonalizable, while in the latter paper, Jia and Stewart remove the constraint that A is diagonalizable and give a further analysis on the convergence of refined Ritz vectors for a general matrix A, which may have a cluster of close eigenvalues or may be defective. The results show that the refined Ritz vectors converge to the eigenvectors of A if only the Ritz values do.

Jia [18] investigates an important property of the refined Arnoldi method, polynomial characterizations of the refined Ritz vectors, and applies the implicitly restarting scheme proposed by Sorensen to the refined Arnoldi method successfully. The roots of these polynomials are used as shifts, called refined shifts, within an implicitly restarted refined Arnoldi algorithm. The numerical experiments also show that implicitly restarting the refined Arnoldi method is much more efficient than Sorensen's implicitly restarted Arnoldi algorithm. In the spirit of [18], Jia and Zhang [22] propose certain refined shifts to implicitly restart the refined harmonic Arnoldi method, and the resulting algorithm can be considerably more efficient than the implicitly restarted harmonic Arnoldi algorithm of Morgan [27]. A great advantage of this algorithm is that it may compute interior eigenpairs of large matrices efficiently without factoring a large matrix explicitly. Jia and Zhang [23] present a refined shift and invert Arnoldi method for large unsymmetric generalized eigenproblems, which leads to better numerical behavior and faster convergence.

In the refined subspace iteration algorithm [20], apart from replacing Ritz vectors by refined counterparts, Jia makes an innovation on updating (restarting) matrix. In it, rather than use Ritz vector or Schur vector matrix as an updating matrix, he suggests to use the matrix whose columns are generated only by the refined Ritz vectors that are supposed to approximate the desired eigenvectors. This is one of the keys to make the algorithm much more efficient.

Jia and his research group pay much efforts in refining the Jacobi-Davidson method and compute an invariant subspace.

Stewart [33] and van der Vorst [34] have given an excellent account of the refined methods. Bai *et al.* [4] have briefly introduced the refined projection methods and some related work.

5. Some problems

The existing theory and algorithms imply by no means that no problems are worth studying further. In fact, many challenging problems need to be considered carefully.

Suppose that the eigenvalues in some rectangular region in the complex plane are required. In this case, a shift-and-invert transformed matrix is generaly involved. This will lead to two problems: first, one must solve a large linear system at each step. If it is feasible to solve it using a direct solver, then it will cause no big problem; however, in general it is impossible to solve the linear system efficiently by a direct solver. One must rely on an iterative solver, which can not guarantee to converge or give a relatively high accuracy at low cost. Also, it is not clear to us nowadays that how the accuracy of the approximate solution of the linear system affects overall performance of a general iterative eigensolver. Second, because there is no inertia law, we do not know how to guarantee to find all the eigenvalues in the given region.

Restarting is crucial to the success and efficiency of a projection algorithm in the non-Hermitian case. For Krylov subspace based algorithms, how to select restarting (block) vectors as good as possible has been under consideration, though there exist some good approaches.

For the biorthogonalization Lanczos method, breakdown or near breakdown problems have been cured successfully both theoretically and numerically when solving a non-Hermitian linear system. However, as far as a matrix eigenproblem is concerned, such problems are by far from solved numerically. How small a number is can be treated to be a numerical breakdown? It is not yet known to the community. This is why the method has been very popular for solving a linear system but much less used to solve eigenproblems.

In contrast to orthogonal and oblique projection methods, refined projection methods are only used to compute individual eigenvectors. How to use them to compute an invariant subspace is interesting and worths considering.

References

[1] W.E. Arnoldi, The principle of minimized iteration in the solution of the matrix eigenvalue problem, *Quart. Appl. Math.*, **9** 17-29 (1951).

[2] J. Baglama, D. Calvetti and L. Reichel, Iterative methods for the computation of a few eigenvalues of a large symmetric matrix, *BIT*, **36** 400-421 (1996).

[3] Z. Bai, Error analysis of the Lanczos algorithm for the nonsymmetric eigenvalue problem, *Math. Comput.*, **62** 209-226 (1994).

[4] Z. Bai, J. Demmel, J. Dongarra, A. Ruhe and H.A. van der Vorst, *Templates for the Solution of the Algebraic Eigenvalue Problem: A Practical Guide*, SIAM, Philadelphia, 2000.

[5] M. Clint and A. Jenning, A simultaneous iteration method for the unsymmetric eigenvalue problem, *J. Inst. Math. Appl.*, **8** 111-121 (1971).

[6] J. Cullum and R.A. Willoughby, *Lanczos Algorithms for Large Symmetric Eigenvalue Computation, Vol. 1 Theory*, Birkhauser, Boston (1985).

[7] J. Cullum and R.A. Willoughby, A practical procedure for computing eigenvalues of large sparse nonsymmetric matrices, In *Large Scale Eigenvalue Problem*, (Edited by J. Cullum and R.A. Willoughby), pp. 199-240, Elsevier, Amsterdam, (1986).

[8] E.R. Davidson, The iterative calculation of a few of the lowest eigenvalues and corresponding eigenvectors of large real symmetric matrices, *J. Comput. Phys.*, **17** 87-94 (1975).

[9] M.H. Gutknecht, A completed theory of the unsymmetric Lanczos process and related algorithms, Part I, *SIAM J. Matrix Anal. Appl.*, **13** 594-639 (1992).

[10] M.H. Gutknecht, A completed theory of the unsymmetric Lanczos process and related algorithms, Part II, *SIAM J. Matrix Anal. Appl.*, **15** 15-58 (1994).

[11] A. Jenning and W.J. Stewart, A simultaneous iteration algorithm for real matrices, *ACM Trans. Math. Soft.*, **7** 185-198 (1981).

[12] Z. Jia, Some numerical methods for large unsymmetric eigenproblems, Ph.D. Thesis, Department of Mathematics, University of Bielefeld, Germany, 1994.

[13] Z. Jia, The convergence of generalized Lanczos methods for large unsymmetric eigenproblems, *SIAM J. Matrix Anal. Appl.*, **16** 843-862 (1995).

[14] Z. Jia, A block incomplete orthogonalization method for large nonsymmetric eigenproblems, *BIT*, **34** 519-536 (1995).

[15] Z. Jia, Refined iterative algorithms based on Arnoldi's process for large unsymmetric eigenproblems, *Linear Algebra Appl.*, **259** 1-23 (1997).

[16] Z. Jia, A refined iterative algorithm based on the block Arnoldi process for large unsymmetric eigenproblems, *Linear Algebra Appl.*, **270** 171-189 (1998).

[17] Z. Jia, Generalized block Lanczos methods for large unsymmetric eigenproblems, *Numer. Math.*, **80** 239-266 (1998).

[18] Z. Jia, Polynomial characterizations of the approximate eigenvectors by the refined Arnoldi method and an implicitly restarted refined Arnoldi algorithm, *Linear Algebra Appl.*, **287** 191-214 (1999).

[19] Z. Jia, Composite orthogonal projection methods for large eigenproblems, Science in China (Series A), **42** 577-585 (1999).

[20] Z. Jia, A refined subspace iteration algorithm for large sparse eigenproblems, *Appl. Numer. Math.*, **32** 35-52 (2000).

[21] Z. Jia and G.W. Stewart, An analysis of the Rayleigh-Ritz method for approximating eigenspaces, *Math. Comput.*, **70** 637-647 (2001).

[22] Z. Jia and Y. Zhang, The refined harmonic Arnoldi method and an implicitly restarted algorithm for computing interior eigenpairs of large matrices, *Appl. Numer. Math.*, to appear.

[23] Z. Jia and Y. Zhang, A refined shift-and-invert Arnoldi algorithm for large generalized eigenproblems, *Comput. Math. Appl.*, to appear.

[24] C. Lanczos, An iteration method for the solution of the eigenvalue problem of linear differential and integral operators, *J. Res. Natur. Bur. Stand.*, *45* 225-280 (1950).

[25] R.B. Morgan, On restarting the Arnoldi method for large nonsymmetric eigenvalue problems, *Math. Comput.*, **65** 1213-1230 (1996).

[26] R.B. Morgan and M. Zeng, Harmonic projection methods for large non-symmetric eigenvalue problems, *Numer. Linear Algebra Appl.*, **5** 33-55 (1998).

[27] R.B. Morgan, Implicitly restarted GMRES and Arnoldi methods for nonsymmetric systems of equations, *SIAM J. Matrix Anal. Appl.*, **21** 1112-1135 (2000).

[28] A. Ruhe, Rational Krylov sequence methods for eigenvalue computation, *Linear Algebra Appl.*, **58** 391-405 (1984).

[29] Y. Saad, *Numerical Methods for Large Eigenvalue Problems*, Manchester University Press in Algorithms and Architecture for Advanced Scientific Computing (1992).

[30] M. Sadkane, Block-Arnoldi and Davidson methods for unsymmetric large eigenvalue problems, *Numer. Math.*, **64** 195-211 (1993).

[31] G.L.G. Sleijpen and H.A. Van der Vorst, A Jacobi-Davidson iteration method for linear eigenvalue problems, *SIAM J. Matrix Anal. Appl.*, **17** 401-425 (1996).

[32] D.C. Sorensen, Implicit application of polynomial filters in a k-step Arnoldi method, *SIAM J. Matrix Anal. Appl.*, **13** 357-385 (1992).

[33] G.W. Stewart, *Matrix Algorithms: Vol.II, Eigensystems*, SIAM, Philadephia, PA, 2001.

[34] H.A. van der Vorst, *Computational Methods for Large Eigenvalue Problems*, Elsevier-North Holland, 2001.

[35] Q. Ye, A convergence analysis for nonsymmetric Lanczos algorithms, *Math. Comput.*, **56** 677-691 (1991).

GLOBAL PROPAGATION OF REGULAR NONLINEAR HYPERBOLIC WAVES

Tatsien Li
Department of Mathematics, Fudan University, Shanghai 200433, China
dqli@fudan.edu.cn

1. Introduction

In this work we shall consider the nonlinear hyperbolic waves described by the following Cauchy problem for first order quasilinear hyperbolic systems

$$\begin{cases} \dfrac{\partial u}{\partial t} + A(u)\dfrac{\partial u}{\partial x} = 0, & (1) \\ t = 0: \quad u = \varphi(x), & (2) \end{cases}$$

where $u = (u_1, \cdots, u_n)^T$ is the unknown vector function of (t, x), $A(u) = (a_{ij}(u))$ is an $n \times n$ matrix with suitably smooth entries $a_{ij}(u)$ $(i, j = 1, \cdots, n)$ and $\varphi(x) = (\varphi_1(x), \cdots, \varphi_n(x))^T$ is a C^1 vector function of x with bounded C^1 norm.

By definition of *hyperbolicity*, for any given u on the domain under consideration, the matrix $A(u)$ possesses n real eigenvalues $\lambda_1(u), \cdots, \lambda_n(u)$ and a complete set of left (resp. right) eigenvectors $l_1(u), \cdots, l_n(u)$ (resp. $r_1(u), \cdots, r_n(u)$): for $i = 1, \cdots, n$,

$$l_i(u)A(u) = \lambda_i(u)l_i(u) \qquad (\text{resp. } A(u)r_i(u) = \lambda_i(u)r_i(u)). \tag{3}$$

Without loss of generality, we may suppose that

$$l_i(u)r_j(u) \equiv \delta_{ij} \quad (i = 1, \cdots, n) \tag{4}$$

and

$$r_i^T(u)r_i(u) \equiv 1, \tag{5}$$

where δ_{ij} stands for the Kronecker's symbol.

In particular, if the matrix $A(u)$ possesses n distinct real eigenvalues

$$\lambda_1(u) < \lambda_2(u) < \cdots < \lambda_n(u), \tag{6}$$

system (1) is called to be *strictly hyperbolic*.

If the matrix A is independent of u, we meet linear hyperbolic waves given by

$$\begin{cases} \frac{\partial u}{\partial t} + A\frac{\partial u}{\partial x} = 0, & (7) \\ t = 0: \quad u = \varphi(x). & (8) \end{cases}$$

The acoustic wave is a typical example of linear hyperbolic waves. In the scalar case, we have, for instance, the Cauchy problem

$$\begin{cases} \frac{\partial u}{\partial t} + \frac{\partial u}{\partial x} = 0, & (9) \\ t = 0: \quad u = \varphi(x). & (10) \end{cases}$$

The wave speed is constant: $\frac{dx}{dt} = 1$ and the wave always keeps its shape in the course of propagation. In the general case, there are n linear waves given by (7)–(8) with constant speeds

$$\frac{dx}{dt} = \lambda_i \qquad (i = 1, \cdots, n) \tag{11}$$

respectively. Each wave keeps its shape in the propagation and the interaction among waves is only a linear superposition. It is the reason that we can hear and distinguish many persons speaking at the same time. Otherwise, our life will be very complicated.

The situation for nonlinear hyperbolic waves is totally different. In the scalar case, let us consider, for instance, the Cauchy problem for Burger's equation

$$\begin{cases} \frac{\partial u}{\partial t} + u\frac{\partial u}{\partial x} = 0, & (12) \\ t = 0: \quad u = \varphi(x). & (13) \end{cases}$$

The wave speed depends on u : $\frac{dx}{dt} = u$ and then the wave can not keep its shape in the course of propagation. Generically, there will be a distortion of wave shape such that the wave steepens and finally blows up in a finite time. In the general case, there are n nonlinear hyperbolic waves given by (1)–(2) with speeds

$$\frac{dx}{dt} = \lambda_i(u) \quad (i = 1, \cdots, n) \tag{14}$$

respectively and there are nonlinear interactions among these waves such that the situation is much more complicated.

As a conclusion, Cauchy problem (1)–(2) always admits a unique C^1 solution $u = u(t, x)$ at least for a short time $0 \leq t \leq \delta$ (cf. [1] and the references therein), however, generically speaking, the C^1 solution $u = u(t, x)$ to Cauchy problem (1)-(2) exists only locally in time and the singularity may occur in a finite time, i.e., there exist $t_0 > 0$ such that

$$\|u(t, .)\|_0 + \|u_x(t, .)\|_0 \to +\infty \quad \text{as } t \uparrow t_0, \tag{15}$$

no matter how smooth and how small the initial data are (cf. [2] and the references therein).

Therefore, it is of great importance in both theory and application to study the following two problems:

(1) Under what conditions does Cauchy problem (1)–(2) admit a unique global C^1 solution $u = u(t, x)$ on $t \geq 0$ or for all $t \in \mathbb{R}$?

(2) Under what conditions does the C^1 solution to Cauchy problem (1)-(2) blow up in a finite time? What is the sharp estimate on the life-span of C^1 solution, i.e., on the maximum length of existence t-interval? What is the mechanism of the formation of singularity and what is the character of singularity?

When $n = 1$ or 2, the answer to these two problems is relatively simple (cf. [2] and the references therein).

For the general system (1) of n equations, the first result in this direction was given by F. John [3]. Suppose that in a neighbourhood of $u = 0$, $A(u) \in C^2$, system (1) is strictly hyperbolic and *genuinely nonlinear* (GN) in the sense of P. D. Lax: for $i = 1, \cdots, n$,

$$\nabla \lambda_i(u) r_i(u) \neq 0. \tag{16}$$

Suppose furthermore that $\varphi(x) \in C^2$ has a compact support:

$$\text{Supp } \varphi \subseteqq [\alpha_0, \beta_0]. \tag{17}$$

F.John proved that if

$$\theta \triangleq (\beta_0 - \alpha_0)^2 \sup_{x \in \mathbb{R}} |\varphi''(x)| \tag{18}$$

is small enough, then the first order derivatives of the C^2 solution $u = u(t, x)$ to Cauchy problem (1)-(2) must blow up in a finite time.

T.P.Liu [4] generalized F.John's result to the case that in a neighbourhood of $u = 0$, a nonempty part of characteristics is genuinely nonlinear, while the other part of characteristics is *linearly degenerate* (LD) in the sense of P. D. Lax: for the corresponding indices i,

$$\nabla \lambda_i(u) r_i(u) \equiv 0. \tag{19}$$

Under the additional hypothesis "linear waves do not generate nonlinear waves", he got the same result as in F.John [3] for a quite large class of initial date. His result can be applied to the system of one-dimensional gas dynamics with convexity.

L.Hörmander [5, 6] reproved F.John's result and, when $\varphi(x) = \varepsilon\psi(x)$, where $\varepsilon > 0$ is a small parameter. He obtained the following asymptotic behaviour of the life-span $\tilde{T}(\varepsilon)$:

$$\lim_{\varepsilon \downarrow 0}\{\varepsilon \tilde{T}(\varepsilon)\} = M_0, \tag{20}$$

where M_0 is a positive constant independent of ε, defined by

$$M_0 = \Big(\max_{i=1,\cdots,n} \sup_{x\in\mathbb{R}}\{-(\nabla\lambda_i(0)r_i(0))l_i(0)\psi'(x)\}\Big)^{-1}. \tag{21}$$

Thus, there exist two positive constants c and C independent of ε, such that the life-span $\tilde{T}(\varepsilon)$ satisfies the following optimal estimate:

$$c\varepsilon^{-1} \leq \tilde{T}(\varepsilon) \leq C\varepsilon^{-1}, \tag{22}$$

denoted by

$$\tilde{T}(\varepsilon) \approx \varepsilon^{-1}. \tag{23}$$

On the other hand, A.Bressan [7] gave a result on the global existence of classical solution as follows: Suppose that system (1) is strictly hyperbolic and linearly degenerate in the sense of P.D.Lax: (19) holds for $i = 1.\cdots, n$. Suppose furthermore that the initial data φ have a compact support. If the total variation of φ is small enough:

$$TV\{\varphi\} \ll 1, \tag{24}$$

then Cauchy problem (1)-(2) admits a unique global classical solution $u = u(t,x)$ for all $t \in \mathbb{R}$.

All the previous results are obtained under the following three hypotheses on system (1):

(1) The system is strictly hyperbolic.

(2) (a) The system is genuinely nonlinear, i.e., all the characteristics are genuinely nonlinear; or

(b) A nonempty part of characteristics is genuinely nonlinear, while the other part of characteristics is linearly degenerate; or

(c) The system is linearly degenerate, i.e., all the characteristics are linearly degenerate.

(3) In case (2) (b), "linear waves do not generate nonlinear waves".

In order to explain that these three hypotheses restrict the applications, we give the following examples.

Ex.1. The *system of nonlinear elasticity* can be written as

$$\begin{cases} \dfrac{\partial v}{\partial t} - \dfrac{\partial w}{\partial x} = 0, \\ \dfrac{\partial w}{\partial t} - \dfrac{\partial K(v)}{\partial x} = 0, \end{cases} \tag{25}$$

where $K(v)$ is a suitably smooth function of v such that

$$K'(0) > 0 \tag{26}$$

and

$$K''(0) = 0, \qquad \text{but} \quad K''(v) \not\equiv 0. \tag{27}$$

In a neighbourhoods of $(v, w) = (0, b)$, where b is an arbitrarily given constant, (26) implies that system (25) is strictly hyperbolic. However, noting (27), system (25) is neither genuinely nonlinear nor linearly degenerate.

Ex.2. The *system of one-dimensional gas dynamics* can be written in Lagrangian representation as

$$\begin{cases} \dfrac{\partial \tau}{\partial t} - \dfrac{\partial u}{\partial x} = 0, \\ \dfrac{\partial u}{\partial t} + \dfrac{\partial p(\tau, S)}{\partial x} = 0, \\ \dfrac{\partial S}{\partial t} = 0, \end{cases} \tag{28}$$

where $p = p(\tau, S)$ is the equation of state satisfying

$$p_\tau < 0, \quad \forall \tau > 0, \tag{29}$$

which implies that system (28) is strictly hyperbolic. One characteristic is linearly degenerate, however, the other two characteristics are neither genuinely nonlinear nor linearly degenerate except when $p = p(\tau, S)$ is a strictly convex or concave function with respect to τ.

Ex.3. The *system of the motion of elastic strings* can be written as

$$\begin{cases} \dfrac{\partial u}{\partial t} - \dfrac{\partial v}{\partial x} = 0, \\ \dfrac{\partial v}{\partial t} - \dfrac{\partial}{\partial x}\left(\dfrac{T(r)}{r} u\right) = 0 \end{cases} \tag{30}$$

(see [8]–[10]), where $u = (u_1, \cdots, u_n)^T$, $v = (v_1, \cdots, v_n)^T$, $r = |u| = \sqrt{u_1^2 + \cdots + u_n^2}$ (in practice $n = 2$ or 3) and $T = T(r)$ is a suitably smooth function of r such that

$$T'(r) > \frac{T(r)}{r} > 0. \tag{31}$$

Under hypothesis (31), system (30) is hyperbolic but not strictly hyperbolic except for $n = 2$. There are two linearly degenerate characteristics $\pm\sqrt{\frac{T(r)}{r}}$ with multiplicity $n-1$, while, two simple characteristics $\pm\sqrt{T'(r)}$ are neither genuinely nonlinear nor linearly degerate except when $T''(r) \equiv 0$, namely, $T = T(r)$ is an affine function of r. Moreover, for system (30), "linear waves *do* generate nonlinear waves". Thus, all the three hypotheses mentioned above fail.

Ex.4. The *system of finite amplitude plane elastic waves for hyperelastic materials* can be written as

$$\frac{\partial u}{\partial t} + A(u)\frac{\partial u}{\partial x} = 0 \tag{32}$$

(see [3]) with $u = (u_1, \cdots, u_6)^T$ and

$$A(u) = \begin{pmatrix} 0 & -I \\ -V'' & 0 \end{pmatrix}, \tag{33}$$

where $V'' = V''(u_1, u_2, u_3)$ is a 3×3 matrix determined by the material.

For instance, for the material of Ciarlet-Geymomat (cf. [11]), it is easy to see that system (32) has two linearly degenerate characteristics with multiplicity 2 and two simple genuinely nonlinear characteristics, then (32) is a non-strictly hyperbolic system.

Actually, many authors have pointed out the necessity of studying the hyperbolic system with general characteristics. For instance, A.Majda has proposed in [12] the open problem "Investigate shock formation in non-genuinely nonlinear systems for initial data of compact support". He has also mentioned two specially interesting systems in nonlinear elasticity, namely, our examples 3 and 4.

The aim of this talk is to establish a complete theory on both global existence and blow-up phenomenon of C^1 solution to the Cauchy problem for general quasilinear hyperbolic systems with small initial data with compact support or more generally with small and decaying initial data.

2. Weak linear degeneracy

In what follows we first consider the strictly hyperbolic case.

In order to present our results, it is necessary to introduce a new concept—–the weak linear degeneracy (cf. [13]).

Definition 1: The i-th characteristic $\lambda_i(u)$ is called to be *weakly linearly degenerate* (WLD), if, along the i-th characteristic trajectory $u = u^{(i)}(s)$ passing through $u = 0$ in the u-space, defined by

$$\begin{cases} \dfrac{du}{ds} = r_i(u), & (34) \\ s = 0: \quad u = 0, & (35) \end{cases}$$

we have

$$\nabla\lambda_i(u)r_i(u) \equiv 0, \quad \forall \text{ small } |u|, \tag{36}$$

namely,

$$\lambda_i(u^{(i)}(s)) \equiv \lambda_i(0), \quad \forall \text{ small } |s|. \ \square \tag{37}$$

Obviously, if $\lambda_i(u)$ is linearly degenerate, then $\lambda_i(u)$ is weakly linearly degenerate; while, if $\lambda_i(u)$ is genuinely nonlinear, then $\lambda_i(u)$ is not weakly linearly degenerate.

By definition, if $\lambda_i(u)$ is not weakly linearly degenerate, then $\lambda_i(u^{(i)}(s))$ is not identically equal to a constant for small $|s|$, therefore, either there exists an integer $\alpha_i \geq 0$ such that

$$\left.\frac{d^l\lambda_i(u^{(i)}(s))}{ds^l}\right|_{s=0} = 0 \quad (l = 1, \cdots, \alpha_i), \ but \ \left.\frac{d^{\alpha_i+1}\lambda_i(u^{(i)}(s))}{ds^{\alpha_i+1}}\right|_{s=o} \neq 0, \tag{38}$$

or

$$\left.\frac{d^l\lambda_i(u^{(i)}(s))}{ds^l}\right|_{s=0} = 0 \quad (l = 1, 2, \cdots) \tag{39}$$

but (37) fails, denoted by $\alpha_i = +\infty$.

Thus, for each characteristic $\lambda_i(u)$ we have the following table:

$$\lambda_i(u)$$

<table>
<tr><td colspan="2">non WLD</td><td rowspan="2">WLD</td></tr>
<tr><td>$\alpha_i = \underbrace{\overset{(GN)}{0}, 1, 2, \cdots\cdots}_{\text{finite}}$</td><td>$\alpha_i = +\infty$</td></tr>
</table>

If $\alpha_i = 0$, then in a neighbourhood of $u = 0$, $\lambda_i(u)$ is genuinely nonlinear. Moreover, when α_i increases, $\lambda_i(u)$ is closer and closer to the weakly linearly degenerate case.

Definition 2: System (1) is said to be *weakly linearly degenerate*, if all the characteristics $\lambda_1(u), \cdots, \lambda_n(u)$ are weakly linearly degenerate. □

Hence, if system (1) is not weakly linearly degenerate, then there exists a nonempty set of indices $J \subseteqq \{1, \cdots, n\}$ such that $\lambda_i(u)$ is not weakly linearly degenerate if and only if $i \in J$.

Let

$$\alpha = \min_i \{\alpha_i \mid i \in J\}. \tag{40}$$

α is an integer ≥ 0 or $+\infty$.

Thus, for any given quasilinear strictly hyperbolic system (1), all possible situations can be shown in the following table:

<table>
<tr><td colspan="2">non WLD</td><td rowspan="2">WLD</td></tr>
<tr><td>$\underbrace{\alpha = 0, 1, 2, \cdots\cdots}_{\text{finite}}$</td><td>$\alpha = +\infty$</td></tr>
</table>

It gives us a complete category.

Let

$$J_1 = \{i \mid i \in J, \quad \alpha_i = \alpha\}. \tag{41}$$

When $\alpha = 0$, then for every $i \in J_1$, $\lambda_i(u)$ is genuinely nonlinear in a neighbourhood of $u = 0$. Furthermore, when α increases, system (1) is closer and closer to a weakly linearly degenerate system.

3. Main results

We now consider the Cauchy problem for system (1) with small and decaying C^1 initial data (2) satisfying that there exists a number $\mu > 0$ such that

$$\theta \triangleq \sup_{x \in \mathbb{R}} \{(1 + |x|)^{1+\mu}(|\varphi(x)| + |\varphi'(x)|)\} < +\infty \tag{42}$$

and θ is small enough.

Theorem 1. *(Global existence of C^1 solution) (see [13]–[14], [17], [19]–[20]): Suppose that in a neighbourhood of $u = 0$, $A(u) \in C^2$, system (1) is strictly hyperbolic and weakly linearly degenerate. Then there exists $\theta_0 > 0$ so small that for any given $\theta \in [0, \theta_0]$, Cauchy problem (1)-(2) admits a unique C^1 solution $u = u(t, x)$ with small C^1 norm for all $t \in \mathbb{R}$.*

Conversely, under the assumption that in a neighbourhood of $u = 0$, $A(u) \in C^1$ and system (1) is strictly hyperbolic, if Cauchy problem (1)-(2) always admits *a unique C^1 solution $u = u(t, x)$ on $t \geq 0$ for any given C^1 initial data $\varphi(x)$ with small θ, then system (1) must be weakly linearly degenerate.* □

Thus, for small θ, the weak linear degeneracy is equivalent to the global existence of C^1 solution to Cauchy problem (1)–(2), hence, if system (1) is not weakly linearly degenerate, then we should meet the blow-up phenomenon.

Remark 1: Theorem 1 fails when $\mu = 0$ (cf. [21]). □

Remark 2: The result of A.Bressan in [7] is still valid if system (1) is only weakly linearly degenerate (see [23]). □

Theorem 2. *(Blow-up phenomenon) (see [13–14], [18], [22]): Suppose that in a neighbourhood of $u = 0$, $A(u)$ is suitably smooth and system (1) is strictly hyperbolic. Suppose furthermore that system (1) is not weakly linearly degenerate and the corresponding index α defined by (40) is a finite integer≥ 0. Suppose finally that $\varphi(x) = \varepsilon\psi(x)$, where $\varepsilon > 0$ is a small parameter and $\psi(x) \in C^1$ satisfies (42). Then, for a large class of initial data, precisely speaking, if there exists $i_0 \in J_1$ such that*

$$l_{i_0}(0)\psi(x) \not\equiv 0, \tag{43}$$

then there exists $\varepsilon_0 > 0$ so small that for any given $\varepsilon \in (0, \varepsilon_0]$, the following conclusions hold:

(a) The first order derivative u_x of C^1 solution $u = u(t, x)$ to Cauchy problem (1)–(2) must blow up in a finite time, while the solution itself remains bounded and small. Moreover, the life-span $\tilde{T}(\varepsilon)$ of C^1 solution possesses the following asymptotic property:

$$\lim_{\varepsilon \downarrow 0}(\varepsilon^{\alpha+1}\tilde{T}(\varepsilon)) = M_0, \tag{44}$$

where M_0 is a positive constant independent of ε, given by

$$M_0 = \Big(\max_{i \in J_1} \sup_{x \in \mathbb{R}}\Big\{ -\frac{1}{\alpha!}\frac{d^{\alpha+1}\lambda_i(u^{(i)}(s))}{ds^{\alpha+1}}\Big|_{s=0} \cdot (l_i(0)\psi(x))^{\alpha} l_i(0)\psi'(x)\Big\}\Big)^{-1}, \tag{45}$$

where $u = u^{(i)}(s)$ is defined by (34)-(35). Hence, there exist two positive constants c and C independent of ε, such that

$$c\varepsilon^{-(\alpha+1)} \leq \tilde{T}(\varepsilon) \leq C\varepsilon^{-(\alpha+1)}, \tag{46}$$

denoted by

$$\tilde{T}(\varepsilon) \approx \varepsilon^{-(\alpha+1)}. \tag{47}$$

(b) The singularity occurs at the beginning of the envelope of characteristics of the same family, i.e., at the point with minimum t-value on the envelope.

(c) For every $i \notin J_1$, the i-th family of characteristics does not generate any envelope on the domain $0 \le t \le \tilde{T}(\varepsilon)$. In particular, every family of weakly linearly degenerate characteristics and then every family of linearly degenerate characteristics do not generate any envelope on the domain $0 \le t \le \tilde{T}(\varepsilon)$.

(d) Let (t_0, x_0) ($t_0 \triangleq \tilde{T}(\varepsilon)$) be a blow-up point. There exists $i_0 \in J_1$ such that along the i_0-th characteristic passing through (t_0, x_0), the blow-up rate is given by

$$u_x(t,x) = O((t_0 - t)^{-1}), \qquad \forall t < t_0, \tag{48}$$

which is independent of the index α.

(e) On the line $t = \tilde{T}(\varepsilon)$, the set of blow-up points can not possess a positive (even very small) measure. □

Remark 3: Theorem 2 implies all the previous results given by of F.John, T.P.Liu and L.Hörmander. □

Remark 4: When $\mu = 0$, Theorem 2 is still valid (cf. [21]). □

We now consider the critical case that system (1) is not weakly linearly degenerate, but the corresponding index α is equal to $+\infty$. By Theorem 1, in this case we still have the blow-up phenomenon. However, the situation on the life-span should be much better. In fact, we have

Theorem 3. *(see [14]): Under the assumptions of Theorem 2, but instead of assuming that the index α is an finite integer ≥ 0, we suppose that α is equal to $+\infty$, for any given integer $N \ge 1$, there exists $\varepsilon_N > 0$ small enough and a positive constant C_N independent of ε such that for any given $\varepsilon \in (0, \varepsilon_N]$ we have*

$$\tilde{T}(\varepsilon) \ge C_N \varepsilon^{-N}. \square \tag{49}$$

However, even in the scalar case

$$\begin{cases} u_t + \lambda(u) u_x = 0, & (50)\\ t = 0: \quad u = \varepsilon\psi(x), & (51)\end{cases}$$

where $\lambda(u) \in C^\infty$, $\lambda'(u) \not\equiv 0$ with

$$\lambda^{(l)}(0) = 0 \quad (l = 1, 2, \cdots), \tag{52}$$

we may choose $\lambda(u)$ in different ways such that

$$\exp\{c\varepsilon^{-p}\} \le \tilde{T}(\varepsilon) \le \exp\{C\varepsilon^{-p}\}, \quad \forall p > 0 \tag{53}$$

or

$$\exp\{c(\ln\varepsilon)^2\} \le \tilde{T}(\varepsilon) \le \exp\{C(\ln\varepsilon)^2\} \tag{54}$$

etc., where c and C are positive constants independent of ε (see [14]). Therefore, it is impossible to get a unified sharp estimate on the life-span in the critical case $\alpha = +\infty$. Fortunately, up to now we have never encountered this case in applications.

4. Normalized coordinates

The proof of Theorems 1–3 is very long and quite technical. However, the basic idea is as follows: Since the initial data are small and decay as $|x| \to +\infty$, n waves should be essentially separated from each other in a finite time and the interaction among n waves can be controlled to be relatively small, thus for every wave the problem can be essentially reduced to the scalar case.

Noting that the definition of weak linear degeneracy depends on the characteristic trajectories passing through $u = 0$, $u = u^{(i)}(s)$ $(i = 1, \cdots, n)$, the key point in the proof of Theorems 1–3 is to find new coordinates $\tilde{u} = \tilde{u}(u)$ $(\tilde{u}(0) = 0)$ such that in the $\tilde{u}$-space the characteristic trajectories passing through $\tilde{u} = 0$ can be expressed in a simple way. We have

Theorem 4. *(see [13]): Suppose that in a neighbourhood of* $u = 0$, *system (1) is strictly hyperbolic and* $A(u) \in C^k$, *where* k *is an integer* ≥ 1. *Then there exists a* C^{k+1} *diffeomorphism* $u = u(\tilde{u})$ $(u(0) = 0)$ *such that in the* $\tilde{u}$*-space, for each* $i = 1, \cdots, n$, *the i-th characteristic trajectory passing through* $\tilde{u} = 0$ *coincides with the* $\tilde{u}_i$*-axis at least for small* $|\tilde{u}_i|$, *namely,*

$$\tilde{r}_i(\tilde{u}_i e_i) // e_i, \quad \forall \, small \, |\tilde{u}_i| \ (i = 1, \cdots, n), \tag{55}$$

where $\tilde{r}_i(\tilde{u})$ *denotes the corresponding i-th right eigenvector in the* $\tilde{u}$*-space and* $e_i = (0, \cdots, 0, \overset{(i)}{1}, 0, \cdots, 0)^T$. □

We refer to the diffeomorphism given by Theorem 4 as the *normalized transformation*, and the corresponding variables $\tilde{u} = (\tilde{u}_1, \cdots, \tilde{u}_n)$ are called the *normalized variables* or *normalized coordinates.*

In normalized coordinates $\tilde{u}$, the i-th characteristic $\tilde{\lambda}_i(\tilde{u}) = \lambda_i(u(\tilde{u}))$ is weakly linearly degenerate if and only if

$$\tilde{\lambda}_i(\tilde{u}_i e_i) \equiv \tilde{\lambda}_i(0), \quad \forall \, small \, |\tilde{u}_i|, \tag{56}$$

while, if $\lambda_i(\tilde{u})$ is not weakly linearly degenerate, then, either there exists an integer $\alpha_i \geq 0$ such that

$$\frac{d^l \tilde{\lambda}_i(\tilde{u}_i e_i)}{d\tilde{u}_i^l}\Big|_{\tilde{u}_i=0} = 0 \quad (l = 1, \cdots, \alpha_i), \qquad but \ \frac{d^{\alpha_i+1} \tilde{\lambda}_i(\tilde{u}_i e_i)}{d\tilde{u}_i^{\alpha_i+1}}\Big|_{\tilde{u}_i=0} \neq 0, \tag{57}$$

or

$$\frac{d^l \tilde{\lambda}_i(\tilde{u}_i e_i)}{d\tilde{u}_i^l}\Big|_{\tilde{u}_i=0} = 0 \quad (l = 1, 2, \cdots) \tag{58}$$

but (56) fails, denoted by $\alpha_i = +\infty$.

The system in normalized coordinates can be regarded as a standard form of strictly hyperbolic system. The proof of Theorems 1–3 are taken in normalized coordinates.

Some further properties of normalized coordinates can be found in [19]-[20].

5. Non-strictly hyperboric case

Up to now, all results are presented in the strictly hyperbolic case. In this Section we consider the non-strictly hyperbolic system only in some physically meaningful and interesting cases.

Consider the quasilinear hyperbolic system of conservation laws

$$\frac{\partial u}{\partial t} + \frac{\partial f(u)}{\partial x} = 0, \tag{59}$$

where $u = (u_1, \cdots, u_n)^T$ and $f(u) = (f_1(u), \cdots, f_n(u))^T$. Moreover, we suppose that every eigenvalue of $A(u) = \bigtriangledown f(u)$ has a constant multiplicity. Without loss of generality, we may suppose that

$$\lambda(u) \triangleq \lambda_1(u) \equiv \cdots \equiv \lambda_p(u) < \lambda_{p+1}(u) < \cdots < \lambda_n(u), \tag{60}$$

where $1 \leq p \leq n$. When $p = 1$, system (59) is strictly hyperbolic; while, when $p > 1$, (59) is a non-strictly hyperbolic system of conservation laws with characteristics with constant multiplicity. According to the result given in [24]–[25], for the hyperbolic system of conservation laws, every characteristic with constant multiplicity $p > 1$ must be linearly degenerate:

$$\bigtriangledown\lambda(u)r_i(u) \equiv 0 \qquad (i = 1, \cdots, p), \tag{61}$$

then weakly linearly degenerate. It turns out that all previous results for the strictly hyperbolic case can be similarly extended to the present situation (see [15]-[16]).

6. Applications

Now we can apply the previous results to solve the Cauchy problem with small and decaying initial data for the four physical examples given in Section 1 in a complete manner on both the global existence and the blow-up phenomenon except in the critical case $\alpha = +\infty$.

For instance, we consider the following Cauchy problem for the system of the motion of elastic strings

$$\begin{cases} \dfrac{\partial u}{\partial t} - \dfrac{\partial v}{\partial x} = 0, \\ \dfrac{\partial v}{\partial t} - \dfrac{\partial}{\partial r}(\dfrac{T(r)}{r}u) = 0, \end{cases} \tag{62}$$

$$t = 0: \quad u = \tilde{u}_0 + \varepsilon u_0(x), \qquad v = \tilde{v}_0 + \varepsilon v_0(x), \tag{63}$$

where $u = (u_1, \cdots, u_n)^T$, $v = (v_1, \cdots, v_n)^T$, $r = |u| = \sqrt{u_1^2 + \cdots + u_n^2}$, $T(r)$ is a suitably smooth function of $r > 1$ such that

$$T'(\tilde{r}_0) > \frac{T(\tilde{r}_0)}{\tilde{r}_0} > 0, \tag{64}$$

where $\tilde{r}_0 = |\tilde{u}_0| > 1$, $\tilde{u}_0$ and $\tilde{v}_0$ are constant vectors, $\varepsilon > 0$ is a small parameter and $(u_0(x), v_0(x)) \in C^1$ satisfies (42).

By Theorems 1–2, we get (see [14]-[15])

Theorem 5. *Suppose that*

$$T''(r) \equiv 0, \qquad \forall r > 1. \tag{65}$$

System (62) is linearly degenerate, then weakly linearly degenerate, and Cauchy problem (62)–(63) always admits a unique global C^1 solution $(u(t,x), v(t,x))$ for all $t \in \mathbb{R}$, provided that $\varepsilon > 0$ is small enough. □

Theorem 6. *Suppose that there exists an integer $\alpha \geq 0$ such that*

$$T''(\tilde{r}_0) = \cdots = T^{(\alpha+1)}(\tilde{r}_0) = 0, \quad but \quad T^{(\alpha+2)}(\tilde{r}_0) \neq 0. \tag{66}$$

If

$$\left(\tilde{u}_0^T u_0(x), \ \tilde{u}_0^T v_0(x)\right) \not\equiv 0, \tag{67}$$

namely, $\tilde{u}_0$ is not simultaneously orthogonal to $u_0(x)$ and $v_0(x)$ for all $x \in \mathbb{R}$, then for $\varepsilon > 0$ small enough, the first order derivative $(u_x(t,x),\ v_x(t,x))$ of the C^1 solution $(u(t,x), v(t,x))$ to Cauchy problem (62)-(63) must blow up in a finite time and the life-span

$$\tilde{T}(\varepsilon) \approx \varepsilon^{-(\alpha+1)}. \tag{68}$$

The formation of singularity is due to the envelope of characteristics of the first or last family. Moreover, along the first or last characteristic passing through a blow-up point $(\tilde{T}(\varepsilon), x_0)$, the blow-up rate is given by

$$(u_x, v_x)(t,x) = O((\tilde{T}(\varepsilon) - t)^{-1}), \qquad \forall t < \tilde{T}(\varepsilon). \ \square \tag{69}$$

7. Generalized null Condition

The null condition was introduced in the study of nonlinear wave equations for getting the global existence of classical solutions with small initial data (cf. [26]–[27]). For the first order quasilinear strictly hyperbolic system (1) we can similarly introduce the following

Definition 3: Strictly hyperbolic system (1) is said to satisfy the *null condition*, if every small plane wave solution $u = u(s)$ $(u(0) = 0)$, where $s = ax + bt$, a *and* b being constants, to the corresponding linearalized system

$$u_t + A(0)u_x = 0 \tag{70}$$

is always a solution to the original quasilinear system (1), namely

$$u_t + A(0)u_x = (A(0) - A(u))u_x \triangleq B(u)u_x.\square \tag{71}$$

Without loss of generality, we may suppose that

$$A(0) = diag\{\lambda_1(0), \cdots, \lambda_n(0)\}. \tag{72}$$

Thus, the general solution to system (70) is

$$u_i = u_i(x - \lambda_i(0)t) \quad (i = 1, \cdots, n), \tag{73}$$

where $u_i = u_i(s)$ $(i = 1, \cdots, n)$ are arbitrarily given smooth functions of s. Hence, noting the strict hyperbolicity, for each plane wave solution u to system (70), there exists an index $i \in \{1, \cdots, n\}$ such that

$$u = u_i(s)e_i, \tag{74}$$

where

$$s = x - \lambda_i(0)t \tag{75}$$

and

$$e_i = (0, \cdots, 0, \overset{(i)}{1}, 0, \cdots, 0)^T. \tag{76}$$

Therefore, we get

$$\begin{gathered}
\text{The null condition for system (1)} \\
\Updownarrow \\
B(u_i(s)e_i)u_i'(s)e_i \equiv 0, \ \forall \, small \ u_i(s) \ (u_i(0) = 0) \quad (i = 1, \cdots, n) \\
\Updownarrow \\
B(u_ie_i)e_i \equiv 0, \ \forall \, small \ |u_i| \quad (i = 1, \cdots, n) \\
\Updownarrow \\
A(u_ie_i)e_i \equiv \lambda_i(0)e_i, \ \forall \, small \ |u_i| \quad (i = 1, \cdots, n) \\
\Updownarrow \\
\lambda_i(u_ie_i)e_i \equiv \lambda_i(0), \ \forall \, small \ |u_i| \quad (i = 1, \cdots, n),
\end{gathered}$$

$$r_i(u_ie_i)//e_i, \ \forall \ small \ |u_i| \quad (i = 1, \cdots, n)$$
$$\Updownarrow$$
$$\text{system(1) is weakly linearly degenerate and}$$
$$u = (u_1, \cdots, u_n) \text{ are normalized coordinates.}$$

Moreover, since (1) is a system with n unknown variables, we may introduce the following

Definition 4: System (1) is said to satisfy the *generalized null condition*, if there exists a local C^2 diffeomorphism $\tilde{u} = \tilde{u}(u)$ $(\tilde{u}(0) = 0)$ such that the corresponding system for $\tilde{u}$ satisfies the null condition. □

In Definition 4, $\tilde{u} = (\tilde{u}_1, \cdots, \tilde{u}_n)$ are normalized coordinates and $\tilde{u} = \tilde{u}(u)$ is noting but a normalized transformation. Thus, system (1) satisfies the generalized null condition simply means that system (1) satisfies the null condition in normalized coordinates.

Noting that for every $i = 1, \cdots, n$, $\nabla\lambda_i(u)r_i(u)$ is an invariant under any given invertible C^2 transformation $\tilde{u} = \tilde{u}(u)$, we have

Proposition 1 (see [13]): System (1) is weakly linearly degenerate if and only if system (1) satisfies the generalized null condition. □

References

[1] Li Ta-tsien, Yu Wen-ci, *Boundary Value Problems for Quasilinear Hyperbolic Systems*, Duke University Mathematics Series V, 1985.

[2] Li Ta-tisen, *Global Classical Solutions for Quasilinear hyperbolic Systems*, Research in Applied Mathematics 32, Masson/ John Wiley, 1994.

[3] F. John, *Formation of singularities in one-dimensional nonlinear wave propagation*, Comm. Pure Appl. Math., 27(1974), 377-405.

[4] T. P. Liu, *Development of singularities in the nonlinear waves for quasilinear hyperbolic partial differential equations*, J. Diff. Equations, 33(1979), 92-111.

[5] L. Hörmander, *The life span of classical solutions of nonlinear hyperbolic equations*, Report No.5, Institute Mittag-Leffler, 1985.

[6] L. Hörmander, *Lectures on Nonlinear Hyperbolic Differerntial Equations*, Mathématiques & Applications, Vol.26, Springer, Berlin, 1997.

[7] A. Bressan, *Contrative metrics for nonlinear hyperbolic systems*, Indiana University Mathematics Journal, 37(1988), 409-420.

[8] C. Carasso, M. Rascle, D. Serre, *Etude d'un modèle hyperbolique en dynamique des câbles*, Modélisation Mathématique et Analyse Numérique, 19(1985), 573-599.

[9] Li Ta-tsien,, D. Serre, Zhang Hao, *The generalized Riemann problem for the motion of elastic strings*, SIAM J. Math. Analysis, 23(1992), 1189-1203.

[10] Li Ta-tsien, Peng Yue-jun, *Problème de Riemann généralisé pour une sorte de système des cables*, Portugaliae Mathematica, 50(1993), 407-437.

[11] P. G. Ciarlet, *Mathematical Elasticity, Vol. I: Three-Dimensional Elasticity*, North-Holland, Amsterdam, 1988.

[12] A. Majda, *Compressible Fluid Flow and System of Conservation Laws in Several Space Variables*, Applied Mathematical Sciences 53, Springer-Verlag, 1984.

[13] Li Ta-tsien, Zhou Yi, Kong De-xing, *Weak linear degeneracy and global classical solutions for general quasilinear hyperbolic systems*, Commun. in Partial Differential Equations, 19(1994), 1263-1317.

[14] Li Ta-tsien, Zhou Yi, Kong De-xing, *Global classical solutions for general quasilinear hyperbolic systems with decay initial data*, Nonlinear Analysis, Theory, Methods & Applications, 28(1997), 1299-1332.

[15] Li Ta-tsian, Kong De-xing, Zhou Yi, *Global Classical solutions for quasilinear non-strictly hyperbolic systems*, Nonlinear Studies, 3(1996), 203-229.

[16] Li Ta-tsian, Kong De-xing, *Initial value problem for general quasilinear hyperbolic systems with characteristics with constant multiplicity*, J. Partial Diff. Eqa., 10(1997), 299-322.

[17] Li Ta-tsien, Kong De-xing, *Global classical solutions with small amplitude for general quasilinear hyperbolic systems*, New Approachs in Nonlinear Analysis, Edited by Themistocles M. Rassias, Hadronic Press, 1999, pp.203-237.

[18] Li Ta-tsien, Kong De-xing, *Breakdown of classical solutions to quasilinear hyperbolic systems*, Nonlinear Analysis, 40(2000), 407-437.

[19] Li Ta-tsien, *Une remarque sur les coordonnées normalisées et ses applications aux systèmes hyperboliques quasi linéaires*, C. R. Acad. Sci. Paris, 331, Série I (2000), 447-452.

[20] Li Ta-tsien, *A remark on the normalized coordinates and its applications to quasilinear hyperbolic systems*, Optimal Control and Partial Differential Equations, Edited by J. L. Menaldi, E. Rofman and A. Sulem, IOS Press, 2001, pp. 181-187.

[21] Kong De-xing, *Life-span of classical solutions to quasilinear hyperbolic systems with slow decay initial data*, Chin. Ann. of Math., 21B(2000), 413-440.

[22] Kong De-xing, Li Ta-tsien, *A note on blow-up phenomenon of classical solutions to quasilinear hyperbolic systems*, to appear in Nonlinear Analysis.

[23] Yan Ping, *Global classical solutions with small initial total variation for quasilinear hyperbolic systems*, to appear.

[24] G. Boillat, *Chocs caractéristiques*, C. R. Acad. Sci. Paris, 274, Série A (1972), 1018-1021.

[25] H. Freistühler, *Linear degeneracy and shock waves*, Math. Zeit., 207(1991). 583-596.

[26] S. Klainerman, *The null condition and global existance to nonlinear wave equations*, Lectures in Applied Mathematics, 23(1986), 293-326.

[27] D. Christodoulou, *Global solution of nonlinear hyperbolic equations for small initial data*, Comm. Pure Appl. Math., 39(1986), 267-282.

NUMERICAL SIMULATION OF 3D SHALLOW WATER WAVES ON SLOPING BEACH *

Tiejun Li, Pingwen Zhang
School of Mathematical Sciences, Peking University,
Beijing 100871, P.R. China
tieli@pku.edu.cn
pzhang@pku.edu.cn

Abstract In this paper, we solve a special 3D model problem concerning moving waterline of shallow water equations. We adopted two kinds of description for moving waterline formulations, of which the second produces better result and is recommended. To solve the 3D model problem, a stable scheme for smooth solution is given through special treatment of boundary, based on which we present numerical example for the 3D model problem, and the result demonstrates the accuracy and stability of our scheme.

Keywords: 3D shallow water equations, moving waterline, semi-Eulerian

1. Introduction

Shallow-water flows are nearly horizontal, which allows a considerable simplification in the mathematical formulation and numerical solution by describing the average behavior of the fluid flow. The relative simplicity is the reason why such flows have attracted the attention of many mathematicians and hydrodynamicists.

The numerical solution of the shallow water equations (SWE) was one of the early applications of digital computers when these became available in the late 1940's. Simulations were done by Charney et all [1] for atmospheric and Hansen [4] for oceangraphic flows. Now many developments has been achieved in numerics [10].

Study of run-up of ocean waves on sloping beach is one of the classical problems in hydrodynamics because the wave motion around moving waterline is highly nonlinear. There is same nonlinearity for SWE. The moving water-

*Subsidized by the Special Funds for Major State Research Projects G1999032804 and by the Teaching and Research Award Fund for Outstanding Young Teachers in Higher Education Institutions of MOE.

line can also be mathematically considered as the moving boundary conditions for a hyperbolic system. As well known, the most important mathematical and physical quantities of the hyperbolic equations are characteristics, and the wellposedness of the hyperbolic equations is closely related to the propagating direction of the characteristics on the boundary. At the moving waterline, the propagating directions all coalesce into a single one, such that the hyperbolic character of SWE gets lost and the wellposedness is not so straightforward. Numerical study will not stop here. An Eulerian-Lagrangian hybrid method for two dimensional run-up of ocean waves on sloping beach is proposed in [11]. The artificial boundary condition (ABC) is considered for bounded domain problem, and conservative scheme is applied to SWE for overcoming the appearance of hydraulic jump In [6]. In this paper, we solve a special 3D model problem concerning moving waterline of shallow water equations. We adopted two kinds of description of moving waterline formulations, of which the second produces better result for the stability of numerical discretization and is recommended. The 3D model problem, which is periodic in x direction and symmetric in y direction, could be used to avoid the ABC of high dimensional conservation laws. A stable scheme for smooth solution is given through special treatment of the boundary, based on which we present a numerical example for the 3D model problem, and the result demonstrates the accuracy and stability of our scheme. More robust schemes are still wanted for general solution.

The rest of this paper is organized as follows: in section 2 we will introduce the 3D SWEs. Two kinds of moving waterline formulation are described in section 3 and 4 respectively. In section 5 we present some numerical experiments to verify the accuracy and stability of our scheme for smooth solution. The conclusions are drawn in the last section.

2. Three dimensional SWEs

The shallow water equation in three dimension is of the form

$$\begin{cases} (\eta+h)_t + ((\eta+h)u)_x + ((\eta+h)v)_y = 0 \\ ((\eta+h)u)_t + ((\eta+h)u^2 + \frac{1}{2}g(\eta+h)^2)_x + ((\eta+h)uv)_y = g(\eta+h)h_x \\ ((\eta+h)v)_t + ((\eta+h)uv)_x + ((\eta+h)v^2 + \frac{1}{2}g(\eta+h)^2)_y = g(\eta+h)h_y \end{cases} \tag{1}$$

where g is the gravity coefficient, $h(x, y)$ is the undisturbed water depth, $\eta(x, y, t)$ is the water elevation from the undisturbed water surface at time t and $u(x, y, t), v(x, y, t)$ are the x, y component of flow velocities of the wave motion respectively. Thus $\eta(x, y, t) + h(x, y)$ denotes the distance between the surface and the bottom of the river, it is bigger than zero always. The derivation of this system can be seen from [9] or derived from conservation of physical quantities directly.

We simply write this equation as

$$\frac{\partial \mathbf{U}}{\partial t} + \frac{\partial \mathbf{F}}{\partial x} + \frac{\partial \mathbf{G}}{\partial y} = \mathbf{S} \tag{2}$$

where

$$\mathbf{U} = \begin{pmatrix} \eta + h \\ (\eta + h)u \\ (\eta + h)v \end{pmatrix}, \quad \mathbf{F} = \begin{pmatrix} (\eta + h)u \\ (\eta + h)u^2 + \frac{1}{2}g(\eta + h)^2 \\ (\eta + h)uv \end{pmatrix},$$

$$\mathbf{G} = \begin{pmatrix} (\eta + h)v \\ (\eta + h)uv \\ (\eta + h)v^2 + \frac{1}{2}g(\eta + h)^2 \end{pmatrix}, \quad \mathbf{S} = \begin{pmatrix} 0 \\ g(\eta + h)h_x \\ g(\eta + h)h_y \end{pmatrix}.$$

We will use the conservative formulation of these equations in the following, from now on, we will use u, v, w instead of the height function $\eta + h$, the momentum in x-direction $(\eta + h)u$ and the momentum in y-direction $(\eta + h)v$ respectively. Then we have

$$\mathbf{U} = \begin{pmatrix} u \\ v \\ w \end{pmatrix}, \quad \mathbf{F} = \begin{pmatrix} v \\ \frac{v^2}{u} + \frac{1}{2}gu^2 \\ \frac{vw}{u} \end{pmatrix},$$

$$\mathbf{G} = \begin{pmatrix} w \\ \frac{vw}{u} \\ \frac{w^2}{u} + \frac{1}{2}gu^2 \end{pmatrix}, \quad \mathbf{S} = \begin{pmatrix} 0 \\ guh_x \\ guh_y \end{pmatrix}.$$

3. Fully Lagrangian moving waterline equations

We can extend moving waterline equations from two dimension [11] to three dimension directly. The formulation is based on the fully Lagrangian. We describe the equations as follows:

$$\begin{cases} h(X(t), Y(t)) + \eta(X(t), Y(t), t) = 0 \\ \dfrac{dX}{dt} = u(X(t), Y(t), t) = U(t) \\ \dfrac{dY}{dt} = v(X(t), Y(t), t) = V(t) \\ \dfrac{dU}{dt} = u_t + u_x U + u_y V = -g\eta_x(X(t), Y(t), t) \\ \dfrac{dV}{dt} = v_t + v_x U + v_y V = -g\eta_y(X(t), Y(t), t) \end{cases} \tag{3}$$

where $X(t), Y(t)$ denote the time-varying position of the fluid particles at the waterline, and $U(t), V(t)$ are the Lagrangian velocities. The first equation is the definition of moving waterline, and the second and third are kinematical equations. The fourth and fifth are dynamical equations which are Lagrangian description of the momentum conservation of SWE, where the water depth is zero. The evolution equation (1) and (3) completely specify the motion of the system. The Lagrangian formulation will lead to domain changing with the evolution of the moving waterline, which will cause numerical difficulties. Actually, we designed schemes for this formulation, and the sensitivity related to the moving domain leads to numerical instability seriously!

4. Semi-Eulerian moving waterline equations

We assume a time dependent curve in x-y plane can be parameterized by x, and the particles on this curve move with the speed (u, v). The curve could be described as: $y = Y(x, t)$. Then we have

$$Y_t = V - UY_x \tag{4}$$

where $U(x,t) = u(x, Y(x,t), t), V(x,t) = v(x, Y(x,t), t)$.

Lemma 1. *For arbitrary time dependent Lagrangian curve* $(X(\xi,t), Y(\xi,t))$, *if we assume that the material derivative of the coordinates satisfies*

$$\begin{cases} \frac{dX}{dt} = U, \\ \frac{dY}{dt} = V, \end{cases}$$

and a function defined on this curve satisfies $\frac{df}{dt} = g$, *then we have the relation:*

$$D_t f = g - UD_x f. \tag{5}$$

Proof: Using the curve expression, we have

$$Y_t - V + UY_x = 0$$

then we get

$$\begin{aligned} D_t f = D_t f(x, Y(x,t), t) &= f_t + f_y Y_t = f_t + f_y(V - UY_x) \\ &= g - Uf_x - Vf_y + Vf_y - Uf_yY_x = g - U(f_x + f_yY_x) = g - UD_x f \end{aligned}$$

This ends the proof.

Taking advantage of above lemma, the semi-Eulerian moving waterline equations could be written:

$$\begin{cases} \eta + h = 0 \\ \dfrac{DY}{Dt} = V - UY_x \\ \dfrac{DU}{Dt} = -g\eta_x - UU_x \\ \dfrac{DV}{Dt} = -g\eta_y - UV_x \end{cases} \tag{6}$$

We call the system as semi-Eulerian formulation because here only y-direction is Lagrangian. The domain will not move in x-direction if we use the semi-Eulerian formulation.

5. Numerical methods

Here we assume the moving waterline is a function of x along the beach. We will perform a coordinate transformation to the equations in order to keep the computational domain fixed. The calculation will be done in the fixed domain. The strategy is as follows: define coordinate transformation:

$$\begin{cases} t = t' \\ x = x' \\ y = \frac{Y(x',t')}{Y_0} y' \end{cases}$$

this transformation changes the moving waterline $(x, Y(x,t))$ into a fixed straight line (x', Y_0), where x', y', t' are transformed variables. Y_0 is a constant. We have the following relations:

$$\begin{pmatrix} \frac{\partial}{\partial t'} \\ \frac{\partial}{\partial x'} \\ \frac{\partial}{\partial y'} \end{pmatrix} = \mathbf{T} \begin{pmatrix} \frac{\partial}{\partial t} \\ \frac{\partial}{\partial x} \\ \frac{\partial}{\partial y} \end{pmatrix} \tag{7}$$

where

$$\mathbf{T} = \begin{pmatrix} 1 & 0 & \frac{y'}{Y_0}\frac{\partial Y}{\partial t'} \\ 0 & 1 & \frac{y'}{Y_0}\frac{\partial Y}{\partial x'} \\ 0 & 0 & \frac{Y}{Y_0} \end{pmatrix}$$

and X, Y are defined as before.

After simple calculation, we may get

$$\mathbf{T}^{-1} = \begin{pmatrix} 1 & 0 & a \\ 0 & 1 & b \\ 0 & 0 & c \end{pmatrix}$$

where

$$a = \frac{y'}{Y}\frac{\partial Y}{\partial t'}, \quad b = \frac{y'}{Y}\frac{\partial Y}{\partial x'}, \quad c = \frac{Y_0}{Y}$$

Under this transformation, the equations become (all the primes of the variables are omitted):

$$\mathbf{U}_t + a\mathbf{U}_y + \mathbf{F}_x + b\mathbf{F}_y + c\mathbf{G}_y = \mathbf{S}$$

where a, b, c are defined as above.

We rewrite the transformed equations in conservative form:

$$\mathbf{U}_t + \mathbf{F}_x + (a\mathbf{U} + b\mathbf{F} + c\mathbf{G})_y = (a_y\mathbf{U} + b_y\mathbf{F} + \mathbf{S})$$

where

$$\begin{cases} a_y &= -\frac{1}{Y}\frac{\partial Y}{\partial t} \\ b_y &= \frac{1}{Y}\frac{\partial Y}{\partial x} \end{cases}$$

Define

$$\tilde{\mathbf{U}} = \mathbf{U}, \quad \tilde{\mathbf{F}} = \mathbf{F}, \quad \tilde{\mathbf{G}} = a\mathbf{U} + b\mathbf{F} + c\mathbf{G},$$
$$\tilde{\mathbf{S}} = a_y\mathbf{U} + b_y\mathbf{F} + \mathbf{S}$$

we have the following system:

$$\tilde{\mathbf{U}}_t + \tilde{\mathbf{F}}_x + \tilde{\mathbf{G}}_y = \tilde{\mathbf{S}} \tag{8}$$

To solve this nonhomogeneous conservation laws, we apply third order WENO scheme to space and TVD-RK3 to time [8]. As well known, the third order WENO is a five point stencil in each direction, which is not enough for the points neighboring to the boundary. Mixed scheme will cause numerical instability seriously!

For smooth solution, in principle, we can use Lax-Wendroff scheme. Unfortunately, 2D Lax-Wendroff scheme is still a five point stencil in each direction [7], which will lead numerical instability seriously. Using centered difference to space, and TVD-RK3 to time [3], whose stable region covers one segment of imaginary axis, and fully coupling the boundary equations and SWEs inside, we obtain a stable scheme for smooth solution.

Remark: TVD-RK3 is:

$$\begin{cases} u^{(1)} &= u^n + \triangle t L(u^n) \\ u^{(2)} &= \frac{3}{4}u^n + \frac{1}{4}u^{(1)} + \frac{1}{4}\triangle t L(u^{(1)}) \\ u^{n+1} &= \frac{1}{3}u^n + \frac{2}{3}u^{(2)} + \frac{2}{3}\triangle t L(u^{(2)}) \end{cases} \tag{9}$$

The stable region is the same as the classical third order RK method.

6. Numerical results

We choose the domain as $[-\pi, \pi] \times [0, \pi]$, in which x direction is period and y direction is symmetric. The height function is also periodic in x direction and symmetric in y direction. The initial value is chosen as:

$$\eta(x, y, 0) = \eta_0(\cos(x) + 1)(\cos(y) + 1)$$

and the bottom function is chosen as one dimensional shape:

$$h(x, y) = 1 - \frac{y^2}{\pi^2}$$

we set $\eta_0 = 0.0025$. The physical time covers from $t = 0$ to $t = 100$. The numerical results are stable and smooth with very small diffusion. The evolution surfaces in different times are shown in Figure 1:

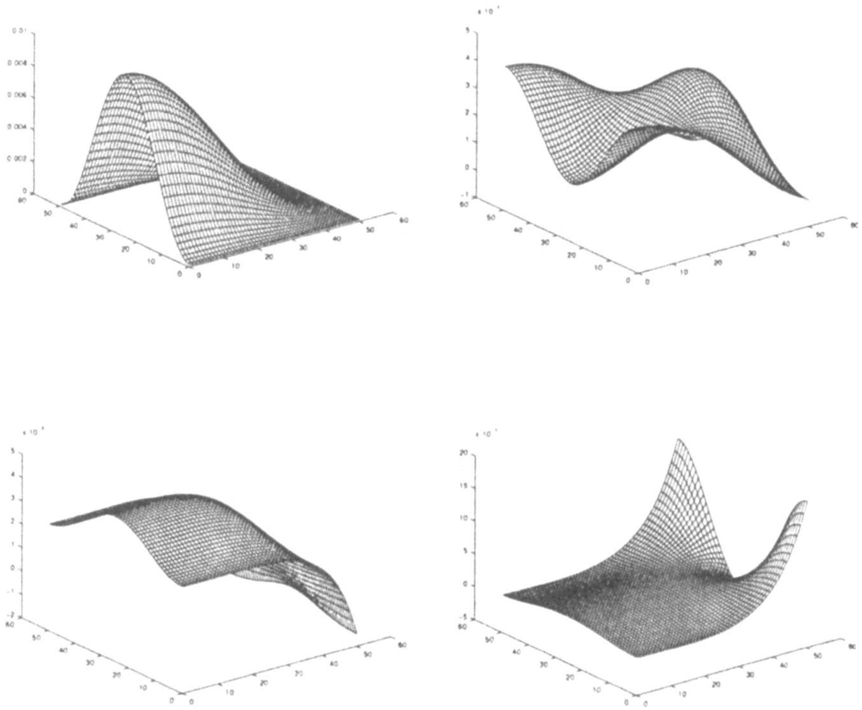

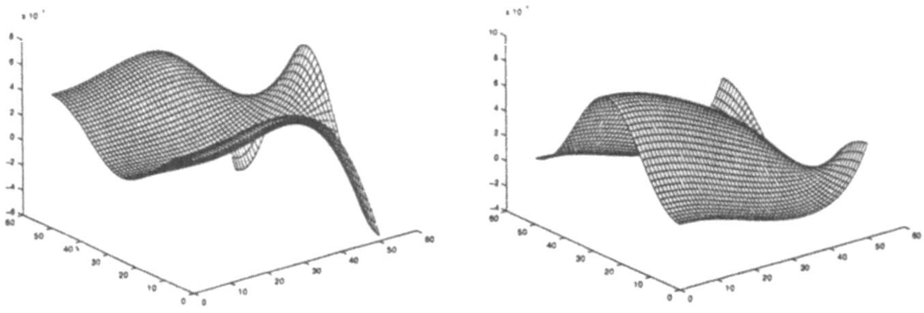

Figure 1. shallow water wave evolution

The evolution of moving waterline are shown in Figure 2.

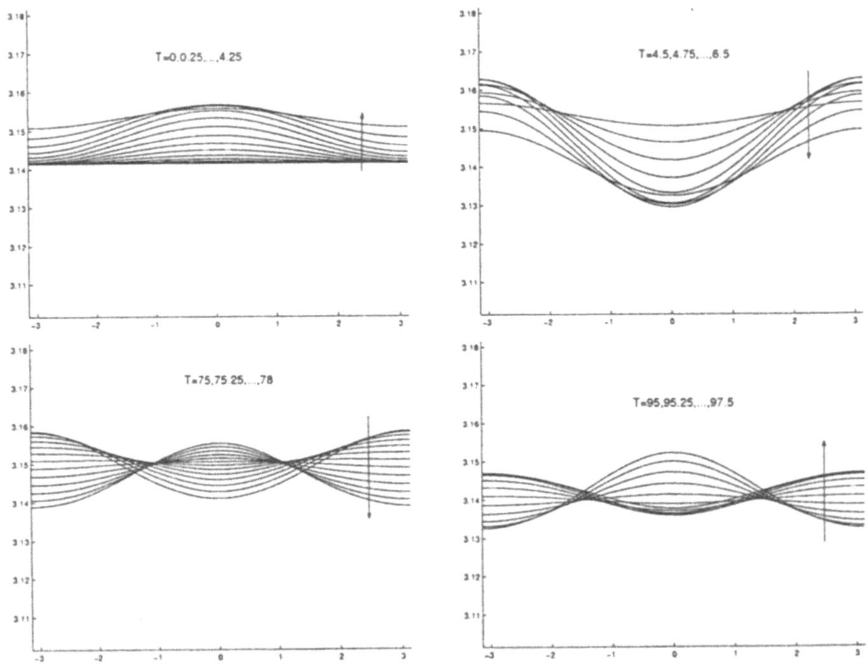

Figure 2. evolution of moving waterline

In order to estimate the order of the scheme, we perform computation with subdivision 50×50, 100×100, 200×200, and 400×400. we consider the numerical result in finest mesh as accurate solution, then we can do L^∞

estimate numerically. The results are shown in Figure 3, two order accuracy was observed.

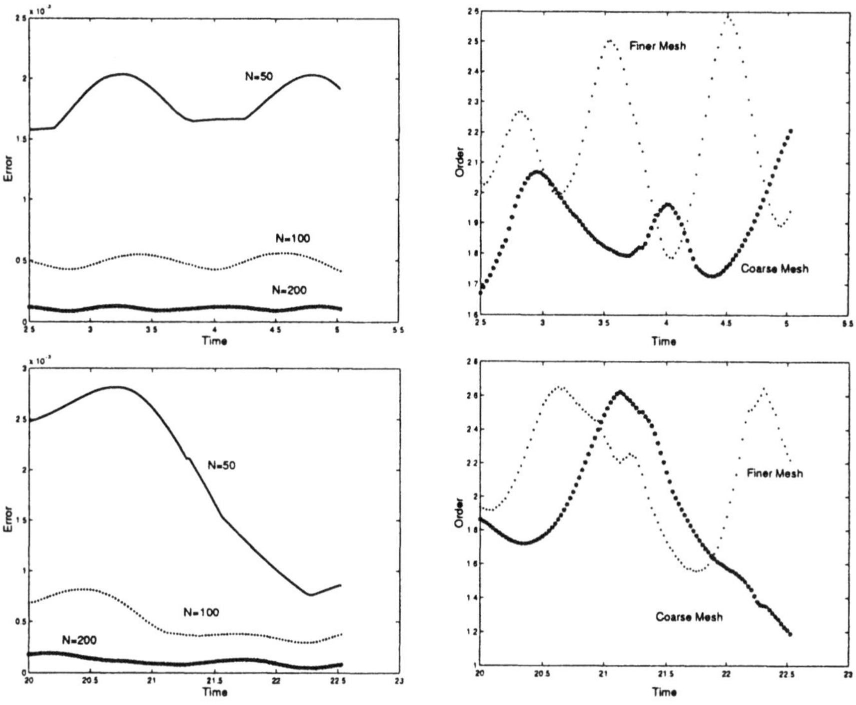

Figure 3. L^∞ error and order estimates

7. Conclusions

In this paper, we solve a special 3D model problem concerning the moving waterline of shallow water equations. A stable scheme for smooth solution is given through special treatment of boundary. The application is limited. We are looking for more robust scheme, such as modifying standard WENO schemes or discontinuous Galerkin methods etc, and hope new schemes can be used for general solution. These results will be reported elsewhere.

References

[1] J. G. Charney, R. Fjörtoft and J. von Neumann, Numerical integration of the barotropic vorticity equation, *Tellus* **2** (1950), pp. 237-254.

[2] J. Crank, Free and moving boundary problems, Oxford University Press, 1987.

[3] W. E and J. Liu, Vorticity boundary condition and related issues for finite difference schemes, *J. Comp. Phys.* **124** (1996), pp. 368-382.

[4] W. Hansen, Theorie zur Errechnung des Wasserstandes und der Strömungen in Randmeeren nebst Anwendungen, *Tellus* **8**, 3 (1956), pp. 187-300.

[5] R. J. LeVeque and H. C. Yee, A study of numerical methods for hyperbolic conservation laws with stiff source terms, *J. Comput. Phys.* **86** (1990), 187-210.

[6] T. Li and P. Zhang, Numerical studies of shallow water waves on sloping beach with articial boundary, accepted by *J. Comp. Math.*.

[7] R. D. Richtmyer and K. W. Morton, Difference Methods for Initial Value Problems, Interscience, New York, 1967.

[8] C.-W.Shu, Essentially non-oscilatory and weighted essentially non-oscilarory schemes for hyperbolic conservation laws, Lecture Notes in Mathematics **1697**, pp. 329-432.

[9] J. J. Stoker, Water waves: the mathematical theory with applications, John Wiley & Sons, U.S.A., 1992, 2nd edition.

[10] C. B. Vreugdenhil, Numerical Methods for Shallow Water Flow, Kluwer Academic Publishers, Netherlands, 1994.

[11] J. E. Zhang, T. Y. Wu and T. Y. Hou, Coastal hydrodynamics of ocean waves on beach, in: Adv. App. Mech. **37** (1999), pp. 1-50.

HIGH PERFORMANCE FINITE ELEMENT METHODS

Qun Lin
Institute of Computational Mathematics
Chinese Academy of Sciences, Beijing, China
qlin@staff.iss.ac.cn

Abstract We examine expansions of FEMs. Although this topic has been discussed for above 20 years some progress has been made in 2001. We first introduce the main progress by Jiafu Lin and Lutz Tobiska, et al. for the solution problem of PDEs, and we then also study about the application of the error expansion method to an eigenvalue problem.

1. Part I Solution Problem

Reserachers are interested in the "simple FEM, coarse mesh and high accuracy". We do not discuss directly in this part about which elements are high performance but discuss about how the extrapolation technique supported by "postprocessing FE" will produce a high performance. The theoretical base is the "transitional" error expansion. There are some survey papers or books concerning with such an expansion, e.g., Rannacher [16], Shen [19] or Brunner [3], Chen and Huang [3], Krizek and Neittaanmaki [6], Liu and Zhu [15], Lin and Yan [14], Shaidurov [18] and Sloan [20], including a complete state and art before 2000 and the related references. We do not repeat them here but introduce the progress in 2001.

First, Jiafu Lin in his postdoctoral thesis [7] investigated systematically the error expansions for conforming FEs, including 5 kinds of triangular elements (linear, quadratic, Hood-Taylor, P_1-P_1, Raviart-Thomas) and 6 kinds of rectangular elements (bilinear, biquadratic, Hood-Taylor, high degree Hood-Taylor, Q_2^+ - Q_1, Nedelec). Although some results have been known (e.g., in [14]) Jiafu Lin's proof is more mechanical and systematical, especially in constructing postprocessed FE solution.

Let us consider the Poisson model on a rectangular domain Ω: find solution $u \in H_0^1(\Omega)$:

$$a(u, v) = (f, v) \quad \forall v \in H_0^1(\Omega),$$

$$a(u, v) = \int_\Omega u_x v_x + u_y v_y$$

($dxdy$ is omitted in this article). We construct the rectangulation $T_h = \{e\}$, introduce the FE space $V_h \subset H^1(\Omega)$ and $V_{h,0} = V_h \cap H_0^1(\Omega)$, and find the FE solution $u_h \in V_{h,0}$:

$$a(u_h, v) = (f, v) \quad \forall v \in V_{h,0}.$$

We cannot have the error expansion straightly for the FE solution itself but

i) A transitional expansion

Compare an interpolation $u_I \in V_h$: for $v \in V_{h,0}$,

$$a(u_h - u_I, v) = a(u - u_I, v) = \int_\Omega (u - u_I)_x v_x + (u - u_I)_y v_y.$$

The integrals are then expanded into a dominant term and a higher order remainder:

$$ch^k + o(h^k)|v|_j$$

for some integers k and j. From which we can get the transitional expansion: there exists a function ϕ, independent of h, such that in appropriate norms

$$u_h - u_I = \phi \cdot h^k + o(h^k).$$

We want, however, to approximate u rather than approximate u_I. For this we need

ii) Postprocessing

This is a high degree interpolation on a coarse mesh, see, e.g., the interpolations Π_{3h}^6 and Π_{2h}^4 for biquadratic and Adini elements below, respectively. Construct postprocessed solution $\Pi_{mh}^l u_h$ (e.g., $m = 3$, $l = 6$), we have in appropriate norms

$$\Pi_{mh}^l u_h - u = \phi_1 \cdot h^k + o(h^k).$$

iii) Extrapolation

Taking $\widetilde{u}_h = \Pi_{mh}^l u_h$, then we get in appropriate norms

$$\frac{2^k \widetilde{u}_{h/2} - \widetilde{u}_h}{2^k - 1} = u + o(h^k).$$

Therefore, we may concentrate ourself on the expansions for the integrals, like

$$\int_e (u - u_I)_x v_x$$

where $e \in T_h$: $e = [x_e - h_e, x_e + h_e] \times [y_e - k_e, y_e + k_e]$.

1.1 Biquadratic element

Let us introduce the biquadratic polynomial space

$$V_h = \{v \in C(\bar{\Omega}); v \mid_e \in \mathrm{Q}_2 \ \forall e \in T_h\},$$

$$\mathrm{Q}_2 = \mathrm{span}\{1, x, y, x^2, xy, y^2, x^2y, xy^2, x^2y^2\}.$$

Let the biquadratic interpolation u_I be defined by the point-line-plane condition:

$$u_I(Z_i) = u(Z_i), \int_{l_i} (u_I - u)ds = 0, \ \ i = 1,2,3,4, \int_e (u_I - u) = 0,$$

where Z_i and l_i are four vertices and four edges, respectively, of the element e.

Lemma 1. *If* $v \in V_h$, *then*

$$\int_e (u - u_I)_x v_x = -\frac{k_e^4}{45} \int_e u_{xyyy} v_{xy} + O(h^4)|u|_{5,e}|v|_{1,e}, \tag{1}$$

$$\int_e (u - u_I)_x v_x = -\frac{k_e^4}{45} \int_e u_{xyyy} v_{xy} + \frac{32k_e^6}{4725} \int_e u_{xyyyy} v_{xyy} + O(h^6)|u|_{6,e}|v|_{2,e}. \tag{2}$$

Proof E.g., the last expansion can be checked, by the Bramble-Hilbert Lemma, on the standard reference element $\hat{e} = [-1,1] \times [-1,1]$ for the polynomial $\hat{u} \in P_5(\hat{e})$:

$$\int_{\hat{e}} (\hat{u} - \hat{u}_I)_{\hat{x}} \hat{v}_{\hat{x}} d\hat{x}d\hat{y} + \frac{1}{45} \int_{\hat{e}} \hat{u}_{\hat{x}\hat{y}\hat{y}\hat{y}} \hat{v}_{\hat{x}\hat{y}} d\hat{x}d\hat{y} - \frac{32}{4725} \int_{\hat{e}} \hat{u}_{\hat{x}\hat{y}\hat{y}\hat{y}\hat{y}} \hat{v}_{\hat{x}\hat{y}\hat{y}} d\hat{x}d\hat{y} = 0$$

where $\hat{u}_I = \hat{u}$ when $\hat{u} \in Q_2(\hat{e})$, otherwise

$\hat{u}$	$\hat{x}^3$	$\hat{y}^3$	$\hat{x}^4$	$\hat{x}^3\hat{y}$	$\hat{x}\hat{y}^3$	$\hat{y}^4$
$\hat{u}_I$	$\hat{x}$	$\hat{y}$	$\frac{6\hat{x}^2-1}{5}$	$\hat{x}\hat{y}$	$\hat{x}\hat{y}$	$\frac{6\hat{y}^2-1}{5}$
$\hat{u}$	$\hat{x}^5$	$\hat{x}^4\hat{y}$	$\hat{x}^3\hat{y}^2$	$\hat{x}^2\hat{y}^3$	$\hat{x}\hat{y}^4$	$\hat{y}^5$
$\hat{u}_I$	$\hat{x}$	$(\frac{6\hat{x}^2-1}{5})\hat{y}$	$\hat{x}\hat{y}^2$	$\hat{x}^2\hat{y}$	$\hat{x}(\frac{6\hat{y}^2-1}{5})$	$\hat{y}$

Theorem 1. *If* $v \in V_{h,0}$ *and* T_h *is uniform:* $k_e \equiv h_2$ $(\forall e \in T_h)$, *then*

$$\int_\Omega (u - u_I)_x v_x \ = \ -\frac{h_2^4}{45} \int_\Omega u_{xxyyyy} v + O(h^5)|u|_6|v|_1.$$

Proof Integration by parts in the right-hand side of (2), leads to

$$\sum_e \int_e u_{xyyy}v_{xy} = -\sum_e \int_e u_{xxyyy}v_y = \sum_e \int_e u_{xxyyyy}v,$$

$$\sum_e \int_e u_{xyyyy}v_{xyy} = -\sum_e \int_e u_{xxyyyy}v_{yy}.$$

Theorem 1 now follows from Lemma 1.

From expansion (2) and Theorem 1 we can get an extrapolation algorithm as follows.

First, we construct on the coarse element τ consisted of 9 elements e_i $(i = 1, \cdots, 9)$ the postprocessing interpolation $\Pi_{3h}^6 : C(\tau) \to Q_6(\tau)$ such that

$$\Pi_{3h}^6 v(Z_i) = v(Z_i), \ \ i = 1, \dots, 16, \ \ \int_{l_i} (\Pi_{3h}^6 v - v) ds = 0, \ \ i = 1, \dots, 24,$$
$$\int_{e_i} (\Pi_{3h}^6 v - v) = 0, i = 1, \dots, 9$$

where Z_i and l_i are 16 nodal points and 24 edges, respectively, of τ. We then have an error expansion for postprocessed FE solution in the L^2 norm:

$$\Pi_{3h}^6 u_h - u = \phi_1 h^4 + O(h^6)$$

which leads to a new extrapolation in the L^2 norm:

$$\widetilde{u}_h = \Pi_{3h}^6 u_h, \qquad \frac{2^4 \widetilde{u}_{h/2} - \widetilde{u}_h}{2^4 - 1} = u + O(h^6).$$

A question arises: what is the result when the biquadratic interpolation u_I is defined by the standard nodal point condition: $u_I(Z_i) = u(Z_i)$ $(i = 1, \cdots, 9)$?

Lemma 2. *If $v \in V_h$ and u_I is the nodal point interpolation, then*

$$\int_e (u - u_I)_x v_x = -\frac{k_e^4}{45} \int_e u_{xyyy} v_{xy} + \frac{h_e^4}{180} \int_e u_{xxxx} v_{xx} + O(h^4)|u|_{5,e}|v|_{1,e} \tag{3}$$

with one more term than (1).

The proof is similar to Lemma 1: checking on $\hat{e}$ for $\hat{u} \in P_4(\hat{e})$ the expansion

$$\int_{\hat{e}} (\hat{u} - \hat{u}_I)_{\hat{x}} \hat{v}_{\hat{x}} d\hat{x} d\hat{y} + \frac{1}{45} \int_{\hat{e}} \hat{u}_{\hat{x}\hat{y}\hat{y}\hat{y}} \hat{v}_{\hat{x}\hat{y}} d\hat{x} d\hat{y} - \frac{1}{180} \int_{\hat{e}} \hat{u}_{\hat{x}\hat{x}\hat{x}\hat{x}} \hat{v}_{\hat{x}\hat{x}} d\hat{x} d\hat{y} = 0$$

where $\hat{u}_I = \hat{u}$ when $\hat{u} \in Q_2(\hat{e})$, otherwise

$\hat{u}$	$\hat{x}^3$	$\hat{y}^3$	$\hat{x}^4$	$\hat{x}^3\hat{y}$	$\hat{x}\hat{y}^3$	$\hat{y}^4$
$\hat{u}^I$	$\hat{x}$	$\hat{y}$	$\hat{x}^2$	$\hat{x}\hat{y}$	$\hat{x}\hat{y}$	$\hat{y}^2$

The additional term in (3) cannot achieve a high order. Even if T_h is uniform,

$$\frac{h_1^4}{180}\sum_e \int_e u_{xxxx}v_{xx} = \frac{h_1^4}{180}\sum_e (\int_{l_1} - \int_{l_3})u_{xxxx}v_x dy + O(h^4)|u|_5|v|_1, \quad (4)$$

where l_1 and l_3 are the right and left edges, respectively, of the element e. Since v_x is not continuous across the edges l_1 and l_3, the above edge integrals cannot be cancelled. So that the nodal point interpolation makes the integral to be of a low order:

$$\int_\Omega (u - u_I)_x v_x = O(h^3)|u|_4|v|_1. \quad (5)$$

Therefore, the point-line-plane interpolation is better than the nodal point one.

1.2 Bicubic rectangular element

Let, for $k \geq 3$,

$$V_h = \{v \in C(\bar{\Omega}); v\,|_e \in Q_k \;\; \forall e \in T_h\},$$

$$Q_k = \text{span}\{x^i y^j; 0 \leq i \leq k, 0 \leq j \leq k\}$$

and the interpolation $u_I \in V_h$ defined by the point-line-plane condition:

$$u_I(Z_i) = u(Z_i), \quad \int_{l_i}(u_I - u)vds = 0 \;\; \forall v \in P_{k-2}(l_i), i = 1,2,3,4,$$

$$\int_e (u_I - u)v = 0 \;\; \forall v \in Q_{k-2}(e).$$

Lemma 3. *If $v \in V_h$, then*

$$\int_e (u - u_I)_x v_x =$$

$$-\frac{k_e^{2k}}{(2k-1)!!(2k+1)!!}\int_e \partial_x\partial_y^{k+1}u \cdot \partial_x\partial_y^{k-1}v + O(h^{k+3})|u|_{k+3,e}|v|_{2,e}.$$

Proof The expansion can be checked on $\hat{e}$ for $\hat{u} \in P_{k+2}(\hat{e})$, where $\hat{u}_I = \hat{u}$ when $\hat{u} \in Q_k(\hat{e})$, otherwise,

$\hat{u}$	$\hat{x}^{k+1}$	$\hat{y}^{k+1}$	$\hat{x}^{k+2}$	$\hat{x}^{k+1}\hat{y}$	$\hat{x}\hat{y}^{k+1}$	$\hat{y}^{k+2}$
$\hat{u} - \hat{u}_I$	$p_k(\hat{x})$	$p_k(\hat{y})$	$p_{k+1}(\hat{x})$	$p_k(\hat{x})\hat{y}$	$\hat{x}p_k(\hat{y})$	$p_{k+1}(\hat{y})$

where

$$p_k(t) = \frac{(k+1)!}{(2k)!}\frac{d^{k-1}(t^2-1)^k}{dt^{k-1}}.$$

When $k = 3$ we get superclose estimate

$$\int_\Omega (u - u_I)_x v_x = O(h^5)|u|_6|v|_1. \tag{6}$$

On the contrary to $k \leq 2$ we cannot have the extrapolation estimate for $k \geq 3$, although we can construct by (6) the postprocessed solution like $\Pi^5_{2h} u_h$ which is superconvergent to u.

A question arises: what is the result when the interpolation u_I returns to be the standard nodal point condition rather than our point-line-plane conditions?

Consider the case $k = 3$. Set $x_1 = -1$, $x_2 = a$, $x_3 = b$, $x_4 = 1$, $-1 < a < b < 1$. The interpolation $u_I \in V_h$ is defined by the nodal point condition: on the element e

$$(u_I - u)(x_e + x_i h_e, y_e + x_j k_e) = 0, \quad i, j = 1, 2, 3, 4.$$

Lemma 4. *If $v \in V_h$, then*

$$\int_e (u - u_I)_x v_x = \tag{7}$$

$$\frac{h_e^4}{12}\left[\frac{2}{5} + \frac{ab-1}{3}\right]\int_e u_{xxxx}v_{xx} - (a+b)\frac{h_e^5}{180}\int_e u_{xxxx}v_{xxx} + O(h^4)|u|_{5,e}|v|_{1,e}.$$

The proof is similar to Lemma 3, but when $\hat{u} = \hat{x}^4$,

$$\hat{u}_I(t) = (a+b)t^3 - (ab-1)t^2 - (a+b)t + ab.$$

Case 1. Take the equidistance nodal points: $a = -1/3$, $b = 1/3$. Then (7) leads to

$$\int_e (u - u_I)_x v_x = \frac{h_e^4}{405}\int_e u_{xxxx}v_{xx} + O(h^4)|u|_{5,e}|v|_{1,e}.$$

Like (4), the integration by parts will lead to a low order, and hence we only have (5).

Case 2. To make the two dominant terms of (7) to be zero we may select

$$\frac{2}{5} + \frac{ab-1}{3} = 0, \quad a + b = 0$$

or $a = -b$, $a^2 = 1/5$, which are unequidistance nodal points. Then,

$$\int_e (u - u_I)_x v_x = O(h^4)|u|_{5,e}|v|_{1,e}, \tag{8}$$

a better result than the case 1. To make clear if (8) is optimal we need to further expand the integral:

$$\int_e (u-u_I)_x v_x = \frac{2h_e^6}{7875}\int_e u_{xxxxx}v_{xx} - \frac{k_e^6}{1575}\int_e u_{xyyyy}v_{xyy} + O(h^5)|u|_{6,e}|v|_{1,e}. \tag{9}$$

This can be checked on $\hat{e}$ for $\hat{u} \in P_5(\hat{e})$, where $\hat{u}_I = \hat{u}$ when $\hat{u} \in Q_3(\hat{e})$, otherwise,

$\hat{u}$	$\hat{x}^4$	$\hat{y}^4$	$\hat{x}^5$	$\hat{x}^4\hat{y}$	$\hat{x}\hat{y}^4$	$\hat{y}^5$
$\hat{u}_I$	$p(\hat{x})$	$p(\hat{y})$	$\hat{x}p(\hat{x})$	$\hat{y}p(\hat{x})$	$\hat{x}p(\hat{y})$	$\hat{y}p(\hat{y})$

where $p(t) = \dfrac{6t^2-1}{5}$. We can see that the second term in the right-hand side of (9) cannot achieve a higher order like (6).

Therefore, different interpolations lead to different orders for the integral $\int_\Omega (u-u_I)_x v_x$ and the point-line-plane interpolation is the best one.

1.3 Adini element for the biharmonic problem

Let us consider the biharmonic problem and the nonconforming Adini element

$$V_h = \{v \in H^1(\Omega), v|_e \in \tilde{Q}_3(e) \forall e \in T_h, v_x, v_y \text{ are continuous at vertices of } e\},$$

$$\tilde{Q}_3 = \text{span}\{1, x, y, x^2, y^2, xy, x^3, y^3, x^2y, xy^2, x^3y, xy^3\},$$

$$V_{h,00} = \{v \in V_h; v, v_x, v_y = 0 \text{ on nodal point of the edges }\}.$$

Let $u_I \in V_h$ be the interpolation defined on e by

$$u_I(Z_i) = u(Z_i),\ (u_I)_x(Z_i) = u_x(Z_i),\ (u_I)_y(Z_i) = u_y(Z_i),\ i = 1,2,3,4.$$

Let Adini FE solution be defined by

$$u_h \in V_{h,00}: \quad A_h(u_h, v) = (\Delta^2 u, v) \quad \forall v \in V_{h,00},$$

$$A_h(u,v) = \sum_e \int_e (u_{xx}v_{xx} + 2u_{xy}v_{xy} + u_{yy}v_{yy}),$$

$$A_h(u-u_h, v) = \sum_e \Big(\int_{l_1} - \int_{l_3}\Big)(u_{xx}v_x + u_{xy}v_y)dy + \sum_e \Big(\int_{l_2} - \int_{l_4}\Big)(u_{xy}v_x + u_{yy}v_y)dx,$$

$$A_h(u_h - u_I, v) = A_h(u - u_I, v) - A_h(u - u_h, v).$$

The last two terms can be expanded as follows.

Lemma 5. *If $v \in V_h$, then*

$$\int_e (u - u_I)_{xx} v_{xx} = -\frac{k_e^2}{3} \int_e u_{xxyy} v_{xx} + \frac{4k_e^4}{45} \int_e u_{xxyyy} v_{xxy} + O(h^4)|u|_{6,e}|v|_{3,e}.$$

Proof The expansion can be checked on $\hat{e}$ for $\hat{u} \in P_4(\hat{e})$, where $\hat{u}_I = \hat{u}$ when $\hat{u} \in \tilde{Q}_3(\hat{e})$, otherwise

$\hat{u}$	$\hat{x}^4$	$\hat{x}^2\hat{y}^2$	$\hat{y}^4$
$\hat{u}_I$	$2\hat{x}^2 - 1$	$\hat{x}^2 + \hat{y}^2 - 1$	$2\hat{y}^2 - 1$

Lemma 6. *If $v \in V_h$, then*

$$\int_e (u - u_I)_{xy} v_{xy} = O(h^4)|u|_{5,e}|v|_{3,e}.$$

Proof Integration by parts on e and its boundary,

$$\int_e (u - u_I)_{xy} = 0.$$

Consider the bilinear functional on $\hat{e}$:

$$B(\hat{u}, \hat{v}) = \int_{\hat{e}} (\hat{u} - \hat{u}_I)_{\hat{x}\hat{y}} \hat{v}_{\hat{x}\hat{y}} d\hat{x} d\hat{y}.$$

Then, $\forall \hat{p} \in P_4(\hat{e})$ and $\hat{q} \in P_2(\hat{e})$,

$$B(\hat{u} + \hat{p}, \hat{v} + \hat{q}) = 0 \quad \forall \hat{u} \in H^5(e), \forall \hat{v} \in \tilde{Q}_3(\hat{e}).$$

Lemma 6 follows from the Bramble-Hilbert Lemma.

Let $\|\cdot\|_{k,h} = (\sum_e \|\cdot\|_{k,e}^2)^{\frac{1}{2}}$.

Theorem 2. *(Lin and Luo [11]) If $v \in V_{h,00}$ and $u \in H^4(\Omega)$, then*

$$\sum_e \int_e (u - u_I)_{xx} v_{xx} = \sum_e \frac{k_e^2}{3} \int_e u_{xxxyy} v_x + O(h^4)\|u\|_6 \|v\|_{3,h},$$

$$\sum_e \int_e (u - u_I)_{xy} v_{xy} = O(h^4)|u|_5 |v|_{3,h}.$$

Proof See [11] or [14].

Theorem 3. *(Lin and Luo [11]) If* $v \in V_{h,00}$, *then*

$$A_h(u-u_h,v) = \frac{1}{3}\sum_e \int_e (h_e^2 u_{yyyy}v_{xx} + k_e^2 u_{xxxx}v_{yy}) + O(h^3)\|u\|_4\|v\|_{4,h}.$$

Proof See [11] or [14], p. 130 and p. 132:

$$A_h(u-u_h,v) = \sum_e \int_e (u_{xxx}F(y)v_{xyy} + u_{yyy}E(x)v_{yxx} + O(h^3)\|u\|_4\|v\|_{4,h},$$

$$\begin{aligned}
&\int_e u_{xxx}F(y)v_{xyy} \\
&= -\frac{1}{6}\int_e \left(F^2(y)\right)'\left(u_{xxxy}v_{xyy} + u_{xxx}v_{xyyy}\right) - \frac{1}{3}k_e^2\int_e u_{xxx}v_{xyy} \\
&= -\frac{1}{3}k_e^2\int_e u_{xxx}v_{xyy} + O(h^3)\|u\|_4\|v\|_{4,h}.
\end{aligned}$$

From Theorems 2 and 3 we can get an extrapolation algorithm.

First, we construct on the coarse element τ consisted of 4 elements e the postprocessing interpolation $\Pi_{2h}^4 : C(\tau) \to Q_4(\tau) + \mathrm{span}\{x^5y, xy^5\}$ such that

$$\Pi_{2h}^4 v(Z_i) = v(Z_i),\ (\Pi_{2h}^4 v)_x(Z_i) = v_x(Z_i),$$
$$(\Pi_{2h}^4 v)_y(Z_i) = v_y(Z_i),\ i = 1,\dots,9$$

where Z_i are the all nodal points on the 4 elements. We then have, if $u \in H^6(\Omega)$ and T^h is uniform, postprocessing expansion in H^1 norm:

$$\Pi_{2h}^4 u_h - u = \phi_1 h^2 + O(h^4)$$

and postprocessing extrapolation in H^1 norm:

$$\widetilde{u}_h = \Pi_{2h}^4 u_h,\quad \frac{4\widetilde{u}_{h/2} - \widetilde{u}_h}{3} = u + O(h^4).$$

As a by-produce of Theorem 3, we have the following error expansion for eigenvalues with almost the same dominant term as that in Theorem 3 but twice:

$$\begin{aligned}
\lambda_h - \lambda &= \frac{2}{3}\sum_e \int_e \left(h_e^2 u_{yyyy}(u_h)_{xx} + k_e^2 u_{xxxx}(u_h)_{yy}\right) + O(h^3) \\
&= -\frac{2}{3}h_1^2\int_\Omega u_{xyy}^2 - \frac{2}{3}h_2^2\int_\Omega u_{xxy}^2 + O(h^3).
\end{aligned}$$

(Yang [24] also obtained,by using our Theorem 3, the same result).

Another progress is made by Tobiska, et al. [12] for simple and typical nonconforming elements, i.e. Rannacher-Turek element [17] and its generalization [12]. See [12] for details.

Postprocessed extrapolation connects with postprocessed superconvergence. Superconvergence of various types has been discussed by, e.g., Babuska, et al. [1], Chen and Huang [3], Krizek, et al. [5] [6], Zhu [15], Lu [9] [10], Schatz, Sloan and Wahlbin [22], Sloan [20], Xu [23]. See a recent survey by Brandts and Krizek [2], and a new paper by Brandts and Krizek: *Linear splines and their derivatives on uniform simplicial partitions of polytopes*.

2. Part II Eigenvalue Problem

We try now to answer the questions: which FEs are high performance? Which one is better, bilinear or linear element, conforming or nonconforming element? Do we have a mathematical way to give a judgment?

Let us recall an elementary experience (see Rannacher [16] and Shen [19]), the calculation of π approximated by the perimeter of $n-$sided regular polygon π_n inscribed in a unit circle. To characterize the error we can use the expression:

$$\pi_n = n \sin \frac{\pi}{n}$$

and the Taylor expansion:

$$\pi_n - \pi = -\frac{\pi^3}{6} n^{-2} + O(n^{-4}).$$

Such an expansion contains a lot of information. E.g., from the sign of the dominant error term we can see that π_n gives a lower bound for π, and from the coefficient of the dominant error term we can see how good the approximation is. Furthermore, the expansion leads to a high order algorithm, i.e., the extrapolation algorithm.

It is surprising that such an elementary experience can be generalized to the calculation of differential eigenvalue λ approximated by the FE eigenvalue λ_h. See the early paper by Lin and Lu [9] [10] and the development by Chen and Huang [3], Rannacher [16] and Shaidurov [18], see also Krizek and Neittaanmaki [6]. Briefly speaking, they also expanded the eigenvalue error into a dominant term and a high order reminder, and the expansion was then used to derive the extrapolation algorithm.

2.1 Outline

In this article we return the early work but with a more careful analysis on the coefficient and the sign in the error expansion for some simple and typical conforming and nonconforming elements. From the coefficients of dominant error terms of different FEs we can judge which FE gives a better eigenvalue bound. In the nonconforming case, from the sign, positive or negative of the dominant error term we can see which FE gives an upper or a low bound for the eigenvalues. For example, we can see that

(1) Rannacher-Turek [17], rotated multilinear element Q_1^{rot} gives a good upper bound for square meshes;

(2) its generalization [12] called GQ_1^{rot}, for uniform rectangular meshes, gives a good lower bound for the eigenvalues;

(3) bilinear element, called Q_1, gives a bad upper bound;

(4) linear element, called P_1, gives a better upper bound.

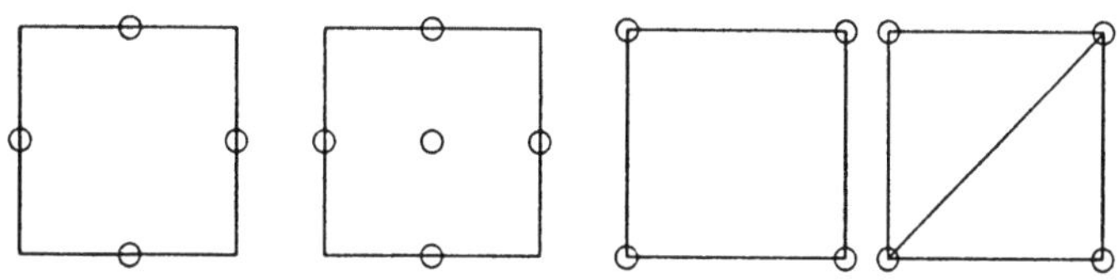

(1): upper (2): lower (3): upper (4): upper

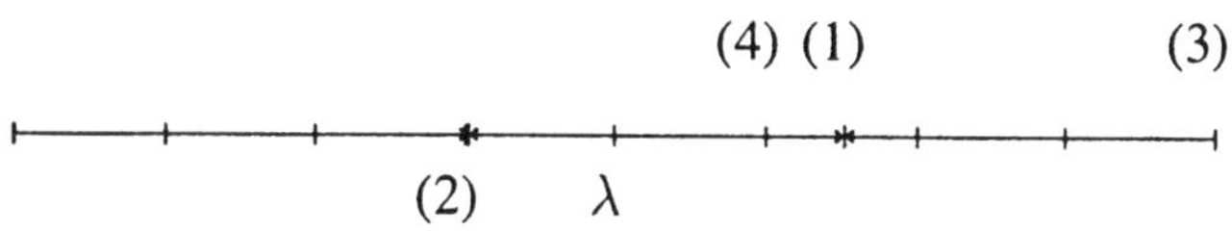

$$(1) \doteq \frac{(2)+(3)}{2}, \quad (4) \doteq \text{Refined } (3)$$

Compare the degrees of freedom: GQ_1^{rot} is more than Q_1^{rot} (but less than Wilson element). Compare the approximate property: Q_1^{rot} (for square meshes) gives a upper bound while GQ_1^{rot} (for uniform rectangular meshes) gives a lower bound. Both of them are sharper than Q_1 (and, we guess, also sharper than Wilson element). Furthermore, Q_1^{rot} gives almost the exact eigenvalue for the simplest differential model (i.e., $\rho \equiv 1$ in the problem (10) below). This is a superconvergence phenomenon in the eigenvalue problem.

Our conclusion: the error expansion method is neat and efficient for selecting the high performance FEs for the eigenvalue problems on a rectangular domain. And the extrapolation leads to a higher order accuracy.

Open problems: The most serious one is the answer of Krizek problem [6], i.e., what is the necessary condition for meshes to obtain the error expansions?

2.2 Text

Consider the eigenvalue problem of Laplace operator on a rectangular domain Ω: find eigenpair $(\lambda, u) \in \mathbf{R} \times H_0^1(\Omega)$ and $\|u\|_\rho = 1$ such that

$$a(u, v) = \lambda(\rho u, v) \quad \forall v \in H_0^1(\Omega), \tag{10}$$

$$a(u, v) = \int_\Omega (u_x v_x + u_y v_y), \quad \|u\|_\rho = (\rho u, u)^{\frac{1}{2}},$$

where $\rho(x, y) \geq C > 0$ is a smooth function. We construct the FE rectangulation $T_h = \{e\}$, introduce the FE space V_h and its subspace $V_{h,0}$ satisfying the zero boundary condition, and find the FE eigenpair $(\lambda_h, u_h) \in \mathbf{R} \times V_{h,0}$ and $\|u_h\|_\rho = 1$ such that

$$a_h(u_h, v) = \lambda_h(\rho u_h, v) \quad \forall v \in V_{h,0}, \tag{11}$$

$$a_h(u, v) = \sum_e \int_e (u_x v_x + u_y v_y).$$

Let us consider the $k-$th eigenvalue λ of the continuous problem (10). Take the case where λ is simple. In this case, λ is associated with one eigenfunction u. Find the $k-$th eigenvalue λ_h of FE problem (11). Since the eigenvalues of the FEM converge to the continuous problem, we can assert that λ_h is also simple. Thus λ_h is associated with only one eigenfunction u_h. See Shaidurov [18] for details.

Let us introduce now the FE projection $R_h \in H_0^1(\Omega) \to V_{h,0}$: for $u \in H_0^1(\Omega)$, let

$$a(u, v) = (f, v) \quad \forall v \in H_0^1(\Omega)$$

(or $-\Delta u = f$ when u is an eigenfunction), $R_h u$ is defined by the equation

$$a_h(R_h u, v) = (f, v) \quad \forall v \in V_{h,0}.$$

In particular, for u being an eigenfunction, we have

$$a_h(R_h u, v) = (-\Delta u, v) = a_h(u, v) - \sum_e \int_{\partial e} \frac{\partial u}{\partial n} v ds \quad \forall v \in V_{h,0},$$

where ∂e is the boundary of e. So, the projection R_h is not orthogonal in general:

$$a_h(u - R_h u, v) = \sum_e \int_{\partial e} \frac{\partial u}{\partial n} v ds \neq 0. \tag{12}$$

This is the nonconforming error, c.f. Strang [21].

Theorem 4. *The error of FE eigenvalue reduces into the errors of the interpolation u_I and the projection $R_h u$:*

$$\begin{aligned}\lambda_h - \lambda &= \lambda(u - u_I, \rho u) - a_h(u - u_I, R_h u) + a_h(u - R_h u, R_h u) \\ &\quad -\lambda \|u - u_h\|_\rho^2 + |R_h u - u_h|_{1,h}^2,\end{aligned}$$

$$|R_h u - u_h|_{1,h} \leq C(|\lambda_h - \lambda| + \|u_h - u\|_0), \quad \|\cdot\|_{k,h} = (\sum_e \|\cdot\|_{k,e}^2)^{\frac{1}{2}}.$$

The proof is based on simple algebraic operations. See Lin and Lin [8].

By the above theorem, in which the last two terms are higher order, we can concentrate ourself on the expansions for the integrations of interpolation error and projection error:

$$a_h(u - R_h u, R_h u), \quad a_h(u - u_I, R_h u), \quad (u - u_I, \rho u). \tag{13}$$

In this article we mainly consider two nonconforming elements:

(1) $\mathrm{Q}_1^{\mathrm{rot}}$ element for square meshes:

$$\begin{aligned}V_h &= \left\{ v \in L^2(\Omega), v|_e \in \mathrm{span}\{1, x, y, x^2 - y^2\}, \right. \\ &\qquad \left. \int_l v|_{e_1} ds = \int_l v|_{e_2} ds \text{ if } e_1 \cap e_2 = l \right\}, \\ V_{h,0} &= \left\{ v \in V_h, \int_l v|_e ds = 0 \text{ if } e \cap \partial\Omega = l \right\}.\end{aligned}$$

For $u \in H^1(\Omega)$, the interpolation $u_I \in V_h$ is defined as follows: on the four edges l_i of the element e

$$\int_{l_i} (u_I - u) ds = 0, \quad i = 1, 2, 3, 4.$$

where l_i are the four edges of the element e.

(2) Generalized $\mathrm{Q}_1^{\mathrm{rot}}$ element for rectangular meshes:

$$\begin{aligned}V_h &= \left\{ v \in L^2(\Omega), v|_e \in \mathrm{span}\{1, x, y, x^2, y^2\}, \right. \\ &\qquad \left. \int_l v|_{e_1} ds = \int_l v|_{e_2} ds \text{ if } e_1 \cap e_2 = l \right\}, \\ V_{h,0} &= \left\{ v \in V_h, \int_l v|_e ds = 0 \text{ if } e \cap \partial\Omega = l \right\}.\end{aligned}$$

For $u \in H^1(\Omega)$, the interpolation $u_I \in V_h$ is defined by one more condition:

$$\int_{l_i} (u_I - u)ds = 0, \quad i = 1, 2, 3, 4; \qquad \int_e (u_I - u) = 0.$$

We first list the dominant terms of the expansions for the integrations in (13), and the corresponding dominant terms for the errors of eigenvalues of FEMs (1), (2), (3) and (4) in Section 1 on square meshes.

FE	$a_h(u - R_h u, R_h u)$	$a_h(u - u_I, R_h u)$	$(u - u_I, \rho u)$
Q_1^{rot}	$-\frac{2h^2}{3}\int u_{xy}^2$	0	$-\frac{h^2}{6}\int (u_{xx} + u_{yy})\rho u$
GQ_1^{rot}	$-\frac{h_1^2 + h_2^2}{3}\int u_{xy}^2$	0	0
Q_1	0	$\frac{1}{3}\sum_e (h_e^2 + k_e^2)\int_e u_{xx}u_{yy}$	$-\frac{1}{3}\sum_e \int_e (h_e^2 u_{xx} + k_e^2 u_{yy})\rho u$

FE	Dominant term of $\lambda_h - \lambda$
Q_1^{rot}	$-\frac{2h^2}{3}\int u_{xy}^2 + \frac{h^2}{6}\int (u_{xx} + u_{yy})^2 = \frac{h^2}{6}\int (u_{xx} - u_{yy})^2$
GQ_1^{rot}	$-\frac{2h^2}{3}\int u_{xy}^2 < 0$
Q_1	$-\frac{2h^2}{3}\int u_{xy}^2 + \frac{h^2}{3}\int (u_{xx} + u_{yy})^2 = \frac{h^2}{3}\int (u_{xx}^2 + u_{yy}^2)$
P_1	$\frac{h^2}{12}\int (u_{xx}^2 + u_{yy}^2)$

Notice that Q_1^{rot} could leads to $\lambda_h \doteq \lambda$ when $\rho \equiv 1$ (i.e., $u_{xx} = u_{yy}$).

We now start to prove the above integral table and the corresponding eigenvalue error expansions. We want to first expand for both Q_1^{rot} and GQ_1^{rot} the projection error:

$$a_h(u - R_h u, v) = \sum_e \int_{\partial e} \frac{\partial u}{\partial n} v ds = \sum_e (\int_{l_1} - \int_{l_3}) u_x v dy + (\int_{l_2} - \int_{l_4}) u_y v dx,$$

where l_1, l_3 and l_2, l_4 are the right, left and above, below edges of e, respectively. Since $v \in V_h$, we have $\int_l v|_{e_1} ds = \int_l v|_{e_2} ds$ for $e_1 \cap e_2 = l$ and therefore

$$a_h(u - R_h u, v) = \sum_e (\int_{l_1} - \int_{l_3}) u_x (v - \bar{v}) dy + \sum_e (\int_{l_2} - \int_{l_4}) u_y (v - \bar{v}) dx$$

where $\bar{v}|_l = \int_l v ds/|l|$ and thus for $v \in V_h$ (i.e., $v \in \mathrm{Q}_1^{\mathrm{rot}}$ or $\mathrm{GQ}_1^{\mathrm{rot}}$)

$$(v - \bar{v})\Big|_{l_i} = (y - y_e)v_y - \frac{v_{yy}}{2}\left((y - y_e)^2 + \frac{k_e^2}{3}\right), \quad i = 1, 3 \tag{14}$$

since $v\,|_{l_i} = \{1, y - y_e, (y - y_e)^2\}$.

Let $e \in T_h$: $e = [x_e - h_e, x_e + h_e] \times [y_e - k_e, y_e + k_e]$.

Lemma 7. *If $v \in V_h$, then*

$$(\int_{l_1} - \int_{l_3})u_x(v - \bar{v})dy = \frac{k_e^2}{3}\int_e u_{xxy}v_y - \frac{4k_e^4}{45}\int_e u_{xxyy}v_{yy} + O(h^4)|u|_{5,e}|v|_{1,e}.$$

Proof By (14)

$$\begin{aligned}
&(\int_{l_1} - \int_{l_3})u_x(v - \bar{v})dy \\
= \; & (\int_{l_1} - \int_{l_3})u_x\left((y - y_e)v_y - \frac{v_{yy}}{2}\left((y - y_e)^2 + \frac{k_e^2}{3}\right)\right)dy \\
= \; & \int_e u_{xx}\left((y - y_e)v_y - \frac{v_{yy}}{2}\left((y - y_e)^2 + \frac{k_e^2}{3}\right)\right)dxdy. \qquad (15)
\end{aligned}$$

Here we have used integration by parts and that $v_{xy} = v_{xyy} = 0$. Denote the right-hand side of (15) by I. We can use the Bramble-Hilbert Lemma to check the expansion

$$I = \frac{k_e^2}{3}\int_e u_{xxy}v_y - \frac{4k_e^4}{45}\int_e u_{xxyy}v_{yy} + O(h^4)|u|_{5,e}|v|_{1,e}. \tag{16}$$

In fact, the expansion can be checked on the standard reference element $\hat{e} = [-1, -1] \times [-1, 1]$ for the polynomial $\hat{u} \in P_4(\hat{e})$:

$$\int_{\hat{e}} \hat{u}_{\hat{x}\hat{x}}\left(\hat{y}\hat{v}_{\hat{y}} - \left(\hat{y}^2 + \frac{1}{3}\right)\frac{\hat{v}_{\hat{y}\hat{y}}}{2}\right)d\hat{x}d\hat{y} - \frac{1}{3}\int_{\hat{e}} \hat{u}_{\hat{x}\hat{x}\hat{y}}\hat{v}_{\hat{y}}d\hat{x}d\hat{y} + \frac{4}{45}\int_{\hat{e}} \hat{u}_{\hat{x}\hat{x}\hat{y}\hat{y}}\hat{v}_{\hat{y}\hat{y}}d\hat{x}d\hat{y} = 0.$$

Lemma 8. *If $v \in V_h$ and T_h is uniform, then*

$$a_h(u - R_h u, v) = -\frac{h_1^2 + h_2^2}{3}\int_\Omega u_{xxyy}v + O(h^4)\|u\|_5\|v\|_{2,h}.$$

Proof Since T_h is uniform: $h_e \equiv h_1, k_e \equiv h_2$. Integration by parts in the first term of the right-hand side of (16):

$$\sum_e \int_e u_{xxy}v_y = -\int_\Omega u_{xxyy}v + \sum_e(\int_{l_2} - \int_{l_4})u_{xxy}v dx.$$

Using the expansion for v:

$$(v-\bar{v})\Big|_{l_i} = (x-x_e)v_x - \frac{v_{xx}}{2}\left((x-x_e)^2 + \frac{h_e^2}{3}\right), \quad i=2,4$$

and integration by parts,

$$\begin{aligned}
\sum_e \Big(\int_{l_2} - \int_{l_4}\Big) u_{xxy} v dx &= \sum_e \Big(\int_{l_2} - \int_{l_4}\Big) u_{xxy}(v-\bar{v}) dx \\
&= \sum_e \Big(\int_{l_2} - \int_{l_4}\Big) u_{xxy}\left((x-x_e)v_x - \frac{v_{xx}}{2}\left((x-x_e)^2 + \frac{h_e^2}{3}\right)\right) dx \\
&= \sum_e \int_e u_{xxyy}\left((x-x_e)v_x - \frac{v_{xx}}{2}\left((x-x_e)^2 + \frac{h_e^2}{3}\right)\right) dxdy \\
&= \sum_e \int_e u_{xxyy} E'(x) v_x + O(h^2)|u|_4 |v|_{2,h} \\
&= -\sum_e \int_e E(x)(u_{xxxyy} v_x + u_{xxyy} v_{xx}) + O(h^2)|u|_4 |v|_{2,h} \\
&= O(h^2)\|u\|_5 \|v\|_{2,h},
\end{aligned}$$

where $E(x) = ((x-x_e)^2 - h_e^2)/2$. From (2.2), we obtain

$$\sum_e \int_e u_{xxy} v_y = -\int_\Omega u_{xxyy} v + O(h^2)\|u\|_5 \|v\|_{2,h}.$$

By the same reason

$$\sum_e \int_e u_{xyy} v_x = -\int_\Omega u_{xxyy} v + O(h^2)\|u\|_5 \|v\|_{2,h}.$$

Consequently, Lemma 8 follows.

Taking $v = R_h u$ in Lemma 8, we have

$$\begin{aligned}
a_h(u - R_h u, R_h u) &= -\frac{h_1^2 + h_2^2}{3} \int_\Omega u_{xxyy} R_h u + O(h^4)\|u\|_5 \|R_h u\|_{2,h} \\
&= -\frac{h_1^2 + h_2^2}{3} \int_\Omega u_{xxyy} u + O(h^4).
\end{aligned}$$

Here we have used $\|R_h u - u\|_0 \leq Ch^2$ and $\|R_h u\|_{2,h} \leq C$. Thus, we obtained for both Q_1^{rot} and GQ_1^{rot} the same expansion for the projection error in (13).

For expanding the interpolation errors in the first two terms of (13) we need, however, a separate study for GQ_1^{rot} and Q_1^{rot}.

For GQ_1^{rot} we have

Lemma 9. *If V_h is GQ_1^{rot} space, then $u_I \in V_h$ is an orthogonal interpolation of u under the piecewise H^1 inner product:*

$$a_h(u - u_I, v) = 0. \tag{17}$$

And, u_I is an almost orthogonal interpolation of u in L_2:

$$\int_e (u - u_I)v =$$
$$-\frac{h_e^4}{45}\int_e u_{xxx}v_x - \frac{k_e^4}{45}\int_e u_{yyy}v_y + O(h^4)|u|_{4,e}|v|_{0,e} = O(h^4)\|u\|_4\|v\|_{1,h}.$$

Proof To prove the first formula we need integration by parts and use the definition of u_I and that $v_{xx}|_e, v_x|_{l_i}$ $(i = 1, 3)$ are constant:

$$\int_e (u - u_I)_x v_x = -\int_e (u - u_I)v_{xx} + (\int_{l_1} - \int_{l_3})(u - u_I)v_x dy = 0.$$

To prove the second formula, by the Bramble-Hilbert Lemma we only need to check the expansion on the standard reference element for the cubic polynomial:

u	xy	x^3	x^2y	xy^2	y^3
u_I	0	x	$\frac{1}{3}y$	$\frac{1}{3}x$	y

Taking $v = (\rho u)_I$ in the second formula and noting that

$$\|(\rho u)_I\|_{1,h} \le \|\rho u - (\rho u)_I\|_{1,h} + \|\rho u\|_1 \le C\|\rho u\|_2,$$

we obtain the expansion:

$$\begin{aligned}(u - u_I, \rho u) &= (u - u_I, (\rho u)_I) + (u - u_I, \rho u - (\rho u)_I) \\ &= O(h^4)\|u\|_4\|(\rho u)_I\|_{1,h} + O(h^4)\|\rho u\|_2^2 = O(h^4).\end{aligned}$$

By Theorem 4 and the above expansions we obtain

Theorem 5. *If V_h is GQ_1^{rot} space and T_h is uniform, then*

$$\lambda_h - \lambda = -\frac{h_1^2 + h_2^2}{3}\int_\Omega u_{xxyy}u + O(h^4) = -\frac{h_1^2 + h_2^2}{3}\int_\Omega u_{xy}^2 + O(h^4).$$

Lemma 10. *If V_h is Q_1^{rot} space and T_h is a square mesh, then (17) holds too.*

And, u_I is not an orthogonal interpolation of u in L_2:

$$\begin{aligned}\int_e (u-u_I)v &= -\frac{h^2}{6}\int_e (u_{xx}+u_{yy})v + \frac{h^4}{30}\int_e (u_{xxx}v_x + u_{yyy}v_y)\\ &+\frac{h^4}{18}\int_e (u_{xxy}v_y + u_{yyx}v_x) + O(h^4)|u|_{4,e}|v|_{0,e}.\end{aligned}$$

Proof To prove the first formula we need integration by parts and use the definition of u_I and that $\frac{\partial v}{\partial n}|_{l_i}$ is a constant and $\triangle v = 0$:

$$\int_e \nabla(u-u_I)\nabla v = \int_{\partial e}(u-u_I)\frac{\partial v}{\partial n}ds - \int_e (u-u_I)\triangle v = 0.$$

To prove the second formula, by the Bramble-Hilbert Lemma we only need to check the expansion on the standard reference element for the cubic polynomial:

u	x^2	xy	y^2	x^3	x^2y	xy^2	y^3
u_I	$\frac{2}{3}+\frac{1}{2}(x^2-y^2)$	0	$\frac{2}{3}-\frac{1}{2}(x^2-y^2)$	x	$\frac{1}{3}y$	$\frac{1}{3}x$	y

The second formula leads to

$$(u-u_I,v) = -\frac{h^2}{6}\int_e (u_{xx}+u_{yy})v + O(h^4)\|u\|_3\|v\|_{1,h}.$$

Taking $v=(\rho u)_I$ and noting that

$$\|\rho u - (\rho u)_I\|_0 \le Ch^2\|\rho u\|_2, \quad \|(\rho u)_I\|_0 \le C\|\rho u\|_2,$$

we obtain the expansion

$$\begin{aligned}(u-u_I,\rho u) &= (u-u_I,(\rho u)_I) + (u-u_I,\rho u-(\rho u)_I)\\ &= -\frac{h^2}{6}\int_e (u_{xx}+u_{yy})\rho u + O(h^4).\end{aligned}$$

By Theorem 4 and the above expansions we obtain

Theorem 6. *If V_h is Q_1^{rot} space and T_h is a square mesh, then*

$$\begin{aligned}\lambda_h-\lambda &= -\frac{2h^2}{3}\int_\Omega u_{xxyy}u - \frac{h^2}{6}\lambda\int_\Omega (u_{xx}+u_{yy})\rho u + O(h^4)\\ &= \frac{h^2}{6}\int_\Omega (u_{xx}-u_{yy})^2 + O(h^4).\end{aligned}$$

About eigenfunctions: the error expansion method (or extrapolation) is not efficient for FE eigenfunction itself but efficient for postprocessing FE eigenfunction.

Acknowledgment

The author wishes to thank to Professor Michal Krizek for valuable discussions, suggestions and corrections.

References

[1] I. Babuska, et al., *Computer-based proof of the existance of superconvergence points in the finite element method*, Numer. Methods PDEs, 12:347-392, 1996.

[2] J. Brandts and M. Krizek *History and future of superconvergence in three-dimensional FEMs*, GAKUTO International Series, Math. Sci. Appl. Vol. 15, Tokyo, 2001, 24-35.

[3] H. Brunner, *Collocation Methods for Volterra Integral and Related Functional Differential Equations*, to be published by Cambridge University Press.

[4] C. Chen and Y. Huang, *High Accuracy Theory for Finite Element Methods*. Hunan Sci. Tech. Press, China, 1995.

[5] M. Krizek, Q. Lin and Y. Huang, *A nodal superconvergence arising from combination of linear and bilinear elements*, Syst. Sci. Math. Sci. 1:191-197, 1988.

[6] M. Krizek and P. Neittaanmaki, *Finite Element Approximation of Variational Problems and Applications*, Pitman Monograpgs and Surveys in Pure and Applied Mathematics, vol. 50, 1990.

[7] J. Lin, *A systematic study for postprocessing FE solutions*, Postdoc. thesis, Inst. Sys. Sci., CAS, 2001.

[8] Q. Lin and J. Lin, *High Performance FEMs*, to appear in China Sci. Tech. Press, 2002.

[9] Q. Lin and T. Lu, *Asymptotic expansions for finite element eigenvalues and finite element solution*, Proceedings 6 Int. Conf. on Comp. Meth. Appl. Sci. Eng. Versailles, 1983.

[10] Q. Lin and T. Lu, *Asymptotic expansions for finite element approximation of elliptic problem of polygonal domains*, Bonn. Math. Schrift, 1984.

[11] Q. Lin and P. Luo, *Error expansions and extrapolations for Adini nonconforming finite element*, Beijing Mathematics, 1:2, 65-83, 1995.

[12] Q. Lin, L. Tobiska and A. Zhou, *On the superconvergence of nonconforming low order finite elements applied to the Poisson equation*, Preprint, 2001.

[13] Q. Lin and R. Xie, *Some advances in the study of error expansion for finite elements*, J. Comp. Math., Vol. 4:368-382, 1986.

[14] Q. Lin and N. Yan, *Construction and analysis for effective FEMs*, Hebei University Press, China, 1996.

[15] Q. Lin and Q. Zhu, *Preprocessing and Postprocessing for FEMs*, Shanghai Sci. & Tech. Press, China, 1994.

[16] R. Rannacher, *Extrapolation techniques in the FEM(A survey), In Summer school on Numerical Analysis*, MATC7. Helsinki, Univ. of Tech., pp.80-113, 1988.

[17] R. Rannacher and S. Turek, *Simple nonconforming quadrilateral Stokes element*, Numer. Methods PDEs, 8:97-111, 1992.

[18] V. Shaidurov, *Multigrid Methods for Finite Elements*, Math. and its Appl., Vol. 318, Kluwer Academic Publisher, Dordrecht, 1995.

[19] X. Shen, *Extrapolation: from calculation of π to finite element method of PDEs*, Appl. Math. Reviews, Vol. 1, World Sci. Publishing, River Edge, NJ, 2000, pp. 537-558.

[20] I. Sloan, *Superconvergence and the Galerkin method for integral equations of the second kind*, Acad. Press, New York, London, 197-207, 1982.

[21] G. Strang and G. Fix, *An Analysis of the Finite Element Method*, Prentice-Hall Series in Automatic Comp., Prentice-Hall, Englewood Cliffs, NJ, 1973.

[22] L.B. Wahlbin, *Superconvergence in Galerkin Finite Element Methods*, Lecture Notes in Mathematics,1605, Springer-Verlag, Berlin, 1995.

[23] J. Xu, *The error analysis and the improved algorithms for the infinite element method*, Sci. Press, Beijing, 1985, pp. 326-331. J. Comp. Math. Univ. 4, 1985.

[24] Y. Yang, *A posteriori error estimates in Adini finite element for eigenvalue problems*, J. Comp. Math., 4:413-418, 2000.

ALGEBRAIC MULTIGRID METHOD ON LATTICE BLOCK MATERIALS *

Shi Shu
Institute for Computational and Applied Mathematics
Xiangtan University, Hunan, China
shushi@xtu.edu.cn

Jinchao Xu
Department of Mathematics and Center for Computational Mathematics and Applications
Pennsylvania State University, University Park, PA, USA
xu@math.psu.edu

Yingxiong Xiao
Institute for Computational and Applied Mathematics
Xiangtan University, Hunan, China
xyx610@yahoo.com

Ludmil Zikatanov
Department of Mathematics and Center for Computational Mathematics and Applications
Pennsylvania State University, University Park, PA, USA

Abstract In this paper, we construct and numerically analyze a class of algebraic multigrid methods applied to discrete mathematical models for lattice block materials. Some extensive numerical experiments and comparison results are presented. They clearly demonstrate that the constructed algebraic mutigrid method is uniformly convergent with respect to the size of the lattice and some crucial parameters.

Keywords: Lattice Block Materials, Algebraic Multigrid, ILU, PCG

*Supported by Special Funds for Major State Basic Research Projects of China G1999032804.

1. Introduction

In this paper we consider models of periodic, elastic lattice block materials. In the last four decades such materials have been used in many engineering applications. A lattice block material is a composition of connected thin beams which can be arranged in various different ways forming cells and trusses. As a consequence, the designed materials can have different elastic properties. Certain advantageous properties of these materials are: they are light and can withstand considerable forces. An important question in the design of such light but strong materials is: How to arrange the beams in order to achieve desired strength and flexibility behavior. One effective way of providing an answer to such question is given by appropriate mathematical and numerical models for these materials. In the recent years, there have been developed models describing the elastic properties of lattice block materials (see for example Babuška [5, 7]). In this paper, we shall deal with models based on linear elasticity and describing the displacements in a lattice block material. Such model usually derived from the Hooke's law for a single beam and then a superposion principle is applied to combine all beam equations in a global *coefficient matrix*. Thus, finding the displacements due to stretching, bending and twisting forces in a lattice block material is reduced to the solution of a linear algebraic problem of a huge size. Typically, the number of beams in a practical application will be in the range $10^7 - 10^9$ and the solution of these systems of equations requires considerable amount of computing resources. We focus on the development of uniformly convergent iterative methods for the solution of the algebraic problem.

We shall report some numerical results on the application of an algebraic multigrid method for the solution of the resulting algebraic system. The reason we choose the algebraic multigrid roots mainly in the fact that such method can be tuned up to cope with wide range of the elastic parameters and thus is applicable to many practical situations. The multigrid technique we shall focus on is based on the operator dependent prolongations, which in turn are defined by using various energy minimizing techniques (see [6, 8, 10, 11, 25, 27, 35]).

The algebraic problems corresponding to the lattice material models are in fact similar to the solution of systems of elliptic partial differential equations. In this work, we have extended the multigrid techniques to the case of equation groups, and we have developed a class of AMG methods called AMV (using the V-cycle as iterator) and APCG (preconditioned conjugate gradient with V-cycle as preconditioner). The numerical experiments we have performed, convincingly show that these methods converge independently of the size of the problem or the parameters considered. In addition, the methods presented in this paper are also efficient for general truss materials which are not always of periodic structures.

The rest of the paper is organized as follows. In section 2, we set up the discrete models on lattice block materials. In section 3, using some common iterative methods, we report on some numerical results for the the discrete models. In section 4, we present two possible constructions of iterative methods for the lattice materials. and present some numerical data which shows the robustness of the method. In section 5, we make further analysis and propose some approximate, "homogenized" continuous differential models, corresponding to the discrete algebraic equations.

2. Discrete model on lattice block materials

Following the models described in [2, 5, 7], we view the lattice material as a composition of periodic trusses (cells). One important parameter, will be the number of vertices in such periodic cell (we denote this number with q throughout the paper). To introduce some formality in the description, we associate with the lattice material a graph G. The periodicity then means, that we can recover G by translating of a subgraph $G_{sub} \subset G$ which has q vertices. This subgraph is called *cell* and its vertices are called *reference nodes*. The lattices with $q = 1, 3$ and 4 are shown in figures 1–3. A little bit more precise definition is as follows.

Definition2.1 Given graph G=(V,E), let $\mathcal{G} = \{G_i = (V_i, E_i)\}, i = 1 : N$ be a collection of subgraphs of G such that $\cup_{i=1}^{N} V_i = V$. We define the factor graph $G/\mathcal{G} = (\mathcal{V}, \mathcal{E})$ in the following way

(1) $\mathcal{V} = \{G_i\}, i = 1 : N.$

(2) $(G_i, G_j) \in \mathcal{E}$ iff there exists $v_i \in V_i$ and $v_j \in V_j$ such that $(v_i, v_j) \in E$.

Definition 2.2 We call graph $G = (V, E)$ a periodic lattice in $\mathbb{R}^d$ if and only if there exists a subgraph of G such that a finite number of its translations cover the vertex set of G and the corresponding factor graph is isomorphic to the usual integer lattice in $\mathbb{R}^d$. The mathematical model describing the elastic forces in such a lattice is derived by using a one dimensional elastic beam model, written for every edge. To describe this model, we consider a horizontal beam and a rotated beam with the angle ψ as shown in figure 4. We assume that all beams have modulus of elasticity E, cross section S and moment of inertia I. Let L denote the length of a beam, $[x_j, y_j, \theta_j]^t$ F_j^x, F_j^y and M_j denote the local deflections vector and nodal internal forces at node j, $(j = 1, 2)$, respectively. Let us set $\mathbf{X} = [x_1, y_1, \theta_1, x_2, y_2, \theta_2]^t$, $\mathbf{F} = [F_1^x, F_1^y, M_1, F_2^x, F_2^y, M_2]^t$. and describe the relations between the deflection vector $\mathbf{X}$ and the internal force vector $\mathbf{F}$. We consider a horizontal beam as shown in figure 4(a) and assume that the deformation of the beam satisfies the principle of superposition and the beam is an undetermined one. In the right-handed coordinate system, the displacement is counted as positive and consistent with the coordinate axis

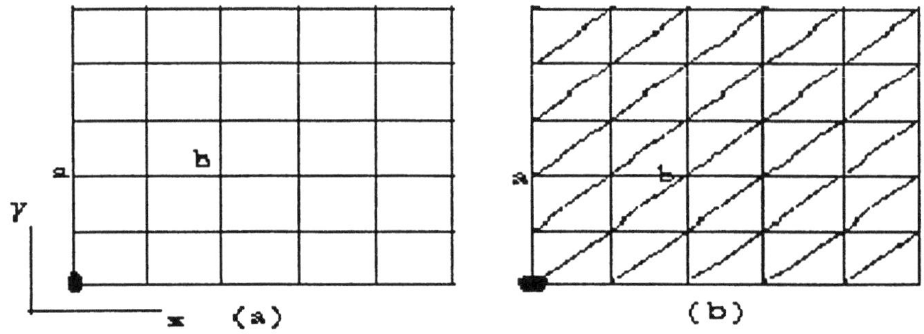

Figure 1. Square lattices $q = 1$

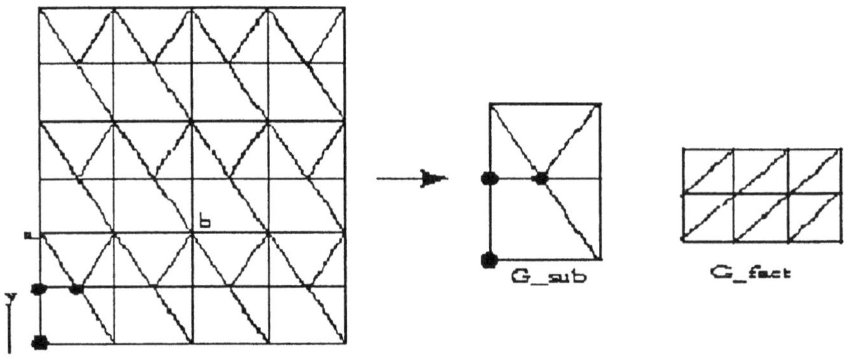

Figure 2. Graph factorization and periodicity–square lattice, $q = 3$

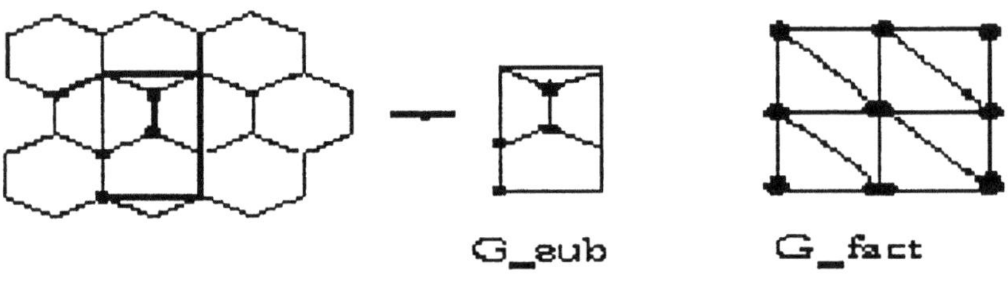

Figure 3. Graph factorization and periodicity–honeycomb lattice, $q = 4$

direction and the twist is measured in counterclockwise direction. Similar conventions apply to the internal forces as well. For a beam which is fixed at its right endpoint the deformation can then be decomposed using three basic deformations as shown in figure 5. If $X = (x_1, 0, 0, 0, 0, 0)^t$, as illustrated in figure 5(a), we then have by Hooke's law that

$$[F_1^x, F_1^y, M_1, F_2^x, F_2^y, M_2]^t = [\frac{SE}{L}, 0, 0, -\frac{SE}{L}, 0, 0]^t x_1 \qquad (2.1)$$

Further, we suppose $X=(0, y_1, 0, 0, 0, 0)^t$, as shown in figure 5(b). We can get the deflection function $V(x) = y_1(\frac{x}{L} - 1)^2(2\frac{x}{L} + 1)$. It is known that the beam has an upward shearing force $F_1^y=EIV'''(x)=\frac{12EI}{L^3}$ at its left endpoint. According to the equilibrium of forces, there is a downward shearing force $F_2^y=-EIV'''(x)=-\frac{12EI}{L^3}$ at the right endpoint. These forces, produce a pair of moments rotating clockwise. To reach the balance of the moments, there should be a pair of moments acting on the beam namely $M_1=EIV''(0)=\frac{6EI}{L^2}y_1$, $M_2=EIV''(L)=\frac{6EI}{L^2}y_1$. Accordingly, we have

$$[F_1^x, F_1^y, M_1, F_2^x, F_2^y, M_2]^t = [0, \frac{12EI}{L^3}, \frac{6EI}{L^2}, 0, -\frac{12EI}{L^3}, \frac{6EI}{L^2}]^t y_1 \qquad (2.2)$$

Finally, we consider $X = (0, 0, \theta_1, 0, 0, 0)^t$, as illustrated in figure 5(c). Again we solve for the displacement to obtain that $V(x) = \theta_1 x(\frac{x}{L} - 1)^2$. By using the moment $M_1=EIV''(0) = \frac{4EI}{L}\theta_1$ at the left endpoint of the beam and the same moment $M_2=EIV''(L) = \frac{2EI}{L}\theta_1$ at the right endpoint and the balance equations for the moments, we obtain the shearing forces $F_1^y=EIV'''(0)=\frac{6EI}{L^2}\theta_1$ and $F_2^y=EIV'''(L)=-\frac{6EI}{L^2}\theta_1$. Therefore, we have that

$$[F_1^x, F_1^y, M_1, F_2^x, F_2^y, M_2]^t = [0, \frac{6EI}{L^2}, \frac{4EI}{L}, 0, -\frac{6EI}{L^2}, \frac{2EI}{L}]^t \theta_1 \qquad (2.3)$$

In analogous way, we can derive the relations between the internal forces acting

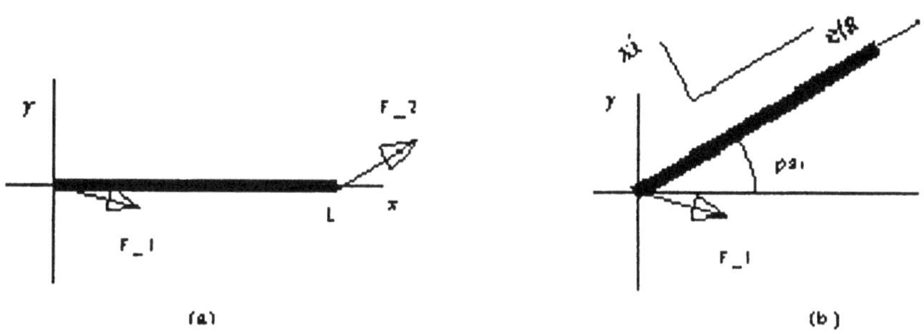

Figure 4.

at a lattice node and the displacements of the beam fixed at its left endpoint under the three basic deformations

$$[F_1^x, F_1^y, M_1, F_2^x, F_2^y, M_2]^t = x_2\left[-\frac{SE}{L}, 0, 0, \frac{SE}{L}, 0, 0\right]^t \qquad (2.4)$$

$$[F_1^x, F_1^y, M_1, F_2^x, F_2^y, M_2]^t = y_2\left[0, -\frac{12EI}{L^3}, -\frac{6EI}{L^2}, 0, \frac{12EI}{L^3}, -\frac{6EI}{L^2}\right]^t \qquad (2.5)$$

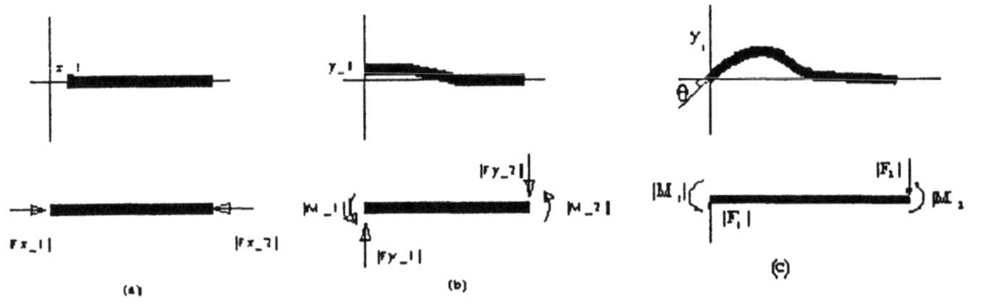

Figure 5.

$$[F_1^x, F_1^y, M_1, F_2^x, F_2^y, M_2]^t = \theta_2 \left[0, \frac{6EI}{L^2}, \frac{2EI}{L}, 0, -\frac{6EI}{L^2}, \frac{6EI}{L}\right]^t \quad (2.6)$$

From equations (2.1)–(2.6), we obtain

$$\begin{bmatrix} SE/L & 0 & 0 & -SE/L & 0 & 0 \\ 0 & 12EI/L^3 & 6EI/L^2 & 0 & -12EI/L^3 & 6EI/L^2 \\ 0 & 6EI/L^2 & 4EI/L & 0 & -6EI/L^2 & 2EI/L \\ -SE/L & 0 & 0 & SE/L & 0 & 0 \\ 0 & -12EI/L^3 & -6EI/L^2 & 0 & 12EI/L^3 & -6EI/L^2 \\ 0 & 6EI/L^2 & 2EI/L & 0 & -6EI/L^2 & 4EI/L \end{bmatrix} \begin{bmatrix} x_1 \\ y_1 \\ \theta_1 \\ x_2 \\ y_2 \\ \theta_2 \end{bmatrix} = \begin{bmatrix} F_1^x \\ F_1^y \\ M_1 \\ F_2^x \\ F_2^y \\ M_2 \end{bmatrix} \quad (2$$

To simplify the notation, let us introduce two parameters $\alpha = \frac{12I}{SL_0^2}$ and $\lambda = \frac{L_0}{L}$, where L_0 is the length of the shortest beam in the lattice. For example, for a cuboid beam, α is proportional to $\frac{h}{L^2}$, where h and L are the height of cross section and the length of the beam. In such case, small values of α correspond to long and thin beams. By re-scaling in equation (??), we get the following local (for each cell) relation

$$A_e U_e = F_e, \quad (2.8)$$

where

$$A_e = \begin{bmatrix} A^e_{11} & A^e_{12} \\ A^e_{21} & A^e_{22} \end{bmatrix} = \lambda \begin{bmatrix} 1 & 0 & 0 & -1 & 0 & 0 \\ 0 & \alpha\lambda^2 & \frac{1}{2}\alpha\lambda & 0 & -\alpha\lambda^2 & \frac{1}{2}\alpha\lambda \\ 0 & \frac{1}{2}\alpha\lambda & \frac{1}{3}\alpha & 0 & -\frac{1}{2}\alpha\lambda & \frac{1}{6}\alpha \\ -1 & 0 & 0 & 1 & 0 & 0 \\ 0 & -\alpha\lambda^2 & -\frac{1}{2}\alpha\lambda & 0 & \alpha\lambda^2 & -\frac{1}{2}\alpha\lambda \\ 0 & \frac{1}{2}\alpha\lambda & \frac{1}{6}\alpha & 0 & -\frac{1}{2}\alpha\lambda & \frac{1}{3}\alpha \end{bmatrix},$$

and the the corresponding local displacement and force vectors are then as follows

$$\begin{aligned} U_e = [U^e_1, U^e_2]^t &= [\frac{x_1}{L_0}, \frac{y_1}{L_0}, \theta_1, \frac{x_2}{L_0}, \frac{y_2}{L_0}, \theta_2]^t, \\ F_e = [F^e_1, F^e_2]^t &= \frac{1}{SE}[F^x_1, F^y_1, \frac{M_1}{L_0}, F^x_2, F^y_2, \frac{M_2}{L_0}]^t. \end{aligned}$$

Following analogy with the finite element method, w shall call A_e and F_e the *element stiffness matrix* and the *element load vector* of horizontal beam, respectively. Let us now consider rotated beam as shown in Figure 4(b). We have that $A_e\tilde{U}_e = \tilde{F}_e$, where $\tilde{U}_e$, $\tilde{F}_e$ are element stiffness matrix and element load vector in the local coordinate system $o\xi\eta$, respectively. Using the rotation $\tilde{U}^e_j$=$U_\psi U^e_j$ we obtain

$$A_{\psi,e}U_e = F_{\psi,e}, \tag{2.9}$$

where

$$U_\psi = \begin{bmatrix} \cos\psi & \sin\psi & 0 \\ -\sin\psi & \cos\psi & 0 \\ 0 & 0 & 1 \end{bmatrix}, \qquad A_{\psi,e} = \begin{bmatrix} U^t_\psi A^e_{11} U_\psi & U^t_\psi A^e_{12} U_\psi \\ U^t_\psi A^e_{21} U_\psi & U^t_\psi A^e_{22} U_\psi \end{bmatrix},$$

$$F_{\psi,e} = \begin{bmatrix} U^t_\psi F^e_1 \\ U^t_\psi F^e_2 \end{bmatrix}.$$

Let U_x,U_y and U_θ be the deflection vectors in x direction, y direction and θ direction, and F_x,F_y, F_θ are the corresponding load vectors. By assembling together the equations for all beams, we obtain a system of linear algebraic equations as follows

$$AU = F, \quad U = \begin{bmatrix} U_x \\ U_y \\ U_\theta \end{bmatrix}, \quad F = \begin{bmatrix} F_x \\ F_y \\ F_\theta \end{bmatrix}. \tag{2.10}$$

3. Iterative methods for lattice block materials

To design an iterative method for the solution of (2.10), we first split A in the following way

$$A = \begin{bmatrix} A_{11} & A_{12} & A_{13} \\ A_{12}^t & A_{22} & A_{23} \\ A_{13}^t & A_{23}^t & A_{33} \end{bmatrix}$$

We shall focus our attention on a square grids and 1-lattices, i.e. $q = 1$ in Figure 1. Let $NN = n \times m$ denote the total number of cells, where n and m denote the number of cells in x direction and in y direction, respectively. The boundary conditions which we set are, that a given initial displacement $(0.2, 0, 0)^t$ is prescribed at a fixed boundary node and an external force vector $(0, 0, 0.2)^t$ t a fixed internal node. We further assume that the displacements of the rest of the boundary nodes and the external forces are all zero. To apply an iterative method for the solution of the resulting algebraic problem, let us consider the following general iteration towards the solution of $AU = F$ as follows: Given initial guess $U^{(0)}$, $U^{(k+1)}$ is obtained from $U^{(k)}$, by the following relation:

$$U^{(k+1)} = U^{(k)} + B(F - AU^{(k)}) \tag{3.1}$$

where the iterator B is usually taken to be some approximation of A^{-1}. We summarize in table 3.1 some performance comparison for several iterative methods when applied towards the solution of (2.10). The results presented are when B in (3.1)corresponds to block Gauss-Seidel iteration (G-S). We also present results for other iterative methods (not of the form (3.1), such as Conjugate gradients (CG) method and Preconditioned conjugate gradients method (PCG) with incomplete LU decomposition as preconditioner (ILU(0)-PCG).

In table 3.1, the number of iterations for two different types of lattices are shown. We consider square grids without the diagonals, and also square grids with a beam added to each cell–one of the diagonals of the square. For the square grids without the diagonals $\lambda \equiv 1$. Accordingly for a square lattice with diagonals, $\lambda \equiv 1$ for a horizontal beam or a vertical beam, and $\lambda \equiv \frac{1}{\sqrt{2}}$ for the diagonal beams. The stopping criteria of all iterative methods is $\|r_k\|/\|r_0\| \leq 10^{-6}$, where r_k is the residual vector at the k-th iteration and r_0 is the initial residual.

α	NN	G-S method	CG method	ILU(0)-PCG
	64×64	2666	181	60
		2254	183	59
0.9	128×128	6678	323	106
		6483	275	95
	256×256	> 12000	560	178
		> 10000	482	154
	64×64	1940	198	42
		1795	187	49
0.125	128×128	4734	324	69
		5148	326	83
	256×256	9624	564	105
		> 10000	498	143
	64×64	2360	232	18
		1747	221	41
0.01	128×128	6516	383	30
		5089	350	71
	256×256	> 12000	616	47
		> 10000	524	107

Table 3.1. Convergence behavior of different iterative methods.

	8×8	16×16	32×32	64×64
$\alpha = 1.0$	34.146	138.633	560.141	2252.54
$\alpha = \frac{1}{1024}$	5493.94	5971.95	6101.60	6134.44

Table 3.2. Condition numbers for BA for the square grids without diagonals. B corresponds to block Gauss-Seidel iteration.

	8×8	16×16	32×32	64×64	128×128
$\alpha = 0.9$	31.193	127.365	517.715	2090.858	8401.805
$\alpha = 0.125$	37.75	142.752	570.327	2284.245	9141.581
$\alpha = 0.01$	410.319	432.528	607.235	2421.664	9678.807

Table 3.3. Condition numbers for A for the square grids with diagonals.

	8×8	16×16	32×32	64×64	128×128
$\alpha = 0.9$	3.818	12.419	46.978	185.672	741.432
$\alpha = 0.125$	2.376	6.284	21.969	84.253	333.311
$\alpha = 0.01$	1.273	1.845	4.190	13.101	50.772

Table 3.4. Condition number of BA for the square grids without diagonals, where B corresponds to ILU(0).

	8×8	16×16	32×32	64×64	128×128
$\alpha = 0.9$	4.484	14.285	52.520	206.672	825.360
$\alpha = 0.125$	3.516	9.889	35.037	135.589	538.065
$\alpha = 0.01$	3.334	9.022	31.380	120.046	474.553

Table 3.5. Condition numbers for BA for the square grids with diagonals, where B corresponds to ILU(0).

From Tables 3.2 - 3.5, we see that the condition number of BA increases as the number of lattice nodes increases for a given α. Also for a fixed mesh (i.e. fixed number of vertices), when α decreases, the corresponding condition number increases. This leads to the conclusion that the a simple iterative method is not good choice for the solution of (2.10) and in the next section we test some more sophisticated techniques for the solution of (2.10).

4. AMG method and numerical experiments

4.1 Block Gauss-Seidel iteration based on AMG method

As a basic product method iteration, let us consider the following Gauss-Seidel method for the solution of (2.10),

$$\begin{cases} U_x \leftarrow A_{11}^{-1}(F_x - A_{12}U_y - A_{13}U_\theta), \\ U_y \leftarrow A_{22}^{-1}(F_y - A_{12}^t U_x - A_{23}U_\theta), \\ U_\theta \leftarrow A_{33}^{-1}(F_\theta - A_{13}^t U_x - A_{23}^t U_y) \end{cases} \tag{4.1}$$

Obviously, the CPU time consuming part of such iteration is in the computing of the inverses of the diagonal blocks. The first algorithm, which we propose is in fact an algebraic multigrid algorithm for each of the diagonal blocks. Let us briefly describe here the main part of such algorithm, namely the coarsening phase. We shall briefly describe the important steps in such algorithm below only for the matrix A_{11}. For the other two diagonal blocks, the algorithm is analogous.

Algorithm 4.1

Step 1 Define the set of coarse nodes (e.g. pick a maximal independent set of vertices in the graph corresponding to A_{11}). Define the nonzero pattern of the prolongation matrix. and initial prolongation matrix by using simple piece wise constant interpolation.

Step 2 Correct the above initial prolongation operator, by using the techniques developed in [8] or some energy minimization procedure (e.g. minimizing the trace of $P_x^t A_{11} P_x$.

recursion Apply Step 1 and Step 2 recursively to obtain the coarse matrices on all levels.

Applying algorithm 4.1 to to obtain a sequence of coarser subspaces, we obtain a AMG method approximate A_{ii}^{-1}, $i = 1, 2, 3$. The results are summarized in table 4.1 - 4.4

		without diagonals		with diagonals	
α	NN	BAPCG-GS	BAMV-GS	BAPCG-GS	BAMV-GS
0.9	64×64	14	14	18	18
		3	2	4	3
	128×128	13	14	18	18
		4	2	5	3
	256×256	13	14	18	18
		4	2	5	3
0.125	64×64	10	10	13	13
		4	3	4	4
	128×128	10	10	13	13
		4	3	5	4
	256×256	10	10	13	13
		4	3	5	4
0.01	64×64	8	8	11	11
		8	11	5	4
	128×128	8	8	11	11
		8	11	5	4
	256×256	8	8	11	11
		9	11	5	4

Table 4.1. Convergence of block Gauss-Seidel iteration. The first row for each grid is the number of outer iterations, the second is the number of inner iterations. The columns correspond to various methods used for the inner iteration for A_{ii}^{-1}, i.e. BAPCG denotes PCG with V-cycle as preconditioner as an inner iteration and BAMV is just V-cycle as inner iteration.

4.2 Condition numbers of the block Gauss-Seidel iteration

We consider the square 1-lattices with and without diagonals. In this paragraph, we present numerical results related to the conditioning of each of the diagonal blocks and also a table giving the spectral radii of $G = I - BA$, where B in (3.1) corresponds to block Gauss-Seidel method.

	8×8	16×16	32×32	64×64	128×128
$\alpha = 0.9$	25.274	103.087	414.275	1659.344	6638.096
$\alpha = 0.125$	25.274	103.092	413.203	1657.952	6442.095
$\alpha = 0.01$	25.097	102.438	412.413	1674.331	6691.433

Table 4.2. Condition number of A_{11}.

	8×8	16×16	32×32	64×64	128×128
$\alpha = 0.9$	2.788	2.943	2.986	2.995	2.992
$\alpha = 0.125$	2.788	2.944	2.987	2.982	2.979
$\alpha = 0.01$	2.789	2.929	2.939	2.938	2.938

Table 4.3. Condition number of A_{11}.

	8×8	16×16	32×32	64×64	128×128
$\alpha = 0.9$	1.023	1.026	1.051	1.098	1.346
$\alpha = 0.125$	1.021	1.033	1.072	1.127	1.475
$\alpha = 0.01$	1.019	1.256	1.680	2.093	2.567

Table 4.4. Condition number of BA_{11}, B corresponds to the AMG V-cycle in (3.1).

ρ(G)/NN	16×16	32×32	64×64	128×128	256×256
$\alpha = 1.0$	0.480	0.495	0.498	0.499	0.499
$\alpha = 0.5$	0.403	0.450	0.471	0.485	0.492
$\alpha = 0.125$	0.306	0.395	0.443	0.466	0.481
$\alpha = 0.015$	0.133	0.255	0.358	0.424	0.452

Table 4.5. Spectral radii of $G = I - BA$, where B in (3.1) corresponds to the block Gauss-Seidel iteration.

The results in table 4.5 show that the block Gauss-Seidel iteration provides an optimal preconditioner for 1 lattices. But unfortunately, this is not the case for other type of lattices. In table 4.6 the results are shown about 4 lattices and it can be easily seen that such iterative method is not optimal in this case. For example, from the results in table 4.6, we can conclude that for 4-lattices, the number of iterations needed for the PCG with block Gauss-Seidel as a preconditioner depend heavily on the value of the parameter α. When α is very small, the number of iterations makes this method unusable. In the following paragraph, we shall discuss corresponding AMG method which is directly applied to solve the equations (2.10) converges uniformly with respect to α.

iter.num/NN	8×8	16×16	32×32	64×64	128×128
α=0.5	22	23	22	22	22
α=0.1	49	51	51	51	51
α=0.01	357	394	> 1000	> 1000	> 1000

Table 4.6.

4.3 Algebraic multigrid method

The algebraic multigrid method we consider is based on a generalization of algorithm 4.1. The new two level algorithm is as follows

Algorithm 4.2

Step 1 Using algorithm 4.1, we get interpolation operators P_x, P_y and P_θ of all grid levels from three diagonal sub-block matrices of stiffness matrix A in the equations (2.10), respectively.

Step 2 Use P_x, P_y and P_θ in the following way and get the global interpolation operator P

$$P = \begin{bmatrix} P_x & 0 & 0 \\ 0^t & P_y & 0 \\ 0^t & 0^t & P_\theta \end{bmatrix}$$

Step 3 Define the coarse grid matrix as $A_{coarse} = P^t AP$.

The corresponding multilevel algorithm is obtained as a recursive application of algorithm 4.2. In the numerical experiments, we used the algorithm 4.2 to construct a sequence of nested spaces. The corresponding V-cycle algorithms we shall call AMV method and the corresponding PCG iteration with AMV as a preconditioner, we call APCG. The numerical results for $q = 3$ and $q = 4$ are summarized below.

4.3.1. Case $q = 3$**.** We consider lattice block materials with $q = 3$ as shown in Figure 2 with $n \times m$ cells and the same boundary and external force conditions as for the previous case $q = 1$. The corresponding numerical results are given in table 4.10 and table 4.11. From these results we can see that the APCG method is very good choice for the solution of the algebraic system.

4.3.2. Case $q = 4$**.** We now consider the honeycomb materials (see Fig. 3) with the same boundary and force conditions. Assume that there are n macrocells in direction $\vec{x} = \frac{1}{2}(3, \sqrt{3})^t$ and m cells in direction $\vec{y} = \sqrt{3}(0, 1)^t$, respectively. The results are summarized in table 4.10 and 4.12.

		without diagonals			with diagonals		
		APCG method		AMV method	APCG method		AMV method
cline3-8 α	NN	n_I	t_ρ	n_I	n_I	t_ρ	n_I
0.9	32×32	4	0.0192	3	5	0.0558	5
	64×64	4	0.0196	3	5	0.0559	5
	128×128	4	0.0203	3	5	0.0559	5
	256×256	4	0.0204	3	5	0.0560	5
0.125	32×32	4	0.0238	4	4	0.0303	4
	64×64	4	0.0282	4	4	0.0311	4
	128×128	4	0.0292	4	5	0.0414	4
	256×256	4	0.0294	4	5	0.0422	4
0.01	32×32	7	0.1164	12	4	0.0278	5
	64×64	8	0.1538	15	4	0.0275	5
	128×128	8	0.1768	16	4	0.0280	5
	256×256	8	0.1767	16	4	0.0282	5

Table 4.7. 1-lattices, n_I is the number of iterations needed to achieve the desired stopping criteria and t_ρ is the average reduction per iteration.

	8×8	16×16	32×32	64×64	128×128
$\alpha = 0.9$	1.05	1.07	1.10	1.16	1.33
$\alpha = 0.125$	1.03	1.07	1.13	1.18	1.54
$\alpha = 0.01$	1.03	1.32	1.96	2.64	3.43

Table 4.8. 1-lattices, square grid without diagonals. Condition number of preconditioned matrix.

	8×8	16×16	32×32	64×64	128×128
$\alpha = 0.9$	1.06	1.12	1.40	1.42	1.47
$\alpha = 0.125$	1.11	1.22	1.29	1.39	1.51
$\alpha = 0.01$	1.11	1.22	1.30	1.42	1.58

Table 4.9. 1-lattices, square grid with diagonals: condition numbers.

		q=3			q=4		
		APCG method		AMV method	APCG method		AMV method
α	NN	n_I	t_ρ	n_I	n_I	t_ρ	n_I
0.9	16×16	7	0.1078	8	5	0.0372	5
	32×32	7	0.1222	9	5	0.0373	5
	64×64	7	0.1356	9	5	0.0386	5
	128×128	8	0.1531	9	5	0.0388	5
0.125	16×16	8	0.1553	15	8	0.1550	14
	32×32	8	0.1683	15	8	0.1576	14
	64×64	8	0.1752	15	8	0.1586	14
	128×128	9	0.1917	15	8	0.1655	14
0.01	16×16	8	0.1722	22	22	0.5284	141
	32×32	9	0.2085	23	23	0.5453	141
	64×64	9	0.2065	23	24	0.5516	141
	128×128	9	0.2073	23	24	0.5539	141

Table 4.10. 3-lattices, n_I is the number of iterations needed to achieve the desired stopping criteria and t_ρ is the average reduction per iteration.

	before preconditioning				after preconditioning			
	8×8	16×16	32×32	64×64	8×8	16×16	32×32	64×64
$\alpha = 0.9$	121.55	490.54	1967.17	7925.18	1.46	1.78	2.01	2.81
$\alpha = 0.125$	140.56	568.39	2300.07	9240.13	1.79	2.51	2.87	3.14
$\alpha = 0.01$	575.41	638.18	2576.36	10365.74	2.06	3.19	3.90	4.31

Table 4.11 3-lattices: Condition numbers.

	before preconditioning				after preconditioning			
	8×8	16×16	32×32	64×64	8×8	16×16	32×32	64×64
$\alpha = 0.9$	81.26	346.53	1431.43	5821.15	1.15	1.24	1.39	1.56
$\alpha = 0.125$	124.55	528.12	2175.77	8840.46	1.89	2.28	2.81	3.03
$\alpha = 0.01$	1220.51	4536.45	18141.77	72986.7	14.30	20.10	25.94	29.16

Table 4.12. Honeycomb lattices: Condition numbers.

		G-S	CG	ILU(0)-CG	APCG	AMV
α	NN	CPU-time	CPU-time	CPU-time	CPU-time	CPU-time
0.9	32×32	11.81	1.16	0.66	0.61	0.71
	64×64	118.80	5.71	3.41	2.36	1.98
	128×128	> 500.00	39.00	21.53	9.34	7.08
0.125	32×32	8.84	1.10	0.44	0.77	0.83
	64×64	85.02	6.20	2.36	2.37	2.36
	128×128	> 500.00	39.16	21.86	9.17	8.95
0.01	32×32	10.00	1.49	0.22	1.04	1.59
	64×64	104.68	7.03	1.04	3.90	6.32
	128×128	> 500.00	46.30	5.98	14.94	26.53

Table 4.13. 1-lattices with diagonals: CPU time comparison.

		G-S	CG	ILU(0)-CG	APCG	AMV
α	NN	CPU-time	CPU-time	CPU-time	CPU-time	CPU-time
0.9	32×32	13.40	1.20	0.77	0.71	0.76
	64×64	140.61	7.74	4.34	2.69	2.64
	128×128	> 500.00	44.54	27.25	10.05	10.05
0.125	32×32	12.14	1.37	0.66	0.60	0.60
	64×64	111.94	7.97	3.63	2.31	2.31
	128×128	> 500.00	52.62	24.66	9.94	8.23
0.01	32×32	11.48	1.97	0.60	0.72	0.99
	64×64	109.47	9.17	3.08	2.36	2.63
	128×128	> 500.00	57.56	20.38	8.57	9.72

Table 4.14. 3-lattices: CPU time comparison.

		G-S	CG	ILU(0)-CG	APCG	AMV
α	NN	CPU-time	CPU-time	CPU-time	CPU-time	CPU-time
0.9	32×32	81.51	5.27	2.31	3.08	3.68
	64×64	700.03	34.06	15.27	11.31	13.79
	100×100	> 1000.00	116.44	70.69	31.91	38.78
0.125	32×32	64.15	5.82	1.97	3.40	5.61
	64×64	581.05	35.75	11.81	13.13	22.30
	100×100	> 1000.00	122.43	40.64	38.56	57.39
0.01	32×32	63.88	6.92	1.64	3.73	8.02
	64×64	601.21	40.59	9.66	14.56	32.74
	100×100	> 1000.00	132.37	41.58	39.00	84.53

Table 4.15. Honeycomb lattices: CPU time comparison.

From these numerical results, it can be concluded, that the APCG method is efficient and robust with respect to the number of lattice beams and the parameter α. From the results presented in Table 4.12, it can also be seen that the condition numbers of the coefficient matrix rapidly increase as the size of problem increases and α becomes smaller. The preconditioned system behaves however in a different way, and the condition numbers corresponding to the APCG method are uniformly bounded for larger values of the parameter α, but they increase for smaller α as has been numerically observed in Table 4.11. Our guess is that such behaviour for small values of α is due to the fact that the point Gauss-Seidel relaxation, is not a good smoother in this case.

5. Conclusions and some remarks on the corresponding continuous models

To justify the convergence of the constructed methods in a general situation is a rather complicated task. But in some special cases, it is possible to derive a "homogenized" differential model. In these cases, the standard techniques for proving the convergence of the multigrid method can be applied. For example it can be easily derived that on square grids, the coefficient matrix A, up to an order L^2 is the finite element matrix, corresponding to the discretization of the following system of partial differential equations:

$$\left\{\begin{array}{l} -\frac{\partial^2 u}{\partial x^2} - \alpha\frac{\partial^2 u}{\partial y^2} - \alpha\frac{\partial \theta}{\partial y} = f_1 \\ -\alpha\frac{\partial^2 v}{\partial x^2} - \frac{\partial^2 v}{\partial y^2} + \alpha\frac{\partial \theta}{\partial x} = f_2 \\ \alpha\frac{\partial u}{\partial y} - \alpha\frac{\partial v}{\partial x} + 2\alpha\cdot\theta = f_3. \end{array}\right. \tag{5.1}$$

Using this continuous model, a new preconditioner can be obtained, by using the discretization of these equations with finite element method. The corresponding numerical results are given in Table 5.1.

iter.num/NN	16×16	32×32	64×64	128×128	256×256
α=0.9	13	14	13	13	13
	3	3	4	4	5
α=0.125	9	10	10	10	10
	3	4	4	5	5
α=0.01	6	7	8	8	8
	5	7	8	8	9

Table 5.1. Numerical results for (5.1).

For square grids with diagonals, another approximate continuous model can be obtained. The corresponding system of differential equations is as follows.

$$\left\{\begin{array}{l} -\frac{8+2\sqrt{2}+\sqrt{2}\alpha}{8}\frac{\partial^2 u}{\partial x^2} - \frac{2\sqrt{2}+(8+\sqrt{2})\alpha}{8}\frac{\partial^2 u}{\partial y^2} - \frac{2\sqrt{2}(2+\alpha)}{8}\frac{\partial^2 u}{\partial x\partial y} \\ -\frac{(2-\alpha)\sqrt{2}}{8}\left(\frac{\partial^2 v}{\partial x^2} + \frac{\partial^2 v}{\partial y^2} + 2\frac{\partial^2 v}{\partial x\partial y}\right) - \frac{\sqrt{2}\alpha}{4}\frac{\partial \theta}{\partial x} - \frac{4+\sqrt{2}}{4}\alpha\frac{\partial \theta}{\partial y} = f_1 \\ \\ -\frac{2\sqrt{2}+(8+\sqrt{2})\alpha}{8}\frac{\partial^2 v}{\partial x^2} - \frac{8+2\sqrt{2}+\sqrt{2}\alpha}{8}\frac{\partial^2 v}{\partial y^2} - \frac{2\sqrt{2}(2+\alpha)}{8}\frac{\partial^2 v}{\partial x\partial y} \\ -\frac{(2-\alpha)\sqrt{2}}{8}\left(\frac{\partial^2 u}{\partial x^2} + \frac{\partial^2 u}{\partial y^2} + 2\frac{\partial^2 u}{\partial x\partial y}\right) + \frac{4+\sqrt{2}}{4}\alpha\frac{\partial \theta}{\partial x} + \frac{\sqrt{2}\alpha}{4}\frac{\partial \theta}{\partial y} = f_2 \\ \\ \frac{\sqrt{2}\alpha}{4}\frac{\partial u}{\partial x} + \frac{4+\sqrt{2}}{4}\alpha\frac{\partial u}{\partial y} - \frac{4+\sqrt{2}}{4}\alpha\frac{\partial v}{\partial x} - \frac{\sqrt{2}\alpha}{4}\frac{\partial v}{\partial y} + \frac{4+\sqrt{2}}{2}\alpha\cdot\theta = f_3. \end{array}\right. \tag{5.2}$$

We hope that some quantitative convergence results related to the convergence of Gauss-Seidel preconditioners and the algebraic multigrid methods can be obtained by using these continuous models. Such areas will be a subject of future research.

References

[1] Hua Yun-long and Yu Tong-xi, *M*echanical behavior of cellular solids, ADVANCES IN MECHANICS, Vol. 21, No. 4(1991), pp. 457-569.

[2] I. Babuška. Approximation by Hill functions. Comment Math. Univ. Carolimae, **11**, (1970), pp. 787-811.

[3] R. E. Bank and R. K. Smith. *T*he incomplete factorization multigraph algorithm. to appear, 1998.

[4] J. W. Ruge and K. Stüben. *A*lgebraic mutigrid, Mutigrid methodsM, SIAM, 1987.

[5] Lorna J. Gibson and Michael F. Ashby, *C*ellular Solids Structure and properties, Cambridge University Press, 1997.

[6] T. F. Chan, J. Xu, and L. Zikatanov, *A*n agglomeration multigrid method for unstructured grids. Proceedings of 10-th International conference on Domain Decomposition methods, edit by J. Mandel, C. Farhat, and X. J. Cai, AMS, Providence, RI, to appear.

[7] I. Babuška and S. A. Sauter, *M*athematical description of periodic trusses, in preparation.

[8] J. Mandel, M. Brezina, P. Vaněk, *E*nergy optimization of algebraic multigrid bases. Computing, to appear.

[9] J. H. Bramble, J. E. Pasciak, and J. Xu, *Convergence estimates for multigrid algorithms without regularity assumptions*, Math. Comp., 57, pp. 23-45, 1991.

[10] T. F. Chan, B. F. Smith, and W. L. Wan, *An energy-minimizing interpolation for multigrid methods*. Presentation at the 10th International Conference on Domain Decomposition, Boulder, CO, August 1997.

[11] P. Vaněk, J. Mandel, and M. Brezina, *Algebraic multigrid on unstructured meshes*, UCD/CCM Report 34, Center for Computational Mathematics, University of Colorado at Denver, December 1994.

[12] W. L. Wan, *An energy-minimizing interpolation for multigrid methods*, UCLA CAM Report 97-18, Department of mathematics, UCLA, April 1997.

[13] J. Mandel, S. McCormick and J. Ruge, *An agebraic theory for multigrid methods for variational problems*, SIAM J. Numer. Anal., 25(1988), pp. 91-110.

[14] Jinchao Xu, *Iterative methods by space decomposition and subspace correction*, SIAM Rev., 34(1992), pp. 581-613.

[15] Jinchao Xu, Jun Zou, *Some nonoverlapping domain decomposition methods*, SIAM Rev., Vol. 40, No. 4, pp. 581-613, December 1998.

[16] Q. S. Chang, Y. S. wong and H. Fu, *On the Algebraic Multigrid MethodsJ*, J. Comput. Phys., 125 (1996), pp. 279-292.

[17] R. H. chan, Q. S. Chang, Y. S. wong and H. W. Sun, *Multigrid Method for ill-conditioned Symmetric ToeplitzJ*, SIAM J. Sci. Comput., 1998, 19(2), pp. 516-529.

[18] Q. S. Chang, S. Q. Ma and G. Y. Lei, *Algebraic Multigrid for Queuing NetworksJ*, Inter. J. of Computer Math., 70(1999), pp. 539-552.

[19] J. Mandel, M. Brezina, P. Vaněk, *Energy optimization of algebraic multigrid bases*. Computing, to appear.

[20] T. Y. Hou, X. H. Wu, and Z. Q. Cai, *Convergence of a multiscale finite element method for elliptic problems with rapidly oscillating coefficients*. Submitted to Math. Comp..

[21] J. E. Jones, and P. S. Vassilevski, *AMGE based on element agglomeration*, to appear.

[22] J. H. Bramble, J. E. Pasciak, and J. Xu, *Convergence estimates for multigrid algorithms without regularity assumptions*, Math. Comp., 57, pp. 23-45, 1991.

[23] T. Chan, B. Smith, *Domain decomposition and multigrid algorithms for eliptic problems unstructured meshes*, Electronic Transactions in Numerical Analysis, 2, pp. 171-182, 1994.

[24] P. Vaněk, J. Mandel, and M. Brezina, *Algebraic multigrid on unstructured meshes*, UCD/CCM Report 34, Center for Computational Mathematics, University of Colorado at Denver, December 1994.

[25] W. L. Wan, T. F. Chan, and B. Smith, *An energy-minimizing interpolation for robust multigrid methods*, UCLA CAM Report 98-6, Department of Mathematics, UCLA, February 1998.

[26] R. Bank, and J. Xu, *An algorithm for coarsing unstructured meshes*, Numer. Math., 73, No. 1, pp. 1-36, 1996.

[27] W. Hackbusch, *Multigrid methods and applications*, vol. 4 of Computational Mathematics, Spring-Verlag, Berlin, 1985.

[28] A. Brandt, *Algebraic multigrid theory: The sysmmetric case*, Appl. Math. Computing., 19, pp. 23-56, 1986.

[29] D. Braess, *T*owards algebraic multigrid for elliptic problems of second order, Computing, 55, pp. 379-393, 1995.

[30] P. Vaněk, M. Brezina, and R. Tezaur, *T*wo-grid method for linear elasticity on unstructured meshes, to appear in SIAM J. Sci. Comp. , 1998.

[31] P. Vaněk, J. Mandel, and M. Brezina, *A*lgebraic multigrid by smoothed aggregation for second and fourth order elliptic problems, Computing, 56, pp. 179-196, 1996.

[32] J. Bramble, *M*ultigrid methods, Pitman, Notes on Mathematics, 1994.

[33] P. Vaněk, M. Brezina, and J. Mandel, *C*onvergence of algebraic multigrid based on smoothed aggregation, UCD/CCM Report 126, February 1998.

[34] P. Vaněk, J. Mandel, and M. Brezina, *A*lgebraic multi-grid by smoothed aggregation for second and forth order elloptic problems, Computing, 56, pp. 179-196, 1996.

ON THE EXISTENCE OF SYMMETRIC THREE DIMENSIONAL FINGER SOLUTIONS

Jianzhong Su and Bao Loc Tran
Department of Mathematics, The University of Texas at Arlington, Arlington, Texas 76019, USA

Abstract In this note, the existence problem of symmetric 3-dimenensional finger solutions of Mullins-Sekerka equation is studied. The finger solutions are traveling wave solutions whose finger-shaped interfaces are moving along a certain direction at a constant speed within a cylindrical domain. The existence of finger solutions is shown through a fixed point argument of the Hilbert Transformation.

1. Introduction

The fingering phenomena arise from a variety of physical processes. These fingers typically occur on the interface between two immisible fluids in a porous medium flow or on the interface of two different phases in phase transition problems. Under external conditions (such as a pressure, the gravitational force or a heat source from far field), the interface between the fluids(or phases) tends to develop into the shape of a finger, and it penetrates into the region of the other fluid (or other phase). The fluid-fingering phenomenon is meant to be the entire evolutionary process involving splitting and merging of finger-shaped interfaces, emerging of new fingers, vanishing of existing fingers and so on.

This type of problems is interesting and important for both practical and theoretical purposes. There is a wide range of applications from the secondary oil production processes where water is injected into oil reservoir [7,15-16] to crystal growth problems [19]. In the physical setting of Hele-Shaw apparatus [14], the process commonly referred as "Saffman-Taylor Instability." The theory was formulated by Saffman and Taylor [26] and Chouke [7], although a simplified one-dimensional analysis appeared earlier by Hill [15]. There are very extensive studies and discussions on this subject in the literature. We refer to Homsy [16] and Kessler [19] and their hundreds of references. Recent research activies are accounted by Tanveer in [31].

The mathematical analyses of two phase fluid or heat flow were considered by Duchon and Robert [9], Constantin and Pugh [8], Otto and E [24], Chen [5], Bazalli [3], Escher and Simonett [11-12] for the existence of general solutions of Hele-Shaw Equation with surface tension. Similar problems in higer dimensions are studied by Chen, Hong and Yi [6]. Tanveer [30] provided the analytic solutions for the small surface tension cases. Tian [32] and Nie and Tian [23] studied the zero surface tension cases and indicated the formation of singularities for a non-zero surface tension. Almgren [2] considered the non-smooth crystalline evolution for Hele-Shaw flow. There were earlier works available in the direction of finger solutions. They are either through numerical simulations [2,20,26,32 and etc] or through singular perturbation methods [2,17,20,26,28-29 and ect]. The remaining studies [18,26 and ect] were about exact solutions under the assumption that the surface tension of the interfaces is zero. The existence of at least one finger for any surface tension is given by Su [27]. When the surface tension is small, Xie and Tanveer [34] showed there is a sequence of solutions with their width in a discreet set.

The fingering problem considered in this paper is in the two-phase heat flow in a cynlindrical domain where the thermal diffusivity is near infinity. The dynamics of the interface motion in such a setting is governed by a mathematical equation called "Mullins-Sekerka equation" to be described in detail below. The Gibbs-Thompson relation is assumed on the interface.

The main purpose of this article is to provide a rigorous proof of the existence of the solutions of Mullins-Sekerka equation whose finger-shaped interfaces are moving at a constant velocity along a certain direction.

Our finger solutions are smooth and indeed analytic for $0 \leq t < \infty$. Previously only the short time existence was known for all initial interfaces except the near-circular cases [1,8-9,23,30,32]. Further, sicne Mullins-Sekerka equation has been shown to be a singular limit of Cahn-Hillard equation [1] and Phase-field equation [4] in the sense that there is one-to-one relation between the two, the knowledge of the fingering solutions in Mullins-Sekerka equation will hep to understand the dendritic solification phenomena in the phase transition problems [10,19] which is very important area of study.

2. The Mathematical Model

We assume that our situation occurs with a two-phase heat flow in an infinitely long cylindrical region in 3-dimensional space. The flows in each phase satisfy heat equations with a no-flux condition on the wall of the cylinder. The key physical law is Gibbs-Thompson relation on the sharp interface that is similar to those in relations in [9,14,16,26,30]. We model the motions of the interface as follows. The value of the temperature functions at the interface is proportional to the curvature of the interface at every point. Meanwhile the interface is moving according to the jump of normal derivatives of the temperature function along the interface.

Consider the region $\Omega = \{(x, y, z) \in \mathbb{R}^3, -\infty < x < \infty, y^2 + z^2 < R^2\}$ to be the infinitely long cylinder with radius R. Assume $R = 1$ from now on. Denote $u = u(x, y, z, t), t \geq 0, (x, y, z) \in \Omega$ to be the temperature function, $\widetilde{\Gamma} = \cup_{t \geq 0}(\Gamma_t \times \{t\})$ to be free interface. The equation which governs the motion can be rewritten as in ([5,9])

$$u_t(z, y, z, t) - D\Delta u(x, y, z, t) \equiv u_t(x, y, z, t) - D(u_{xx} + u_{yy} + u_{zz}) = 0$$

$$(x, y, z) \in \Omega \backslash \Gamma_t, t \geq 0,$$

$$\frac{\partial u}{\partial \mathrm{n}} = 0 \qquad (x, y, z) \in \partial\Omega, t \geq 0,$$

$$u(x, y, z, t) = \mu \mathbf{K}(\Gamma_t)(x, y, z) \qquad (x, y, z) \in \Gamma_t, t \geq 0, \tag{2.1}$$

$$[\frac{\partial u}{\partial \mathrm{n}}]_{\Gamma_t} = \mathbf{V}(x, y, z) \qquad (x, y, z) \in \Gamma_t, t \geq 0$$

where n is the outer normal of the boundary, D is the heat diffusivity constant, μ is a surface tension constant,$[\frac{\partial u}{\partial \mathrm{n}}]_\Gamma = \frac{\partial u}{\partial \mathrm{n}}|_{\Gamma-} - \frac{\partial u}{\partial \mathrm{n}}|_{\Gamma+}$ represents the jump from the region connected to negative infinity to the region connected to positive infinity, $\mathbf{K}(\Gamma_t)(x, y, z)$ symbols the mean curvature of Γ_t at the point (x, y, z) and $\mathbf{V}(x, y, z)$ is the normal velocity at which Γ_t propagates. Further, we assume the diffusivity is near infinity so that the heat flow reaches the equilibrium at an infinitesimally short time and the governing equations over each of the phase become Laplacian equation. Then Eq. (2.1) becomes

$$\Delta u(x, y, z, t) \equiv u_{xx} + u_{yy} + u_{zz} = 0 \qquad (x, y, z) \in \Omega \backslash \Gamma_t, t \geq 0,$$

$$\frac{\partial u}{\partial n} = 0 \qquad (x, y, z) \in \partial\Omega, t \geq 0,$$

$$u(x, y, z, t) = \mu \mathbf{K}(\Gamma_t)(x, y, z) \qquad (x, y, z) \in \Gamma_t, t \geq 0, \tag{2.2}$$

$$[\frac{\partial u}{\partial \mathrm{n}}]_{\Gamma_t} = \mathbf{V}(x, y, z) \qquad (x, y, z) \in \Gamma_t, t \geq 0$$

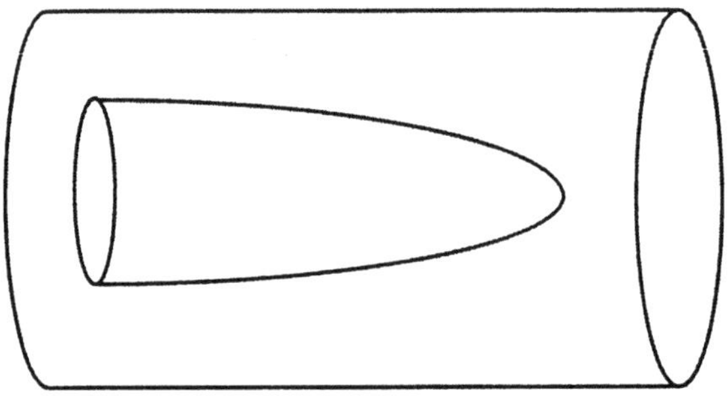

Figure 1. The interface of two fluids in 3-dimension is finger-shaped and is moving towards x-direction at a constant speed $\mathbf{V}$ for every point in the interface. It is better known as a finger solution or a "steady state" finger.

Eq. (2.2) is commomly know as Mullins-Sekerka Equation [22].

We are interested in the solutions of Eq. (2.2) with finger-shaped interfaces, and those interfaces are propagating with time. We consider the so called "steady state case" where the interface is moving at a constant speed $\mathbf{V}$ towards the positive x-axis. See figure 1. Thus the interface can be expressed as $\Gamma_t = (x_t, y_t, z_t)$ for $x_t = x_0 + \mathbf{V}t, y_t \equiv y_0, z_t \equiv z_0$. We let $x' = x - x_t, y' - y, z' = z$ be the new independent variables but still be denoted as (x, y, z). Let Γ be the interface in the new coordinates. Along with the appropriate physical conditions in this scenario, the Eq. (2.2) becomes

$$\Delta u(x,y,z,t) \equiv u_{xx} + u_{yy} + u_{zz} = 0 \qquad (x,y,z) \in \Omega \backslash \Gamma,$$
$$\frac{\partial u}{\partial \mathrm{n}} = 0 \qquad (x,y,z) \in \partial\Omega,$$
$$u(x,y,z) \to 0 \qquad x \to \infty,$$
$$u(x,y,z)_x \to \mathbf{V} \qquad x \to -\infty,$$
$$u(x,y,z,t) = \mu \mathbf{K}(\Gamma)(x,y,z) \qquad (x,y,z) \in \Gamma, \tag{2.3}$$
$$[\frac{\partial u}{\partial \mathrm{n}}]_\Gamma = \mathbf{V}\cos\theta(x,y,z) \qquad (x,y,z) \in \Gamma,$$
$$\int_{\Gamma_s} \mathbf{K}_1(s) \quad ds = \pi$$

where θ is the angle between the normal direction of the interface and the positive x-axis. The curve $\Gamma_s = \Gamma \cap \mathbf{P}(a,b) = \Gamma \cap \{(x,y,z)|ay + bz = 0, a \in$

$\mathbb{R}, b \in \mathbb{R}\}$ is the intersection of Γ with any plan through the x-axis. The function $\mathbf{K}_1(s)$ is the section curvature Γ along Γ_s.The condition $\int_{\Gamma_s} \mathbf{K}_1(s) \quad ds = \pi$ is from the geometric conditions of the channel which require the surface Γ to be parallel to the wall as it goes to infinity. The condition of temperature u at the negative infinity is corresponding to the heat source at left far field, and the condition of u at the positive infinity represents the terminal temperature there.

If the solution u is symmetric about the x-axis, then $u = u(x, r)$ for $r \equiv \sqrt{x^2 + y^2}$ and the last equation in Eq. (2.2) becomes just one constraint because of the symmetry.

The natural question is certainly about the existence of the solutions of Eq. (2.2) or (2.3). We resolve the existence problem of at least one symmetric 3-d finger solution affirmatively in the following way.

Theorem 2.1. *Let the assumptions stated above stand. There exists at least one solution $u(x, y, z) \in \mathbf{C}^2(\Omega \backslash \Gamma)$ of the equation (2.2) or (2.3) such that (1) both the solution and the corresponding curvature of the interface Γ are symmetric with respect to the x-axis and (2) $\mathbf{K}(r)$ is at least $\mathbf{C}^0$.*

3. The Formulations

The Hilbert transformation formulation of Eq. (2.2) or Eq. (2.3) was considered by several authors, particularly in [5] and [17]. For the sake of completeness as well as the needs for properties of Green's functions, we derive the formula here.

Write the function

$$\mathbf{G}(p, q) = \frac{1}{4\pi}(|p - q|^{-1} + h(p, q)) \tag{3.1a}$$

where

$$\Delta_q h(p, q) = 0 \qquad q \in \Omega, \tag{3.1b}$$

$$\frac{\partial h}{\partial \mathrm{n_q}} = -\frac{\partial |p - q|^{-1}}{\partial \mathrm{n_q}} \qquad q \in \partial\Omega.$$

Such a function $h(p, q)$ for this geometric formation Ω can be expressed directly as follows. Let $p = (x, y, z), q = (\xi, \eta, \chi)$. Then $h(p, q) = |p^* - q|^{-1}$ where p^* is the conjugate point of p with respect to the cylinder. We see $p^* = (x, \frac{y}{y^2+z^2}, \frac{z}{y^2+z^2})$. Therefore $h(p, q)$ has the form

$$h(p,q) = |(x, \frac{y}{y^2+z^2}, \frac{z}{y^2+z^2}) - (\xi,\eta,\chi)|^{-1}. \tag{3.2}$$

Eq. (3.2) is obtained by the mirror image method. We find for each point $p = (x,y,z) \in \Omega$ the images and its image p^* about $r = 1$. Since

$$\frac{\partial}{\partial y}(|(x,y,z) - (\xi,\eta,\chi)|^{-1} + |(x, \frac{y}{r^2}, \frac{z}{r^2} - (\xi,\eta,\chi)|^{-1})|_{r=1} = 0,$$

we can verify that $h(p,q)$ satisfies Eq. (3.1b).

Let $\Omega_M = \Omega \cap \{(x,y,z), |x| \le M\}$. Since $G(\cdot,p) \in C^2(\Omega\backslash\{p\}) \cap C^1(\partial\Omega \cup \Omega)$ and any solution of Eq. (2.2) has $u(\cdot) \in C^2(\Omega\backslash\Gamma) \cap C^1(\partial\Omega \cup \Omega\backslash\Gamma) \cap C^0(\Omega)$, we have

$$\int_{\Omega_M\backslash(B_\epsilon(p)\cup\Gamma_\epsilon)} (u(q)\Delta_q G(p,q) - G(p,q)\Delta_q u(q))\, d\, V_q \tag{3.3}$$

$$= \int_{\partial\Omega_M - (\partial B_\epsilon(p)\cup\partial\Gamma_\epsilon)} (u(q)\frac{\partial G(p,q)}{\partial n_q} - G(p,q)\frac{\partial u(q)}{\partial n_q})\, d\, S_q$$

where $B_\epsilon(p) \equiv \{q, |p-q| = \epsilon\}, \Gamma_\epsilon \equiv \{q, |q-\Gamma| = \epsilon\}$. Letting $\epsilon \to 0^+$ and $M \to \infty$, using the infinity condition of u, we derive

$$u(p) = \int_{\partial\Omega} (u(q)\frac{\partial G(p,q)}{\partial n_q} - G(p,q)\frac{\partial u(q)}{\partial n_q})\, d\, S_q \tag{3.4}$$

$$+ \int_\Gamma G(p,q)[\frac{\partial u(q)}{\partial n_q}]|_\Gamma\, d\, S_q + C$$

where C is an arbitrary constant to be determined by the infinity conditions

Since $\frac{\partial G(p,q)}{\partial n_q} = 0$ on $\partial\Omega$ which is from the construction of $h(p,q)$ and $\frac{\partial u(q)}{\partial n_q} = 0$ on $\partial\Omega$, the equation (3.4) reduces to

$$u(p) = \int_\Gamma G(p,q)[\frac{\partial u(q)}{\partial n_q}]|_\Gamma\, d\, S_q + C \tag{3.5}$$

In the new cylindrical coordinates (x,r,ϕ) where $y = r\cos\phi, z = r\sin\phi$, we can express the interface Γ as $\{(x,y,z), x = -f(r)\}$ and its mean curvature

$\mathbf{K} = \mathbf{K}(r)$ will be also be independent of the angle ϕ. The normal direction of the interface Γ is expressed as

$$n = \frac{(1, f'(r)\cos\phi, f'(r)\sin\phi)}{\sqrt{1 + f'^2(r)}} \tag{3.6}$$

and the angle between the normal direction and the x-axis satisfies

$$\cos\theta = \frac{1}{\sqrt{1 + f'^2(r)}}. \tag{3.7}$$

The surface element on the interface can also be similarly represented

$$d\,A = \sqrt{1 + f_y{}^2 + f_z{}^2}\, dy\, dz = \sqrt{1 + f'^2(r)}\, rdr\, d\phi. \tag{3.8}$$

Through a direct calculation, we have the curvature formula expressed as

$$\mathbf{K}(r) = \Big(\frac{f'(r)}{\sqrt{1 + f'^2(r)}}\Big)' + \frac{1}{r}\Big(\frac{f'(r)}{\sqrt{1 + f'^2(r)}}\Big). \tag{3.9}$$

If we let $p \in \Gamma$, then we formulate l Eq. (2.2) into an integral equation of the mean curvature function $\mathbf{K} \equiv \mathbf{K}(r)$ where $r = 0$ designated as the tip of the finger,

$$\mu\mathbf{K}(r) = \mathbf{H}(\mathbf{K}(r)) \equiv \textstyle\int_{\Gamma(\lambda)} \mathbf{G}(s,\tau)\mathbf{V}\cos\theta(\tau)\, d\mathbf{A}(\tau) + \mathbf{C}, \tag{3.10a}$$

$$\mathbf{K}(r) = \Big(\tfrac{f'(r)}{\sqrt{1+f'^2(r)}}\Big)' + \tfrac{1}{r}\Big(\tfrac{f'(r)}{\sqrt{1+f'^2(r)}}\Big). \tag{3.10b}$$

$$\cos\theta(r) = \tfrac{1}{\sqrt{1+f'^2(r)}} \tag{3.10c}$$

$$\lim_{r\to 1}\theta(r) = \tfrac{\pi}{2}, \tag{3.10d}$$

Eq(3.10d) can be satisfied by choosing an appropriate constant $\mathbf{C} = \mathbf{C}(\pi)$ in (3.10a). Thus Eq. (3.10a-3.10d) can be reduced to Eq. (3.10a-3.10c) with $\mathbf{C} = \mathbf{C}(\pi)$.

4. The Laray-Schauder Argument

We study the family of equations

$$\mu\mathbf{K}(r) = \mathbf{H}(\mathbf{K}(r)) \equiv \textstyle\int_{\Gamma(\lambda)} \mathbf{G}(s,\tau)\mathbf{V}\cos\theta(\tau)\, d\mathbf{A}(\tau) + \mathbf{C}(\lambda), \quad (4.1a)$$

$$\mathbf{K}(r) = \left(\tfrac{f'(r)}{\sqrt{1+f'^2(r)}}\right)' + \tfrac{1}{r}\left(\tfrac{f'(r)}{\sqrt{1+f'^2(r)}}\right). \quad (4.1b)$$

$$\cos\theta(r) = \tfrac{1}{\sqrt{1+f'^2(r)}} \quad (4.1c)$$

$$\lim_{r\to 1}\theta(r) = \tfrac{\lambda}{2}, \quad (4.1d)$$

for $0 \le \lambda < \pi$. We drop the dependency to λ in further writing unless the parameter is needed to distinguish different subjects.

Let us discuss the physical problems corresponding to Eq. (4.1a)-(4.1d). Since the total change of angles is smaller than π, Γ is of finite area and intersects $\partial\Omega$. These solutions are not exactly finger solutions in the classical sense, but closely related to them as mentioned by McLean and Saffman [20]. The finger solutions will be obtained in this article by letting $\lambda \to \pi^-$ in Section 5.

Lemma 1. *If* $\mathbf{K}(r)$ *is a symmetric* $\mathbf{C}^0$ *- solution of Eq. (4.1a)-(4.1d) and* $0 \le \lambda \le \pi - \epsilon < \pi$ *, then there* $\exists \mathbf{M} \equiv \mathbf{M}(\mu, \mathbf{V})$ *and* $\mathbf{M}_0 \equiv \mathbf{M}_0(\mu, \mathbf{V})$ *such that* $\mathbf{K}(r) \le \mathbf{M}$ *and* $|\frac{d\mathbf{K}(r)}{ds}| \le \mathbf{M}_0$ *for the derivative respect to the arc length s of* Γ_s.

Proof. From the Eq. (4.1c), we derive

$$\frac{f'(r)}{\sqrt{1+f'^2(r)}} = \mathbf{F}(r) = \frac{1}{r}\int_0^r \tau\mathbf{K}(\tau)\, d\,\tau \quad (4.2)$$

Then (4.1d) can be changed to

$$\int_{\Gamma_s} \tau\mathbf{K}(\tau)\, d\,\tau = \sin\frac{\lambda}{2} \quad (4.3)$$

that can be satisfied by choosing the right constant in (4.1a).

We now first estimate the size of the mean curvature function $\mathbf{K}(r)$ by dividing into two regions:

$$|\mathbf{K}(r)| \le \frac{1}{4\pi\mu}\Big(\int_{\Gamma(\lambda)} (|p(r) - q(\tau)|^{-1} + |p^*(r) - q(\tau)|^{-1})\mathbf{V}|\cos\theta(\tau)|\, d\,\mathbf{A}(\tau)$$

$$+|\mathbf{C}(\lambda)|)$$

$$= \frac{1}{4\pi\mu}\Big(\int_{\Gamma(\lambda),|p-q|\le\sigma}(|p(r)-q(\tau)|^{-1}+|p^*(r)-q(\tau)|^{-1})\mathbf{V}|\cos\theta(\tau)|\,d\,\mathbf{A}(\tau)$$

$$+\frac{1}{4\pi\mu}\Big(\int_{\Gamma(\lambda),|p-q|\ge\sigma}(|p(r)-q(\tau)|^{-1}+|p^*(r)-q(\tau)|^{-1})\mathbf{V}|\cos\theta(\tau)|\,d\,\mathbf{A}(\tau)$$

$$+|\mathbf{C}(\lambda)|). \tag{4.4}$$

We notice the relation

$$|\cos\theta(\tau)|\,d\,\mathbf{A}(\tau) = \,dy\,dz. \tag{4.5}$$

Then for $|p-q| > \sigma$,

$$\int_{\Gamma(\lambda),|p-q|\ge\sigma}(|p(r)-q(\tau)|^{-1}+|p^*(r)-q(\tau)|^{-1})\mathbf{V}|\cos\theta(\tau)|\,d\,\mathbf{A}(\tau) \tag{4.6}$$

$$\le \frac{2}{\sigma}\int_{\Gamma(\lambda),|p-q|\ge\sigma}\mathbf{V}dy\,dz \le 2\mathbf{V}\frac{\pi}{\sigma}.$$

For $|p-q| \le \sigma$, we divide into two cases. If $p \notin \partial\Omega$, then $|p-q|^{-1}$ is the only singular function. If $p \in \partial\Omega$, then $p = p^*$, $|p-q|^{-1} = |p^*-q|^{-1}$. In either case, let $\rho = |(y(r), z(r)) - (y(\tau), z(\tau))|$. Then

$$\int_{\Gamma(\lambda),|p-q|\le\sigma}(|p(r)-q(\tau)|^{-1}+|p^*(r)-q(\tau)|^{-1})\mathbf{V}|\cos\theta(\tau)|\,d\,\mathbf{A}(\tau) \tag{4.7}$$

$$\le \int_{\Gamma(\lambda),|p-q|\le\sigma}2|p(r)-q(\tau)|^{-1}\mathbf{V}|\cos\theta(\tau)|\,dy\,dz$$

$$\le \frac{2}{\sigma}\int_{\Gamma(\lambda),|p-q|\le\sigma}\frac{1}{\rho}\rho\,d\rho\,d\phi \le 4\mathbf{V}\pi.$$

The estimates for $\mathbf{K}_s(s)$, the derivative of the curvature with respect to the arc length of Γ_s can be obtained in terms of Cauchy Principal Value [21] such that

$$||\mathbf{K}_s(s)|| \le \mathbf{M}_0(\mu, \mathbf{V}).$$

The derivation however is lengthy and technical. We will have the details in a seperate technical paper later.

Remark: We note that the curvature condition (4.3) on the curvature is equivalent to that of the section curvature in (2.3).

Let $\mathbf{K}(r) > 0$. We define

$$\Gamma(\mathbf{K}) \equiv \{(x, y, z) \in \Omega, x = -f(r), (\frac{f'(r)}{\sqrt{1+f'^2(r)}})' + \frac{1}{r}\frac{f'(r)}{\sqrt{1+f'^2(r)}} = \mathbf{K}(r), \tag{4.8}$$

$$f(0) = f'(0) = 0, r \leq 1\},$$

and

$$\theta(\mathbf{K}) = \cos^{-1}(\frac{1}{\sqrt{1+f'^2}}). \tag{4.9}$$

The curvature $\Gamma(\mathbf{K})$ is well defined if $0 < \int_{\Gamma_s(\mathbf{K})} \mathbf{K}_1(\tau)\, ds(\tau) \leq \pi - \epsilon$.

Lemma 2. *Let* $\mu\mathbf{T}(\mathbf{K}) \equiv \int_{\Gamma(\mathbf{K})} \mathbf{G}(s, \tau)\mathbf{V}\cos\theta(\mathbf{K})\ d\mathbf{A}(\tau) + \mathbf{C}(\lambda)$. *The constant* $\mathbf{C}(\lambda)$ *is chosen so that* $\int_{\Gamma_s(\mathbf{K})} r\mathbf{T}(\mathbf{K}(s(r)))\ dr = \sin(\lambda/2)$. *Then for* $0 \leq \lambda \leq \pi - \epsilon$, *the mapping* $\mathbf{T} : \mathbf{C}^0[0,1] \to \mathbf{C}^0[0,1]$ *is a compact mapping which satisfies* $|\mathbf{T}(\mathbf{K})| \leq \mathbf{M}$ *for any* $\mathbf{K}(r) \in \mathbf{C}_0[0,1]$ *with* $0 < \int_{\Gamma_s(\mathbf{K})} \mathbf{K}_1(r)\, ds(r) = \lambda \leq \pi - \epsilon$.

Proof The two estimates in Lemma 1 imply that $|\mathbf{T}(\mathbf{K})(s(y))| \leq \mathbf{M}$ and $|\mathbf{T}(\mathbf{K})_s(s(y))| \leq \mathbf{M}_0$. The compactness follows from the fact that space $\mathbf{C}^1[0,1]$ embeds compactly into $\mathbf{C}^0[0,1]$.

Proposition 1. *Let* $0 \leq \lambda \leq \pi - \epsilon < \pi$. *Then there exists* $\mathbf{K}(r)$, *a symmetric* $\mathbf{C}^0$ *- solution of Eq. (4.1a)-(4.1d). Further, the solution* $\mathbf{K}(s)$ *satisfies the properties that* $\exists \mathbf{M} \equiv \mathbf{M}(\mu, \mathbf{V}), \mathbf{M}_0 \equiv \mathbf{M}_0(\mu, \mathbf{V})$ *such that* $|\mathbf{K}(r)| \leq \mathbf{M}$ *and* $|\frac{d\mathbf{K}(r)}{ds}| \leq \mathbf{M}_0$ *for derivative with respect to the arc length s.*

Remark. 1) The uniform bound for $\mathbf{K}_s(s)$ is mainly for the compactness in Proposition 1. But its importance will be see when $\lambda \to \pi^-$ in Section 5 below for the uniform convergence. Although a Lipschitz bound for $\mathbf{K}(s)$ should suffice, the derivation of such bound will involve similar difficulty. 2) It is not clear to us if $\mathbf{K}(r) \geq 0$ or Γ is convex. In 2-D case, with the help of Maximum

Principle [25], we are able to show $\mathbf{K}(s) \geq 0$ for any point. The argument does not extend itself to 3-D case.

Proposition 1 follows directly from the estimates in Lemmas 1 and 2. The existence of $\mathbf{K}(s)$ can then be derived by using Leray-Schauder Fixed Point Theorem [13,35] in the function space $\mathbf{C}^0[0,1]$. The proof is a standard argument and hence omitted.

5. A Proof of Theorem 2.1

Proof of Theorem 2.1. The solutions $\mathbf{K}(\lambda, s)$ of Eq. (4.1a)-(4.1d) can be expressed by

$$\mu\mathbf{K}(r) = \mathbf{H}(\mathbf{K}(\lambda, s)) \equiv \int_{\Gamma(\lambda)} \mathbf{G}(s,\tau)\mathbf{V}\cos\phi(\tau)\, d\mathbf{A}(\tau) + \mathbf{C}(\lambda) \tag{5.1a}$$

$$\mathbf{K}(r) = \Big(\frac{f'(r)}{\sqrt{1+f'^2(r)}}\Big)' + \frac{1}{r}\Big(\frac{f'(r)}{\sqrt{1+f'^2(r)}}\Big). \tag{5.1b}$$

$$\cos\theta(r) = \frac{1}{\sqrt{1+f'^2(r)}} \tag{5.1c}$$

$$\lim_{r\to 1}\theta(r) = \frac{\lambda}{2}, \tag{5.1d}$$

From the boundedness of $\mathbf{K}_s(r)$, there exists a subsequence $\lambda(n) \to \pi$ such that

$$\mathbf{K}(\lambda(n), s) \to \mathbf{K}(s) \quad unif. \quad in \quad \mathbf{C}^0 - norm, \tag{5.2}$$

$$\theta_{\lambda(n)}(s) \to \theta(s) \quad unif. \quad in \quad \mathbf{C}^1 - norm \tag{5.3}$$

and

$$\Gamma(\mathbf{K}(\lambda(n), s)) \to \Gamma \quad unif. \quad in \quad \mathbf{C}^2 - norm. \tag{5.4}$$

Thus by letting $\lambda(n) \to \pi^-$ in Eq. (5.1a)-(5.1d), we derive the existence of solutions of Eq. (2.3) by the mean of the limiting functions. The uniform bound of $\mathbf{K}(\lambda, s)$ implies the same bound for $\mathbf{K}(s)$. Therefore, given any small positive number $\mathbf{M}_1$, there exists $r_1 = r_1(\mathbf{M}_1) > 0$ such that $|f(r)| \leq \mathbf{M}_1$ for $|r| \leq r_1$. Thus the limiting solutions are not degenerate.

Acknowledgments

The first author was partial supported by the Research Enhancement Program in University of Texas at Arlington and the Texas ARP grant No. 003656-0009-1999. The authors thank some interesting discussions with Saleh Tanveer.

References

[1] Alikabos, N. D., Bates, P. W. and Chen, X., *Convergence of the Cahn-Hillard equation to the Hele-Shaw model*, Arch Rational Mech. Anal. 128(1994), 165-205.

[2] Almgren, R. F., *Crystalline Saffman-Taylor fingers*, SIAM J. Appl. Math. 55(1995), 1511-1535.

[3] Bazilli, B. V., *Steffan Problem for the Laplace equation with regard for the curvature of the free boundary,*, Ukrain. Math. J. 49 (1997), 1465-1484.

[4] Caginalp, G,, *Steffan and Hele-Shaw type models as asymptotic limits of teh phasse field equations*, Phys. Rev. A 39 (1989), 5887-5896.

[5] Chen, X., *Hele-Shaw problem and area-preserving curve shorting motion*, Arch. Rational Mech. Anal. 123 (1993, 117-151.

[6] Chen, X., Hong J. X. and Yi, F. H., *Existence, uniqueness, and regularity of classical solutions of the Mullins-Sekerka problem*, Comm. Partial Diff. Eq. 21 (1996), 1705-1727.

[7] Chouke, R. L., van Muers, P and van der Poel, C., *The instablility of slow immiscible viscous liquid - liquid displacements in permeable media*, Trans AIME 216 (1959), 188-194.

[8] Constantin P. and Pugh. M., *Global solutions for small data to the Hele-Shaw equation*, Nonlinearity 6 (1993), 393-415.

[9] Duchon, J. and Robert, R., *Evolution d'une interface par capilarite et diffusion de volume I. existence locale en temps*, Ann. Inst. H. Poincare, Analyses Non Lineaire 1 (1984), 361-378.

[10] Elliott, C. M. and Ockendon, J. R., *Weak and variational methods for moving boundary problems*, Pitman Advanced Publishing Program, 1982.

[11] Escher, J. and Simonett, G., *On Hele-Shaw models with surface tension*, Math. Res. Lett. 3 (1996), 467-474.

[12] Escher, J. and Simonett, G., *Classical solutions to Hele-Shaw models with surface tension*, Adv. Differential Equation 2 (1997), 439-459.

[13] Gillarg D. and Trudinger N. S., *Elliptic partial differential equations of second order*, Spring-Verlag, 1983.

[14] Hele-Shaw, H. J. S., *On the motion of a viscous fluid between two parallel plates*, Nature 58 (1898), 34-36.

[15] Hill, S., *Channeling in packed columns*, Chem. Eng. Sci 1 (1952), 247-253.

[16] Homsy, G. M., *Viscous fingering in porous media*, Ann. Rev. Fluid Mech. 19 (1987), 271-311.

[17] Hong, D.C. and Langer, J. S., *Analytic theory of the selection mechanism in the Staffman-Taylor problem*, Phys. Rev. Lett. 56 (1986), 2032-2035.

[18] Howinson, S. D., *Cusp development in Hele-Shaw flow with a free surface*, SIAM J. of Appl. Math 46 (1986), 20-26.

[19] Kessler, D. A., Koplik, J, and Levine, H., *Pattern Selection in fingered growth phenomena*, Advance in Physics 39 (1988), 255-329.

[20] Mclean J. W. and Saffman, P. G., *The effect of surface tension on the shape of fingers in Hele-Shaw cell*, J. Fluid Mech 102 (1981), 455-469.

[21] Mikhlin S. G., and Prossdorf S., *Singular integral operators [translated from German by Albrecht Bottcher, Reinhard Lehmann]*, Springer-Verlag, 1986.

[22] Mullins, W. W. and Sekerka, R. F., *Morphological stability of a particle growing by diffusion of heat flow*, Journal of Applied Physics 34 (1963), 323-328.

[23] Nie Q. and Tian F. R., *Singularities in Hele-Shaw flows*, SIAM J. Appl Math 58 (9998), 34-54.

[24] Otto F. and E, W., *Thermodynamically driven incompressible fluid mixtures*, Journal of Chemical Physics 107 (1997), 10177-10184.

[25] Protter M. H. and Weinberger H. F., *Maximum principles in differential equations*, Prentice Hall, 1967.

[26] Saffman, P. G. and Taylor, G. I., *The penetration of a fluid into a porous medium or Hele-Shaw cell containing a more viscous liquid*, Proc. R. Soc. London, Ser. A 245 (1958), 312-329.

[27] Su, J. *On the existence of finger solutions in Hele-Shaw Equation*, Nonlinearity 14 (2001), 153-166.

[28] Tanveer, S., *Analytic theory for the selection of symmetric Saffman-Taylor finger*, Phys. Fluids 30 (1987), 1589-1605.

[29] Tanveer, S., *Analytic theory for the selection of Saffman-Taylor finger in the presence of thin film effects*, Proc. R. Soc. Lond. A A 428 (1990), 511-545.

[30] Tanveer, S., *Evolution of Hele-Shaw interface for small surface tension*, Phil. Trans. R. Soc. Lond. A 343 (1993), 155-204.

[31] Tanveer, S., *Surprises in viscous fingering*, J. Fluid Mech. 409 (2000), 273-308.

[32] TianF. R. *A Cauchy integral approach to Hele-Shaw problems with a free boundary: The case of zero surface tension*, Arch. Rational Mech. Anal 135 (1996), 175-196.

[33] Tryggvason, G. and Aref, H., *Numerical experiments on Hele-Shaw flow with a sharp interface*, J. Fluid Mech. 139 (1983), 1-30.

[34] Xie, X. and Tanveer, S., *Rigorous results in steady finger selection in viscous fingering*, Preprint, Ohio State University (2001), 1-91.

[35] Zeidler E., *Nonlinear functional analysis and its applications [translated by Peter R. Wadsack]*, vol. 1, Springer-Verlag, 1985.

A COMBINED MIXED FINITE ELEMENT AND DISCONTINUOUS GALERKIN METHOD FOR MISCIBLE DISPLACEMENT PROBLEM IN POROUS MEDIA

Shuyu Sun, Béatrice Rivière and Mary F. Wheeler

The Center for Subsurface Modeling, Texas Institute of Computational and Applied Mathematics, The University of Texas, Austin, TX 78712, USA

shuyu@ticam.utexas.edu

riviere@ticam.utexas.edu

mfw@ticam.utexas.edu

Abstract A combined method consisting of the mixed finite element method for flow and the discontinuous Galerkin method for transport is introduced for the coupled system of miscible displacement problem. A "cut-off" operator $\mathcal{M}$ is introduced in the discontinuous Galerkin formular in order to make the combined scheme converge. Optimal error estimates in $L^2(H^1)$ for concentration and in $L^\infty(L^2)$ for velocity are derived.

Keywords: discontinuous Galerkin method, mixed finite element method, miscible displacement, error estimate

1. Introduction

Numerical modeling of miscible displacement in porous media is important and interesting in oil recovery and enviromental pollution problem. The miscible displacement problem is described by a coupled system of non-linear partial differential equations. The need for accurate solutions to the coupled equations challenges numerical analysts to design new methods.

The mixed finite element methods [1, 6] gained great popularity in the last two decades for the reasons that they provide very accurate approximations of the primary unknown and its flux and they conserve mass locally. The discontinuous Galerkin method gained even greater popularity recently for at least four reasons [10, 11, 12, 13]: 1) the flexibility inherent to it allows more general meshes construction and degree of non uniformity than permitted by the more conventional finite element method; 2) it also conserves mass locally on any element; 3) it has, in general, less numerical diffusion and provides more accurate local approximations for problems with rough coefficients; 4) it is easy to implement. Traditional numerical methods were studied for solving the miscible displacement problem by Darlow, Ewing, Wheeler and Douglas [3, 4, 5, 7, 9, 14]. The formulation of discontinuous Galerkin for both of flow and transport subproblems is given by B. Rivière [11].

In this paper, a combined method with mixed finite element method for flow and discontinuous Galerkin method for transport is introduced and analyzed. This paper consists of four additional sections. Problem definition is given in section 2 and the formulation of the combined method is described in section 3. In section 4, the results and proofs of error estimates for the subproblems and coupled system are given. Conclusions are described in the last section.

2. .Governing Equations

The displacement of one incompressible fluid by another in porous media is considered in this paper. Detailed discussion on physical theories of miscible displacement in porous media can be found in [2] or [8].

Let Ω denote a bounded domain in $\mathbb{R}^d$, $(n = 2, 3)$ and Let J denote the time interval $(0, T_f]$. The classical equations governing the miscible displacement in porous media is as follows.

- Continuity equation

$$\nabla \cdot \mathbf{u} = q \qquad (x,t) \in \Omega \times J \tag{1}$$

- Transport equation

$$\phi \frac{\partial c}{\partial t} + \nabla \cdot (\mathbf{u}c - \mathbf{D}(\mathbf{u})\nabla c) = qc^* \qquad (x,t) \in \Omega \times J \tag{2}$$

- Darcy velocity

$$\mathbf{u} = -\frac{K}{\mu}\nabla p \qquad (x,t) \in \Omega \times J \tag{3}$$

- Dispersion/diffusion tensor

$$\mathbf{D}(\mathbf{u}) = d_m\mathbf{I} + |\mathbf{u}| \{\alpha_l \mathbf{E}(\mathbf{u}) + \alpha_t (\mathbf{I} - \mathbf{E}(\mathbf{u}))\} \tag{4}$$

where,

$$d = \phi\tau D_m$$

- Constitutive relation

$$\mu = \mu(c)$$

where the dependent variables are p, the pressure in the fluid mixture, and $\mathbf{u}$, the Darcy velocity of the mixture (volume flowing across a unit across-section per unit time), and c, the concentration of interested species measured in amount of species per unit volume of the fluid mixture. The permeability K of the medium measures the conductivity of the medium to fluid flow; the viscosity μ of the fluid measures the resistance to flow of the fluid mixture; ρ is the density of fluid mixture; the porosity ϕ is the fraction of the volume of the medium occupied by pores; $\mathbf{D}(\mathbf{u})$ is the dispersion/diffusion tensor, which has contributions from molecular diffusion and mechanical dispersion, and it can be calculated by equation (4), where $\mathbf{E}(\mathbf{u})$ is the tensor that projects onto the $\mathbf{u}$ direction, whose (i,j) component is $(\mathbf{E}(\mathbf{u}))_{i.j} = \frac{u_i u_j}{|\mathbf{u}|^2}$; τ is the tortuosity coefficient; D_m is the molecular diffusivity; α_l and α_t are the longitudinal and transverse dispersivities, respectively. The commonly used constitutive relation is the quarter-power mixing rule $\mu(c) = \left(c\mu_s^{-0.25} + (1-c)\mu_o^{-0.25}\right)^{-4}$, but we consider $\mu(c)$ to be a general nonlinear relation in this paper. The imposed external total flow rate q is a sum of sources (injection) and sinks (extraction), c^* is the injected concentration c_w if $q > 0$ and is the resident concentration c if $q < 0$.

The continuity equation (1) can be obtained by the mass conservation for the whole fluid mixture and the equation (3) is a formulation of Darcy's law. Combination of equations (1) and (3) will give the flow equation.

$$-\nabla \cdot \left(\frac{K}{\mu(c)}\nabla p\right) = q \qquad (x,t) \in \Omega \times J \qquad (5)$$

The flow equation (5) governs the fluid flow and gives the pressure field and Darcy velocity field if the concentration is given. It is elliptic if the concentration is considered to be given.

The transport equation (2) can be obtained by the mass conservation of the interested species. It governs the convection-diffusion transport process and gives the concentration profile provided the velocity field is given. It is parabolic but normally convection-dominated.

We assume Ω is a bounded domain with Lipschitz boundary $\partial\Omega = \Gamma_N = \overline{\Gamma}_{in} \cup \overline{\Gamma}_{out}$, where Γ_N is the Neumann boundary for flow subproblem; Γ_{in} is the inflow boundary and Γ_{out} is the outflow/noflow boundary condition, defined as follows.

$$\begin{aligned}\Gamma_{in} &= \Gamma_{in}(t) = \{x \in \partial\Omega : \mathbf{u}(t)\cdot\nu < 0\}\\ \Gamma_{out} &= \Gamma_{out}(t) = \{x \in \partial\Omega : \mathbf{u}(t)\cdot\nu \geq 0\}\end{aligned}$$

where, ν denotes the unit outward normal vector to $\partial\Omega$. Though Γ_{in} and Γ_{out} can be time-dependent for some physical problem, we assume they are fixed at all the time in J for simplicity.

We consider following boundary condition for this problem.

$$\mathbf{u}\cdot\nu = u_B \qquad (x,t) \in \Gamma_N \times J \tag{6}$$

$$(\mathbf{u}c - \mathbf{D}\nabla c)\cdot\nu = c_B\mathbf{u}\cdot\nu \qquad t \in J,\ x \in \Gamma_{in}(t) \tag{7}$$

$$(-\mathbf{D}\nabla c)\cdot\nu = 0 \qquad t \in J,\ x \in \Gamma_{out}(t) \tag{8}$$

We know the flow problem with above boundary conditions has a solution for pressure, which is unique up to an additive constant, thus has a unique solution for velocity, provided that $\int_\Omega f = \int_{\partial\Omega} u_B$ is satisfied and the viscosity is given.

The initial concentration is specified in the following way.

$$c(x,0) = c_0(x) \qquad x \in \Omega \tag{9}$$

In this paper we are only interested in the convergence result for velocity $\mathbf{u}$ and concentration c.

3. Discontinuous Galerkin/Mixed Finite Element Scheme

3.1 Assumption

We consider the scheme of mixed finite element (MFE) method for flow subproblem and discontinuous Galerkin (DG) with interior penalty term for transport subproblem. The scheme used for transport subproblem is referred to as the Non-symmetric Interior Penalty Galerkin (NIPG) [11].

For simplicity, we consider only two or three dimensional rectangular domain $\Omega = \prod_{i=1}^{d}(0, L_i)$, $d = 2$, or 3 and we only consider rectangular mesh. However, the results can be directly extended to logically rectangular domain/mesh by conforming mapping. Though we can choose separate domain partitions for flow and for transport problem, the same rectangular domain partition $\mathcal{T}_h$ is considered here for both of flow and transport equations. We also assume the permeability tensor K is invertible and is uniformly positive definite and uniformly bounded above.

3.2 Notation

Let $\mathcal{T}_{h>0}$ be a quasi-uniform family of rectangular partition of Ω such that no element crosses the boundaries of $\Gamma_D, \Gamma_N, \Gamma_{in}$, or Γ_{out}, where h is the maximal element diameter. The set of all interior edges (for 2 dimensional domain) or faces (for 3 dimensional domain) for $\mathcal{T}_h$ are denoted by E_h. On each edges or faces $e \in E_h$, a unit normal vector ν_e is arbitrarily fixed. The set of all edges or faces on Γ_{out} and on Γ_{in} for $\mathcal{T}_h$ are denoted by $E_{h,out}$ and $E_{h,in}$, respectively, for which the normal vector ν_e coincides with the outward unit normal vector.

For $s \geq 0$, define,

$$H^s(\mathcal{T}_h) = \{\phi \in L^2(\Omega): \ \phi|_R \in H^s(R), \ R \in \mathcal{T}_h\} \tag{10}$$

We now define the average jump for $\phi \in H^s(\mathcal{T}_h)$, $s > 1/2$. Let $R_i, R_j \in H^s(\mathcal{T}_h)$ and $e = \partial R_i \cap \partial R_j \in E_h$ with ν_e exterior to R_i. Denote

$$\langle \phi \rangle = \frac{1}{2}\left(\phi|_{R_i}\right)\Big|_e + \frac{1}{2}\left(\phi|_{R_j}\right)\Big|_e \tag{11}$$

$$[\phi] = \left(\phi|_{R_i}\right)\Big|_e - \left(\phi|_{R_j}\right)\Big|_e \tag{12}$$

The usual Sobolev norm on Ω is denoted by $\|\cdot\|_{m,\Omega}$. The broken norms are defined, for positive integer m, as

$$|||\phi|||_m^2 = \sum_{R \in \mathcal{T}_h} \|\phi\|_{m,\Omega}^2 \tag{13}$$

The finite element space is taken to be

$$\mathcal{D}_r(\mathcal{T}_h) \equiv \{\phi \in L^2(\Omega): \ \phi|_R \in P_r(R), \ R \in \mathcal{T}_h\} \tag{14}$$

where $P_r(R)$ denotes the space of polynomials of (total) degree less than or equal to r on R.

Define

$$V \equiv H(\Omega; \mathrm{div}) \equiv \left\{\mathbf{u} \in \left(L^2(\Omega)\right)^d : \ \mathrm{div}\mathbf{u} \in L^2(\Omega)\right\} \tag{15}$$

$$W \equiv L^2(\Omega) \tag{16}$$

Let V^0 and V^N be the subspaces of V consisting of functions with normal trace on Γ_N (weakly) equal to zero and $\mathbf{u}_B$, respectively.

Let the approximating subspace $V_k(\mathcal{T}_h) \times W_k(\mathcal{T}_h)$ of $V \times W$ be the k-th ($k \geq 0$) order Raviart-Thomas space (RT_k) of the partition $\mathcal{T}_h$. For example, for three dimensional domain Ω, it is defined as

$$\begin{aligned} V_k(\mathcal{T}_h) &= \{\mathbf{v} \in H(\Omega; \mathrm{div}) : \mathbf{v}|_R \in Q_{k+1,k,k}(R) \times Q_{k,k+1,k}(R) \\ &\qquad \times Q_{k,k,k+1}(R),\ R \in \mathcal{T}_h\} \\ W_k(\mathcal{T}_h) &= \{w \in L^2(\Omega) : w|_R \in Q_{k,k,k}(R),\ R \in \mathcal{T}_h\} \end{aligned}$$

where, we denote by $Q_{i,j,k}(R)$ the space of polynomials of degree less than or equal to i (j, k) in the first (second, third) variable restricted to R.

Corresponding to V^0 and V^N, define their subspaces $V_k^0(\mathcal{T}_h) = V_k(\mathcal{T}_h) \cap V^0$ and

$$V_h^N(\mathcal{T}_h) = \left\{ \mathbf{v} \in V_k(\mathcal{T}_h) : (\mathbf{v} \cdot \nu, \lambda)_{\Gamma_N} = 0 \quad \forall \lambda \in \Lambda_h \right\} \tag{17}$$

where $\Lambda_h \subset L^2(\partial\Omega)$ is the corresponding hybrid space of Lagrange multipliers for the pressure restricted to $\partial\Omega$, and we have, $\Lambda_h = V_h \cdot \nu|_{\partial\Omega}$.

The inner product in $\left(L^2(\Omega)\right)^d$ or $L^2(\Omega)$ is indicated by $(\cdot, \cdot)$ and the inner product in boundary function space $L^2(\Gamma)$ is indicated by $(\cdot, \cdot)_\Gamma$. Denote

$$|\mathbf{u}| = |\mathbf{u}|_2 = \sqrt{\sum_{i=1}^{d} (\mathbf{u})_i^2} \tag{18}$$

$$\|\mathbf{u}\|_{(L^2(\Omega))^d} = \|(|\mathbf{u}|_2)\|_{L^2(\Omega)} \tag{19}$$

$$\|\mathbf{u}\|_{(L^\infty(\Omega))^d} = \|(|\mathbf{u}|_2)\|_{L^\infty(\Omega)} \tag{20}$$

3.3 Continuous in time scheme

Let us define the bilinear form $B(c, \psi; \mathbf{u})$ and the linear functional $L(\psi; \mathbf{u})$ as follows.

$$\begin{aligned} B(c, \psi; \mathbf{u}) &= \sum_{R \in \mathcal{T}_h} \int_R (\mathbf{D}(\mathbf{u})\nabla c - c\mathbf{u}) \cdot \nabla\psi - \sum_{e \in E_h} \int_e \langle \mathbf{D}(\mathbf{u})\nabla c \cdot \nu_e \rangle [\psi] \\ &+ \sum_{e \in E_h} \int_e \langle \mathbf{D}(\mathbf{u})\nabla\psi \cdot \nu_e \rangle [c] + \sum_{e \in E_h} \int_e c^* \mathbf{u} \cdot \nu_e [\psi] \\ &+ \sum_{e \in E_{h,out}} \int_e c\mathbf{u} \cdot \nu_e \psi - \int_\Omega c q^- \psi + J_0^{\sigma,\beta}(c, \psi) \end{aligned} \tag{21}$$

where, $c^*|_e$ is the upwind value of concentration,

$$c^*|_e = \begin{cases} c|_{R_1} & \text{if } \mathbf{u} \cdot \nu_e > 0 \\ c|_{R_2} & \text{if } \mathbf{u} \cdot \nu_e < 0 \end{cases}$$

for $e = \partial R_1 \cap \partial R_2$ and ν_e is the outward unit normal vector to R_1. Notice $\mathbf{u} \cdot \nu_e$ is continuous on the direction of ν_e, thus has well-defined value at the interface.

q^+ is the injection part of source term and q^- is the extraction part of source term,

$$\begin{aligned} q^+ &= \max(q, 0) \\ q^- &= \min(q, 0) \end{aligned}$$

Of course, we have $q = q^+ + q^-$.

$J_0^{\sigma,\beta}(c, \psi)$ is the interior penalty term,

$$J_0^{\sigma,\beta}(c, \psi) = \sum_{e \in E_h} \frac{\sigma_e}{h_e^\beta} \int_e [c]\,[\psi] \tag{22}$$

where, σ is a discrete positive function that takes constant value σ_e on the edge or face e, and is bounded below by $\sigma_* > 0$ and above by σ^*, h_e denotes the size of e and $\beta \geq 0$ is a real number.

The linear functional $L(\psi)$ is defined as

$$L(\psi; \mathbf{u}) = \int_\Omega c_w q^+ \psi - \sum_{e \in E_{h,in}} \int_e c_B \mathbf{u} \cdot \nu_e \psi \tag{23}$$

The continuous in time DG/MFE scheme for approximating the solution of the equations (1), (2) and (3) is as follows.

Finding $\mathbf{u}_h \in L^\infty(J, V_k^N(\mathcal{T}_h)), p_h \in L^\infty(J, W_k(\mathcal{T}_h)), c_h \in L^\infty(J, \mathcal{D}_r(\mathcal{T}_h))$ such that,

$$(\mu(c_h) K^{-1}\mathbf{u}_h, \mathbf{v}) - (\nabla \cdot \mathbf{v}, p_h) = -(p_B, \mathbf{v} \cdot \nu)_{\Gamma_D} \tag{24}$$

$$\forall \mathbf{v} \in V_k^0(\mathcal{T}_h) \quad \forall t \in J$$

$$(\nabla \cdot \mathbf{u}_h, w) = (q, w) \qquad \forall w \in W_k(\mathcal{T}_h) \quad \forall t \in J \tag{25}$$

$$\left(\phi \frac{\partial c_h}{\partial t}, \psi\right) + B(c_h, \psi; \mathbf{u}_h^M) = L(\psi; \mathbf{u}_h^M) \tag{26}$$

$$\forall \psi \in \mathcal{D}_r(\mathcal{T}_h) \quad \forall t \in J$$

$$(c_h, \psi) = (c_0, \psi) \qquad \forall \psi \in \mathcal{D}_r(\mathcal{T}_h) \quad t = 0 \tag{27}$$

where the $\mathbf{u}_h^M$ is defined as,

$$\mathbf{u}_h^M = \min\left(|\mathbf{u}_\mathbf{h}|, M\right) \frac{\mathbf{u}_\mathbf{h}}{|\mathbf{u}_\mathbf{h}|} \tag{28}$$

where, M is a fixed positive real number and $|\mathbf{u}_\mathbf{h}| = |\mathbf{u}_\mathbf{h}|_2 = \sqrt{\sum_{i=1}^{d} (\mathbf{u}_\mathbf{h})_i^2}$.

The reason for using $\mathbf{u}_h^M$ rather than $\mathbf{u}_h$ for approximation of transport equation will be clear in the next section.

4. Error Estimates for Combined DG/MFE Approximation for Miscible Displacement Problem

4.1 Notations

Throughout this paper, C denotes a generic constant whose value may change with different occurrences.

We first define three projection operators and their approximation properties. Let P_h denote L^2-projection of W onto $W_h = W_k(\mathcal{T}_h)$: for $p \in W$, $P_h p \in W_h$ is defined by

$$(P_h p - p, w) = 0, \qquad \forall w \in W_h \tag{29}$$

Let Π_h denote the usual Raviart-Thomas projection $\Pi_h : V \to V_h$ satisfies the following properties [6],

$$(\nabla \cdot (\mathbf{u} - \Pi_h \mathbf{u}), w) = 0, \qquad \forall w \in W_h \tag{30}$$

$$\|\mathbf{u} - \Pi_h \mathbf{u}\|_{(L^2(\Omega))^d} \leq C \|\mathbf{u}\|_{(H^j(\Omega))^d} h^j \qquad 1 \leq j \leq k+1 \tag{31}$$

$$\nabla \cdot \Pi_h = P_h \nabla \cdot \tag{32}$$

where, k is the order of the RT spaces.

Furthermore, we have [6],

$$\|P_h p - p\|_{L^2(\Omega)} \leq C \|p\|_{H^j(\Omega)} h^j \qquad 0 \leq j \leq k+1 \tag{33}$$

$$(\nabla \mathbf{v}, P_h p - p) = 0, \qquad \forall \mathbf{v} \in V_h \tag{34}$$

Let $\widehat{P}_h$ be the L^2-projection of $H^s(\mathcal{T}_h)$ to $\mathcal{D}_r(\mathcal{T}_h)$ defined by,

$$\left(\widehat{P}_h c - c, \psi\right) = 0, \qquad \forall \psi \in \mathcal{D}_r(\mathcal{T}_h) \tag{35}$$

We have,

$$\|\widehat{P}_h c - c\|_0 \leq Ch^j \|c\|_j \qquad 0 \leq j \leq r+1 \tag{36}$$

Define the interpolation errors for velocity, pressure and concentration as

$$E_{\mathbf{u}}^I = \Pi_h \mathbf{u} - \mathbf{u} \tag{37}$$
$$E_p^I = P_h p - p \tag{38}$$
$$E_c^I = \widehat{P}_h c - c \tag{39}$$

Define the finite element solution error for velocity, pressure and concentration as

$$E_{\mathbf{u}} = \mathbf{u} - \mathbf{u}_h \tag{40}$$
$$E_p = p - p_h \tag{41}$$
$$E_c = c - c_h \tag{42}$$

Define the auxiliary error for velocity, pressure and concentration as

$$E_{\mathbf{u}}^A = E_{\mathbf{u}}^I + E_{\mathbf{u}} = \Pi_h \mathbf{u} - \mathbf{u}_h \tag{43}$$
$$E_p^A = E_p^I + E_p = P_h p - p_h \tag{44}$$
$$E_c^A = E_c^I + E_c = \widehat{P}_h c - c_h \tag{45}$$

4.2 *a priori* error estimate for flow subproblem

In this section, we derive the error estimate for the MFE approximation of flow subproblem. We assume that the error of concentration is given.

Theorem 4.1. (Error estimate for flow) *Assume that the equations (1), (2) and (3) with boundary conditions (6) through (8) and initial condition (9) has a solution. Also assume the permeability tensor K is invertible and is uniformly positive definite and uniformly bounded above; $\mu(c)$ is uniformly Lipschitz continuous with respect to c and $\mu(c)$ is uniformly bounded below and uniformly bounded above. Let k be the order of Raviart-Thomas space (RT_k) as defined in above notation subsection.*

Then, there exists a constant $C > 0$ independent of finite element size h and independent of exact solution $(\mathbf{p}, \mathbf{u}, c)$ such that the following inequality hold for any $t \in (0, T_f]$,

$$\|E_{\mathbf{u}}\|_{(L^2(\Omega))^d} \leq C \|\mathbf{u}\|_{(L^\infty(\Omega))^d} \|E_c\|_{L^2(\Omega)} + Ch^j \|\mathbf{u}\|_{(H^j(\Omega))^d} \tag{46}$$

where, $1 \leq j \leq k+1$.

- *Moreover, we have, for any* $t \in (0, T_f]$,

$$\|\nabla \cdot E_{\mathbf{u}}\|_{L^2(\Omega)} \leq Ch^j \|\nabla \cdot \mathbf{u}\|_{H^j(\Omega)} \tag{47}$$

where, $1 \leq j \leq k+1$.

Proof. It is clear that if $(\mathbf{p}, \mathbf{u}, c)$ is solution of the equations (1), (2) and (3) with boundary conditions (6) through (8) and initial condition (9), then it satisfies the following formulation for any $t \in J$.

$$\left(\mu(c) K^{-1}\mathbf{u}, \mathbf{v}\right) - (\nabla \cdot \mathbf{v}, p) = -\left(p_B, \mathbf{v} \cdot \nu\right)_{\Gamma_D} \qquad \forall \mathbf{v} \in V_k^0(\mathcal{T}_h)$$
$$(\nabla \cdot \mathbf{u}, w) = (q, w) \qquad \forall w \in W_k(\mathcal{T}_h)$$

Subtracting above equations by equations (24) and (25), respectively, we have,

$$\left(\mu(c) K^{-1}\mathbf{u} - \mu(c_h) K^{-1}\mathbf{u}_h, \mathbf{v}\right) - (\nabla \cdot \mathbf{v}, p - p_h) = 0 \quad \forall \mathbf{v} \in V_k^0(\mathcal{T}_h)$$
$$(\nabla \cdot (\mathbf{u} - \mathbf{u}_h), w) = 0 \quad \forall w \in W_k(\mathcal{T}_h)$$

or

$$\left(\mu(c_h) K^{-1}(\mathbf{u} - \mathbf{u}_h), \mathbf{v}\right) - (\nabla \cdot \mathbf{v}, P_h p - p_h) = \left((\mu(c_h) - \mu(c)) K^{-1}\mathbf{u}, \mathbf{v}\right) \qquad \forall \mathbf{v} \in V_k^0(\mathcal{T}_h)$$

$$(\nabla \cdot (\Pi_h \mathbf{u} - \mathbf{u}_h), w) = 0 \qquad \forall w \in W_k(\mathcal{T}_h)$$

Now, let $\mathbf{v} = E_{\mathbf{u}}^A = \Pi_h \mathbf{u} - \mathbf{u}_h$ and $w = E_p^A = P_h p - p_h$, and add above two equations, we have

$$\left(\mu(c_h) K^{-1} E_{\mathbf{u}}^A, E_{\mathbf{u}}^A\right) = \left((\mu(c_h) - \mu(c)) K^{-1}\mathbf{u}, E_{\mathbf{u}}^A\right) + \left(\mu(c_h) K^{-1} E_{\mathbf{u}}^I, E_{\mathbf{u}}^A\right) \tag{48}$$

Let us bound the left hand side of equation (48) from below:

$$\left(\mu(c_h) K^{-1} E_{\mathbf{u}}^A, E_{\mathbf{u}}^A\right) \geq \frac{\mu_*}{k^*} \left\|E_{\mathbf{u}}^A\right\|_{(L^2(\Omega))^d}^2$$

where, we have used the fact that uniform positive definiteness and upper boundedness of K implies the uniform definiteness and upper boundedness of K^{-1}

provided K^{-1} exists everywhere. $\mu_* > 0$ is the positive constant such that $\mu(c) \geq \mu_*$ for all c; $1/k^*$ is the constant for uniform positive definiteness of K^{-1}.

Let us bound the right hand side of equation (48) from above:

$$\begin{aligned} ((\mu(c_h) - \mu(c))K^{-1}\mathbf{u}, E_{\mathbf{u}}^A) &\leq \frac{1}{k_*} \|\mathbf{u}\|_{(L^\infty(\Omega))^d} \|\mu(c_h) - \mu(c)\|_{L^2(\Omega)} \|E_{\mathbf{u}}^A\|_{L^2(\Omega)} \\ &\leq \frac{r_\rho}{k_*} \|\mathbf{u}\|_{(L^\infty(\Omega))^d} \|c_h - c\|_{L^2(\Omega)} \|E_{\mathbf{u}}^A\|_{L^2(\Omega)} \\ (\mu(c_h) K^{-1} E_{\mathbf{u}}^I, E_{\mathbf{u}}^A) &\leq \frac{\mu^*}{k_*} \|E_{\mathbf{u}}^I\|_{(L^2(\Omega))^d} \|E_{\mathbf{u}}^A\|_{(L^2(\Omega))^d} \end{aligned}$$

where, $r_\rho \geq 0$ is the Lipschitz constant for $\mu(c)$, i.e. $|\mu(c_1) - \mu(c_2)| \leq r_\rho |c_1 - c_2|$; μ^* is the upper boundedness constant for $\mu(c)$; $1/k_*$ is the upper boundedness constant for K^{-1}.

Combining above bounds for the left-hand side and right-hand side, we obtain

$$\|E_{\mathbf{u}}^A\|_{(L^2(\Omega))^d} \leq \frac{k^* r_\rho}{k_* \mu_*} \|\mathbf{u}\|_{(L^\infty(\Omega))^d} \|E_c\|_{L^2(\Omega)} + \frac{k^* \mu^*}{k_* \mu_*} \|E_{\mathbf{u}}^I\|_{(L^2(\Omega))^d}$$

Using approximation properties, we have,

$$\|E_{\mathbf{u}}^A\|_{(L^2(\Omega))^d} \leq C \|\mathbf{u}\|_{(L^\infty(\Omega))^d} \|E_c\|_{L^2(\Omega)} + Ch^j \|\mathbf{u}\|_{(H^j(\Omega))^d}$$

The first result follows by triangle inequality.

To show the second result, we only need to notice the following equality from error equation,

$$\nabla \cdot (\Pi_h \mathbf{u} - \mathbf{u}_h) = 0$$

The second result follows by using the approximation properties. ■

4.3 *a priori* error estimate for transport subproblem

Before presenting the error estimate for the transport subproblem, let us study the property of "cut-off" operator $\mathcal{M}$ defined as

$$\mathcal{M}(\mathbf{u})(x) = \min(|\mathbf{u}(x)|, M) \frac{\mathbf{u}(x)}{|\mathbf{u}(x)|} \tag{49}$$

where, M is a fixed positive real number and $|\mathbf{u}| = |\mathbf{u}|_2 = \sqrt{\sum_{i=1}^{d} (\mathbf{u})_i^2}$.

The notation $\mathbf{u}_h^M$ used in last section is equivalent to $\mathbf{u}_h^M = \mathcal{M}(\mathbf{u}_h)$. Similarly, we denote $\mathbf{u}^M = \mathcal{M}(\mathbf{u})$. The "cut-off" operator $\mathcal{M}$ is uniformly Lipschitz continuous in the following sense.

Lemma 4.2. (Property of operator $\mathcal{M}$) *The "cut-off" operator $\mathcal{M}$ defined as in equation (49) is uniformly Lipschitz continuous,*

$$\|\mathcal{M}(\mathbf{u}) - \mathcal{M}(\mathbf{v})\|_{(L^\infty(\Omega))^d} \leq \|\mathbf{u} - \mathbf{v}\|_{(L^\infty(\Omega))^d} \tag{50}$$

Proof. We notice that for all $x \in \Omega$,

$$|\mathcal{M}(\mathbf{u}) - \mathcal{M}(\mathbf{v})|_2 (x) \leq |\mathbf{u} - \mathbf{v}|_2 (x)$$

which can be shown by separately studying the three cases for fixed x: 1) $|\mathbf{u}|_2 (x) \leq M, |\mathbf{v}|_2 (x) \leq M$; 2) $|\mathbf{u}|_2 (x) \leq M, |\mathbf{v}|_2 (x) > M$; 3) $|\mathbf{u}|_2 (x) > M, |\mathbf{v}|_2 (x) > M$.

Taking the essential superium on both sides of above equation, we get the result. ■

Thus we have,

$$|\mathbf{u}_h^M - \mathbf{u}^M| \leq |\mathbf{u}_h - \mathbf{u}|$$

and

$$\|\mathbf{u}_h^M\|_{(L^\infty(\Omega))^d} \leq M$$

If the exact solution u is bounded, i.e. $\mathbf{u} \in (L^\infty(\Omega))^d$, we can pick M large enough such that $M \geq \|\mathbf{u}\|_{(L^\infty(\Omega))^d}$, then $\mathbf{u}^M = \mathbf{u}$.

Let us state and prove three lemmas for the properties of dispersion/diffusion tensor, which are needed to derive the error estimate for transport subproblem.

Lemma 4.3. (Uniform positive definiteness of $\mathbf{D}(\mathbf{u})$) *Let $\mathbf{D}(\mathbf{u})$ defined as in equation (4), where, $d_m(x) \geq 0$, $\alpha_l(x) \geq 0$ and $\alpha_t(x) \geq 0$ are nonnegative functions of $x \in \Omega$.*

Then

$$\mathbf{D}(\mathbf{u})\nabla c \cdot \nabla c \geq (d_m + \min(\alpha_l, \alpha_t) |\mathbf{u}|) |\nabla c|^2 \tag{51}$$

In particular, if $d_m(x) \geq d_{m,*} > 0$ uniformly in the domain Ω, then $\mathbf{D}(\mathbf{u})$ is uniformly positive definite and for all $x \in \Omega$, we have,

$$\mathbf{D}(\mathbf{u})\nabla c \cdot \nabla c \geq d_{m,*} |\nabla c|^2 \tag{52}$$

Proof. Notice that

$$\begin{aligned}
&\mathbf{D}(\mathbf{u})\nabla c\cdot\nabla c\\
&= d_m\nabla c\cdot\nabla c+|\mathbf{u}|\left\{\alpha_l\mathbf{E}(\mathbf{u})+\alpha_t\left(\mathbf{I}-\mathbf{E}(\mathbf{u})\right)\right\}\nabla c\cdot\nabla c\\
&= d_m\left|\nabla c\right|^2+|\mathbf{u}|\left|\nabla c\right|^2\alpha_l\cos^2(\theta)+|\mathbf{u}|\left|\nabla c\right|^2\alpha_t\left(1-\cos^2(\theta)\right)\\
&\geq \left(d_m+\min\left(\alpha_l,\alpha_t\right)|\mathbf{u}|\right)\left|\nabla c\right|^2
\end{aligned}$$

where θ is the angle between $\mathbf{u}$ and ∇c, i.e.

$$\cos(\theta)=\frac{\mathbf{u}\cdot\nabla c}{|\mathbf{u}|\left|\nabla c\right|}$$

■

Lemma 4.4. (Uniform Lipschitz continuousness of $\mathbf{D}(\mathbf{u})$) *Let $\mathbf{D}(\mathbf{u})$ defined as in equation (4), where, $d_m(x)\geq 0$, $\alpha_l(x)\geq 0$ and $\alpha_t(x)\geq 0$ are nonnegative of domain $x\in\Omega$, and the dispersivity α_l and α_t is uniformly bounded, i.e. $\alpha_l(x)\leq\alpha_l^*$ and $\alpha_t(x)\leq\alpha_t^*$.*
Then

$$\|\mathbf{D}(\mathbf{u})-\mathbf{D}(\mathbf{v})\|_{(L^2(\Omega))^{d\times d}}\leq k_D\,\|\mathbf{u}-\mathbf{v}\|_{(L^2(\Omega))^d} \tag{53}$$

where, $k_D=(4\alpha_t^*+3\alpha_l^*)\,d^{3/2}$ is a fixed number ($d=2$ or 3 is the dimension of domain Ω).

Proof. Notice that

$$\begin{aligned}
&\left|\mathbf{D}(\mathbf{u})-\mathbf{D}(\mathbf{v})\right|_1\\
&= \sum_{i=1}^{d}\max_{j=1,\cdots,d}\left|(\mathbf{D}(\mathbf{u}))_{i,j}-(\mathbf{D}(\mathbf{u}))_{i,j}\right|\\
&= \sum_{i=1}^{d}\max_{j=1,\cdots,d}\left|\alpha_t\delta_{ij}\left(|\mathbf{u}|_2-|\mathbf{v}|_2\right)+\left(\alpha_l-\alpha_t\right)\left(\frac{u_iu_j}{|\mathbf{u}|_2}-\frac{v_iv_j}{|\mathbf{v}|_2}\right)\right|\\
&\leq d\alpha_t\left||\mathbf{u}|_2-|\mathbf{v}|_2\right|+3d\left|\alpha_l-\alpha_t\right|\left|\mathbf{u}-\mathbf{v}\right|_2\\
&\leq \left(\alpha_t+3\left|\alpha_l-\alpha_t\right|\right)d\left|\mathbf{u}-\mathbf{v}\right|_2
\end{aligned}$$

Thus,

$$\begin{aligned}
\left|\mathbf{D}(\mathbf{u})-\mathbf{D}(\mathbf{v})\right|_2 &\leq \sqrt{d}\left|\mathbf{D}(\mathbf{u})-\mathbf{D}(\mathbf{v})\right|_1\\
&\leq \left(\alpha_t+3\left|\alpha_l-\alpha_t\right|\right)d^{3/2}\left|\mathbf{u}-\mathbf{v}\right|_2
\end{aligned}$$

where, we have used the property of matrix norm: for any matrix $A \in \mathbb{R}^{m\times n}$,

$$\frac{1}{\sqrt{m}}\|A\|_1 \leq \|A\|_2 \leq \sqrt{n}\,\|A\|_1$$

The result follows by integration. ■

We need a trace estimate inequality, which is stated in the following Lemma.

Lemma 4.5. *Let $\Omega = \prod_{i=1}^{d}(0, L_i)$ ($d = 2$, or 3), $\mathcal{T}_{h>0}$ and $H^s(\mathcal{T}_h)$ ($s > 1/2$) defined as in the notation section. Then there exist a constant C (C depends only on the domain Ω) such that*

$$\sum_{e\subset\partial\Omega}\|\phi\|_{L^2(e)}^2 \leq \sum_{e\in E_h}\int_e [\phi]^2 + \varepsilon|\!|\!|\nabla\phi|\!|\!|_0^2 + \frac{C}{\varepsilon}|\!|\!|\phi|\!|\!|_0^2$$

holds for any $\varepsilon \in (0,1)$ and any $\phi \in H^s(\mathcal{T}_h)$.

Proof. Denote $\Gamma_{i,+}$ and $\Gamma_{i,-}$ be the boundary faces of domain Ω such that the unit outward normal vector coincide with the positive and negative x_i direction, respectively. That is,

$$\Gamma_{i,+} = \{x \in \partial\Omega : \nu(x) = \mathbf{e}_i\} \qquad i = 1,\cdots,d$$

$$\Gamma_{i,-} = \{x \in \partial\Omega : \nu(x) = -\mathbf{e}_i\} \qquad i = 1,\cdots,d$$

We have,

$$\overline{\partial\Omega} = \bigcup_{i=1}^{d}\left(\overline{\Gamma_{i,+}} \cup \overline{\Gamma_{i,-}}\right)$$

Similarly, denote $E_{h,i}$ the set of interior edges (faces) e with the unit normal vector ν_e being the positive or negative x_i direction. That is,

$$E_{h,i} = \{e \in E_h : \nu_e = \mathbf{e}_i \text{ or } \nu_e = -\mathbf{e}_i\} \qquad i = 1,\cdots,d$$

Obviously, $E_h = \bigcup_{i=1}^{d} E_{h,i}$.

Define a subspace of $H^s(\mathcal{T}_h)$ as

$$C^\infty(\mathcal{T}_h) = \{\phi \in L^2(\Omega) : \phi|_R \in C^\infty(R) \cap H^s(R),\ R \in \mathcal{T}_h\}$$

Pick an arbitrary $\phi \in C^\infty(\mathcal{T}_h)$, let us bound the term $\|\phi\|^2_{L^2(\Gamma_{1,+})}+\|\phi\|^2_{L^2(\Gamma_{1,-})}$. Fix a point $(0,\zeta_2,\cdots,\zeta_d)\in\Gamma_{1,-}$ such that

$$(0,\zeta_2,\cdots,\zeta_d)\notin\bigcup_{e\in E_h}\bar{e}$$

We know $(L_1,\zeta_2,\cdots,\zeta_d)\in\Gamma_{1,+}$ and

$$(L_1,\zeta_2,\cdots,\zeta_d)\notin\bigcup_{e\in E_h}\bar{e}$$

thus $\phi(0,\zeta_2,\cdots,\zeta_d)$ and $\phi(L_1,\zeta_2,\cdots,\zeta_d)$ have well-defined values.

Define ϕ_0 as the average value:

$$\phi_0=\frac{1}{L_1}\int_0^{L_1}\phi\left(\zeta_1,\zeta_2,\cdots,\zeta_d\right)d\zeta_1$$

We know there exist at least one value $\chi\in(0,L_1)$ such that,

$$\phi\left(\chi_-,\zeta_2,\cdots,\zeta_d\right)\le\phi_0\le\phi\left(\chi_+,\zeta_2,\cdots,\zeta_d\right)$$

or

$$\phi\left(\chi_+,\zeta_2,\cdots,\zeta_d\right)\le\phi_0\le\phi\left(\chi_-,\zeta_2,\cdots,\zeta_d\right)$$

where, $\phi\left(\chi_-,\zeta_2,\cdots,\zeta_d\right)$ and $\phi\left(\chi_+,\zeta_2,\cdots,\zeta_d\right)$ are understood as

$$\phi\left(\chi_-,\zeta_2,\cdots,\zeta_d\right)=\lim_{\delta\to0+}\phi\left(\chi-\delta,\zeta_2,\cdots,\zeta_d\right)$$

$$\phi\left(\chi_+,\zeta_2,\cdots,\zeta_d\right)=\lim_{\delta\to0+}\phi\left(\chi+\delta,\zeta_2,\cdots,\zeta_d\right)$$

Integrating ϕ^2 along the line connecting $(0,\zeta_2,\cdots,\zeta_d)$ and $(\chi,\zeta_2,\cdots,\zeta_d)$ and the line connecting $(\chi,\zeta_2,\cdots,\zeta_d)$ and $(L_1,\zeta_2,\cdots,\zeta_d)$, we find,

$$\begin{aligned}&\phi^2(0,\zeta_2,\cdots,\zeta_d)+\phi^2(L_1,\zeta_2,\cdots,\zeta_d)\\ \le\;& 2\phi_0^2+\int_0^{L_1}\left|\frac{\partial}{\partial\zeta_1}\phi^2\left(\zeta_1,\zeta_2,\cdots,\zeta_d\right)\right|d\zeta_1\\ &+\sum_{\zeta\in(0,L_1)}\left|\phi^2\left(\zeta_+,\zeta_2,\cdots,\zeta_d\right)-\phi^2\left(\zeta_-,\zeta_2,\cdots,\zeta_d\right)\right|\end{aligned}$$

but,

$$\begin{aligned}
&\int_0^{L_1}\left|\frac{\partial}{\partial\zeta_1}\phi^2(\zeta_1,\zeta_2,\cdots,\zeta_d)\right|d\zeta_1\\
=\;& 2\int_0^{L_1}|\phi(\zeta_1,\zeta_2,\cdots,\zeta_d)\,\phi_{\zeta_1}(\zeta_1,\zeta_2,\cdots,\zeta_d)|\,d\zeta_1\\
\le\;& 2\left(\int_0^{L_1}\phi^2(\zeta_1,\zeta_2,\cdots,\zeta_d)\,d\zeta_1\int_0^{L_1}\phi^2_{\zeta_1}(\zeta_1,\zeta_2,\cdots,\zeta_d)\,d\zeta_1\right)^{1/2}\\
\le\;& \frac{1}{\varepsilon}\int_0^{L_1}\phi^2(\zeta_1,\zeta_2,\cdots,\zeta_d)\,d\zeta_1+\varepsilon\int_0^{L_1}\phi^2_{\zeta_1}(\zeta_1,\zeta_2,\cdots,\zeta_d)\,d\zeta_1
\end{aligned}$$

$$2\phi_0^2\le\frac{2}{L_1}\int_0^{L_1}\phi^2(\zeta_1,\zeta_2,\cdots,\zeta_d)\,d\zeta_1$$

Thus,

$$\begin{aligned}
&\phi^2(0,\zeta_2,\cdots,\zeta_d)+\phi^2(L_1,\zeta_2,\cdots,\zeta_d)\\
\le\;&\left(\frac{2}{L_1}+\frac{1}{\varepsilon}\right)\int_0^{L_1}\phi^2(\zeta_1,\zeta_2,\cdots,\zeta_d)\,d\zeta_1+\varepsilon\int_0^{L_1}\phi^2_{\zeta_1}(\zeta_1,\zeta_2,\cdots,\zeta_d)\,d\zeta_1\\
&+\sum_{\zeta\in(0,L_1)}|\phi^2(\zeta_+,\zeta_2,\cdots,\zeta_d)-\phi^2(\zeta_-,\zeta_2,\cdots,\zeta_d)|
\end{aligned}$$

Notice that above inequality holds for a.e. $(\zeta_2,\cdots,\zeta_d)\in\prod_{i=2}^d(0,L_i)$. Now integrating above inequality on $\int_0^{L_2}d\zeta_2\cdots\int_0^{L_d}d\zeta_d$, we have

$$\|\phi\|^2_{L^2(\Gamma_{1,-})}+\|\phi\|^2_{L^2(\Gamma_{1,+})}\le\left(\frac{2}{L_1}+\frac{1}{\varepsilon}\right)|||\phi|||_0^2+\varepsilon|||\phi_{\zeta_1}|||_0^2+\sum_{e\in E_{h,1}}\int_e[\phi]^2$$

Similarly, for $i=1,\cdots,d$, we can have

$$\|\phi\|^2_{L^2(\Gamma_{i,-})}+\|\phi\|^2_{L^2(\Gamma_{i,+})}\le\left(\frac{2}{L_i}+\frac{1}{\varepsilon}\right)|||\phi|||_0^2+\varepsilon|||\phi_{\zeta_i}|||_0^2+\sum_{e\in E_{h,i}}\int_e[\phi]^2$$

Summing above inequality for $i=1,\cdots,d$, we obtain,

$$\sum_{e\subset\partial\Omega}\|\phi\|^2_{L^2(e)}\le\sum_{e\in E_h}\int_e[\phi]^2+\varepsilon|||\nabla\phi|||_0^2+\left(\frac{d}{\varepsilon}+\sum_{i=1}^d\frac{2}{L_i}\right)|||\phi|||_0^2$$

Let

$$C = d + \sum_{i=1}^{d} \frac{2}{L_i}$$

C is fixed constant depending only on the size of domain Ω.

Notice that $\frac{d}{\varepsilon} + \sum_{i=1}^{d} \frac{2}{L_i} \leq \frac{C}{\varepsilon}$ for any $\varepsilon \in (0,1)$, we obtain that

$$\sum_{e \subset \partial\Omega} \|\phi\|_{L^2(e)}^2 \leq \sum_{e \in E_h} \int_e [\phi]^2 + \varepsilon |\!|\!| \nabla\phi |\!|\!|_0^2 + \frac{C}{\varepsilon} |\!|\!| \phi |\!|\!|_0^2$$

holds for any $\varepsilon \in (0,1)$ and any $\phi \in C^\infty(\mathcal{T}_h)$.

Using the fact that $C^\infty(\mathcal{T}_h)$ is dense in $H^s(\mathcal{T}_h)$, the lemma follows by density argument. ■

Now, we can obtain the error estimate for transport subproblem. Let us first derive the error equation without any assumptions.

It is clear that if $(\mathbf{p}, \mathbf{u}, c)$ is solution of the equations (1), (2) and (3) with boundary conditions (6) through (8) and initial condition (9), then it satisfies the following formulation for any $t \in J$.

$$\left(\phi \frac{\partial c}{\partial t}, \psi\right) + B(c, \psi; \mathbf{u}) = L(\psi; \mathbf{u}) \qquad \forall \psi \in \mathcal{D}_r(\mathcal{T}_h)$$

Denote $\widehat{c} = \widehat{P_h c}$ and $\widehat{\mathbf{u}} = \Pi_h \mathbf{u}$. Notice that $[\widehat{c} - c] = [\widehat{c}]$ on any interior edge (face) $e \in E_h$ and that $\mathbf{u}^M = \mathbf{u}$ if we picked M large enough, then above equation can be written as, $\forall \psi \in \mathcal{D}_r(\mathcal{T}_h)$,

$$\begin{aligned}
&\left(\phi \frac{\partial \widehat{c}}{\partial t}, \psi\right) + \sum_{R \in \mathcal{T}_h} \int_R \left(\mathbf{D}(\mathbf{u}_h^M)\nabla\widehat{c}\right) \cdot \nabla\psi + J_0^{\sigma,\beta}(\widehat{c}, \psi) \\
&= \sum_{R \in \mathcal{T}_h} \int_R \widehat{c}\mathbf{u}_h^M \cdot \nabla\psi + \sum_{e \in E_h} \int_e \langle \mathbf{D}(\mathbf{u}_h^M)\nabla\widehat{c} \cdot \nu_e \rangle [\psi] \\
&\quad - \sum_{e \in E_h} \int_e \langle \mathbf{D}(\mathbf{u}_h^M)\nabla\psi \cdot \nu_e \rangle [\widehat{c}] - \sum_{e \in E_h} \int_e \widehat{c}^* \mathbf{u}_h^M \cdot \nu_e [\psi] - \sum_{e \in E_{h,out}} \int_e \widehat{c}\mathbf{u}_h^M \cdot \nu_e \psi \\
&\quad + \int_\Omega \widehat{c} q^- \psi + \int_\Omega c_w q^+ \psi - \sum_{e \in E_{h,in}} \int_e c_B \mathbf{u}_h^M \cdot \nu_e \psi + \left(\phi \frac{\partial \widehat{c} - c}{\partial t}, \psi\right) \\
&\quad + \sum_{R \in \mathcal{T}_h} \int_R \left(\mathbf{D}(\mathbf{u}_h^M) - \mathbf{D}(\mathbf{u}^M)\right) \nabla\widehat{c} \cdot \nabla\psi + \sum_{R \in \mathcal{T}_h} \int_R \mathbf{D}(\mathbf{u}^M)\nabla(\widehat{c} - c) \cdot \nabla\psi
\end{aligned}$$

$$+ J_0^{\sigma,\beta}(\widehat{c}-c,\psi) - \sum_{R\in\mathcal{T}_h}\int_R \widehat{c}\left(\mathbf{u}_h^M - \mathbf{u}^M\right)\cdot\nabla\psi$$

$$- \sum_{R\in\mathcal{T}_h}\int_R (\widehat{c}-c)\,\mathbf{u}^M\cdot\nabla\psi - \sum_{e\in E_h}\int_e \left\langle\left(\mathbf{D}(\mathbf{u}_h^M)-\mathbf{D}(\mathbf{u}^M)\right)\nabla\widehat{c}\cdot\nu_e\right\rangle[\psi]$$

$$- \sum_{e\in E_h}\int_e \left\langle\mathbf{D}(\mathbf{u}^M)\nabla(\widehat{c}-c)\cdot\nu_e\right\rangle[\psi] + \sum_{e\in E_h}\int_e \left\langle\mathbf{D}(\mathbf{u}_h^M)\nabla\psi\cdot\nu_e\right\rangle[\widehat{c}-c]$$

$$+ \sum_{e\in E_h}\int_e \widehat{c}^*\left(\mathbf{u}_h^M-\mathbf{u}^M\right)\cdot\nu_e[\psi] + \sum_{e\in E_h}\int_e (\widehat{c}-c)^*\,\mathbf{u}^M\cdot\nu_e[\psi]$$

$$+ \sum_{e\in E_{h,out}}\int_e \widehat{\mathbf{c}}\left(\mathbf{u}_h^M-\mathbf{u}^M\right)\cdot\nu_e\psi + \sum_{e\in E_{h,out}}\int_e (\widehat{c}-c)\,\mathbf{u}^M\cdot\nu_e\psi$$

$$- \int_\Omega (\widehat{c}-c)\,q^-\psi + \sum_{e\in E_{h,in}}\int_e c_B\left(\mathbf{u}_h^M-\mathbf{u}^M\right)\cdot\nu_e\psi$$

Subtracting above equation by equation (26) and set $\psi = E_c^A$, we have,

$$\left(\phi\frac{\partial E_c^A}{\partial t}, E_c^A\right) + \sum_{R\in\mathcal{T}_h}\int_R \left(\mathbf{D}(\mathbf{u}_h^M)\nabla E_c^A\right)\cdot\nabla E_c^A + J_0^{\sigma,\beta}\left(E_c^A, E_c^A\right) \quad (54)$$

$$= \sum_{R\in\mathcal{T}_h}\int_R E_c^A\mathbf{u}_h^M\cdot\nabla E_c^A - \sum_{e\in E_h}\int_e \left(E_c^A\right)^*\mathbf{u}_h^M\cdot\nu_e\left[E_c^A\right]$$

$$- \sum_{e\in E_{h,out}}\int_e E_c^A\mathbf{u}_h^M\cdot\nu_e E_c^A + \int_\Omega E_c^A q^- E_c^A + \sum_{i=1}^{15} T_i$$

where,

$$\begin{aligned}
T_1 &= \left(\phi\frac{\partial\widehat{c}-c}{\partial t}, E_c^A\right)\\
T_2 &= \sum_{R\in\mathcal{T}_h}\int_R \left(\mathbf{D}(\mathbf{u}_h^M)-\mathbf{D}(\mathbf{u}^M)\right)\nabla\widehat{c}\cdot\nabla E_c^A\\
T_3 &= \sum_{R\in\mathcal{T}_h}\int_R \mathbf{D}(\mathbf{u}^M)\nabla(\widehat{c}-c)\cdot\nabla E_c^A\\
T_4 &= J_0^{\sigma,\beta}\left(\widehat{c}-c, E_c^A\right)\\
T_5 &= -\sum_{R\in\mathcal{T}_h}\int_R \widehat{c}\left(\mathbf{u}_h^M-\mathbf{u}^M\right)\cdot\nabla E_c^A
\end{aligned}$$

$$T_6 = -\sum_{R\in\mathcal{T}_h}\int_R (\widehat{c}-c)\,\mathbf{u}^M\cdot\nabla E_c^A$$

$$T_7 = -\sum_{e\in E_h}\int_e \left\langle \left(\mathbf{D}(\mathbf{u}_h^M)-\mathbf{D}(\mathbf{u}^M)\right)\nabla\widehat{c}\cdot\nu_e\right\rangle \left[E_c^A\right]$$

$$T_8 = -\sum_{e\in E_h}\int_e \left\langle \mathbf{D}(\mathbf{u}^M)\nabla(\widehat{c}-c)\cdot\nu_e\right\rangle \left[E_c^A\right]$$

$$T_9 = \sum_{e\in E_h}\int_e \left\langle \mathbf{D}(\mathbf{u}_h^M)\nabla E_c^A\cdot\nu_e\right\rangle [\widehat{c}-c]$$

$$T_{10} = \sum_{e\in E_h}\int_e \widehat{c}^*\left(\mathbf{u}_h^M-\mathbf{u}^M\right)\cdot\nu_e\left[E_c^A\right]$$

$$T_{11} = \sum_{e\in E_h}\int_e (\widehat{c}-c)^*\,\mathbf{u}^M\cdot\nu_e\left[E_c^A\right]$$

$$T_{12} = \sum_{e\in E_{h,out}}\int_e \widehat{\mathrm{c}}\left(\mathbf{u}_h^M-\mathbf{u}^M\right)\cdot\nu_e E_c^A$$

$$T_{13} = \sum_{e\in E_{h,out}}\int_e (\widehat{c}-c)\,\mathbf{u}^M\cdot\nu_e E_c^A$$

$$T_{14} = -\int_\Omega (\widehat{c}-c)\,q^- E_c^A$$

$$T_{15} = \sum_{e\in E_{h,in}}\int_e c_B\left(\mathbf{u}_h^M-\mathbf{u}^M\right)\cdot\nu_e E_c^A$$

The above error equation is difficult to analyze in general boundary condition and we thus assume that only Neumann boundary condition for flow subproblem is used. In this way, we can know the normal velocity on the boundary by exact value. The boundary condition for transport problem can still be either inflow or outflow/noflow.

Theorem 4.6. (Error estimate for transport) *Assume that the equations (1), (2) and (3) with boundary conditions (6) through (8) and initial condition (9) has a solution. Assume that only Neumann boundary condition is imposed for flow subproblem; the exact solution* $\mathbf{u}$ *and* c *are smooth enough:*

$$\mathbf{u}\in C_B\left(\overline{\Omega}\times(0,T_f]\right)\bigcap L^\infty\left((0,T_f],H^{l+\frac{1}{2}}(\Omega)\right) \tag{55}$$

$$c\in C_B\left((0,T_f],W^{1,\infty}(\Omega)\right)\bigcap L^2\left((0,T_f],H^m(\Omega)\right) \tag{56}$$

and

$$\frac{\partial c}{\partial t} \in L^2\left((0,T_f], H^n(\Omega)\right) \tag{57}$$

We also assume that porosity ϕ is time-dependent and is uniformly bounded below and above; the parameter β in interior penalty term for DG formulation is assume to be $\beta = 1$; the extraction part of source term satisfies $q^- \in \left(\Pi_{R\in\mathcal{T}_h} W^{s,1}(R)\right)'$ for all of the partitions $\mathcal{T}_h$ used, where $0 \leq s < 1$ (see remark 4.7 for the explanation of this assumption). Assume M is picked large enough such that $M \geq \|\mathbf{u}\|_{(L^\infty(\Omega))^d}$ for all $t \in (0,T_f]$.

Then, there exist a constant $C > 0$ independent of finite element size h and a constant $h_0 > 0$ such that the following inequality hold for any $\tau \in (0,T_f]$ and for any $h \leq h_0$,

$$\begin{aligned} &|\!|\!| E_c |\!|\!|_0^2(\tau) + \int_0^\tau |\!|\!| \nabla E_c |\!|\!|_0^2(t)dt \\ \leq\ & C\int_0^\tau |\!|\!| E_c |\!|\!|_0^2 + C\int_0^\tau \|E_{\mathbf{u}}\|^2_{(L^2(\Omega))^d} + Ch^{\min(2k+2,2l,2r,2m-2,2n)} \end{aligned} \tag{58}$$

where, $k \geq 0$, $r \geq 1$ are the order of Raviart-Thomas space and discontinuous space, respectively, defined in above notation subsection; l, m, n describe the regularity order of solution $\mathbf{u}, c, \partial c/\partial t$ respectively, as defined in equations (55), (56) and (57).

Proof. Let us first relax the assumption $\beta = 1$ for a while so that we can also show that $\beta = 1$ is indeed the optimal choice for the value of parameter β. Using the lemma (4.3), let us bound the left hand side of the error equation (54) from below,

$$\begin{aligned} &\left(\phi\frac{\partial E_c^A}{\partial t}, E_c^A\right) + \sum_{R\in\mathcal{T}_h}\int_R \left(\mathbf{D}(\mathbf{u}_h^M)\nabla E_c^A\right)\cdot\nabla E_c^A + J_0^{\sigma,\beta}\left(E_c^A, E_c^A\right) \\ \geq\ & \frac{1}{2}\frac{\partial}{\partial t}|\!|\!|\sqrt{\phi}E_c^A|\!|\!|_0^2 + d_{m,*}|\!|\!|\nabla E_c^A|\!|\!|_0^2 + J_0^{\sigma,\beta}\left(E_c^A, E_c^A\right) \end{aligned}$$

where, we have used the uniform positive definiteness of the dispersion/diffusion tensor from Lemma (4.3).

Let us bound from above the right hand side of the error equation (54).

The first term is straightforward.

$$
\begin{aligned}
\sum_{R\in\mathcal{T}_h}\int_R E_c^A \mathbf{u}_h^M \cdot \nabla E_c^A &\leq M\sum_{R\in\mathcal{T}_h}\int_R |E_c^A|\,|\nabla E_c^A| \\
&\leq M\sum_{R\in\mathcal{T}_h} \|E_c^A\|_{L^2(R)}\,\|\nabla E_c^A\|_{(L^2(R))^d} \\
&\leq \frac{C}{\varepsilon}\sum_{R\in\mathcal{T}_h}\|E_c^A\|^2_{L^2(R)} + \varepsilon\sum_{R\in\mathcal{T}_h}\|\nabla E_c^A\|^2_{(L^2(R))^d} \\
&= \frac{C}{\varepsilon}\|\!|E_c^A|\!\|_0^2 + \varepsilon\|\!|\nabla E_c^A|\!\|_0^2
\end{aligned}
$$

where, ε is a small positive constant.

The second term is a little tricky.

$$
\begin{aligned}
&-\sum_{e\in E_h}\int_e \left(E_c^A\right)^* \mathbf{u}_h^M \cdot \nu_e \left[E_c^A\right] \\
&\qquad\leq M\left|\sum_{e\in E_h}\int_e \left(E_c^A\right)^*\left[E_c^A\right]\right| \\
&\qquad\leq M\sum_{e\in E_h}\left\|\left(E_c^A\right)^*\right\|^2_{L^2(e)}\left\|\left[E_c^A\right]\right\|^2_{L^2(e)} \\
&\qquad\leq M\sum_{e\in E_h}\left(\frac{\varepsilon}{M}\frac{\sigma_e}{h_e^\beta}\left\|\left[E_c^A\right]\right\|^2_{L^2(e)} + \frac{Ch_e^\beta}{\varepsilon}\left\|\left(E_c^A\right)^*\right\|^2_{L^2(e)}\right) \\
&\qquad\leq \varepsilon J_0^{\sigma,\beta}\left(E_c^A, E_c^A\right) + \frac{Ch^\beta}{\varepsilon}\sum_{R\in\mathcal{T}_h} h^{-1}\left\|E_c^A\right\|^2_{L^2(R)} \\
&\qquad\leq \varepsilon J_0^{\sigma,\beta}\left(E_c^A, E_c^A\right) + \frac{C}{\varepsilon}\|\!|E_c^A|\!\|_0^2
\end{aligned}
$$

where, we have used the assumption $\beta \geq 1$.

The third term can be bounded by using Lemma (4.5).

$$
\begin{aligned}
-\sum_{e\in E_{h,out}}\int_e E_c^A \mathbf{u}_h^M \cdot \nu_e E_c^A &\leq M\sum_{e\in E_{h,out}}\left\|E_c^A\right\|^2_{L^2(e)} \\
&\leq \frac{Mh^\beta}{\sigma_*}J_0^{\sigma,\beta}\left(E_c^A, E_c^A\right) + \varepsilon\|\!|\nabla E_c^A|\!\|_0^2 + \frac{C}{\varepsilon}\|\!|E_c^A|\!\|_0^2
\end{aligned}
$$

The fourth term comes from the extraction wells, and it can be bounded as follows.

$$\begin{aligned}
\int_\Omega E_c^A q^- E_c^A &\leq \|q^-\|_{(\Pi_{R\in\mathcal{T}_h} W^{s,1}(R))'} \sum_{R\in\mathcal{T}_h} \left\|(E_c^A)^2\right\|_{W^{s,1}(R)} \\
&\leq C \sum_{R\in\mathcal{T}_h} \|E_c^A\|^2_{H^s(R)} \leq C \sum_{R\in\mathcal{T}_h} \|E_c^A\|^{2(1-s)}_{L^2(R)} \|E_c^A\|^{2s}_{H^1(R)} \\
&\leq \sum_{R\in\mathcal{T}_h} \left(\frac{C}{\varepsilon^{s/(1-s)}} \|E_c^A\|^2_{L^2(R)} + \varepsilon \|E_c^A\|^2_{H^1(R)} \right) \\
&\leq \frac{C}{\varepsilon^{s/(1-s)}} \|E_c^A\|_0^2 + \varepsilon \|\nabla E_c^A\|_0^2
\end{aligned}$$

where, we have used the fact,

$$a^{1-s} b^s \leq \left(\frac{a}{\varepsilon^{\frac{s}{1-s}}} + b\varepsilon \right) \qquad a > 0 \quad b > 0 \quad 0 \leq s < 1$$

Now let us bound the terms T_1 through T_{15}.

$$\begin{aligned}
T_1 &\leq \phi^* \left\| \frac{\partial E_c^I}{\partial t} \right\|_{L^2(\Omega)} \|E_c^A\|_{L^2(\Omega)} \leq \frac{\phi^*}{2} \left\| \frac{\partial E_c^I}{\partial t} \right\|^2_{L^2(\Omega)} + \frac{\phi^*}{2} \|E_c^A\|^2_{L^2(\Omega)} \\
T_2 &\leq \|\nabla \widehat{c}\|_{(L^\infty(\Omega))^3} \sum_{R\in\mathcal{T}_h} \|\mathbf{D}(\mathbf{u}_h^M) - \mathbf{D}(\mathbf{u}^M)\|_{(L^2(R))^{d\times d}} \|\nabla E_c^A\|_{(L^2(R))^d} \\
&\leq C \|\nabla c\|_{(L^\infty(\Omega))^3} \sum_{R\in\mathcal{T}_h} \|\mathbf{u}_h^M - \mathbf{u}^M\|_{(L^2(R))^d} \|\nabla E_c^A\|_{(L^2(R))^d} \\
&\leq \sum_{R\in\mathcal{T}_h} \left(\frac{C}{\varepsilon} \|E_\mathbf{u}\|^2_{(L^2(R))^d} + \varepsilon \|\nabla E_c^A\|^2_{(L^2(R))^d} \right) \\
&\leq \frac{C}{\varepsilon} \|E_\mathbf{u}\|^2_{(L^2(\Omega))^d} + \varepsilon \|\nabla E_c^A\|_0^2
\end{aligned}$$

where, we have used the facts of $\|\nabla \widehat{c}\|_{(L^\infty(\Omega))^3} \leq C \|\nabla c\|_{(L^\infty(\Omega))^3} \leq C$ and Lemmas (4.2) and (4.4) for bounding the term T_2.

$$\begin{aligned}
T_3 &\le C\sum_{R\in\mathcal{T}_h}\int_R \nabla E_c^I\cdot\nabla E_c^A \\
&\le \sum_{R\in\mathcal{T}_h}\left(\frac{C}{\varepsilon}\left\|\nabla E_c^I\right\|^2_{(L^2(R))^d}+\varepsilon\left\|\nabla E_c^A\right\|^2_{(L^2(R))^d}\right)\\
&\le \frac{C}{\varepsilon}|||\nabla E_c^I|||_0^2+\varepsilon|||\nabla E_c^A|||_0^2\\
T_4 &= \sum_{e\in E_h}\frac{\sigma_e}{h_e^\beta}\int_e\left[E_c^I\right]\left[E_c^A\right]\\
&\le \sum_{e\in E_h}\frac{\sigma_e}{h_e^\beta}\left(\varepsilon\left\|\left[E_c^A\right]\right\|^2_{L^2(e)}+\frac{C}{\varepsilon}\left\|\left[E_c^I\right]\right\|^2_{L^2(e)}\right)\\
&\le \varepsilon J_0^{\sigma,\beta}\left(E_c^A,E_c^A\right)+\frac{C}{\varepsilon h^\beta}\sum_{e\in E_h}\left\|\left[E_c^I\right]\right\|^2_{L^2(e)}\\
&\le \varepsilon J_0^{\sigma,\beta}\left(E_c^A,E_c^A\right)+\frac{C}{\varepsilon h^\beta}\sum_{R\in\mathcal{T}_h}\left(h^{-1}\left\|E_c^I\right\|^2_{L^2(R)}+h\left\|\nabla E_c^I\right\|^2_{(L^2(R))^d}\right)\\
&\le \varepsilon J_0^{\sigma,\beta}\left(E_c^A,E_c^A\right)+\frac{C}{\varepsilon h^{\beta+1}}|||E_c^I|||_0^2+\frac{C}{\varepsilon h^{\beta-1}}|||\nabla E_c^I|||_0^2
\end{aligned}$$

T_5 can be bounded similarly as the term T_3,

$$\begin{aligned}
T_5 &\le \|\widehat{c}\|_{L^\infty(\Omega)}\sum_{R\in\mathcal{T}_h}\left\|\mathbf{u}_h^M-\mathbf{u}^M\right\|_{(L^2(R))^d}\left\|\nabla E_c^A\right\|_{(L^2(R))^d}\\
&\le \frac{C}{\varepsilon}\left\|E_{\mathbf{u}}\right\|^2_{(L^2(\Omega))^d}+\varepsilon|||\nabla E_c^A|||_0^2
\end{aligned}$$

where, we have used that fact $\|\widehat{c}\|_{L^\infty(\Omega)}\le C\,\|c\|_{L^\infty(\Omega)}\le C$.

$$\begin{aligned}
T_6 &\le M\sum_{R\in\mathcal{T}_h}\left\|E_c^I\right\|^2_{L^2(R)}\left\|\nabla E_c^A\right\|^2_{(L^2(R))^d}\\
&\le \sum_{R\in\mathcal{T}_h}\left(\frac{C}{\varepsilon}\left\|E_c^I\right\|^2_{L^2(R)}+\varepsilon\left\|\nabla E_c^A\right\|^2_{(L^2(R))^d}\right)\\
&\le \frac{C}{\varepsilon}|||E_c^I|||_0^2+\varepsilon|||\nabla E_c^A|||_0^2
\end{aligned}$$

The term T_7 is quite tricky,

$$
\begin{aligned}
T_7 &= \|\nabla\widehat{c}\|_{(L^\infty(\Omega))^d} \sum_{e\in E_h} \left\|\mathbf{D}(\mathbf{u}_h^M) - \mathbf{D}(\mathbf{u}^M)\right\|_{(L^2(e))^{d\times d}} \left\|\left[E_c^A\right]\right\|_{L^2(e)} \\
&\le \sum_{e\in E_h} \left(\varepsilon\frac{\sigma_e}{h_e^\beta} \left\|\left[E_c^A\right]\right\|_{L^2(e)}^2 + \frac{Ch^\beta}{\varepsilon} \left\|\mathbf{D}(\mathbf{u}_h^M) - \mathbf{D}(\mathbf{u}^M)\right\|_{(L^2(e))^{d\times d}}^2\right) \\
&\le \varepsilon J_0^{\sigma,\beta}\left(E_c^A, E_c^A\right) + \frac{Ch^\beta}{\varepsilon} \sum_{e\in E_h} \|\mathbf{u}_h - \mathbf{u}\|_{(L^2(e))^d}^2 \\
&\le \varepsilon J_0^{\sigma,\beta}\left(E_c^A, E_c^A\right) + \frac{Ch^\beta}{\varepsilon} \sum_{e\in E_h} \|\mathbf{u}_h - \widehat{\mathbf{u}}\|_{(L^2(e))^d}^2 \\
&\quad + \frac{Ch^\beta}{\varepsilon} \sum_{e\in E_h} \|\widehat{\mathbf{u}} - \mathbf{u}_h\|_{(L^2(e))^d}^2 \\
&\le \varepsilon J_0^{\sigma,\beta}\left(E_c^A, E_c^A\right) + \frac{Ch^{\beta-1}}{\varepsilon} \sum_{R\in\mathcal{T}_h} \|\mathbf{u}_h - \widehat{\mathbf{u}}\|_{(L^2(R))^d}^2 \\
&\quad + \frac{C}{\varepsilon} h^{\min(2k+\beta+1,2l+\beta-1)} \\
&\le \varepsilon J_0^{\sigma,\beta}\left(E_c^A, E_c^A\right) + \frac{Ch^{\beta-1}}{\varepsilon} \sum_{R\in\mathcal{T}_h} \left(\|E_{\mathbf{u}}\|_{(L^2(R))^d}^2 + \left\|E_{\mathbf{u}}^I\right\|_{(L^2(R))^d}^2\right) \\
&\quad + \frac{C}{\varepsilon} h^{\min(2k+\beta+1,2l+\beta-1)} \\
&\le \varepsilon J_0^{\sigma,\beta}\left(E_c^A, E_c^A\right) + \frac{Ch^{\beta-1}}{\varepsilon} \|E_{\mathbf{u}}\|_{(L^2(\Omega))^d}^2 + \frac{C}{\varepsilon} h^{\min(2k+\beta+1,2l+\beta-1)}
\end{aligned}
$$

where, we have used the facts of $\|\nabla\widehat{c}\|_{(L^\infty(\Omega))^d} \le C\,\|\nabla c\|_{(L^\infty(\Omega))^d} \le C$ and the approximation properties for $\mathbf{u}$.

The boundedness of T_8 can be shown by using penalty term,

$$
\begin{aligned}
T_8 &\le C\sum_{e\in E_h} \|\nabla\left(\widehat{c} - c\right)\|_{(L^2(e))^d} \left\|\left[E_c^A\right]\right\|_{L^2(e)} \\
&\le \varepsilon J_0^{\sigma,\beta}\left(E_c^A, E_c^A\right) + \frac{C}{\varepsilon} h^{\beta-1} \|\nabla E_c^I\|_0^2
\end{aligned}
$$

$$
\begin{aligned}
T_9 &\le C\sum_{e\in E_h} \left\|\nabla E_c^A\right\|_{(L^2(e))^d} \|[\widehat{c} - c]\|_{L^2(e)} \\
&\le \varepsilon\|\nabla E_c^A\|_0^2 + \frac{C}{\varepsilon h^2}\|E_c^I\|_0^2 + \frac{C}{\varepsilon}\|\nabla E_c^I\|_0^2
\end{aligned}
$$

The term T_{10} is similar as T_7

$$
\begin{aligned}
T_{10} &\le \|\widehat{c}\|_{L^\infty(\Omega)} \sum_{e\in E_h} \|\mathbf{u}_h - \mathbf{u}\|^2_{(L^2(e))^d} \left\| \left[E_c^A\right] \right\|_{L^2(e)} \\
&\le \varepsilon J_0^{\sigma,\beta}\left(E_c^A, E_c^A\right) + \frac{Ch^{\beta-1}}{\varepsilon} \|E_{\mathbf{u}}\|^2_{(L^2(\Omega))^d} + \frac{C}{\varepsilon} h^{\min(2k+\beta+1, 2l+\beta-1)} \\
T_{11} &\le C \sum_{e\in E_h} \left\| \left(E_c^I\right)^* \right\|_{L^2(e)} \left\| \left[E_c^A\right] \right\|_{L^2(e)} \\
&\le \varepsilon J_0^{\sigma,\beta}\left(E_c^A, E_c^A\right) + \frac{Ch^{\beta-1}}{\varepsilon} \|E_c^I\|_0^2 + \frac{Ch^{\beta+1}}{\varepsilon} \|\nabla E_c^I\|_0^2
\end{aligned}
$$

Since only Neumann boundary conditions for flow subproblem are assumed, the terms T_{12} and T_{15} vanish if we use the exact value of $\mathbf{u}$ on the boundary in the DG formulation.

$$
\begin{aligned}
T_{13} &\le C \sum_{e\in E_{h,out}} \int_e (\widehat{c} - c)\, E_c^A \le C \sum_{e\in E_{h,out}} \left\|E_c^I\right\|_{L^2(e)} \left\|E_c^A\right\|_{L^2(e)} \\
&\le \sum_{e\in E_{h,out}} \left(\frac{C}{h} \left\|E_c^I\right\|^2_{L^2(e)} + h \left\|E_c^A\right\|^2_{L^2(e)} \right) \\
&\le \sum_{R\in \mathcal{T}_h} \frac{C}{h^2} \left(\left\|E_c^I\right\|^2_{L^2(R)} + h^2 \left\|\nabla E_c^I\right\|^2_{L^2(R)} \right) + C \sum_{R\in \mathcal{T}_h} \left\|E_c^A\right\|^2_{L^2(R)} \\
&\le \frac{C}{h^2} \|E_c^I\|_0^2 + C\|\nabla E_c^I\|_0^2 + C\|E_c^A\|_0^2
\end{aligned}
$$

The term T_{14} can be bounded similarly as the term $\int_\Omega E_c^A q^- E_c^A$ above,

$$
\begin{aligned}
T_{14} &\le \left| \int_\Omega (\widehat{c} - c)\, q^- E_c^A \right| \\
&\le \frac{C}{\varepsilon^{s/(1-s)}} \|E_c^A\|_0^2 + \varepsilon \|\nabla E_c^A\|_0^2 + C\|E_c^I\|_0^2 + C\|\nabla E_c^I\|_0^2
\end{aligned}
$$

Combining the above bounds for both hand sides of the error equation (54), choosing ε small enough, we have, for $\forall h \le \left(\frac{\sigma_*}{3M}\right)^{1/\beta}$,

$$
\begin{aligned}
&\frac{1}{2}\frac{\partial}{\partial t}\left\|\sqrt{\phi}E_c^A\right\|^2 + \frac{d_{m,*}}{2}\|\nabla E_c^A\|_0^2 + \frac{1}{2}J_0^{\sigma,\beta}\left(E_c^A, E_c^A\right)\\
\leq\;& C\|E_c^A\|_0^2 + \left(C + Ch^{\beta-1}\right)\|E_{\mathbf{u}}\|^2_{(L^2(\Omega))^d}\\
&+C\|\partial E_c^I/\partial t\|_0^2 + \left(C + \frac{C}{h^{\beta+1}} + \frac{C}{h^2} + Ch^{\beta-1}\right)\|E_c^I\|_0^2\\
&+\left(C + \frac{C}{h^{\beta-1}} + Ch^{\beta-1} + Ch^{\beta+1}\right)\|\nabla E_c^I\|_0^2 + Ch^{\min(2k+\beta+1, 2l+\beta-1)}
\end{aligned}
$$

We can find the best choice of β is indeed $\beta = 1$, then,

$$
\begin{aligned}
&\frac{1}{2}\frac{\partial}{\partial t}\|\sqrt{\phi}E_c^A\|_0^2 + \frac{d_{m,*}}{2}\|\nabla E_c^A\|_0^2 + \frac{1}{2}J_0^{\sigma,\beta}\left(E_c^A, E_c^A\right)\\
\leq\;& C\|E_c^A\|_0^2 + C\|E_{\mathbf{u}}\|^2_{(L^2(\Omega))^d}\\
&+C\|\partial E_c^I/\partial t\|_0^2 + \frac{C}{h^2}\|E_c^I\|_0^2 + C\|\nabla E_c^I\|_0^2 + Ch^{\min(2k+2, 2l)}
\end{aligned}
$$

Now, we integrate with respect to time between 0 to τ ($0 \leq \tau \leq T_f$) and we have,

$$
\begin{aligned}
&\|\sqrt{\phi}E_c^A\|_0^2(\tau) + \frac{d_{m,*}}{2}\int_0^\tau \|\nabla E_c^A\|_0^2(t)dt + \frac{1}{2}\int_0^\tau J_0^{\sigma,\beta}\left(E_c^A, E_c^A\right)\\
\leq\;& \|\sqrt{\phi}E_c^A\|_0^2(0) + C\int_0^\tau \|E_c^A\|_0^2 + C\int_0^\tau \|E_{\mathbf{u}}\|^2_{(L^2(\Omega))^d}\\
+\;& C\int_0^\tau \|\partial E_c^I/\partial t\|_0^2 + \frac{C}{h^2}\int_0^\tau \|E_c^I\|_0^2 + C\int_0^\tau \|\nabla E_c^I\|_0^2 + CT_f h^{\min(2k+2, 2l)}
\end{aligned}
$$

Notice that $\left\|\sqrt{\phi}E_c^A\right\|^2(0) = \left\|\sqrt{\phi}E_c^I\right\|^2(0)$, and that ϕ is uniformly bounded below and above, using the approximation results for $c \in \mathcal{D}_r(\mathcal{T}_h)$, we have,

$$
\begin{aligned}
&\|E_c^A\|_0^2(\tau) + \int_0^\tau \|\nabla E_c^A\|_0^2(t)dt + \int_0^\tau J_0^{\sigma,\beta}\left(E_c^A, E_c^A\right)\\
\leq\;& C\int_0^\tau \|E_c^A\|_0^2 + C\int_0^\tau \|E_{\mathbf{u}}\|^2_{(L^2(\Omega))^d} + Ch^{\min(2k+2, 2l, 2r, 2m-2, 2n)}
\end{aligned}
$$

The theorem follows by the triangle inequality. ■

Remark 4.7. *In the above theorems, we do not make assumption about the regularity of injection part of source term q^+ and injection concentration c_w,* *which can have singular value such as point source (the delta function). But we assume for the extraction part of source term* $q^- \in \left(\Pi_{R\in T_h} W^{s,1}(R)\right)'$ *for all of the partitions* $\mathcal{T}_h$ *used* $(0 \leq s < 1)$. *This means that* q^- *can be more singular than* $L^2(\Omega)$, *but cannot be too singular as the delta function. Roughly speaking, it is like* $q^- \in \left(W^{s,1}(\Omega)\right)'$ *except that the position for singularity of* q^- *cannot be on the interior and boundary edges or faces for any partition* $\mathcal{T}_h$.

Remark 4.8. *We do not have assumption on the inflow boundary concentration* c_B, *which can have singular value.*

4.4 *A priori* error estimate for the coupled system of flow and transport

We now state and prove our final result, which estimates the coupled system of flow and transport.

Theorem 4.9. (Error estimate for coupled system of flow and transport) *Let the assumption in Theorems (4.1) and (4.6) holds.*

Then, there exist a constant $C > 0$ *independent of finite element size* h *and a constant* $h_0 > 0$ *such that the following inequality hold for any* $h \leq h_0$,

$$\|E_{\mathbf{u}}\|^2_{L^\infty((0,T_f);(L^2(\Omega))^d)} \leq Ch^{\min(2k+2,2l,2r,2m-2,2n)} \tag{59}$$

$$|||E_c|||^2_{L^\infty((0,T_f);L^2(\Omega))} + |||\nabla E_c|||^2_{L^2((0,T_f);(L^2(\Omega))^d)} \leq Ch^{\min(2k+2,2l,2r,2m-2,2n)} \tag{60}$$

where, $k \geq 0$, $r \geq 1$ *are the order of Raviart-Thomas space and discontinuous space, respectively, defined in above notation subsection;* l, m, n *describe the regularity order of solution* $\mathbf{u}$, c, $\partial c/\partial t$ *respectively, as defined in equations (55), (56) and (57).*

Proof. Combining Theorems (4.1) and (4.6), we have,

$$|||E_c|||_0^2(\tau) + \int_0^\tau |||\nabla E_c|||_0^2(t)dt \leq C\int_0^\tau |||E_c|||_0^2 + Ch^{\min(2k+2,2l,2r,2m-2,2n)}$$

Using the Gronwall's inequality, we have the error result for E_c.

To get the bound for $E_{\mathbf{u}}$, we substitute the error result for E_c into the following result, which comes from Theorems (4.1).

$$\|E_{\mathbf{u}}\|^2_{(L^2(\Omega))^d} \leq C\,\|E_c\|^2_{L^2(\Omega)} + Ch^{\min(2k+2,2l+1)}$$

■

Remark 4.10. *If we let $r = k + 1$, and if the exact solution $\mathbf{u}, c$ are smooth enough, then Theorem (4.9) gives the optimal $L^2\left(H^1\right)$ rate of convergence for concentration, and also gives optimal $L^\infty\left(L^2\right)$ rate of convergence for velocity.*

Remark 4.11. *The error estimate for the couple system of flow and transport is not applied for the case of having Dirichlet boundary condition for flow. It is difficult to bound the error for the coupled system for the case of having Dirichlet boundary condition for flow, because we do not yet have sharp control on the error of velocity $\mathbf{u}_h$ in the boundary edge or face.*

5. Conclusion

In this paper, we present a combined method with mixed finite element method for flow and discontinuous Galerkin method for transport for the coupled system of miscible displacement problem. The "cut-off" operator $\mathcal{M}$ is introduced in the discontinuous Galerkin scheme in order to make the combined scheme converge. The property of "cut-off" operator $\mathcal{M}$ is given. The optimal choice of penalty parameter β in DG scheme is derived to be $\beta = 1$. The Neumann boundary condition for flow subproblem is assumed for getting the error estimate of the coupled system. Error estimates in $L^2(H^1)$ and $L^\infty(L^2)$ for concentration and error estimate in $L^\infty(L^2)$ for velocity are derived, which are the optimal $L^2\left(H^1\right)$ rate of convergence for concentration, and optimal $L^\infty\left(L^2\right)$ rate of convergence for velocity. The uniform positive definitiveness and uniform Lipschitz continuity of dispersion/diffusion tensor computed by Engineering standard formula are proved. The injection part of source term is allowed to have arbitrarily singular value and the extraction part of source term to have value singular to some degree.

References

[1] Arbogast, T., M. F. Wheeler, and I. Yotov, *Mixed finite elements for elliptic problems with tensor coefficients as cell-centered finite differences*, SIAM J. Numer. Anal., 34, pp. 828-852, 1997

[2] Bear, J., 1972. *Dynamics of Fluids in Porous Media.* Dover Publications, Inc., New York, 764pp.

[3] Darlow, B. L., *A Penalty-Galerkin method for solving the miscible displacement problem.* PhD thesis, Rice University, 1980

[4] Douglas, J., R. E. Ewing, and M. F. Wheeler. *The approximation of the pressure by a mixed method in the simulation of miscible displacement.* R.A.I.R.O. Numerical Analysis, 17(1):17-33, 1983

[5] Douglas, R. E. Ewing, and M. F. Wheeler. *A time-discretization procedure for a mixed finite element approximation of miscible displacement in porous media.* R. A. I. R. O. Numerical Analysis, 17(3):249-265, 1983

[6] Douglas, J. Jr. and J. E. Roberts. *Global Estimates for Mixed Methods for Second Order Elliptic Equations*, Math. Comp., Vol. 44, No. 169. pp. 39-52, 1985

[7] Douglas, J., M. F. Wheeler, B. L. Darlow, and R. P. Kendall. *Self-adaptive finite element simulation of miscible displacement in porous media.* Computer methods in applied mechanics and engineering, 47:131-159, 1984

[8] Dullien, F. A. L., *Porous media fluid transport and pore structure*, Academic press, Inc., New York, 1979

[9] Ewing, R. E. and M. F. Wheeler, *Galerkin methods for miscible displacement problems in porous media.* SIAM J. Numer. Anal, 17(3): 351-365, June 1980

[10] Oden J. T., Babuska I., Baumann C. E., *A discontinuous hp finite element method for diffusion problems.* J. Compu. Phys. 146 (1998) 495-519

[11] B. Rivière, *Discontinuous Galerkin methods for solving the miscible displacement problem in porous media*, PhD thesis, The University of Texas at Austin, 2000

[12] B. Rivière and M.F. Wheeler, *Discontinuous Galerkin methods for flow and transport problems in porous media.* Communications in Numerical Methods in Engineering, 2001, to appear.

[13] B. Rivière, M.F. Wheeler and V. Girault, *A priori error estimates for finite element methods based on discontinuous approximation spaces for elliptic problems*, SIAM J. Numer. Anal. vol 39 no 3, 902–931 (2001).

[14] Wheeler, M. F. and B. L. Darlow, *Interior penalty Galerkin procedure for miscible displacement problems in porous media.* Computational Methods in Nonlinear Mechanics, pages 485-506, 1980.

NUMERICAL SIMULATION AND COARSE-GRAINING OF LARGE PARTICLE SYSTEMS

Shlomo Ta'asan *

Department of Mathematical Sciences, Carnegie Mellon University, Pittsburgh PA 15213

shlomo@andrew.cmu.edu

Abstract We discuss the use of combined analysis and simulation to tackle the problem of bridging between the scales in large systems of interacting particles (agents). On the microscopic level we consider several models: Hamiltonian systems such as those that arise in atomistic models for fluids or solids; networks of grain boundaries modeled by evolution PDEs, as well as several stochastic processes. Coarse-graining may involve the passage to large scales in time, space or both. The larger scale models, whether at the mesoscopic or macroscopic levels, involve stochastic processes, stochastic differential equations, ordinary and partial differential equations. The passage between scales is done using two approaches. One involves the construction of a Markov process followed by probabilistic methods combined with simulation. The other follows continuum mechanics arguments combined with simulation. The role of simulation in both approaches is to bridge gaps in the analytical steps, suggesting conjecture about the behavior of certain quantities needed for a complete description on a particular scale.

Keywords: Multiscale simulation, coarse graining, microscopic models, macroscopic equations.

1. Introduction

In many areas of science and engineering the bridging between descriptions at small scales to those at large scales is a central problem. In most fields, macroscopic models were derived first, and only years later a detailed study into microscopic aspects of the problem, followed by microscopic models have been completed. Coarse-graining refers to the process of deriving larger scale

*Supported by NSF Grant DMS9805582

models from given models. Major challenges in this direction are found in economics, material science, chemical engineering, and biological sciences.

The need to understand coarse-graining in a mathematical way is obvious by looking at many areas and their contemporary leading questions. Recent years have raised new challenges in material science, in fluids, in nano-technology, in biological systems (immune system, nervous system), and have stimulated the search for connection between models at different scales. A common challenge is to understand how do the laws of physics/chemistry/biology/economics change as we go from the small scales to larger scales in time and space. The context of the problem is very general. In material science, this may refer to the passage from atomistic levels to the level of grains (in polycrystalline materials); in fluids it may refer to passing from atomistic models of mixtures to macroscopic levels; in biology to the passage between molecular level (protein) to networks of molecules sharing a common task, to organelles, to cells, to tissue, ending at the organ level. Many of these challenges are yet to be tackled, and probably each of these challenges will result in new methodologies, new mathematical and computational tools. Studying the problem of coarse-graining in general, examining different fields of application increases the chance of progress. Different fields share common aspects of coarse-graining and progress in one area may help in others. Mathematics is a unified language to bridge between those apparently different problems.

We have considered several areas where the coarse-graining problem is a central one. The following examples, among others, helped us identifying some of the questions to be answered.

Economics. Consider the problem of studying the economic behavior of a country, where the problems of interests are concerned with global quantities, such as inflation, interest rate, unemployment, etc. Assume that the modeling approach is microscopic. Thus, the economy is made up of many 'agents' that interacts with each other; there are many types of agents, among others are consumers, producers (firms), banks, government, federal reserve and more. Other elements in the model are money, commodities, products and labor that are being exchanged between the agents. Behavior at the smaller scales is usually understood better than at the macro level, making this approach a desirable one, provided that it can be carried out. The challenge is not only computational, which is obvious. It is also in the interpretation of such huge systems which exhibit behavior on many scales important for applications. Economic theories use concepts such as aggregate demand, aggregate supply etc, attempting to construct macroscopic laws starting from microscopic level. However, a systematic approach that connect the small scale modeling with larger scales is missing. Probably, this is due to the fact that performing this task using purely analytical techniques is not feasible, and a combined analytical-computational approach may help here.

Material Sciences. Many of the modern challenges in material sciences call for including microstructure description in the modeling. The reason is simple, without it modeling of the macroscopic level is inaccurate or sometimes even wrong. Material strength in polycrystals, for example, is determined by their grains structure, including size, texture and other properties. Dislocations pose another major challenge in integrating microscopic features into macroscopic behavior. Certain questions require atomistic considerations, for others a mesoscale (grain level) is sufficient. To appreciate the richness of models needed for this filed consider a piece of metal, say a small wire, which looks homogeneous on a macroscopic scale. A closer look reveals a granular structure. Each of these grains have an almost perfect crystal structure, with embedded defects of different types; dislocation, voids, impurities and more. The surfaces between grains they move at a rate determined by curvature, and certain properties of the adjacent grains. On even smaller scale these surfaces seems to have random fluctuation, due to thermal effects. The grains, that look almost as perfect crystals are not stationary. The atoms that define them move extremely fast but staying most of the time around some 'center', which we may identify as a lattice point.

Complex Fluids. Fluid mixtures behavior is an interesting problem with many contemporary applications. In most cases fluids are being treated using continuum mechanics, using a set of partial different equations (Navier-Stokes). More recently, advances in technology called for understanding more complex fluids. The problem is that for these fluids the classical equations are not the proper macroscopic description. Here one must go to smaller scales, where the behavior is better understood and to build from there a macroscopic model. An example is a fluid made up of two types of atoms, A and B, such that the interactions AA and BB are attractive and AB is repulsive. Such a fluid (with the proper interactions) will develop 'clusters'. The macroscopic dynamics of such a fluid will depend on the presence of clusters.

These examples, point to some of the issues that need to be considered,

- What is the appropriate type of modeling technique to use? Stochastic versus deterministic, discrete versus continuous.
- Within a given modeling paradigm, what are the correct variables to introduce?
- How to interact across scales, where such interaction is necessary?

To gain an insight into the procedures of coarse graining it is useful to consider examples of microscopic models, macroscopic models and their relations. The different models can be classified as stochastic versus deterministic models or continuous versus discrete models, and subclassifications is possible in both.

2. Microscopic Models

We divide the class of microscopic models studied here into stochastic and deterministic models. The purpose of listing a variety of models of different type is to point out both the need and the possibility of 'mapping' models from different representations. The goal is to approximate certain aspects of the detailed model using a simpler model.

2.1 Stochastic Models

Random Walks: A well studied class of models can be viewed as random walks. Here the basic property under study is a random variable which undergoes a stochastic dynamics specified by a certain distribution function. Particles perform jumps that are independent.

Uniform Media The simplest case of a random walk is that on a uniform lattice, say 1D. The state of a particle is denoted by x_i^n, and transition probabilities are specified,

$$P(x_i^{n+1} = x_i^n \pm h) = p_\pm. \tag{1}$$

Nonuniform Media: A more interesting case, though much more difficult is when the jump probabilities depend on the site, and whose distribution is prescribed. These are less commonly used processes and are more challenging theoretically as well as computationally. See [6] for more information about random walks on nonuniform media.

Random Walks with Waiting Time: A more general class of models allow the jumps to happen at arbitrary times according to a prescribe distribution. Models may involve discrete or continuous time and/or space. We restrict our discussion to the discrete-time discrete-space case.

A probability distribution $\psi(k, n)$ for jumping a distance of n lattice points after staying precisely k time steps in the current position, is given. A common case is when ψ has a representation

$$\psi(k, n) = \omega(k)\lambda(n) \tag{2}$$

for some function ω, λ. A detailed discussion for the continuous time case is given in [7].

One of the interesting features of such models are the long tails behaviors of different averages, for proper choices of the function ψ. The mean square displacement, for example, obeys,

$$< |r(t) - r(0)|^2 > \approx Ct^\alpha \quad \text{for t large} \tag{3}$$

where α depends on the decay rate of $\omega(k)$ at infinity. See [7] for more details.

Monte-Carlo Methods are stochastic process that are govern by a Hamiltonian. Probabilities for moves are related to changes in energy between the possible configurations.

Given a configuration X, and another one X', associated with energies $E(X), E(X')$, respectively, transition probabilities between states can be defined in many ways that obey detailed balance. The Metropolis rule is a common implementation,

$$P(X \longrightarrow X') = \begin{cases} \exp(-\Delta E/T) & \Delta E/T > 0 \\ 1 & \Delta E/T \leq 0 \end{cases}, \tag{4}$$

where ΔE is the change in energy, and T is the temperature in proper units. In these methods the jumps of particles depend on the configuration.

Stochastic Differential Equations are used in many modeling areas. As an example we give the Lengevin's equation. Particles obey the dynamics

$$\begin{aligned} dx &= v\,dt \\ dv &= -\zeta\,dt + dG \end{aligned} \tag{5}$$

where G models a Brownian random noise. One of the well known properties of this model is behavior of mean square displacement,

$$< |x(t) - r(0)|^2 > \approx Ct \quad \text{for t large} \tag{6}$$

2.2 Deterministic Models

Ordinary Differential Equations are used in atomistic modeling, known as molecular dynamics (MD), which in the simple cases take the form,

$$\begin{aligned} \tfrac{dx_j}{dt} &= v_j \\ m_j \tfrac{dv_j}{dt} &= -\textstyle\sum_{i \neq j} \nabla\phi(|x_i - x_j|) \end{aligned} \tag{7}$$

where the potential ϕ is prescribed. Usually, the potential is derived using experiments or detailed computation involving quantum mechanical considerations. The total number of particles may be in the range 10^6 or more.

Partial Differential Equations (PDE). Macroscopic dynamics is usually modeled by differential equation. Here the implicit assumption is that the number of particles involved is so large that the quantities of interest, e.g., averages, do not show any noticeable fluctuations. The equations involved may deal with averages of the microscopic model, such as mass density, or may deal with distribution functions, e.g., particle velocity distribution.

Equation for Averages: Classical examples here deals with the dynamics of fluids or solids in terms of averages (when viewed from a microscopic level).

As an example we mention the compressible Navier-Stokes equations, which can be written in a vector form as

$$\frac{\partial U}{\partial t} + div[F(U)] = 0 \tag{8}$$

for the unknown vector $U = (\rho, \rho V, \rho E)$, and a given flux $F(U)$, see [8]. Macroscopic equations can sometimes be derived from equations for distribution functions.

Equations for Distribution Functions: Markov stochastic processes give rise to the well known Chapman-Kolmogorov equation. This equation, referred in general as a master equation, describes the dynamics of the probability distribution function and can be used to infer equations for averages of interest.

Approximation of the master equation, which in many cases is an integral equation, by a differential equation, is useful in many case. A well known case is the Fokker-Planck equation,

$$\frac{\partial f}{\partial t} - \zeta \frac{\partial (vf)}{\partial v} = D \frac{\partial^2 f}{\partial v^2}. \tag{9}$$

describing the evolution of the velocity distribution for particles following the Lengevin's model presented above.

Other examples for distribution function dynamics were derived from physical principles, e.g., the Boltzmann equation

$$\frac{\partial f}{\partial t} + v \frac{\partial f}{\partial x} = Q(f, f) \tag{10}$$

where $Q(f, f)$ represent the collision term, see [3]. This equation, describing the evolution of the velocity distribution function for dilute gases.

3. Coarse-Graining Approaches

Coarse-graining may be viewed as a mapping between models of different types. A deterministic model may be mapped into a stochastic one, and vice versa, a stochastic model may be mapped into a deterministic one. The goal of such mappings is to approximate certain features of the detailed model, using a much simpler model amenable to analysis and efficient computation. In some cases the simpler model is just a coarse time representation of the problem, while in other cases, it may involve new variables, such as averages. We go briefly over a few approaches.

Kinetic Theory and Macroscopic Equations: One of the classical examples of passing from a microscopic level to macroscopic level is due to Boltzmann. The starting point is an equation for the dynamics of the velocity distribution function (10). By taking moments of this equation one obtains a sequence of equations for the moments of f,

$$M_k = \int v^k f(v) dv \qquad k = 0, 1, 2, \ldots \tag{11}$$

The first two moments are the well known quantities in continuum mechanics, mass and momentum , $M_0 = \rho, M_1 = \rho V$.

This procedure requires the introduction of closure relations. For example, certain second order moments are expressed in terms of derivatives of lower order moments, i.e., velocities.

One role that numerical simulation can serve is to validate or suggest closure relations, or to examine the regime for which a given closure relation holds. In [1], a well known model by Einstein for self diffusion is examined. A closure relation, $\rho V = -D\nabla\rho$, is verified to hold for densities which are not too small. In the other case an extra equation for the variable ρV must be introduced.

This suggests that many closure relations are stationary solutions of evolution equations with time scale much faster than that of the rest of the system. Thus, for times that are not too short, these evolution PDE reduce to algebraic equations between quantities appearing in the problem.

Markov Process Representation: When coarse graining is done some aspects of the system are ignored, and in many case the resulting system has a stochastic character. When temporal coarsening is employed, the details of the dynamics between the larger time increments are lost. The resulting large time 'steps' reveal a non-deterministic dynamics. As an example, consider a gas at equilibrium, where its molecules are interacting via a short range force. If we observe the system at time intervals which are larger than the mean collision time, we observe velocities that do not obey any simple rule; their deflections seem random. Although we have started with a deterministic model, upon temporal coarsening, we have ended up with a stochastic process.

Lets look at another motivation for dealing with stochastic representation of phenomena on large scales. When a large number of 'objects' are interacting, we may be interested in the behavior of populations, and this naturally lead to dealing with distribution functions. A given property of the 'objects' is distributed in the population usually in a nonuniform way, and affect the outcome of the total behavior sought. From probability theory we know that equations for distribution functions are closely related to stochastic processes. For example, a diffusion term in the equation for the distribution function is related to a random walk. Thus, when we discuss equations for distribution functions, we are implicitly discussing some stochastic processes. An example is the DSMC (direct simulation Monte-Carlo) method, see [3], to model dilute gases, whose connection to the Boltzmann equation has been shown years after the method was introduced and used in real world applications.

In view of the above examples we regard the passage to a stochastic process at larger time-space scales as a fundamental step in coarse graining. The starting point in construction of stochastic processes is identifying a proper set of variables. From practical point of view having a Markov process is an advantage, since in this case the history of the 'object' or particle need not be stored.

Proving the Markov property for a stochastic process whose origin is a deterministic dynamics is not an easy task. Numerical simulation can be used to check whether the Markov property holds to a good approximation. This is not a replacement for a proof, but may serve as a selection procedure for the right conjectures.

Once a set of variables for describing the stochastic process have been picked, the following steps follow,

1. Check the Markov assumption for the variables picked. If not Markov look for additional variables.

2. Construct transition probabilities from simulations, check for their time dependence (hope for time independence!).

3. Construct a master equation, Then apply proper approximation (Taylor expansions, for example).

4. Averaging of the kinetic equations analytically, then using numerical closure relations.

The numerical procedure that is invoked to verify that a given dynamics follow a Markov process proceed as follows. First, calculate transition probabilities between states using the microscopic model. This can be done by following all particles that are in a given state, say E, and monitoring the fraction of those particles that end up after k time steps at a new state F. Let

$$\begin{aligned} \mathcal{A}_n(E) &= \{j | x_j^n \in E\} \\ \mathcal{A}_n^m(F|E) &= \{j | x_j^n \in E, \quad x_j^m \in F\}. \end{aligned} \tag{12}$$

An approximation for transition probabilities $p_n^{n+k}(E \to F)$, of being in E at time n and in F at time $n+k$, is given by

$$P_n^{n+k}(E \to F) \approx \frac{|\mathcal{A}_n^{n+k}(F|E)|}{|\mathcal{A}_n(E)} \tag{13}$$

where $|\mathcal{A}|$ stands for the number of elements in $\mathcal{A}$. The particles in a given state E, have reached there from other states, $E'_\alpha \quad \alpha \in \mathcal{J}$. The process satisfies Markov property if the transition probabilities as calculated from the whole population at state E, equals to transition probabilities as calculated from subsets of $\mathcal{A}_n(E)$, corresponding to being in different states at previous times.

A procedure for carrying out step 3 is demonstrated in [5], for an particular example. A general master equation is given, the Chapman-Kolmogorov equation (an integral equation), and using a Taylor approximation the Fokker-Plank and Chandrasekhar equations (differential equations), are derived. These equations are more manageable for analysis as well as computation.

Step 4 above has been done in different models, probably the most significant case is that of deriving the Navier-Stokes equations from kinetic theories and can be found in books on kinetic theories of gases. Recently, we have applied these ideas to coarse graining in modeling of grain growth. This will be presented elsewhere.

Density Representation: In this approach one defines volume averages (and/or time averages) and looks for their equations, which are in general stochastic finite difference equations. For sufficiently large scales, the stochastic parts of the equation diminishes, resulting in finite difference equations corresponding in many cases to some differential equations. Take for example the MD model of section 2.2. Define the following averages

$$M_r^k(x) = \frac{1}{|B(x,r)|} \sum_{x_j \in B(x,r)} m_j v_j^k. \tag{14}$$

For sufficiently large (in microscopic terms) r, these averages are good approximation to the well known quantities from continuum mechanics, e.g., $M_r^0 = \rho(x), M_r^1(x) = \rho V(x), \ldots$.

The dynamics of these densities is constructed by arguments similar to those of fluid dynamics, see for example [4], but carried out on the discrete level. A control volume is defined and the change of each of the moments M^k in this volume is computed. We demonstrate the idea for the balance of mass.

Consider a domain Ω where we would like to evaluate the density ρ as a function of time. Let $m(\Omega, t) = \sum_{x_j(t) \in \Omega} m_j$, be the total mass in Ω at time t. Define an index set $\Lambda(t) = \{j | x_j(t) \in \Omega\}$, thus,

$$m(\Omega, t) = \sum_{j \in \Lambda(t)} m_j. \tag{15}$$

Introducing the notation $\Lambda \equiv \Lambda(t)$ and $\bar{\Lambda} \equiv \Lambda(t + \delta t)$, we have,

$$m(\Omega, t + \delta t) - m(\Omega, t) = \sum_{j \in \bar{\Lambda}} m_j - \sum_{j \in \Lambda} m_j = \sum_{j \in \bar{\Lambda} \setminus \Lambda} m_j - \sum_{j \in \Lambda \setminus \bar{\Lambda}} m_j. \tag{16}$$

The right hand side can be interpreted as mass influx minus mass outflux, that is, net mass flux. The last expression, provided that it has a continuum limit, will give

$$m(\Omega, t + \delta t) - m(\Omega, t) \approx \delta t \int_{\partial \Omega} f_l(s) ds \tag{17}$$

where the function $f_l(s)$, is the mass flux per unit area (length) per unit time. This function is an odd function of the normal $\mathbf{n}$, and a first attempt is to express it as $f_l = J^\rho \cdot \mathbf{n}$.

A numerical study of the net mass flux reveal the following. J^ρ fluctuates both in space and time, and fluctuations depending on the scale chosen. We can interpret it as a random variable whose probability distribution has to be identified.

When partitioning the domain of computation into identical square, and defining the local density in each square $\rho_{i,j} = m_{i,j}/h^2$, where $m_{i,j}$ is the mass in cell (i,j), we get

$$\rho_{i,j}(t+\delta t) - \rho_{i,j}(t) = \tfrac{\delta t}{h}[\ J^\rho_{i+\frac{1}{2},j} \cdot \mathbf{n}_{i+\frac{1}{2},j} + J^\rho_{i+\frac{1}{2},j} \cdot \mathbf{n}_{i+\frac{1}{2},j} + J^\rho_{i,j+\frac{1}{2}} \cdot \mathbf{n}_{i,j+\frac{1}{2}} + J^\rho_{i,j+\frac{1}{2}} \cdot \mathbf{n}_{i,j+\frac{1}{2}}] \tag{18}$$

Using numerical simulation whose details are given in [1] we obtained,

$$J^\rho = -\rho V + \textit{fluctuation terms}. \tag{19}$$

On sufficiently large scale, the fluctuation disappear, and we recover a deterministic equation for the evolution of the mass. In this case we may view the result as an approximation to a continuous analog,

$$\frac{m(\Omega, t+\delta t) - m(\Omega, t)}{\delta t} \approx \int_{\partial\Omega} \rho V \cdot \mathbf{n} ds \tag{20}$$

which holds for an arbitrary domain Ω, leading to the equation

$$\frac{\partial \rho}{\partial t} + div(\rho V) = 0 \tag{21}$$

under proper regularity of the functions.

Thus, in this approach, analysis is combined with simulation to construct approximations for certain fluxes in terms of simpler averages. This is a way of generating closure relations. Of course, in the example given here, we need to define an equation for ρV using a similar approach, starting with balance of mass, finding the expression for its evolution and expressing certain averages in terms of simpler ones.

Temporal Coarse-Graining

Stochastic$\rightarrow$ Stochastic: Consider the simple case of a stochastic process involving independent particles whose master equation can be written as,

$$P^{n+1} = LP^n \tag{22}$$

The random walk described in a previous section satisfies this relation. One question of interest is the transition probability for larger times,

$$P(x_i^{n+k} = x_i^n + mh)$$

This of course, is a well known problem, which has an analytical solution. The question of calculating the above probabilities can be solved by considering

$$\begin{aligned} P^{n+k} &= L^k P^n \\ P^0 &= \delta_0 \end{aligned} \tag{23}$$

This amount to calculating powers of L. Although this example is simple and can be solved analytically, we present it as a model for studying central issues in coarse graining. For example, what features of the original model are preserved upon coarse graining? In what sense, or for what initial data, the coarser model is a good approximation to the detailed model? What numerical schemes can be used to compute the longer time transition probabilities (a task that in some cases cannot be done analytically)?

Deterministic $\longrightarrow$ *Stochastic* A more challenging passage between models is in passing from a deterministic model, such as MD model for fluid, to a stochastic model, such as the Lengevin's equation. For the MD model mentioned above we have,

$$v_j(t+\delta t) = v_j(t) + \frac{1}{m}\int_t^{t+\delta t} \nabla f_j(s)ds \tag{24}$$

where f_j is the force acting on the j-th particle. Its complicated expression involving the interaction with the neighbors is given in equation (8). When δt is large enough to include many 'collisions' it is plausible that the integral on the right hand side can be viewed as a stochastic force. This can be seen by taking snapshots of the MD simulation. Increments in velocities seem random and with a careful study using numerical simulation, are shown to follow a nice distribution function, almost a Gaussian distribution. However, the mean square displacement does not follow a linear behavior in time. This poses an interesting challenge. What stochastic process will result in super linear mean square displacement, and have the velocities follow a Gaussian distribution? A discussion of different attempts to approximate the MD dynamics using stochastic processes will be given elsewhere.

References

[1] Ta'asan S., From molecular dynamics to continuum models, in Multigrid Methods VI, Lecture Notes in Computational Science and Engineering No 14, E. Dick, K. Riemslagh, J. Viendeels, eds.), Springer, 1999.

[2] B. Chough and S. Ta'asan, From Molecular Dynamics to Navier-Stokes and Beyond. Computational Aerosciences in the 21st Century. Kluwer Academic Publishers 2000.

[3] Bird G.A., Molecular Gas Dynamics and the Direct Simulation of Gas Flows. Oxford 1994.

[4] Gurtin M.E., Introduction to Continuum Mechanics. Academic Press 1981.

[5] Chandrasekhar, S. Reviews of Modern Physics, **15**, p. 1, 1943.

[6] Bouchaud J.P, Anomalous diffusion in disordered media: Statistical mechanisms, models and physical applications. Physics Reports 195, Nos 4&5 (1990), 127-293.

[7] Meltzer R. and Klafter J, The random walk's guide to anomalous diffusion: A fractional dynamics approach, Physics Reports (2000) 1-77.

[8] Hirsch C., Numerical Computation of Internal and External Flows, Wiley 1990.

MODELING AND SIMULATIONS FOR ELECTROCHEMICAL POWER SYSTEMS

Jinbiao Wu, Jinchao Xu
Department of Mathematics, The Pennsylvania State University
University Park, PA 16802, USA
wu_j@math.psu.edu
xu@math.psu.edu

Abstract In this paper, we will discuss mathematical models and numerical simulations for electrochemical power systems such as lithium-ion battery. We will address the issue of well-posedness of the underlying system of nonlinear partial differential equations, and introduce numerical methods for battery simulations based on Newton linearization, Krylov subspace iteration and multigrid preconditioning. Finally, we will give a brief introduction to fuel cell modeling and simulations.

1. Introduction

Electrochemical power sources such as lead-acid, nickel-mental hydride (Ni-MH), lithium-ion batteries, as well as various fuel cells, are widely used in consumer applications and electric vehicles. Modeling and simulation of battery and fuel cell systems has been a rapidly expanding field. The analysis of electrochemistry systems, such as a battery during charge, discharge and open-circuit, draws primarily on three fundamental areas: thermodynamics, electrode kinetics and mass transport phenomena. These factors determine system behavior and strongly interact with each other[5, 17].

The majority of advanced battery and fuel cell systems employ porous electrodes because they provide large surface areas and a close proximity of the pore electrolyte or gas (in fuel cells) to the electrode material to facilitate electrochemical reactions. A porous electrode cells considered consist of three phases: a solid matrix (electrode material and separator), and electrolyte (liquid or solid) and a gas phase, with complex interfacial structures. It is usually extremely difficult (if at all possible) to solve the exact equations on a microscopic scale due to the complex interface morphology. Instead, macroscopic cell models are derived by averaging the microscopic (exact) equations over a representative elementary volume that contain all phases.

In this paper we mainly consider the battery model and use lithium-ion battery as an example. Our lithium-ion battery model is based on the so-called micro-macroscopic coupled approach [24] and developed by Gu and Wang [9], see also [6, 4, 3, 14] for related works. The governing equations of the model are given by a system of coupled quasilinear partial differential equations.

We will address the issue of well-posedness of the mathematical model for lithium-ion batteries. By exploring some special structure in this system, we are able to adopt the well-known Moser iteration (which is often used for scalar equation)[19, 20] to establish some crucial a priori maximum norm estimates for the systems. With these a priori estimates, we use the Leray-Schauder theory to establish the existence and uniqueness of a subsystem of elliptic equations that describe the electric potentials in the model. Finally, we employ a Schauder fix point theorem to establish the local (in time) existence for the whole model. We note that, from the physical point of view, global existence is not expected; and give some numerical results to support this argument.

We will also discuss an efficient and robust numerical method for battery simulation. In our numerical simulations, Newton method is applied to linearize the nonlinear system, and preconditioned GMRES to solve the linear system. There are also new and crucial feature of our work, for example, introducing multigrid method in the construction of the preconditioner, which makes our work very efficient in battery simulation.

In the end of the paper, we give a simple introduction on the modeling and simulations for the fuel cells which still remains as a developing and challenging filed.

2. Description of Lithium-ion Model

The lithium ion cells consist of the negative electrode current collector (Cu), carbon negative electrode(Li_xC_6), separator, positive manganese dioxide electrode($Li_yMn_2O_4$), and the positive electrode current collector(Al), as shown in Fig. 1. The electrolyte is a solution of lithium salt in a non-aqueous solvent. Electrochemical reactions occurring at the electrode/electrolyte interfaces are as follow:

Composite positive electrode

$$\mathrm{Li}_{y-x}\mathrm{Mn}_2\mathrm{O}_4 + x\mathrm{Li}^+ + xe^- \underset{\text{charge}}{\overset{\text{discharge}}{\rightleftarrows}} \mathrm{Li}_y\mathrm{Mn}_2\mathrm{O}_4,$$

and omposite negative electrode

$$\mathrm{Li}_x\mathrm{C}_6 \underset{\text{charge}}{\overset{\text{discharge}}{\rightleftarrows}} \mathrm{Li}_0\mathrm{C}_6 + x\mathrm{Li}^+ + xe^-.$$

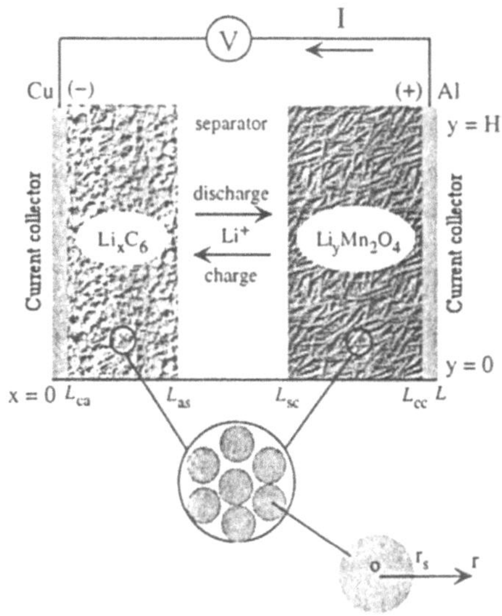

Fig.1. Schematic of a Lithium ion cell and model representation

Lithium-ion inserts into the positive electrode and de-inserts out of the negative electrode during discharge. The change reverses during charge.

Reaction rates. The transfer current from the lithium insertion or de-insertion reaction at the electrode/electrolyte interface that consumes or generates the species Li^+, is assumed to be governed by the Bulter-Volmer equation [17],

$$\mathrm{j} = \mathrm{a}\ \mathrm{i}_{0j}\left[\exp\left(\frac{\alpha_a F}{RT}\eta_j\right) - \exp\left(-\frac{\alpha_c F}{RT}\eta_j\right)\right], \tag{1}$$

where a denotes the specific interfacial area of the manganese dioxide or carbon electrodes; the exchange current density, i_{0j}, is a function of the concentration of lithium ion in the electrolyte phase and the concentrate of lithium in the solid active material phase,

$$\mathrm{i}_{0j} = \mathrm{k}\ (c_e)^{\alpha_a}(c_{s,max} - \bar{c}_{se})^{\alpha_a}(\bar{c}_{se})^{\alpha_c}; \tag{2}$$

$\bar{c}_{se}$ is the area-averaged concentration of lithium at the electrode/electrolyte interface which is determined by

$$\frac{\mathrm{D}_s}{l_{se}}(\bar{c}_{se} - c_s) = -\frac{\mathrm{j}}{\mathrm{a}F}, \tag{3}$$

and the local surface overpotential of reaction j is defined as

$$\eta_j = \Phi_s - \Phi_e - U_j(\bar{c}_{se}, T), \qquad j = 1 \text{ and } j = 2. \tag{4}$$

Here, U_j, a function of $\bar{c}_{se}$ and temperature, stands for the open-circuit (i.e. equilibrium) potential of reaction j.

Effective diffusion coefficient. The effective diffusion coefficient including the effect of tortuosity is evaluated by the Bruggeman relation, i.e.

$$D^{\text{eff}} = D\varepsilon_e^{1.5}, \tag{5}$$

where D is the mass diffusion coefficient of lithium ion in the electrolyte, ε_e is the volume fraction of the electrolyte phase.

Species conservation equations. Let c_e, c_s be the volume-averaged concentration of the lithium in the electrolyte phase and the solid phase, respectively. Conservation of lithium in the electrolyte phase and solid phase gives

$$\frac{\partial(\varepsilon_e c_e)}{\partial t} = \nabla \cdot (D^{\text{eff}} \nabla c_e) + \frac{1 - t_+^0}{F} \mathrm{j}, \tag{6}$$

$$\frac{\partial(\varepsilon_s c_s)}{\partial t} = -\frac{\mathrm{j}}{F}, \tag{7}$$

where t_+^0 is the transference number of the Li^+ with respect to the velocity of the solvent.

Charge conservation equations. For the electrolyte phase, the charge conservation equation takes the following form,

$$\nabla \cdot (\kappa^{\text{eff}} \nabla \Phi_e) - \nabla \cdot (\kappa_D^{\text{eff}} \nabla \ln c_e) + \mathrm{j} = 0. \tag{8}$$

This equation can be used to determine the electrical potential in the electrolyte phase, Φ_e. κ_D^{eff} is the diffusional conductivity given by

$$\kappa_D^{\text{eff}} = \frac{2R\mathrm{T}\kappa^{\text{eff}}}{F}(1 - t_+^0). \tag{9}$$

For the solid phase, the charge conservation equation is expressed by

$$\nabla \cdot (\sigma^{\text{eff}} \nabla \Phi_s) - \mathrm{j} = 0, \tag{10}$$

where s denotes the manganese dioxide or carbon electrode.

Energy conservation equation. Without convection in the cell, the general thermal energy equation reduces to

$$\frac{\partial(\rho c_p \mathrm{T})}{\partial t} = \nabla \cdot (\lambda \nabla \mathrm{T}) + q, \tag{11}$$

where the heat generation rate q is expressed by

$$\begin{aligned} q \;=\; & \mathrm{j}\left(\Phi_s - \Phi_e - U_j + \mathrm{T}\frac{\partial U_j}{\partial \mathrm{T}}\right) + \sigma^{\text{eff}} \nabla \Phi_s \cdot \nabla \Phi_s \\ & + \kappa^{\text{eff}} \nabla \Phi_e \cdot \nabla \Phi_e + \kappa_D^{\text{eff}} \nabla \ln c_e \cdot \nabla \Phi_e. \end{aligned} \tag{12}$$

Initial/Boundary Conditions. Uniform initial conditions are assumed, i.e.

$$c_e = c_{e,0}, \quad c_s = c_{s,0}, \quad \mathrm{T} = \mathrm{T}_0. \tag{13}$$

The computational domain is confined by the two current collects. The electrolyte is confined within the cell and no reaction occurs at the current collector surfaces, giving rise to

$$\frac{\partial c_e}{\partial n} = 0, \quad \frac{\partial c_s}{\partial n} = 0, \quad \text{and } \frac{\partial \Phi_e}{\partial n} = 0, \tag{14}$$

at all boundaries. Current is applied at the tabs on the top and heat is dissipated only through the tabs(basecase)or through the sides as well as through the tabs(see Fig 1.), yielding

At $y = H$,

$$\begin{aligned} -\sigma^{\text{eff}}\frac{\partial \Phi_s}{\partial y} &= I \\ -\lambda\frac{\partial \mathrm{T}}{\partial y} &= h(\mathrm{T} - \mathrm{T}_\alpha) \end{aligned} \quad \text{if } x < L_{ca} \text{ or } x > L_{cc}. \tag{15}$$

At the other boundaries,

$$\begin{aligned} \frac{\partial \Phi_s}{\partial n} &= 0 \\ -\lambda\frac{\partial \mathrm{T}}{\partial n} &= h(\mathrm{T} - \mathrm{T}_\alpha). \end{aligned} \tag{16}$$

3. Local existence

In this section, we will consider the well-posedness of the mathematical model for lithium-ion battery system. The system of partial differential equations we considered are the equations given by (6), (7), (8) and (10) together with initial-boundary value conditions given by (13), (14), (15) and (16). The thermal affect is neglected in these two sections. Among these various equations, the equation (7) is a simple ordinary differential equation. Technically speaking, this equation will not contribute extra difficulty in our analysis given below. Thus, for simplicity of exposition below, we shall drop this equation from the system. In this section, we consider the local (in time) existence of the isothermal lithium-ion model.

3.1 The System of Partial Differential Equations

We will consider the following system of partial differential equations:

$$-\nabla \cdot (\kappa(c)\nabla\Phi_e) + \nabla \cdot (\kappa_D(c)\nabla \ln c) - S_e(\Phi_s - \Phi_e, c) = 0, \quad x \in \Omega, \tag{17}$$

$$-\nabla \cdot (\sigma\nabla\Phi_s) + S_e(\Phi_s - \Phi_e, c) = 0, \quad x \in \Omega', \tag{18}$$

$$\frac{\partial(\varepsilon_e c)}{\partial t} - \nabla \cdot (D\nabla c) - S_c(\Phi_s - \Phi_e, c) = 0, \quad x \in \Omega. \tag{19}$$

Here we have used some simplified notation. For example, we have used c instead of c_e. The "source" term

$$S_e(\Phi_s - \Phi_e, c) = \mathrm{j} = \alpha_0 g(c) \sinh\left(\alpha_2(\Phi_s - \Phi_e - U(c))\right), \tag{20}$$

in Ω' and $S_e = 0$ in Ω_s, where $g(c)(> 0)$ is a smooth function of c; α_0, α_2 are positive constants ($\alpha_1, \alpha(M)$ and so on will also denote some other positive constants in the following). $S_c(\Phi_s - \Phi_e, c) = \frac{1-t_+^0}{F}\mathrm{j}$ where F, $\mathrm{t}_+^0(< 1)$ are positive constants.

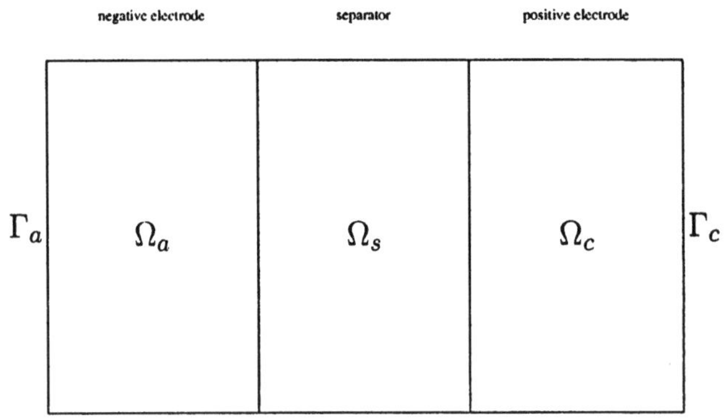

Fig. 2 The domain Ω.

The domain $\Omega = \Omega_a \cup \Omega_s \cup \Omega_c \subset \mathbf{R}^n$ denotes the whole battery domain, where Ω_a, Ω_s, Ω_c denote the negative electrode, the separator, the positive electrode, respectively. Set $\Omega' = \Omega_a \cup \Omega_c$. Eq.(17) and Eq.(19) are defined in the whole domain Ω; Eq.(18) is defined in the domain Ω'. From (9), we have $\kappa_D = \alpha_1 \kappa$. ε_e, D and σ are positive piecewise constants, and κ is a piecewise smooth function of c which satisfies $\kappa(0) = 0$; $\kappa(c) > 0$, if $c > 0$. U, which is defined in Ω', is a known bounded smooth function of c.

We can drop the second term of Eq.(17) by setting $\hat{\Phi}_e = \Phi_e - \alpha_1 \ln c$ and $\hat{U}(c) = U(c) + \alpha_1 \ln c$ (we still use the same notations Φ_e and $U(c)$). For the uniqueness of the system, we impose the following condition:

$$\int_\Omega \Phi_e dx = 0, \tag{21}$$

and define

$$H_*^1(\Omega) = \{u \in H^1(\Omega), \int_\Omega u dx = 0\},$$

$L_*^\infty(\Omega)$ and $C_*^\alpha(\Omega)$ are defined in the similar way.

The boundary condition for Φ_s reads:

$$-\sigma \frac{\partial \Phi_s}{\partial n} = I \quad \text{on} \quad \Gamma_a \cup \Gamma_c,$$

where $\Gamma_a \subset \partial\Omega_a$, $\Gamma_c \subset \partial\Omega_c$. At the other boundaries(note that Φ_s is defined in Ω'),

$$\frac{\partial \Phi_s}{\partial n} = 0;$$

and

$$\frac{\partial \Phi_e}{\partial n} = 0 \text{ on } \partial\Omega, \quad \frac{\partial c}{\partial n} = 0 \text{ on } \partial\Omega,$$

for Φ_e and c. The initial condition for c is

$$c|_{t=0} = c_0,$$

where $c_0 > 0$ and we assume $c_0 \in C^{0,\delta}(\bar{\Omega})$.

3.2 The elliptic system for potential variables

In the subsection, we shall first study the first two elliptic equations given by (17) and (18). Namely, we view the concentration variable c fixed and it satisfies the following condition:

$$\|c\|_{L^\infty} + \|c^{-1}\|_{L^\infty} \leq M,$$

and we consider the following elliptic system:

$$-\nabla \cdot (\kappa(c)\nabla\Phi_e) - S_e = 0, \quad x \in \Omega, \tag{22}$$
$$-\nabla \cdot (\sigma\nabla\Phi_s) + S_e = 0 \quad x \in \Omega'. \tag{23}$$

with same boundary condition for Φ_e, Φ_s described in previous subsection.

Define $X = L^\infty_*(\Omega) \times L^\infty(\Omega')$, and the mapping $Z : X \times [0, 1] \to X$ such that for any $V = (u, v) \in X$, $\delta \in [0, 1]$, $\Phi = (\Phi_e, \Phi_s) = Z(V, \delta)$ is the unique solution in $H^1_* \times H^1$ to the following system:

$$\begin{aligned} -\nabla \cdot (\kappa(c)\nabla\Phi_e) - \delta(S_e(v-u, c) - I(\Omega')(v-u)) \\ -I(\Omega')(\Phi_s - \Phi_e) = 0, \end{aligned} \tag{24}$$
$$-\nabla \cdot (\sigma\nabla\Phi_s) + \delta(S_e(v-u, c) - (v-u)) + (\Phi_s - \Phi_e) = 0, \tag{25}$$

where $I(\Omega') = 1$ in Ω' and $I(\Omega') = 0$ in Ω_s. The boundary conditions are the same as those for the system of Eq.(22) and (23), except on $\Gamma_a \cup \Gamma_c$ the boundary condition for Φ_s reads

$$-\sigma\frac{\partial \Phi_s}{\partial n} = \delta I.$$

We assume that $\Phi = (\Phi_e, \Phi_s) \in X$ satisfies $X = Z(X, \delta)$ for some $\delta \in [0, 1]$. Using Moser iteration[19, 20], we have the following lemma.

Lemma 1. *Let $p > 2$ be an integer, even, and $p \leq \frac{2n}{n-2}$ where n is the dimension of Ω(If $n = 2$, $p < +\infty$; if $n = 3$, $p \leq 6$). Assume that $\|c\|_{L^\infty} + \|1/c\|_{L^\infty} \leq M$, If $(\Phi_e, \Phi_s) \in (H_*^1 \times H^1) \cap (L_*^\infty \times L^\infty)$ is a solution tò the following elliptic system*

$$-\nabla \cdot (\kappa \nabla \Phi_e) - \delta S_e(\Phi_s - \Phi_e, c) - I(\Omega')(1-\delta)(\Phi_s - \Phi_e) = 0, \quad (26)$$

$$-\nabla \cdot (\sigma \nabla \Phi_s) + \delta S_e(\Phi_s - \Phi_e, c) + (1-\delta)(\Phi_s - \Phi_e) = 0, \quad (27)$$

with the same boundary conditions for the system of Eq.(24) and (25). Then Given any $X_0 \in \Omega'$, $R > 0$ such that $R < \operatorname{dist}(X_0, \partial\Omega')$, for any $\theta \in (0,1)$,. we have

$$\operatorname*{ess\ sup}_{B_{\theta R}(X_0)} (|\Phi_e| + |\Phi_s|) \leq \alpha(M)\left(\frac{1}{|B_R(X_0)|} \int_{B_R(X_0)} (|\Phi_e| + |\Phi_s|)^p dx\right)^{1/p}$$

where $B_R(X_0) = \{Y \in \mathbf{R}^n, |Y - X_0| \leq R\}$.

Similar estimates are still valid, if $X_0 \in \Omega_s$ or $X_0 \in \partial\Omega'$. At last, we get the following a priori estimate.

Lemma 2. *Assume that $\|c\|_{L^\infty} + \|1/c\|_{L^\infty} \leq M$. If $\Phi = (\Phi_e, \Phi_s) \in X$ satisfies $\Phi = Z(\Phi, \delta)$ for some $\delta \in [0,1]$; there exists a positive constant $K(M)$ such that*

$$\|\Phi_e\|_{L^\infty(\Omega)} \leq K(M); \quad \|\Phi_s\|_{L^\infty(\Omega')} \leq K(M).$$

The existence of the solution (in $H_*^1 \times H^1$) to the system of Eq.(22) and (23) follows by using the Leray-Schauder theorem:

Theorem 1. *We assume $\|c\|_{L^\infty} + \|c^{-1}\|_{L^\infty} \leq M$; then the system of Eq.(22) and (23) admits a unique solution (Φ_e, Φ_s), and there exists $\alpha(M)$ such that*

$$\|\Phi_e\|_{L^\infty} + \|\Phi_s\|_{L^\infty} \leq \alpha(M).$$

Now we will outline proof of the local in time existence for the system of Eq.(22), (23) and (19). Set

$$M = \|c_0\|_{L^\infty}\| + \|1/c_0\|_{L^\infty} + 1$$

Let $T > 0$ be a small constant to be determined. We want to establish the existence of a solution c in the space,

$$V_T = \{c \in C^0(\Omega \times [0,T]), c \geq 1/M, \|c\|_{C^0} \leq M\}.$$

Let $\hat{c} \in V_T$ be any function. We first define Φ_e and Φ_s by solving the system of Eq.(22) and (23) with c replaced by $\hat{c}$. Once we have (Φ_e, Φ_s), we then can define c by solving (19) in $L^2([0,T], H^1(\Omega))$ (with c in S_c replaced by $\hat{c}$). We define the mapping $\mathbf{T} : \hat{c} \longrightarrow c$. A parabolic a priori estimate gives

$$\|c\|_{C^0} \leq \|c_0\|_{C^0} + T\alpha(M) \leq M,$$

if we take T small. As c is $C^{\alpha,\alpha/2}$ for some $\alpha \in (0,1)$ [15](bounded with bounds depending on M and T). Hence $\mathbf{T}$ is compact.

We can also see that $\mathbf{T}$ is continuous. Hence $\mathbf{T}$ has a fixed point, by the Schauder fixed point theorem.

Theorem 2. *There exists $T > 0$ such that the system of the equations (17), (18) and (19) admits a unique solution (Φ_e, Φ_s, c) where Φ_e satisfies (21).*

3.3 Finite battery lifespan

After the establishment of local existence as done in preceding section, one natural question is whether the local existence result can be extended globally (in time). In this section, we will try to use both physical arguments and numerical experiments to illustrate that a global solution for the system of partial differential equations in our model may not be expected.

The simple (and perhaps nearly naive) physical argument is that a lithium-ion battery can not be run forever and its life expectancy has to be finite. In fact, in the model under our consideration, the battery is either only always being discharged ($I > 0$) or only always being charged ($I < 0$). Hence the battery is expected to be either drawn out (in the former case) or to be blown-up (in the latter case) within a finite amount time. This means, mathematically speaking, that the systems of partial differential equations can not have a global (in time) solution.

Of course, our systems of PDEs is only a model for the reality and the above argument may not be convincing. Next we shall present some numerical examples to support this argument. The major input variable in the model is the current $i = \int_{\Gamma_c} I dA$. Fig .3 and Fig 4. give the numerical results of the cell potential for $i = 0.3C = 0.678A$ and $i = 3C = 6.78A$.

In each of the above, we observe a sharp change of value of cell potential at a critical time and the solution cease to exist beyond this time. As expected, the battery life expectance gets shorter when the current i gets larger.

It would be interesting to mathematically prove (or disprove) that the existence of the critical time (nonexistence of the global solution) with the condition, such as $i = \text{const}$.

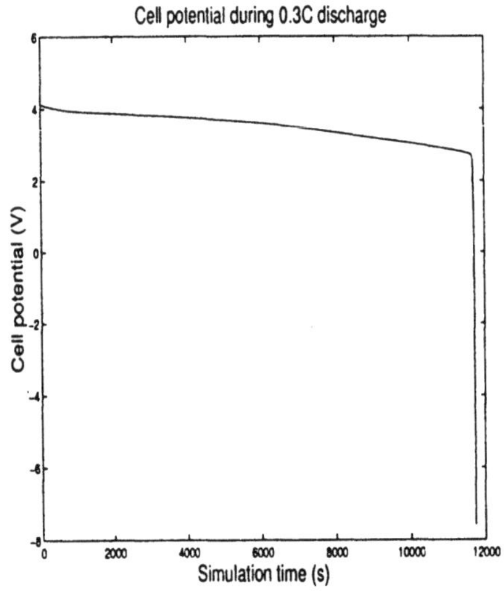

Fig. 3.1

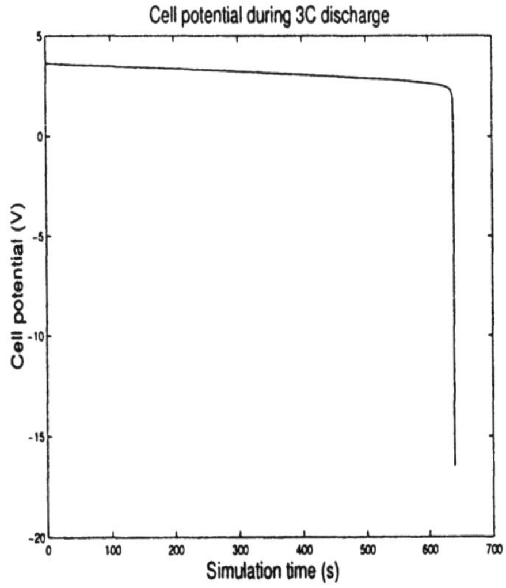

Fig. 3.2

4. Newton-Krylov-Multigrid Method

The intention of this section is to introduce an efficient and robust numerical method for Lithium-ion battery simulation. We use Newton method to linearize the nonlinear system, and use preconditioned GMRES method to solve the linear system. Comparing with the related works [17, 21, 2], there is also new and crucial feature of our work. We use Newton method in different level, Newton method is applied to solved the global linear system, it is also using to decided the value in each vertex (local 1-d problem). For the preconditioner we choose the Block Gauss-Seidel method and use the multigrid method in the construction of the preconditioner.

4.1 Newton Method

The transient terms are discretized by a fully implicit scheme making use of the backward Euler method. After discretizing the spatial terms by finite volume or finite different method, one is faced with the problem of solving a system of nonlinear algebraic equations which we simple express it as

$$f(s) = 0 \tag{28}$$

where f is a vector function, $f : \mathbf{R}^{4n} \rightarrow \mathbf{R}^{4n}$, where n is the number of vertices in the grid. Let $J(s) = \frac{\partial f}{\partial s}$ be the $4n \times 4n$ Jacobian matrix of f.

Algorithm 1. Newton method. Given an initial guess s^0, for $m = 0, 1, \cdots$

1) Solve the linear system

$$J(s^m)\eta^m = -f(s^m). \tag{29}$$

2) Define

$$s^{m+1} = s^m + \eta^m. \tag{30}$$

The convergence of Newton method is locally quadratic and has been found to be particularly fast for the present model problem. The solution to (28) can be obtained in a few iterations using this method.

Remark In fact, j is an implicit function of c_e, c_s, Φ_e, Φ_s (see (1), (2), 4), (3), and (17)). Newton method is also used to get the value of j at each vertex. which is a (local) one-dimensional problem. From our numerical experience, the accuracy of j play a crucial role in the convergence rate of Newton method.

4.2 Preconditioned GMRES

Now the problem is how to solve the linear system (29). We note that the Jacobian matrix of this problem is nonsymmetrical and large. GMRES method introduced by Y.Saad and M.H.Schultz([22]) is a kind of Krylov subspace approximation method, proved to be one of most efficient methods to solve general (large sparse) nonsymmetrical linear systems of equations [8, 12].

The preconditioned GMRES method for solving system (29), is to solve the following equivalent system

$$M^{-1}J(s^m)\eta^m = M^{-1}(-f(s^m)),$$

by GMRES method, where M is called the preconditioner. Among numerous preconditioning technique, we choose Block Gauss-Seidel method and multigrid method in our work.

Write $J = D - L - U$ where D is a block diagonal matrix, and L and U are the strictly lower and upper block triangular parts of $J - D$, respectively. If we assume that $M = D - L$, then M is also called the Block Gauss-Seidel preconditioner.

If we assume that

$$J = \begin{pmatrix} J_{c_e c_e} & J_{c_e \Phi_s} & J_{c_e \Phi_e} & J_{c_e \mathrm{T}} \\ J_{\Phi_s c_e} & J_{\Phi_s \Phi_s} & J_{\Phi_s \Phi_e} & J_{\Phi_s \mathrm{T}} \\ J_{\Phi_e c_e} & J_{\Phi_e \Phi_s} & J_{\Phi_e \Phi_e} & J_{\Phi_e \mathrm{T}} \\ J_{\mathrm{T} c_e} & J_{\mathrm{T}\Phi_s} & J_{\mathrm{T}\Phi_e} & J_{\mathrm{TT}} \end{pmatrix},$$

we need to compute $J_{c_e c_e}^{-1}\nu$ and so on for a given vector ν in the block Gauss-Seidel method. The problem is equivalent to solving the system,

$$J_{c_e c_e} w = \nu,$$

and so on. The matrices $J_{c_e c_e}$, $J_{\Phi_s \Phi_s}$, $J_{\Phi_e \Phi_e}$, J_{TT} are symmetric positive definite. Here, and they will be inverted by the multigrid method. As an iteration method, multigrid method has a very fast convergence speed which is independent on the grid parameter, and the number of whole arithmetic operations needed is only $O(n)$(or $O(n \log n)$), where n is the number of the unknowns. It shows great performance in solving symmetric positive definite problem arise from discretizing elliptic or parabolic partial differential equations [15, 13].

4.3 Results and validation

A general-purpose computational fluid dynamics (CFD) code, which uses Picard-type iteration method, is widely used in this area. Picard-type iteration method solves the governing equations one by one (similar as the nonlinear (block) Guass-Seidel method). The convergence of Picard-type method is very slow for strong coupled problem. Roughly speaking, it is efficient to solve the model problem, the governing equations of which is not strongly coupled. See [4, 21, 2] for related work on Newton iteration for the similar problem.

The following tabular shows CPU times of the two methods during 3C(6.78A) discharge. The case is based on the model in Section 2, and same as that in [9]. Our method is much faster than the Picard-type method, see [26] for more detail.

Tabular 1. CPU times with different grids

2-D grid	CPU time (s) by Picard method	CPU time (s) by Newton method	speed up
45× 32	44958.3	570.899	78.7501
90× 62	335001	2721.63	123.088
178× 122	–	11280.3	–
354× 242	–	47970.2	–

Mathematical modeling is indispensable in the development process for batteries and fuel cells with higher energy density, higher power density and longer cycle life; because a cell model, once validated experimentally, can be used to indentify cell-limiting mechanisms and forecast cell performance for design, scale up, and optimization [24].

The following pictures show the comparison of the numerical result and the experimental data for several batteries (4.1 Lithium/Thionyl Chloride [11], 4.2 4.3 NiMH [10], 4.4 Lead-acid). These battery models are also based on the so called micro-macroscopic coupled approach [24]. The predicted data is fitting perfect well with the experimetal results. Simulation results from a validated model can take the position of experimental data. For example, the optimization of a battery design for a particular application necessarily involved a large amount of time and expermental effort. Computer simulations are very useful in this process because they can potentially lead to a great savings of time and materials.

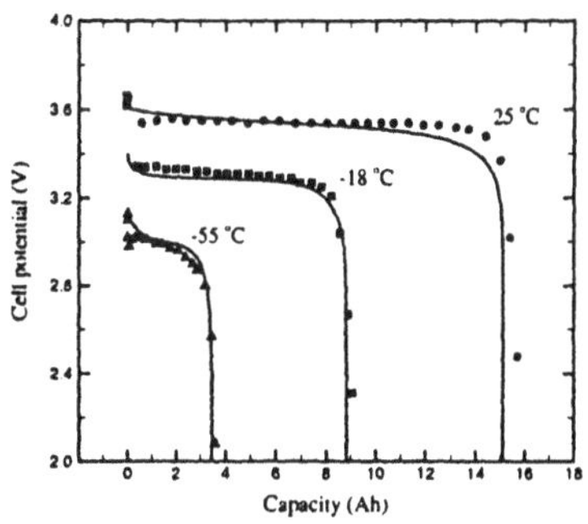

Figure 4.1 Comparison of experimental and predicted discharge curves for a 50 Ω load at -55, -18, and 25 °C. The symbols represent the experimetal data, while the solid lines are for the predicted results.

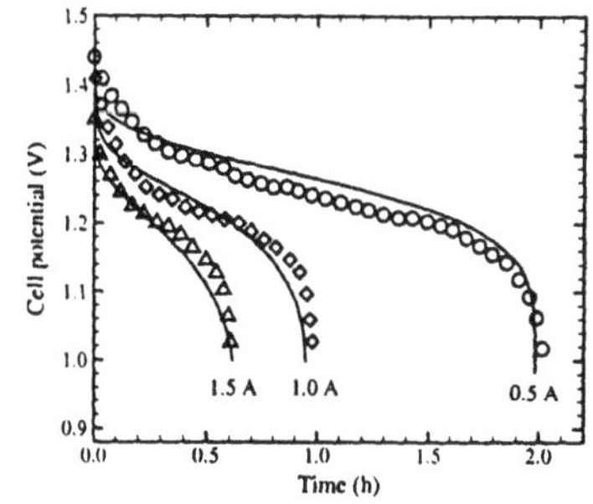

Figure 4.2 Comparison of predicted cell discharge potentials (solid lines) with experimental results (symbols)

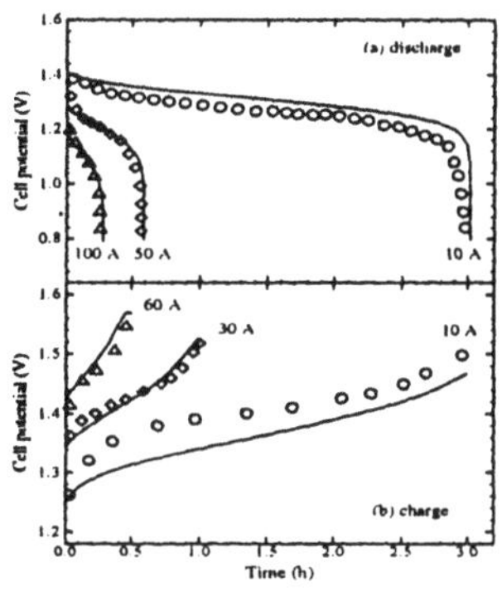

Figure 4.3 Comparison of predicted cell potentials (solid lines) with the experimental results(symbols)

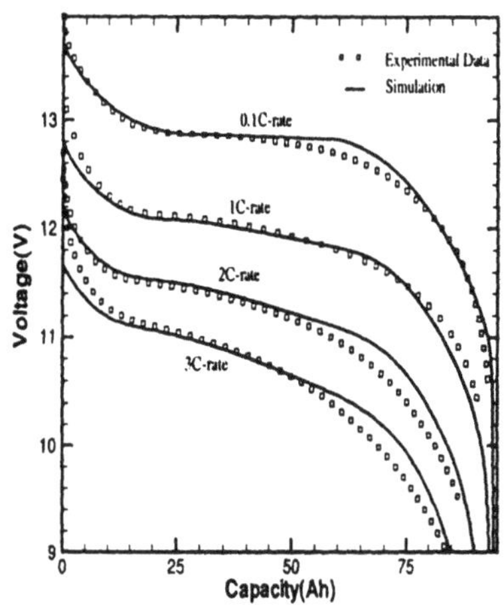

Figure 4.4 Comparison of predicted cell potentials (solid lines) with the experimental results (symbols)

5. Fuel cell – future direction

In this section, we will give a brief description of a different electrochemical power systems, fuel cells; whose mathmatical modeling is related to but more complicated to that of lithium-ion battery. We will only discuss the proton exchange membrance (PEM) fuel cell as an example of fuel cells. Proton exchange membrane fuel cell engines can potentially replace the internal combustion engine because they are clean, energy-efficient, quiet, fuel-flexible, and quickly starting up due to low-temperature operation. Since it usually involves simultaneously electrochemical reaction, hydrodynamics, current distribution, transport and diffusion, heat transfer and mixture and multiple phase materials, a complete model that includes all these effect and their interactions is needed.

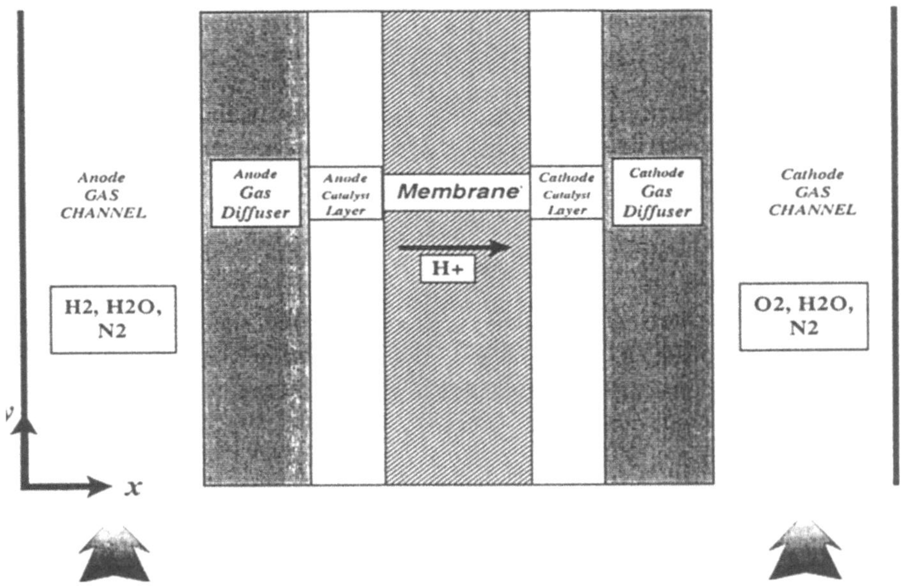

Figure 5. Schematic of PEM Fuel cell

The PEM fuel cell to be modeled is schematically shown in Figure 5 and divided into seven sub-regions: the anode gas channel, gas-diffusion anode, anode catalyst layer, ionomeric membrane, cathode catalyst layer, gas-diffusion cathode, and cathode flow channel. The present model considers the anode feed consisting of hydrogen, water vapor and nitrogen in order to simulate reformate gas (CO will be added in Part IV concerning CO-poisoning effect), whereas humidified air is fed into the cathode channel. Hydrogen oxidation and oxygen reduction reactions are considered to occur only within the active catalyst layers where Pt/C catalysts are intermixed uniformly with recast ionomer. Only a single phase model is given below. The reader should consult Um et al. [27] and Um and Wang [28] for other details of the reformate/air PEM fuel cell model, Mench et al. [18] and Wang et al. [25] for two-phase model and other fuel cell models.

The mathematical model for the PEM fuel cell is derived through different conservation law and balance laws.

Conservation of mass:

$$\frac{\partial(\epsilon\rho)}{\partial t} + \nabla\cdot(\epsilon\rho u) = 0, \tag{31}$$

where ϵ is the porosity, ρ is the density that depends on the concentration of different species. u is the flow velocity.

Conservation of momemtum:

$$\frac{1}{\epsilon}(\frac{\partial(\rho u)}{\partial t} + \nabla(\rho uu)) + \nabla p = \nabla\cdot(\frac{\mu}{\epsilon}\nabla u) + S_u, \tag{32}$$

where p is the pressure, μ is the viscosity and S_u is the fraction drag due that is equal to $-\frac{\mu}{K}u$ in the diffusion layers and zero elsewhere.

Conservation of species:

$$\frac{\partial(\epsilon c_k)}{\partial t} + \nabla\cdot(\epsilon c_k u) = \nabla\cdot(D_k^{eff}\nabla c_k) + S_k, \tag{33}$$

where c_k is the concentration of different species, D_k^{eff} is the effective diffusion constants and S_k is the stoichiometry force that depends on the electric current density j and the current i_e in the catalyst layer and the membrane.

Distribution of charge:

$$\nabla\cdot(K_k^{eff}\nabla\Phi_e) = -S_\Phi, \tag{34}$$

where S_Φ is the force that is equal to the current only on the catalyst layer.

Conservation of heat:

$$\frac{\partial(\epsilon\rho c_p T)}{\partial t} + \nabla\cdot(\rho c_p T u) = \nabla\cdot(K^{eff}\nabla T) + S_T, \tag{35}$$

where T is the temperature, K^{eff} is the effective diffusion coefficient, S_T is only in the catalyst layer and the membrane, depending on j and i_e. The boundary condition of u is that it is zero on the interface between the diffusion layers and the catalyst layers.

Modeling and simulation of fuel cell are more complicated than that of the batteries. Comparision with the battery models, there are following new features for the fuel cell models:

Input and output. For the batteries, all the materials and produces are inside the batteries, and they are reusable during another charge/discharge cycle. But for fuel cells, this is a different story. There are fuel and oxidizer input, and produce output during the working process of fuel cell. No charge is needed for next time using. It is supposed to be powerful enough for future's long time usage, such as automotive engine and so on.

Mass and momentum equations. The flow convection, which is ignored for most battery models, is very important for fuel cells. The mass and momentum equations are coupled with the potential, concentration and heat equations;

so the system of the governing equations are more difficult to analyze and simulate.

Complex geometry and 3-d model. For the modeling of the batteries, 2-d model is mostly enough, and sometimes 1-d model also makes sense. 3-d fuel cell models should be considered for the purpose of validation and prediction, because of much more complicated geometry and phenomenon. One can image that the number of the nodes during simulation for the fuel cell is much larger than that for batteries.

References

[1] R.A. Adams, *Sobolev space*, New York, Academic press, 1975.

[2] P.N. Brown, A.C. Hindmarsh and L.R. Petzold, *Using Krylov Method in the Solution of Large-scale Differential-algebraic Systems*, SIAM J. Sci. Comput. 15 (1994), 1467-1488.

[3] Y. Chen and J.W. Evans, *Thermal analysis of lithium-ion batteries*,J.Electrochem.Soc.,143,2708(1996).

[4] M. Doyle, J. Newman, A.S. Gozdz, C.N. Schmutz, and J.-M. Tarascon, *Comparison of modeling predictions with experimental data from plastic lithium ion cells*, J.Electrochem. Soc.,143,1890(1996).

[5] A. Friedman, *Mathematics in Industrial Problems, Part 7*, 229-240(1995).

[6] T.F. Fuller, M. Doyle and J. Newman, *Simulation and optimization of the dual lithium ion insertion cell*, J.Electrochem. Soc., 141,1(1994).

[7] D. Gilbarg and N.S. Trudinger, *Elliptic Partial Differential equations of Second Order*, Springer-Verlag, Heidelberg, New York, 1977.

[8] A. Greenbaum, *Iterative Methods for Solving Linear Systems*, Frontiers in Applied Mathematics Vol. 17, SIAM, Philadelphia, 1997.

[9] W.B. Gu and C.Y. Wang, *Thermal and Electrochemical Coupled Modeling of a Lithium-Ion Cell*, R.A. Marsh, Z. Oguma, J. Prakash, and S. Surampudi, Editors, The Electrochemical Society Series, PV 99-29, 748 (1999).

[10] W.B. Gu, C.Y. Wang, S.M. Li, M.M. Geng and B.Y. Liaw, *Modeling discharge and charge characteristics of Nickel-Metal hydride batteries*, Electrochimica Acta, Vol.44, pp.4525-4541 (1999).

[11] W.B. Gu, C.Y. Wang, John Weidner and R. Jungst, *Computational Fluid Dynamics Modeling of a Lithium/Thionyl Chloride Battery with Electrolyte Flow*, Journal of Electrochemical Society, Vol.147, pp427-434 (2000).

[12] W. Hackbusch, *Iterative Solution of Large Sparse Systems of Equations*, Springer-verlag, Berlin, New York, 1994.

[13] W. Hackbusch, *Multigrid Methods and Application*, Springer-Verlag, Berlin, 1985.

[14] J.-S. Hong, H. Maleki, S. Al Hallaj, L. Redey and J.R. Selman, *Electrochemical-Calorimetric studies of lithium-ion cells*, J. Eletrochem. Soc., 145, 1489(1998).

[15] O.A. Ladyzenskaja, V.A. Solonnikov and N.N. Ural'tzeva, *Linear and quasilinear equations of parabolic type*, Trans;. Math. Mono., AMS, Vol. 23(1968).

[16] O.A. Ladyzenskaya and N.N. Ural'tseva, *Linear and quasilinear elliptic equations*, English Transl., Academic Press, New York, 1968.

[17] John S. Newman, *Electrochemical Systems*, Second Edition, 1991.

[18] M.M. Mench, C.Y. Wang, and S. Thynell, *An Introduction to Fuel Cells and Related Transport Phenomena*, Accepted for publication in Journal of Transport Phenomena 2001.

[19] J. Moser, *A new proof of de Giorgi's theorem concerning the regularity problem for elliptic differential equations*, Comm. Pure Appl. Math., 13(1960), pp.457-468.

[20] J. Moser, *A Harnack inequality for parabolic differential equations*, Comm. Pure Appl. Math., 17(1964), pp.101-134.

[21] L.R. Petzold, *A description of DASSL: A differential/algebraic system solver*, eds. R.S. stepleman et al., North-Holland, Amsterdam(1983), 65-68.

[22] Y. Saad and M. H. Schultz, *GMRES: A generalized minimal residual algorithm for solving nonsymmetrical linear systems*, SIAM J. Sci. Stat. Comput.,7(1986),865-869.

[23] L.H. Thomas, *Elliptic problems in linear difference equations over a network*, Watson Sci. Comput. Lab. Report, Columbia University, New York, 1949.

[24] C.Y. Wang, W.B. Gu and B.Y. Liaw, *Micro-macroscopic coupled modeling of batteries and fuel cells I. Model development*, J.Eletrochem. Soc., 145, 3407(1998).

[25] Z.H. Wang, C.Y. Wang and K.S. Chen, *Two phase flow and transport in the air cathode of proton exchange membrane fuel cells*, J. Power Sources, Vol. 94 (1), pp. 40-50(2001).

[26] J. Wu, Verkat Srinivasan, J. Xu and C.Y. Wang, *Newton-Krylov-Multigrid algorithms for battery simuation*. to be appear in J. Electrochem. Soc.

[27] S. Um, C. Y. Wang and K. S. Chen, *Computational fluid dynamics modeling of proton exchange membrane fuel cells*, Journal of Electrochemical Society, Vol.147, pp4485-4493(2000).

[28] S. Um and C. Y. Wang, *Three dimensional analysis of transport and reaction in proton exchange membrane fuel cells*, in Proc. of the ASME Heat Transfer Division, Orlando, FL., Nov. 2000.

[29] J. Xu, *An Introduction to Multilevel Methods*, Wavelets, multilevel methods and elliptic PDEs (Leicester, 1996), Oxford Univ. Press, New York, 1997, 213-302.

[30] D. M. Young, *Iterative Solution of Large Linear System*, Academic Press, New York, 1971.

ON THE ERROR ESTIMATES OF THE FULLY DISCRETE NONLINEAR GALERKIN METHOD WITH VARIABLE MODES TO KURAMOTO-SIVASHINSKY EQUATION *

Wu Yu-jiang
Department of Mathematics, Lanzhou University, Lanzhou 730000, China
Institute of Computational Mathematics and Scientific/Engineering Computing, Academy of Mathematics and System Scienses, CAS, Beijing 100080, China
myjaw@lzu.edu.cn
wuyj@lsec.cc.ac.cn

Yang Zhong-hua
School of Mathematical Science, Shanghai Normal University, Shanghai 200234, China

Abstract We investigate the fully discrete schemes of the nonlinear Galerkin method with variable modes for solving the Kuramoto-Sivashinsky equation. We address also the problem of error analysis for the approximate solutions. Theoretical and numerical verification shows that we can change appropriately the number of modes of the small structure components which will lead the discretization to the higher precision than usual.

Keywords: nonlinear Galerkin methods, Kuramoto-Sivashinsky equations, full discretization

1. Introduction

The nonlinear Galerkin methods introduced by Marion and Temam in [7] are well-suited algorithms for the numerical computation of various nonlinear evolution equations over large intervals of time. Many authors make use of these tools to the long-term integration of the Navier-Stokes equations. (See, e.g., [2,6] etc.)

*Project Supported by Foundation for University Key Teacher by the Ministry of Education grant GG-110-73001-1014.

We explore the usage of the nonlinear Galerkin methods for the numerical integration of the Kuramoto-Sivashinsky equations. Taking into account the choice of modes of small structure components, one can obtain a truncation with lower error. This observation motivated the theoretical study and the practical implementation of the nonlinear Galerkin methods with variable modes. We consider two discrete schemes. One is explicit while another is implicit. By virtue of a discrete analogue of Gronwall lemma, we are able to analyze the error of these schemes. The estimates of the fully discrete errors show that the approximate solutions $\{y_m^{(n)} + z_{s(m)}^{(n)}\}$ are far more accurate than what the classical Galerkin methods and the standard nonlinear Galerkin methods provide. Other results, however, especially for the numerical computations with the methods, have been reported elsewhere.

This paper is constructed as follows: Section 2 recalls some preliminaries and the theoretical background. In Section 3 we establish and describe the schemes of the fully discrete system based on our nonlinear Galerkin methods and we provide in Section 4 some useful lemmas. Finally, in Section 5 we present and prove the estimates theorems for errors produced by the two computational schemes, showing a significant gain in the order of precision.

2. Preliminaries and Notations

The one-dimensional Kuramoto-Sivashinsky equations in the primitive formulation are written as:

$$\frac{\partial u}{\partial t} + \nu \frac{\partial^4 u}{\partial x^4} + \frac{\partial^2 u}{\partial x^2} + u\frac{\partial u}{\partial x} = 0 \quad x \in \left(-\frac{l}{2}, \frac{l}{2}\right), \; t > 0 \tag{2.1}$$

$$u(x + l, t) = u(x, t) \tag{2.2}$$

$$u(x, 0) = u_0(x) \tag{2.3}$$

where $\nu > 0$ is an arbitrary constant and $u_0(x)$ is l-periodic and of zero mean.

Define two spaces by $H = \{u | u \in L^2(-\frac{l}{2}, \frac{l}{2}), u \textit{ is odd }\}$, $V = H_p^2(-\frac{l}{2}, \frac{l}{2}) \cap H$. Here the $H_p^2(\cdot,\cdot)$ denotes the 2-order Sobolev space of periodic functions. As usual, we denote H^q the q-order Sobolev space, for any $q \in \mathbf{N}^+$.

For $u \in H$ and $v \in V$ we denote the norms in H and V by $|u| = \left\{\int_{-\frac{l}{2}}^{\frac{l}{2}} |u(x)|^2 dx\right\}^{1/2}$ and $\|v\| = \left\{\int_{-\frac{l}{2}}^{\frac{l}{2}} (D^2 v(x))^2 dx\right\}^{\frac{1}{2}}$ respectively, where $D = \frac{\partial}{\partial x}$. As shown in [11] (Chapter III, Section 4) this definition leads to a norm for l-periodic functions. The corresponding scalar products will be denoted by $(\cdot,\cdot)$ and $((\cdot,\cdot))$ respectively. With these scalar products, both H and V are Hilbert spaces.

As usual let $A = \frac{\partial^4}{\partial x^4}$ be an operator in H with domain $D(A) = H_p^4(-\frac{l}{2}, \frac{l}{2}) \cap H$ and define the bilinear operator $B(\cdot,\cdot)$ by $(B(u,v),w) = \int_{-\frac{l}{2}}^{\frac{l}{2}} u \frac{\partial v}{\partial x} w dx$ $\forall u, v, w \in V$. In particular, we denote $B(u) = B(u,u)$ $\forall u$. Choosing $\phi = \phi(x)$ suitably, we have an $\alpha > 0$ such that

$$((\nu A + C)u, u) \geq \alpha \|u\|^2 \quad \forall u \in D(A) \tag{2.4}$$

where

$$Cu = \begin{cases} \frac{\partial^2 u}{\partial x^2}, & l < 2\pi \\ \frac{\partial^2 u}{\partial x^2} + \phi \frac{\partial u}{\partial x} + \phi' u, & l \geq 2\pi \end{cases} ; \quad f = \begin{cases} 0, & l < 2\pi \\ -\nu\phi^{(4)} - \phi'' - \phi\phi', & l \geq 2\pi \end{cases}$$

Evidently, f is independent of time.

Using the above notation, the system (2.1)-(2.3) is equivalent to the following functional differential equation (see, *e.g.*, [12,16] etc).

$$\frac{d}{dt} u + \nu A u + B(u) + Cu = f \tag{2.5}$$

$$u(0) = u_0 \tag{2.6}$$

For results concerning existence, uniqueness, and regularity of solutions to (2.5),(2.6) the reader is referred to, for instance, [9] and [12].

We recall the following inequalities which are satisfied by $B(u,v)$. (Note that, here and elsewhere in this work, $c_1, c_2, c_3, \ldots$, denote positive absolute constants or nondimensional positive constants that depend at most on l.)

$$|(B(u,v),w)| \leq c_1 |u|^{\frac{1}{2}} \|u\|^{\frac{1}{2}} \|v\| |w|^{\frac{1}{2}} \|w\|^{\frac{1}{2}} \quad \forall u, v, w \in V \tag{2.7}$$

$$|B(u,v)| \leq c_2 |u|^{\frac{1}{2}} \|u\|^{\frac{1}{2}} \|v\|^{\frac{1}{2}} |Av|^{\frac{1}{2}} \quad \forall u \in V, v \in D(A) \tag{2.8}$$

$$|B(u,v)| \leq c_3 |u|^{\frac{1}{2}} |Au|^{\frac{1}{2}} \|v\| \quad \forall u, v \in D(A) \tag{2.9}$$

$$|B(u,v)| \leq c_4 \left(1 + \log \frac{|Au|^2}{\lambda_1 \|u\|^2}\right)^{\frac{1}{2}} \|u\| \|v\| \quad \forall u \in D(A), v \in V \tag{2.10}$$

$$|B(u,v)| \leq c_5 \left(3 - \frac{\lambda_1 \|u\|^2}{\tau |Au|^2}\right)^{\frac{1}{2}} \|u\| \|v\| \quad (\tau > 0) \ \forall u \in D(A), v \in V \tag{2.11}$$

In addition, the operator B enjoys the following property:

$$(B(u,u),u) = 0 \quad \forall u \in V \tag{2.12}$$

It is well-known that A is a linear unbounded self-adjoint positive operator, with A^{-1} compact. Since $D(A) \subset H$ is dense, then H has an orthonormal basis $\{w_j\}_{j=1}^{\infty}$ of eigenvectors of the operator A, $Aw_j = \lambda_j w_j$, $j = 1,2,\ldots$, $0 < \lambda_1 \leq \lambda_2 \leq \ldots$. It is also well-known that there exist constants M_0 and M_1, which depend on ν, $|f|$, and λ_1, such that for every solution $u(t)$ of (2.5), (2.6) there is a time t_* depending on u_0, ν, $|f|$, and λ_1 such that $|u(t)| \leq M_0$, $\|u(t)\| \leq M_1 \ \ \forall t \geq t_*$. In particular if u_0 belongs to the global attractor then these inqualities hold for all $t \in \mathbf{R}$.

3. Nonlinear Galerkin Approximation and Discrete Schemes

Let m be a cut-off value and define

P_m: *the projection operator of H onto* $W_m = \text{span}\{w_1, \ldots, w_m\}$.

We set $s = s(m) \in N$ another integer associated with m. The nonlinear Galerkin method with variable modes is implemented by looking for an approximate solution $y_m + z_{s(m)}$ of the problem (2.5), (2.6) of the form $y_m(t) = \sum_{j=1}^{m} g_{jm}(t) w_j$, $y_m : \mathbf{R}^+ \to W_m$. The function y_m is determined by the resolution of a system involving another function z_s, where

$$z_{s(m)}(t) = \sum_{j=m+1}^{m+s} h_{jm}(t) w_j, \quad z_s : \mathbf{R}^+ \to \tilde{W}_s = \text{span}\{w_{m+1}, w_{m+2}, \ldots, w_{m+s}\}$$

Taking into account the above approximation, the pair (y_m, z_s) verifies the coupled system

$$\frac{dy_m}{dt} + \nu A y_m + P_m(B(y_m, y_m) + B(y_m, z_s) + B(z_s, y_m)) + C y_m = P_m f \tag{3.1}$$

$$\nu A z_s + (P_{s+m} - P_m) B(y_m, y_m) + C z_s = (P_{s+m} - P_m) f \tag{3.2}$$

together with

$$y_m(0) = P_m u_0 \tag{3.3}$$

Define by b the trilinear form on V, $b(u,v,w) = < B(u,v), w >_{V',V}$, $\forall u, v, w \in V$. Therefore, the system (3.1)-(3.3) is equivalent to

$$\frac{d}{dt}(y_m, v) + ((\nu A + C) y_m, v) + b(y_m, y_m, v) + b(y_m, z_s, v) + b(z_s, y_m, v)$$

$$= (f, v), \quad \forall v \in W_m \tag{3.4}$$

$$((\nu A + C)z_s, \tilde{v}) + b(y_m, y_m, \tilde{v}) = (f, \tilde{v}), \quad \forall \tilde{v} \in \tilde{W}_s \tag{3.5}$$

$$(y_m(0), v) = (u_0, v), \quad \forall v \in W_m \tag{3.6}$$

The convergence of this kind of nonlinear Galerkin method has been proved in [16] (with $\nu = 1$). To minimize the error made by the approximation, it is reasonable to choose an approximate number $s = s(m)$ of small wavelengths modes. Usually, we set

$$s = s_m = \gamma \cdot \max\{m[\sqrt{m}], m[m^{\frac{4}{\sigma}}]\} - m \tag{3.7}$$

Our aim in this paper is to study the time discretization of this method. Two schemes of full discretization are used here.

Scheme (I).

$$\frac{y_m^{(n+1)} - y_m^{(n)}}{\tau} + (\nu A + C)y_m^{(n+1)} + P_m(B(y_m^{(n+1)}) + B(y_m^{(n+1)}, z_s^{(n+1)})$$

$$+ B(z_s^{(n+1)}, y_m^{(n+1)})) = P_m f \tag{3.8}$$

$$(\nu A + C)z_s^{(n+1)} + (P_{m+s} - P_m)B(y_m^{(m+1)}) = (P_{m+s} - P_m)f \tag{3.9}$$

Obviously, it is an implicit scheme which needs to be solved by iterative method.

Scheme (II)

$$\frac{y_m^{(n+1)} - y_m^{(n)}}{\tau} + (\nu A + C)y_m^{(n+1)} + P_m\{B(y_m^{(n)}) + B(y_m^{(n)}, z_s^{(n)})$$

$$+ B(z_s^{(n)}, y_m^{(n)})\} = P_m f \tag{3.10}$$

$$(\nu A + C)z_s^{(n+1)} + (P_{m+s} - P_m)B(y_m^{(n+1)}) = (P_{m+s} - P_m)f \tag{3.11}$$

This scheme is an explicit one. One can solve $(y_m^{(n+1)}, z_s^{(n+1)})$ directly

$$\begin{cases} y_m^{(n+1)} = & (I + \tau(\nu A + C))^{-1}(y_m^{(n)} - \tau P_m\{B(y_m^{(n)}) + B(y_m^{(n)}, z_s^{(n)}) \\ & \quad + B(z_s^{(n)}, y_m^{(n)}) - f\}) \\ z_s^{(n+1)} = & z_{s(m)}^{(n+1)} = -(\nu A + C)^{-1}(P_{m+s} - P_m)(B(y_m^{(n+1)}) - f) \end{cases}$$

4. Some Lemmas

In order to give the error estimates of our schemes, we need the following lemmas.

Lemma 4.1 *For the linear operator C, we have*

$$|Cu| \leq c_6\|u\| \quad \forall u \in V. \ \Box \tag{4.1}$$

Lemma 4.2 *For any integer $m' > m$ and $v \in (P_{m'} - P_m)H$, it holds that*

$$\frac{1}{\nu\lambda_{m'} + c_6\lambda_{m'}^{\frac{1}{2}}}|v| \leq |(\nu A + C)^{-1}v| \leq \frac{1}{\alpha\lambda_{m+1}}|v|. \ \square \tag{4.2}$$

Lemma 4.3 *If $Y(t) \geq 0$ $(Y(0) = 0), X(t) \geq 0, g(t) \geq 0$ and $h(t) \geq 0$ satisfy*

$$Y'(t) + X(t) \leq g(t)Y(t) + h(t) \quad \forall t \geq 0 \tag{4.3}$$

then

$$Y(t) + \int_0^t X(\xi)d\xi \leq \int_0^t h(\xi)e^{\int_\xi^t g(\eta)d\eta}d\xi. \ \square \tag{4.4}$$

Lemma 4.4 *Suppose that there are sequences $Y_i > 0,\ g_i > 0,\ h_i > 0$ and a constant $\tau > 0$ such that*

$$Y_n \leq \tau\sum_{i=0}^{n-1}(g_iY_i + h_i) + H_0 \quad n = 1, 2, \ldots. \tag{4.5}$$

If, in addition, $\tau\sum_{i=1}^{\infty} g_i \leq \mu_1, \tau\sum_{i=1}^{\infty} h_i \leq \mu_2$, *then*

$$Y_n \leq (\tau(g_0Y_0 + h_0) + H_0 + \mu_2)e^{\mu_1} \quad n = 1, 2, \ldots. \ \square \tag{4.6}$$

Lemma 4.5 *Suppose that the initial data $u_0, (u_t)_0, (u_{tt})_0$ are in H. Then we have*

$$u, u_t, u_{tt} \in L^\infty(\mathbf{R}^+; H) \cap L^2(\mathbf{R}^+; V), \ \text{for } l < 2\pi \tag{4.7}$$

$$u, u_t, u_{tt} \in L^\infty([0, T]; H) \cap L^2([0, T]; V), \ \text{for } l \geq 2\pi. \ \square \tag{4.8}$$

Lemma 4.6 *Denote*

$$u - (y_m^{(n)} + z_s^{(n)}) = \rho^{(n)} + \theta^{(n)} \tag{4.9}$$

where

$$\rho^{(n)}(x) = u(x, t_n) - P_{m+s}u(x, t_n), \quad \theta^{(n)}(x) = \theta_1^{(n)}(x) + \theta_2^{(n)}(x),$$

$$\theta_1^{(n)}(x) = P_m u(x, t_n) - y_m^{(n)}(x), \theta_2^{(n)}(x) = P_{m+s}u(x, t_n) - P_m u(x, t_n) - z_s^{(n)}(x).$$

Then, for the Scheme (I), the $\theta^{(n)}(x)$ satisfies

$$\frac{\theta_1^{(n+1)} - \theta_1^{(n)}}{\tau} + (\nu A + C)\theta_1^{(n+1)} +$$

$$P_m\{B(u(t_{n+1})) - (B(y_m^{(n+1)}) + B(y_m^{(n+1)}, z_s^{(n+1)}) + B(z_s^{(n+1)}, y_m^{(n+1)}))\}$$
$$= P_m\left\{\frac{u(t_{n+1}) - u(t_n)}{\tau} - u_t(t_{n+1})\right\} \quad (4.10)$$

and

$$\frac{\theta_2^{(n+1)} - \theta_2^{(n)}}{\tau} + (\nu A + C)\theta_2^{(n+1)} + (P_{m+s} - P_m)\{B(u(t_{n+1})) - B(y_m^{(n+1)})\}$$

$$= (P_{m+s} - P_m)\left\{\frac{u(t_{n+1}) - u(t_n)}{\tau} - u_t(t_{n+1})\right\} - \frac{z_s^{(n+1)} - z_s^{(n)}}{\tau}. \square \quad (4.11)$$

Lemma 4.7 *Using the same notation of Lemma 4.6, we know that, for the Scheme (II), the* $\theta^{(n)}(x)$ *satisfies*

$$\frac{\theta_1^{(n+1)} - \theta_1^{(n)}}{\tau} + (\nu A + C)\theta_1^{(n+1)} + P_m\{B(u(t_{n+1})) - (B(y_m^{(n)}) + B(y_m^{(n)}, z_s^{(n)})$$

$$+ B(z_s^{(n)}, y_m^{(n)}))\} = P_m\left\{\frac{u(t_{n+1}) - u(t_n)}{\tau} - u_t(t_{n+1})\right\} \quad (4.12)$$

and

$$\frac{\theta_2^{(n+1)} - \theta_2^{(n)}}{\tau} + (\nu A + C)\theta_2^{(n+1)} + (P_{m+s} - P_m)\{B(u(t_{n+1})) - B(y_m^{(n+1)})\}$$

$$= (P_{m+s} - P_m)\left\{\frac{u(t_{n+1}) - u(t_n)}{\tau} - u_t(t_{n+1})\right\} - \frac{z_s^{(n+1)} - z_s^{(n)}}{\tau}. \square \quad (4.13)$$

5. Error Estimates

For the error estimates of the fully discrete system, we focus, at first, on the case $l < 2\pi$.

5.1 Scheme (I):

Proposition 5.1 *If* u_0 *is in* H, *then we have*

$$\tau \sum_{n=1}^{\infty} (\|y_m^{(n)}\|^2 + \|z_s^{(n)}\|^2) \le \frac{1}{2\alpha}\|y_m^{(0)}\|^2 \quad (5.1)$$

$$\sum_{n=1}^{\infty} |y_m^{(n+1)} - y_m^{(n)}|^2 \le |y_m^{(0)}|^2 \quad (5.2)$$

$$|y_m^{(n)}| \le |y_m^{(0)}| \quad n = 1, 2, \ldots . \square \quad (5.3)$$

Proposition 5.2 *If u_0 is in V, then we have*

$$\tau\sum_{n=1}^{\infty}|Ay_m^{(n)}|^2\le\kappa_1,\ \ \tau\sum_{n=1}^{\infty}|Az_s^{(n)}|^2\le\kappa_2,\ \ \|y_m^{(n)}\|^2\le\kappa_3 \tag{5.4}$$

where κ_1,κ_2 and κ_3 are positive constants depending on α,β and $\|y_m^{(0)}\|^2$. □

Proposition 5.3 *If u_0 is in $H^{\sigma+2}$, then we have*

$$\tau\sum_{n=1}^{\infty}(\|Ay_m^{(n)}\|_{H^\sigma}^2+\|Az_s^{(n)}\|_{H^\sigma}^2)\le\kappa_4,\ \ \|A^{\frac{1}{2}}y_m^{(n)}\|_{H^\sigma}\le\kappa_5,\ \ n=1,2,\ldots \tag{5.5}$$

where κ_4,κ_5 are positive constants depending on α,β and $\|y_m^{(0)}\|_{H^{\sigma+2}}$. □

Proposition 5.4 *If u_0 is in $H^{\sigma+2}$, then we have*

$$\tau\sum_{n=1}^{\infty}\left\|\frac{(y_m^{(n+1)})^2-(y_m^{(n)})^2}{\tau}\right\|_{H^\sigma}^2\le\kappa_6 \tag{5.6}$$

where κ_6 is also a constant depending only on α,β and $\|y_m^{(0)}\|_{H^{\sigma+2}}$. □

The following results tell us the approximate orders of our discretization.

Theorem 5.5 *Let us suppose that u_0 is in $H^{\sigma+2}$. Then, for τ sufficiently small and m sufficiently large, we have*

$$\begin{gathered}|u(t_n)-(y_m^{(n)}+z_s^{(n)})|=O((m+s)^{-\sigma}+\lambda_{m+1}^{-1}m^{-\sigma}+\tau)\\ \textit{uniformly for } n=1,2,\ldots\end{gathered} \tag{5.7}$$

$$\tau\sum_{n=0}^{\infty}\|u(t_n)-(y_m^{(n)}+z_s^{(n)}\|^2=O((m+s)^{-2\sigma}+\lambda_{m+1}^{-2}m^{-2\sigma}+\tau^2) \tag{5.8}$$

Proof. By Lemma 4.6, we take inner product of (4.10) with $\theta_1^{(n+1)}$. It gives

$$\begin{aligned}&\frac{1}{2\tau}(|\theta_1^{(n+1)}|^2-|\theta_1^{(n)}|^2+|\theta_1^{(n+1)}-\theta_1^{(n)}|^2)+((\nu A+C)\theta_1^{(n+1)},\theta_1^{(n+1)})\\
&=-\int_{-\frac{l}{2}}^{\frac{l}{2}}\{B(u(t_{n+1}))-(B(y_m^{(n+1)})+B(y_m^{(n+1)},z_s^{(n+1)})\\
&\quad+B(z_s^{(n+1)},y_m^{(n+1)}))\}\theta_1^{(n+1)}dx\\
&\quad+\int_{-\frac{l}{2}}^{\frac{l}{2}}\left(\frac{u(t_{n+1})-u(t_n)}{\tau}-u_t(t_{n+1})\right)\theta_1^{(n+1)}dx.\end{aligned}$$

By using (2.4) and the Sobolev Imbedding Theorem, we obtain

$$|\theta_1^{(n+1)}|^2-|\theta_1^{(n)}|^2+|\theta_1^{(n+1)}-\theta_1^{(n)}|^2+2\tau\alpha\|\theta_1^{(n+1)}\|^2$$

$$\le \tau \left(\sup_x \{|u(x,t_{n+1}) + y_m^{(n+1)}(x) + z_s^{(n+1)}(x)|\} \right.$$

$$\cdot |u(t_{n+1}) - (y_m^{(n+1)} + z_s^{(n+1)})| \cdot \left| \frac{\partial \theta_1^{(n+1)}}{\partial x} \right|$$

$$\left. + \int_{-\frac{l}{2}}^{\frac{l}{2}} |B(z_s^{(n+1)})||\theta_1^{(n+1)}|dx + 2\left| \frac{u(t_{n+1}) - u(t_n)}{\tau} - u_t(t_{n+1}) \right| \cdot |\theta_1^{(n+1)}| \right)$$

$$\le \alpha\tau \|\theta_1^{(n+1)}\|^2 + \frac{\tau}{\alpha} c_8 \left\{ \|u(t_{n+1}) + y_m^{(n+1)} + z_s^{(n+1)}\|^2 (|\rho^{(n+1)}|^2 \right.$$

$$\left. + |\theta_1^{(n+1)}|^2 + |\theta_2^{(n+1)}|^2) + \|z_s^{(n+1)}\|^2 |z_s^{(n+1)}|^2 + \tau \int_{t_n}^{t_{n+1}} |u_{tt}|^2 dt \right\} \tag{5.9}$$

Upon taking inner product of (4.11) with $\theta_2^{(n+1)}$ and using (2.4) we find

$$|\theta_2^{(n+1)}|^2 - |\theta_2^{(n)}|^2 + |\theta_2^{(n+1)} - \theta_2^{(n)}|^2 + 2\tau\alpha \|\theta_2^{(n+1)}\|^2$$

$$\le \tau \left(\sup_x \{|u(x,t_{n+1}) + y_m^{(n+1)}(x)|\} \cdot |u(t_{n+1}) - y_m^{(n+1)}| \cdot \left| \frac{\partial \theta_2^{(n+1)}}{\partial x} \right| \right.$$

$$\left. + 2\left| \frac{u(t_{n+1}) - u(t_n)}{\tau} - u_t(t_{n+1}) \right| \cdot |\theta_2^{(n+1)}| + 2\left| \int_{-\frac{l}{2}}^{\frac{l}{2}} \left(\frac{z_s^{(n+1)} - z_s^{(n)}}{\tau} \right) \theta_2^{(n+1)} dx \right| \right)$$

$$\le \frac{\alpha\tau}{2} \|\theta_2^{(n+1)}\|^2 + \frac{\tau}{\alpha} c_9 \left\{ \|u(t_{n+1}) + y_m^{(n+1)}\|^2 (|\rho^{(n+1)}|^2 + |\theta_1^{(n+1)}|^2 + |\theta_2^{(n+1)}|^2 \right.$$

$$\left. + |z_s^{(n+1)}|^2) + \tau \int_{t_n}^{t_{n+1}} |u_{tt}|^2 dt \right\} + 2\tau \left| \int_{-\frac{l}{2}}^{\frac{l}{2}} \left(\frac{z_s^{(n+1)} - z_s^{(n)}}{\tau} \right) \theta_2^{(n+1)} dx \right|. \tag{5.10}$$

By Lemma 4.2, we know that (see also [16])

$$\left| \int_{-\frac{l}{2}}^{\frac{l}{2}} \left(\frac{z_s^{(n+1)} - z_s^{(n)}}{\tau} \right) \theta_2^{(n+1)} dx \right|$$

$$= \left| \int_{-\frac{l}{2}}^{\frac{l}{2}} (\nu A + C)^{-1} (P_{m+s} - P_m) \left(\frac{B(y_m^{(n+1)}) - B(y_m^{(n)})}{\tau} \right) \theta_2^{(n+1)} dx \right|$$

$$\le \frac{1}{\alpha\lambda_{m+1}} \left| (P_{m+s} - P_m) \left(\frac{(y_m^{(n+1)})^2 - (y_m^{(n)})^2}{\tau} \right) \right| \left| \frac{\partial \theta_2^{(n+1)}}{\partial x} \right|$$

$$\le \frac{1}{\alpha\lambda_{m+1}} c_9' \left(\frac{1}{m^\sigma} + \frac{1}{(m+s)^\sigma} \right) \left\| \frac{(y_m^{(n+1)})^2 - (y_m^{(n)})^2}{\tau} \right\|_{H^\sigma} \|\theta_2^{(n+1)}\|$$

$$\leq \frac{\alpha}{4}\|\theta_2^{(n+1)}\|^2 + \frac{c_{10}}{\alpha^3\lambda_{m+1}^2 m^{2\sigma}}\left\|\frac{(y_m^{(n+1)})^2 - (y_m^{(n)})^2}{\tau}\right\|_{H^\sigma}^2 \tag{5.11}$$

Note that $\|z_s^{(n+1)}\| = O(\lambda_m^{-1}m^{-\sigma})$. Hence, we add (5.9) and (5.10) and sum n from 0 to k, we obtain

$$|\theta_1^{(k+1)}|^2 + |\theta_2^{(k+1)}|^2 + \alpha\tau\sum_{n=0}^{k}(\|\theta_1^{(n+1)}\|^2 + \|\theta_2^{(n+1)}\|^2)$$

$$+\sum_{n=0}^{k}(|\theta_1^{(n+1)} - \theta_1^{(n)}|^2 + |\theta_2^{(n+1)} - \theta_2^{(n)}|^2)$$

$$\leq c(\alpha)\cdot\tau\sum_{n=0}^{k}\xi_n^2\left(|\rho^{(n+1)}|^2 + |\theta_1^{(n+1)}|^2 + |\theta_2^{(n+1)}|^2 + \frac{1}{\lambda_{m+1}^2 m^{2\sigma}} + \tau^2\right)$$

where $c(\alpha) > 0$ is a constant depending on α, and

$$\xi_n^2 = \max\left\{\|u(t_{n+1}) + y_m^{(n+1)} + z_s^{(n+1)}\|^2,\ \|u(t_{n+1}) + y_m^{(n+1)}\|^2,\right.$$

$$\left.\frac{1}{\tau}\int_{t_n}^{t_{n+1}}|u_{tt}|^2 dt,\ \left\|\frac{(y_m^{(n+1)})^2 - (y_m^{(n)})^2}{\tau}\right\|_{H^\sigma}^2\right\}$$

Clearly, Proposition 5.1 to Proposition 5.4 imply that $\sum_{n=1}^{\infty}\xi_n^2$ (τ sufficiently small) is a convergent series. Now by using Lemma 4.4, and noticing that

$$|\theta^{(n+1)}| \leq \sqrt{|\theta_1^{(n+1)}|^2 + |\theta_2^{(n+1)}|^2}$$

we can easily complete the proof of the theorem. □

Theorem 5.6 *Let us suppose that u_0 is in $H^{\sigma+2}$. Then, for τ sufficiently small and m sufficiently large, we have also*

$$\|u(t_n) - (y_m^{(n)} + z_s^{(n)})\| = O((m+s)^{-\sigma} + \lambda_{m+1}^{-1}m^{-\sigma} + \tau) \tag{5.12}$$

$$\sup_x |u(x,t_n) - (y_m^{(n)}(x) + z_s^{(n)}(x))| = O((m+s)^{-\sigma} + \lambda_{m+1}^{-1}m^{-\sigma} + \tau)$$

uniformly for $n = 1, 2, \ldots$ (5.13)

$$\tau\sum_{n=0}^{\infty}|Au(t_n) - A(y_m^{(n)} + z_s^{(n)})|^2 = O((m+s)^{-2\sigma} + \lambda_{m+1}^{-2}m^{-2\sigma} + \tau^2) \quad □ \tag{5.14}$$

The procedure to prove this theorem can be extended similarly if we begin with the inner product of (4.13) with $A\theta_1^{(n+1)}$ and of (4.14) with $A\theta_2^{(n+1)}$.

Remark 5.1 (i) In the case of full discretization, it is well-known that the main part of the error by usual Galerkin method with m modes is $m^{-\sigma} + \tau$.

(ii) For the nonlinear Galerkin method derived by Marion and Temam[7], the error is measured to be $(2m)^{-\sigma} + \tau$ which coincides with that of the classical Galerkin approximation with $2m$ modes.

(iii) If the number s of small wavelength modes is better chosen, say, $s = s_m$, then the error will reduce to $m^{-(4+\sigma)} + \tau$ which is the lowest one in the same discrete form.

5.2 Scheme (II)

For the values $y_m^{(n)}, z_s^{(n)}, n = 1, 2, \dots$, produced by Scheme (II) at grid-points $t_n = t_0 + \tau n, n = 1, 2, \dots$, the conclusions of Proposition 5.1 to Proposition 5.4 are also true. Using these conclusions, we go directly to the theorems for error.

Theorem 5.7 *Let us suppose that u_0 is in $H^{\sigma+2}$. Then, for τ sufficiently small and m sufficiently large, we have*

$$|u(t_n) - (y_m^{(n)} + z_s^{(n)})| = O((m+s)^{-\sigma} + \lambda_{m+1}^{-1} m^{-\sigma} + \tau) \quad \text{uniformly for } n = 1, 2, \dots \tag{5.15}$$

$$\tau \sum_{n=0}^{\infty} \|u(t_n) - (y_m^{(n)} + z_s^{(n)})\|^2 = O((m+s)^{-2\sigma} + \lambda_{m+1}^{-2} m^{-2\sigma} + \tau^2) \tag{5.16}$$

Proof. By Lemma 4.7, the inner product of (4.12) with $\theta_1^{(n+1)}$ reduces to

$$\frac{1}{2\tau}(|\theta_1^{(n+1)}|^2 - |\theta_1^{(n)}|^2 + |\theta_1^{(n+1)} - \theta_1^{(n)}|^2) + ((\nu A + C)\theta_1^{(n+1)}, \theta_1^{(n+1)})$$

$$= -\int_{-\frac{l}{2}}^{\frac{l}{2}} \{B(u(t_{n+1})) - (B(y_m^{(n)}) + B(y_m^{(n)}, z_s^{(n)}) + B(z_s^{(n)}, y_m^{(n)}))\}\theta_1^{(n+1)} dx$$

$$+ \int_{-\frac{l}{2}}^{\frac{l}{2}} \left(\frac{u(t_{n+1}) - u(t_n)}{\tau} - u_t(t_{n+1})\right) \theta_1^{(n+1)} dx.$$

This leads to

$$|\theta_1^{(n+1)}|^2 - |\theta_1^{(n)}|^2 + |\theta_1^{(n+1)} - \theta_1^{(n)}|^2 + 2\tau\alpha\|\theta_1^{(n+1)}\|^2$$

$$\leq \tau\Bigg(\sup_x \{|u(x,t_n) + y_m^{(n)}(x) + z_s^{(n)}(x)|\} \cdot |u(t_n) - (y_m^{(n)} + z_s^{(n)})| \cdot \left|\frac{\partial\theta_1^{(n+1)}}{\partial x}\right|$$

$$+\sup_x\left\{|u(x,t_{n+1})+u(x,t_n)|\right\}|u(t_{n+1})-u(t_n)|\left|\frac{\partial\theta_1^{(n+1)}}{\partial x}\right|$$

$$+\int_{-\frac{l}{2}}^{\frac{l}{2}}|B(z_s^{(n)})||\theta_1^{(n+1)}|dx\Bigg)+2\tau\left|\frac{u(t_{n+1})-u(t_n)}{\tau}-u_t(t_{n+1})\right|\cdot|\theta_1^{(n+1)}|$$

$$\le\alpha\tau\|\theta_1^{(n+1)}\|^2+\frac{\tau}{\alpha}c_{11}\left\{\|u(t_n)+y_m^{(n)}+z_s^{(n)}\|^2(|\rho^{(n)}|^2+|\theta_1^{(n)}|^2+|\theta_2^{(n)}|^2)\right.$$

$$\left.+\tau\int_{t_n}^{t_{n+1}}|u_{tt}|^2dt+\tau\|u(t_{n+1})+u(t_n)\|^2\int_{t_n}^{t_{n+1}}|u_t|^2dt+\|z_s^{(n)}\|^2|z_s^{(n)}|^2\right\} \tag{5.17}$$

Taking inner product of (4.13) with $\theta_2^{(n+1)}$, we get

$$|\theta_2^{(n+1)}|^2-|\theta_2^{(n)}|^2+|\theta_2^{(n+1)}-\theta_2^{(n)}|^2+2\tau\alpha\|\theta_2^{(n+1)}\|^2$$

$$\le\alpha\tau\|\theta_2^{(n+1)}\|^2+\frac{\tau}{\alpha}c_{12}\left\{\|u(t_{n+1})+y_m^{(n+1)}\|^2(|\rho^{(n+1)}|^2+|\theta_1^{(n+1)}|^2+|\theta_2^{(n+1)}|^2\right.$$

$$\left.+|z_s^{(n+1)}|^2)+\tau\int_{t_n}^{t_{n+1}}|u_{tt}|^2dt\right\}+\frac{\tau c_{13}}{\alpha^3\lambda_{m+1}^2m^{2\sigma}}\left\|\frac{(y_m^{(n+1)})^2-(y_m^{(n)})^2}{\tau}\right\|_{H^\sigma} \tag{5.18}$$

Adding (5.17) and (5.18) and summing n from 0 to k, we obtain

$$|\theta_1^{(k+1)}|^2+|\theta_2^{(k+1)}|^2+\alpha\tau\sum_{n=0}^{k}(\|\theta_1^{(n+1)}\|^2+\|\theta_2^{(n+1)}\|^2)$$

$$+\sum_{n=0}^{k}(|\theta_1^{(n+1)}-\theta_1^{(n)}|^2+|\theta_2^{(n+1)}-\theta_2^{(n)}|^2)$$

$$\le c(\alpha)\cdot\tau\sum_{n=0}^{k}\eta_n^2\left(|\rho^{(n+1)}|^2+|\theta_1^{(n+1)}|^2+|\theta_2^{(n+1)}|^2+\frac{1}{\lambda_{m+1}^2m^{2\sigma}}+\tau^2\right)$$

where $c(\alpha)$ is as before and

$$\eta_n^2=\max\left\{\|u(t_{n+1})+y_m^{(n+1)}+z_s^{(n+1)}\|^2,\ \frac{1}{\tau}\int_{t_n}^{t_{n+1}}|u_{tt}|^2dt,\right.$$

$$\left.\frac{1}{\tau}\int_{t_n}^{t_{n+1}}|u_t|^2dt,\ \|u(t_{n+1})+u(t_n)\|^2,\ \left\|\frac{(y_m^{(n+1)})^2-(y_m^{(n)})^2}{\tau}\right\|_{H^\sigma}^2\right\}$$

By the same token, Proposition 5.1 to Proposition 5.4 guarantee that, for small τ, the series $\tau\sum_{n=1}^{\infty}\eta_n^2$ is convergent. Applying Lemma 4.4, we conclude the proof of Theorem 5.7. □

Theorem 5.8 *Let us suppose that u_0 is in $H^{\sigma+2}$. Then, for τ sufficiently small and m sufficiently large, we have*

$$\|u(t_n) - (y_m^{(n)} + z_s^{(n)})\| = O((m+s)^{-\sigma} + \lambda_{m+1}^{-1} m^{-\sigma} + \tau) \tag{5.19}$$

$$\sup_x |u(x,t_n) - (y_m^{(n)}(x) + z_s^{(n)}(x))| = O((m+s)^{-\sigma} + \lambda_{m+1}^{-1} m^{-\sigma} + \tau) \quad \textit{uniformly for } n = 1,2,\ldots \tag{5.20}$$

$$\tau \sum_{n=0}^{\infty} |Au(t_n) - A(y_m^{(n)} + z_s^{(n)})|^2 = O((m+s)^{-2\sigma} + \lambda_{m+1}^{-2} m^{-2\sigma} + \tau^2) \ \square \tag{5.21}$$

Remark 5.2 (i) Although Scheme (I) is an implicit scheme, and Scheme (II) is an explicit one, the main parts of their errors are the same. However, there seems to be a difference in the stability analysis of the two schemes which we intend to show in other work.

(ii) For the case $l \geq 2\pi$, according to Lemma 4.5 the error estimates about the time t will be naturally weakened. The conclusions are effective only for the finite interval, say, $t \in [0, T]$.

As an example, we list the conclusions for Scheme (I) in the case $l \geq 2\pi$. For Scheme (II) one can almost copy the conclusions verbatimly.

Theorem 5.9 *Let us suppose that u_0 is in $H^{\sigma+2}$. Then, for τ sufficiently small and m sufficiently large, the approximate solutions $y_m^{(n)} + z_s^{(n)}$ satisfies*

$$|u(t_n) - (y_m^{(n)} + z_s^{(n)})| = O((m+s)^{-\sigma} + \lambda_{m+1}^{-1} m^{-\sigma} + \tau) \ \textit{for } n = 1,2,\ldots,M \tag{5.22}$$

$$\tau \sum_{n=0}^{M} \|u(t_n) - (y_m^{(n)} + z_s^{(n)})\|^2 = O((m+s)^{-2\sigma} + \lambda_{m+1}^{-2} m^{-2\sigma} + \tau^2) \tag{5.23}$$

Here, $M = [T/\tau]$ is used. $\square$

Theorem 5.10 *Let us suppose that u_0 is in $H^{\sigma+2}$. Then, for τ sufficiently small and m sufficiently large, the approximate solutions $y_m^{(n)} + z_s^{(n)}$ satisfies also*

$$\|u(t_n) - (y_m^{(n)} + z_s^{(n)})\| = O((m+s)^{-\sigma} + \lambda_{m+1}^{-1} m^{-\sigma} + \tau) \ \textit{for } n = 1,2,\ldots,M \tag{5.24}$$

$$\sup_x |u(x,t_n) - (y_m^{(n)}(x) + z_s^{(n)}(x))| = O((m+s)^{-\sigma} + \lambda_{m+1}^{-1} m^{-\sigma} + \tau) \quad \textit{for } n = 1,2,\ldots,M \tag{5.25}$$

$$\tau \sum_{n=0}^{\infty} |Au(t_n) - A(y_m^{(n)} + z_s^{(n)})|^2 = O((m+s)^{-2\sigma} + \lambda_{m+1}^{-2} m^{-2\sigma} + \tau^2) \ \square \tag{5.26}$$

Acknowledgments

The first author would like to thank Prof. Guo Ben-yu for his constant encouragement. He thanks also Prof. Shi Zhong-ci for his kindness to host a visit to his Institute. The authors also thank Prof. Tang Tao for one of his good suggestion leading to an improved presentation of this article.

References

[1] C. Canuto, M.Y. Hussaini, A. Quarteroni and T.A. Zang, *Spectral Methods in Fluid Dynamics*, Springer-Verlag, Berlin, Heidelberg, 1988.

[2] T. Dubois, F. Jauberteau and R. Temam, The nonlinear Galerkin method for the two and three dimensional Navier-Stokes equations, in: *K.W. Morton, ed., The Proceedings of the Twelfth International Conference on Numerical Methods in Fluid Dynamics*, Lecture Notes in Physics, Springer-Verlag, 1990.

[3] C. Foias, O. Manley and R. Temam, Sur l'interaction des petits et grands troubillons dans des écoulements turbulents, *C.R. Acad. Sc.*, Sér. I, **305** (1987), 497-500.

[4] C. Foias, O. Manley and R. Temam, Modelling of the interaction of small and large eddies in two dimensional turbulent flows, *RAIRO Math. Model. Numer. Anal.*, **22** (1988), 93-118.

[5] C. Foias, O. Manley, R. Temam,and Y. Trève, Asymptic analysis of the Navier-Stokes equations, *Physica D*, **9** (1983), 157-188.

[6] F. Jauberteau, C. Rosier and R. Temam, The nonlinear Galerkin method in computational fluid dynamics, *Appl. Numer. Math.*, **6** (1989-90), 361-370.

[7] M. Marion, and R. Temam, Nonlinear Galerkin methods, *SIAM J. Numer. Anal.*, **26** (1989), 1139-1157.

[8] B. Nicolaenko, B. Scheurer and R. Temam, Some global dynamical properties of the Kuramoto-Sivashinsky equation: nonlinear stability and attractors, *Physica D* **16** (1985), 155-183.

[9] J. Shen, Long time stability and convergence for fully discrete nonlinear Galerkin methods, *Appl. Anal.*, **38** (1990),201-229.

[10] R. Temam, *Navier-Stokes Equations and Nonlinear Functional Analysis*, CBMS-NSF Regional Conference Series in Applied Mathematics, SIAM, Philadelphia, 1983.

[11] R. Temam, *Infinite Dimensional Dynamical Systems in Mechanics and Physics*, Second Edition, Appl. Math. Sci. Ser. *Vol* 68, Springer-Verlag, Berlin, New York, 1997.

[12] R. Temam, Dynamical systems, turbulence and the numerical solution of the Navier-Stokes equations, in: *D. L. Dwoyer & R. Voigt, eds., Proceedings of the Eleventh International Conference on Numerical Methods in Fluid Dynamics*, Lecture Notes in Physics, Springer-Verlag, 1989.

[13] R. Temam, Inertial manifolds and multigrid methods, *SIAM J. Math. Anal.*, **21** (1990), 154-178.

[14] R. Temam, New emerging methods in numerical analysis: applications to fluid mechanics, in: *M. Gunzburger & N. Nicolaides, eds., Incompressible Computational Fluid Dynamics–Trends and Advances*, Cambridge University Press, Cambridge, MA, 1992.

[15] Wu Yu-jiang, Remarks on the nonlinear Galerkin method for Kuramoto-Sivashinsky equation, *Appl. Math. Mech.*, **18** (1997), 1005–1013.

[16] Wu Yu-jiang, A nonlinear Galerkin method with variable modes for Kuramoto-Sivashinsky equation, *J. Comput. Math.*, **17** (1999), 243-256.

[17] Yang Zhong-hua, A. Mahmood and Ye Ruisong, Error estimates of noninear Galerkin methods for Kuramoto-Sivashinsky equation, in: *B. Guo, ed., Proceedings of the 1994 Beijing Symposium on Nonlinear Evolution Equations and Infinite Dimensional Dynamical Systems*, Zhongshan University Press, 1995, 203-208.

A PERTURBED DENSITY-DEPENDENT NAVIER-STOKES EQUATION

Yuelong Xiao
Department of Mathematics, Xiangtan University,
Xiangtan 411105, P.R. China
xiaoyl01@163.com

Abstract For a perturbed density-dependent Navier-Stokes equation, global existence and dynamical behavior of the solutions were investigated and some results were obtained.

Keywords: Navier-Stikes equations, Generalized semi-flow, global attractor, upper Semi-continuity.

1. Introduction

Suppose $\Omega \subset R^2$ is a bounded smooth domain. We consider the following equations

$$\frac{\partial \rho}{\partial t} - div(\rho u) = 0, \quad in\, \Omega, \tag{1}$$

$$\frac{\partial(\rho u)}{\partial t} - div(\rho u \otimes u) - \alpha\Delta\frac{\partial u}{\partial t} - div(2\mu d) + \nabla p = \rho f, \quad in\, \Omega, \tag{2}$$

$$divu = 0, \quad in\, \Omega. \tag{3}$$

$$u = 0, \quad on\, \partial\Omega. \tag{4}$$

where $\alpha > 0$ is a constant, $\mu = \mu(\rho)$ is a positive continuous function, $f \in L^2(\Omega)$, Δ is the Laplacian and $d = \frac{1}{2}(\partial_i u_j + \partial_j u_i)$.

The model comes from the fluid mechanics. In the case when $\alpha = 0$, equations (1)-(4) have been widely studied, for reference, see A.V.Kazhkov [1], A.V.Kazhikov and V.N.Monakhov [2], J.Simon [3, 4, 5], R.Diperna and P.L.Lions [6, 7] and P.L.Lions [8]. It is known that the global in time weak solution exists, but the uniqueness and regularities of the solutions are still not known (see [8]).

In this paper, we investigate the dynamical behavior when $\alpha > 0$. For the Mechanical background of this case, we refer to T.W.Ting [9]. Let us state out our main results. First of all, the global weak solution may be built in a vary similar way of P.L.Lions [8], Namely, we prove

Theorem 1.1. *Under the above hypotheses, then for any* $(\rho_0, u_0) \in X$ *there exists at least one global solution* (ρ, u) *of equation (1)-(4), such that for all* $T > 0$,

$$\rho \in C([0,T], L^q(\Omega)) \cap L^\infty(\Omega \times (0,\infty)),\ 1 \leq q < \infty;$$
$$u \in C([0,T], V),\ \frac{\partial u}{\partial t} \in L^2([0,T], V),\ \frac{\partial(\rho u)}{\partial t} \in L^2(0,T; H^{-1}(\Omega)).$$

Moreover all the solutions has the following properties:

$$\frac{d}{dt}\int_\Omega \beta(\rho)dx = 0,\ \ in\ \Omega, \tag{5}$$

for all $\beta \in C^1(R,R)$;

$$\frac{d}{dt}(\int_\Omega \rho|u|^2 dx + \alpha\|\nabla u\|^2) + \int_\Omega \mu(\partial_i u_j + \partial_j u_i)^2 dx = 2 < \rho f, u > \tag{6}$$

in the scalar distribution sense.
where V *is the subspace of* $H_0^1(\Omega)$*, with the element* u *such that* $divu = 0$, X *is the subspace of* $L^\infty \times V$ *with the element* (ρ, u) *such that* $\rho \geq 0$

It is natural to look for the uniqueness of the solution, but it is also difficult. We do not make deep study here, and we associate the solutions with a generalized semi-flow which is introduced by J. Ball [10] for the systems that may have more than one solutions with given initial data. Let us denote E_q by the subspace of $L^q(\Omega)(1 \leq q < \infty)$with the element $\rho \in L^\infty(\Omega), 0 \leq \rho \leq K$ (for some positive constant K); $X_q = E_q \times V$; G the set of all solutions of equation (1)-(4) with initial data belong to X_q, we have

Theorem 1.2. *G is a generalized semi-flow on* $X = X_q$.

We expect that G would have a global attractor, but, the compactness of ρ depends on its initial data. However, (5) implies the following invariant property: If ρ_0 be such that $\|\rho_0 - a\|_q = \lambda$ for some constants $a > 0, \lambda \geq 0$, then the solution of ρ be the same. So, we denote $S = S(a,\lambda) = \{\rho \in E_q | \|\rho - a\|_q = \lambda\}$, and have

Theorem 1.3. *G restricted on* S *is also a generalized semi-flow which has a global attractor* A_λ.

Note that, when $\lambda \to 0$, the limit system is a density-independent one, We are also interested in the stability, namely, we prove

Theorem 1.4. *The global attractor A_λ is upper semi-continuous at 0.*

Proof of theorem 1.1 has no more idea than that of P.L.Lions [8] in the case $\alpha = 0$, the only difference is that we can get the energy equation (6) since $\frac{\partial u}{\partial t}$ is more regular in our case. The main part of proof of theorem 1.2 is to show G is upper semi-continuous with respect to the initial data. This can be done by a contradiction argument, roughly speaking, if the weak convergent sequence is not strong convergent at some point t_0, then the integration of the energy equation for the limit will be broken at the same point, but the limit is a solution with continuous integration of energy and the integration of the energy equation must be hold for all t. In the proof of theorem 1.3, we developed the energy equation method introduced by I.Moise, R.Rosa and X. Wang [11] to prove the existence of the global attractor for non-compact semi-groups in two aspect: One is that there is no uniqueness in our case, the dynamic is a generalized semi-flow but not a semi-group; Another is the energy equation (6) depending on both ρ and u, the asymptotic compactness of the generalized semi-flow follows from the compactness of ρ and the asymptotic compactness of u. In the proof of theorem 1.4, we modified the abstract result of J.K.Hale and R.Raugel [12] to prove the upper semi-continuity of the attractor for generalized semi-flows.

The outline of this paper is as follows. In section 2, we give some notations and prove the global existence of solutions theorem 1.1. In section 3, we investigate the dynamic behavior of the solutions, and prove theorem 1.2 and theorem 1.3. In section 4, we prove theorem 1.4 the upper-semi-continuity of the attractor.

2. Global weak solution

In this section, we prove the global existence of weak solutions theorem 1.1. We call (ρ, u) is a global weak solution of equation (1)-(4) if for all $T > 0$, $u \in C([0,T]; V)$, $\frac{\partial u}{\partial t} \in L^2(0,T; L^2(\Omega))$, $\rho \in C([0,T]; L^q(\Omega)) \cap L^\infty(\Omega \times (0,T))$ for all $1 \le q < \infty$, $\frac{\partial(\rho u)}{\partial t} \in L^2(0,T; H^{-1}(\Omega))$, and (1) holds in the sense of distribution in $\Omega \times (0,\infty)$, (2)holds in the following sense,

$$- \int_\Omega(\rho(0)u(0) \cdot \phi(0) + \nabla u(0) \cdot \nabla\phi(0))dx + \int\int_{\Omega\times(0,\infty)}(-\rho u \cdot \frac{\partial\phi}{\partial t} - \rho u_i u_j \partial_i \phi_j \\ + \alpha \nabla u \cdot \nabla \frac{\partial \phi}{\partial t} + \frac{1}{2}\mu(\partial_i u_j + \partial_j u_i)(\partial_i\phi_j + \partial_j\phi_i) - \rho f \cdot \phi)dxdt = 0. \tag{7}$$

Now, we look for the global weak solutions, we begin by giving a priori estimates

Lemma 2.1. *Let (ρ, u) be a global weak solution, then we have (6) hold in the scalar distribution sense.*

Proof: Let (ρ, u) be a global weak solution, we denote by $u_\varepsilon = u * \omega_\varepsilon$, where ω_ε is the modifier. Because of (1), we have, for all $T > 0$, $\phi \in D(0,T)$

$$\begin{aligned}
&-\int_0^T \int_\Omega \rho \frac{|u_\varepsilon|^2}{2} dx\phi'(t)dt \\
= &\int_0^T \int_\Omega \frac{\partial}{\partial t}(\rho \frac{|u_\varepsilon|^2}{2}) dx\phi(t)dt \\
= &\int_0^T \int_\Omega (\frac{\partial}{\partial t}(\rho \frac{|u_\varepsilon|^2}{2}) + div(\rho u \frac{|u_\varepsilon|^2}{2}) dx\phi(t)dt \\
= &\int_0^T \int_\Omega (\rho \frac{\partial}{\partial t} + \rho u \cdot \nabla) \frac{|u_\varepsilon|^2}{2}) dx\phi(t)dt \\
= &\int_0^T \int_\Omega (\frac{\partial \rho u_\varepsilon}{\partial t} + div(\rho u \otimes u_\varepsilon) \cdot u_\varepsilon dx\phi(t)dt \qquad (8)
\end{aligned}$$

Since $div(\rho u \otimes u_\varepsilon) \to div(\rho u \otimes u)$ in $L^2(0,T;H^{-1}(\Omega))$ weakly and $\phi u_\varepsilon \to \phi u$ in $L^2(0,T;V)$strongly as $\varepsilon \to \infty$, then, we have

$$\int_0^T \int_\Omega div(\rho u \otimes u_\varepsilon) \cdot u_\varepsilon dx\phi dt \to \int_0^T \int_\Omega div(\rho u \otimes u) \cdot u dx\phi dt. \qquad (9)$$

On the other hand, Since $\frac{\partial u}{\partial t} \in L^2(0,T;V)$, and $\rho u \in L^2(0,T;L^2(\Omega))$, we have

$$\begin{aligned}
&\int_0^T \int_\Omega \frac{\partial(\rho u_\varepsilon)}{\partial t} \cdot u_\varepsilon dx\phi dt = -\int_0^T \int_\Omega \rho u_\varepsilon \cdot \frac{\partial(u_\varepsilon \phi)}{\partial t} dxdt \qquad (10) \\
\to \quad &-\int_0^T \int_\Omega \rho u \cdot \frac{\partial(u\phi)}{\partial t} dxdt,
\end{aligned}$$

as$\varepsilon \to \infty$. Passing to the limit, we get

$$-\int_0^T\int_\Omega \rho \frac{|u|^2}{2} dx\phi' dt = -\int_0^T\int_\Omega \rho u \cdot \frac{\partial(u\phi)}{\partial t} dxdt + \int_0^T\int_\Omega div(\rho u \otimes u) \cdot u dx\phi dt. \qquad (11)$$

Note that $u \in C([0,T];V)$, $\frac{\partial u}{\partial t} \in L^2(0,T;L^2(\Omega)$, $\rho \in L^\infty(\Omega \times (0,T))$ and $\frac{\partial(\rho u)}{\partial t} \in L^2(0,T;H^{-1}(\Omega))$, integrating by part over $[0,T]$, we have

$$\int_0^T \int_\Omega \rho u \cdot \frac{\partial(u\phi)}{\partial t} dxdt = -\int_0^T \int_\Omega \frac{\partial(\rho u)}{\partial t} \cdot u dx\phi dt. \qquad (12)$$

and conclude

$$-\int_0^T \int_\Omega \rho \frac{|u|^2}{2} dx\phi' dt = \int_0^T \int_\Omega (\frac{\partial \rho u}{\partial t} + div(\rho u \otimes u)) \cdot u dx\phi dt. \qquad (13)$$

namely,

$$\frac{d}{dt}\int_\Omega \rho\frac{|u|^2}{2}dx = \int_\Omega (\frac{\partial \rho u}{\partial t} + div(\rho u \otimes u))\cdot u dx. \tag{14}$$

in $D(0,\infty)$.
Finally, multiplying (2) by u and integrating over Ω, from (14), we deduce (6), and the lemma was proved.

Lemma 2.2. *Let (ρ, u) be a weak solution of equation (1)-(4), with $\|\rho^n(0)\|_\infty \le K$, $\|u^n(0)\|_V \le R$ for some positive constants K, R, then there exist positive constants c_1, c_2, c_3 depend on K, R such that*

$$\|\nabla u\|^2 \le c_1, \forall\, t \ge 0, \tag{15}$$

$$\int_0^t \|\nabla \frac{\partial u}{\partial t}\|^2 ds \le c_2 + c_3 t,\ \forall\, t \ge 0, \tag{16}$$

Proof: Since $\|\rho(0)\|_\infty \le K$, then, from (5), it follows that

$$\rho(x,t) \le \|\rho(0)\|_\infty \le K, \tag{17}$$

and

$$| < \rho f, u > | \le \|\rho(0)\|_\infty \|f\| \|u\|^2, \tag{18}$$

also

$$\int_\Omega \rho |u|^2 dx \le \|\rho(0)\|_\infty \|u\|^2. \tag{19}$$

Note that

$$\int_\Omega \mu(\partial_i u_j + \partial_j u_i)^2 dx \ge \min_{0\le s\le K} \mu(s) \int_\Omega (\partial_i u_j + \partial_j u_i)^2 dx, \tag{20}$$

and $V \hookrightarrow L^2(\Omega)$, from the energy equation, we conclude

$$\frac{d}{dt}(\int_\Omega \rho|u|^2 dx + \alpha\|\nabla u\|^2) + c_3(\int_\Omega \rho|u|^2 dx + \alpha\|\nabla u\|^2) \le c_4,\ \ t \ge 0, \tag{21}$$

for some positive constants c_4, c_5. Using the classical Gronwall lemma, we obtain (15).
Since $u \in C([0,T]; V)$, $\frac{\partial u}{\partial t} \in L^2(0,T; L^2(\Omega))$, $\rho \in L^\infty(\Omega \times (0,T))$, $\frac{\partial \rho u}{\partial t} \in L^2(0,T; H^{-1}(\Omega))$, and note that (by using the Galigaliado-Nireberg inequality)

$$|\int_\Omega div(\rho u) u \cdot \phi dx| \le \|\rho\|_\infty \|u\| \|\nabla u\| \|\nabla \phi\|,\ \forall\, \phi \in C_0^\infty(\Omega) \tag{22}$$

By using an approximation argument, we have

$$\frac{\partial(\rho u)}{\partial t} = \frac{\partial \rho}{\partial t}u + \rho\frac{\partial u}{\partial t}, \quad in\ L^2(0,T;H^{-1}(\Omega)). \tag{23}$$

Then, we multiply (2) by $\frac{\partial u}{\partial t}$ and integrate it over Ω to find

$$\begin{aligned}&\int_\Omega \frac{\partial \rho}{\partial t}u \cdot \frac{\partial u}{\partial t}dx + \int_\Omega \rho|\frac{\partial u}{\partial t}|^2 dx + \int_\Omega div(\rho u \otimes u)\cdot \frac{\partial u}{\partial t})dx + \|\nabla \frac{\partial u}{\partial t}\|^2 \\ &+ \int_\Omega \mu(\partial_i u_j + \partial_j u_i)(\partial_i \frac{\partial u_j}{\partial t} + \partial_j \frac{\partial u_i}{\partial t})) =< \rho f, \frac{\partial u}{\partial t} > .\end{aligned} \tag{24}$$

Note that

$$|\int_\Omega div(\rho u \otimes u)\cdot \frac{\partial u}{\partial t}dx| \leq \|\rho\|_\infty \|u\| \|\nabla u\| \|\nabla \frac{\partial u}{\partial t}\|, \tag{25}$$

by using the Cauchy inequality, from (15), (22), (24), and (25) we conclude (16) and the lemma was proved.

Proof of theorem 1.1:

The solutions were built in a vary similar way of P.L.Lions [8] in the case of $\alpha = 0$, to avoid unnecessary repetition, we do not give exact claim, only explain what we can do along this way. First of all, we can solve a similar approximated equation, there is only a better term $\Delta\frac{\partial u}{\partial t}$ extra; secondly, a similar even more elementary estimates of lemma 2.2 may be applied to the approximated solutions; finally, the bounds of the approximated solution allow us to use the compactness result of [8] to pass to the limit and note that $u \in L^\infty(0,T;V), (\frac{\partial u}{\partial t} \in L^2(0,T;V)$ implies u is continuous in V and from equation (2) $\frac{\partial(\rho u)}{\partial t} \in L^2(0,T;H^{-1}(\Omega))$, then we find the solution in the sense above.

3. Generalized semi-flow and attractors

In this section, we investigate the dynamical behavior of the solutions defined in the last section as a generalized semi-flow (see J.Ball [10] for the definition), and prove theorem 1.2 and theorem 1.3.

Proof of theorem 1,2:

Obviously, G satisfies the hypotheses H_1, H_2 and H_3 in the definition of the generalized semi-flow [10]. The rest is to show that it satisfies the hypothesis H_4 the upper semi-continuity of G with respect to the initial data. Let $(\rho_0^n, u_0^n), (\rho_0, u_0) \in X, n = 1, 2, \cdots$, and $(\rho^n, u_0^n) \to (\rho_0, u_0)$ strongly in X; and $(\rho^n(t), u^n(t)) \in G$ with $(\rho^n(0), u^n(0)) = (\rho_0^n, u_0^n)$, From lemma 2.2, we know that $u^n(t)$ remains bound in V for all$t \geq 0$, $\frac{\partial u}{\partial t}$ remains bound in $L^2(0,T;V)$ for all $T > 0$. So, there exists a subsequence also denote by u^n and a function $u \in L^\infty(0,T;V)$, $\frac{\partial u}{\partial t} \in L^2(0,T;L^2(\Omega))$ such that

$$u^n \rightarrow u \text{ } weak \text{ } star \text{ } in \text{ } L^\infty(0,T;V); \tag{26}$$

$$\frac{\partial u^n}{\partial t} \rightarrow \frac{\partial u}{\partial t} \text{ } weakly \text{ } in \text{ } L^2(0,T;V); \tag{27}$$

Note that $\rho_0^n \rightarrow \rho_0$ in E_q, by the compactness result theorem 2.4 of [8], (26) also implies

$$\rho^n \rightarrow \rho, \text{ } in \text{ } C([0,T];E_q), \tag{28}$$

and ρ is the unique solution of equation (1) with respect to u and ρ_0. These convergences allow us to pass to the limit, note that $u \in L^\infty(0,T;V)$, $\frac{\partial u}{\partial t} \in L^2(0,T;L^2(\Omega)$ implies $u \in C([0,T];V)$(see Temam [13,14]), then (ρ, u) is a Global weak solution, namely, $(\rho, u) \in G$. Now, we show

$$u^n \rightarrow u \text{ } strongly \text{ } in \text{ } V, \forall t \geq 0. \tag{29}$$

Indeed, since $(\rho^n, u^n) \in C([0,T];X), \forall T > 0$ and satisfies the energy equation (6), then, we have

$$\begin{aligned} &\int_\Omega \rho^n |u^n|^2 dx + \alpha \|\nabla u^n\|^2 + \int_0^t \int_\Omega \mu(\partial_i u_j^n + \partial_j u_i^n) dx ds \\ &= \int_\Omega \rho_0^n |u_0^n|^2 dx + \alpha \|\nabla u_0^n\|^2 + 2\int_0^t < \rho^n f, u^n > ds, \end{aligned} \tag{30}$$

for all $t \geq 0$. If (29) is not true, then there exists at least a $t_0 > 0$ such that (29) does not hold, note that u^n and u are continuous in V, it follows that

$$\limsup_n \|\nabla u^n(t_0)\|^2 > \|\nabla u(t_0)\|^2 \tag{31}$$

strictly from (26), and we know that(see [8], (2.131))

$$\liminf_n \int_0^t \int_\Omega \mu(\partial_i u_j^n + \partial_j u_i^n)^2 dx ds \geq \int_0^t \int_\Omega \mu(\partial_i u_j + \partial_j u_i)^2 dx ds, \tag{32}$$

and all the other term in (30) are convergent to the limits respect to (ρ, u), we conclude that

$$\begin{aligned} &\int_\Omega \rho(t_0)|u(t_0)|^2 dx + \alpha \|\nabla u(T_0)\|^2 + \int_0^{t_0} \int_\Omega \mu(\partial_i u_j + \partial_j u_i) dx ds \\ &< \int_\Omega \rho_0 |u_0|^2 dx + \alpha \|\nabla u_0\|^2 + 2\int_0^{t_0} < \rho f, u > ds, \end{aligned} \tag{33}$$

strictly. This is a controdiction, since (33) must be an identity for all $t > 0$ by (6) and the continuity of the solutions, and the theorem was proved.

Proof of theorem 1.3:

Without losing generity, we also denote the restriction of G on S by G. According to J.Ball [10], we only need to show G is asymptotic compact. Let $(\rho_0^n, u_0^n) \in S, n = 1, 2, \cdots$ be a bounded Sequence, (ρ^n, u^n) be the corresponding solutions with$(\rho^n(0), u^n(0)) = (\rho_0^n, u_0^n)$. First of all, we claim that ρ_n is compact, namely, there exists a subsequence $\rho^{n'}$ of ρ^nwhich is convergent in $C([0,T]; L^q(\Omega))$for all $T > 0$. Indeed, since (ρ_0^n, u_0^n) is bounded in X, the same argument in the proof of theorem 1.2 shows that (26) also hold, note that ρ_0^n contained on S which is a compact set of $L^q(\Omega) \hookrightarrow L^1(\Omega), 1 \le q < \infty$, then, by using the compactness result of [8] (theorem 2.4.), we deduce (28), namely, ρ^n is compact. Next, we show that u_n is asymptotic compact, namely, for any given $t_n \to \infty$, $u^n(t_n)$ has convergent subsequence. We begin by rewrite the energy equation (6) to the following form

$$\begin{aligned} & \frac{d}{dt}\Big(\int_\Omega \rho^n|u^n|^2dx + \alpha\|\nabla u^n\|^2\Big) + \frac{\bar{\mu}}{\alpha}\Big(\int_\Omega \rho^n|u^n|^2dx + \alpha\|\nabla u^n\|^2\Big) \\ + & \int_\Omega (\mu - \frac{\bar{\mu}}{2})(\partial_i u_j^n + \partial_j u_i^n)^2 dx \\ = & \; 2 < \rho^n f, u^n > + \frac{\bar{\mu}}{\alpha}\int_\Omega \rho^n |u^n|^2 dx. \end{aligned} \tag{34}$$

where $\bar{\mu} = min\{\mu(s); s \in [0, K]\}$. For convenient, we shall denote the subsequence of u^n by u^n below, and remain it in mind that the assertion is to the subsequence if necessary.

Since $u^n(t_n)$ is bounded in V, then

$$u^n(t_n) \to w \; weakly \; in \; V. \tag{35}$$

If the convergence is not strong, then, we have

$$\lim_n \|\nabla u^n(t_n)\|^2 = a > \|\nabla w\|^2. \tag{36}$$

On the other hand, for any given $T > 0$, we also have

$$u^n(t_n - T) \to w_T \; weakly \; in \; V. \tag{37}$$

Note that

$$(\hat{\rho}^n(t), \hat{u}^n(t)) = (\rho^n(t_n - T + t), u^n(t_n - T + t)) \in G, \tag{38}$$

with

$$(\hat{\rho}^n(0), \hat{u}^n(0)) = (\rho^n(t_n - T), u^n(t_n - T)) \tag{39}$$

remains bound in X, similarly as above, it follows that there exists a $(\hat{\rho}, \hat{u}) \in G$ such that

$$\hat{u}^n \rightarrow \hat{u} \; weakly \; \forall t \geq 0; \tag{40}$$

$$\hat{\rho}^n \rightarrow \hat{\rho}, \; in \; C([0,T]; E_q), \; \forall T > 0. \tag{41}$$

and

$$\hat{u}(0) = w_T, \;\; \hat{u}(T) = w. \tag{42}$$

Integrating the energy equation (34) with respect to $(\hat{\rho^n}, \hat{u^n})$ From 0 to T, we have

$$\begin{aligned} & \int_\Omega \hat{\rho}^n(T)|\hat{u}^n(T)|^2 dx + \alpha \|\nabla \hat{u}^n(T)\|^2 \\ + & \int_0^T e^{-\frac{\bar{\mu}}{\alpha}(T-s)} \int_\Omega (\hat{\mu}^n - \frac{\bar{\mu}}{2})(\partial_i \hat{u}_j^n + \partial_j \hat{u}_i^n)^2 dx ds \\ = & (\int_\Omega \hat{\rho}_0^n |\hat{u}_0^n|^2 dx + \alpha \|\nabla \hat{u}_0^n\|^2) e^{-\frac{\bar{\mu}}{\alpha}T} \\ + & \int_0^t e^{-\frac{\bar{\mu}}{\alpha}(T-s)} (\int_\Omega \frac{\bar{\mu}}{\alpha} \hat{\rho}^n |\hat{u}^n|^2 dx + 2 \int_0^t < \hat{\rho}^n f, \hat{u}^n >) ds, \end{aligned} \tag{43}$$

where$\hat{\mu}^n = \mu(\hat{\rho}^n)$.Since $\hat{\mu}^n - \frac{\bar{\mu}}{2} \geq \frac{\bar{\mu}}{2}$, then, we have (see [8],(2.123))

$$\begin{aligned} & \lim_n \int_0^T e^{-\frac{\bar{\mu}}{\alpha}(T-s)} \int_\Omega (\hat{\mu}^n - \frac{\bar{\mu}}{2})(\partial_i u_j^n + \partial_j u_i^n)^2 dx ds \\ \geq & \int_0^T e^{-\frac{\bar{\mu}}{\alpha}(T-s)} \int_0^t \int_\Omega \mu(\hat{\rho})(\partial_i u_j + \partial_j u_i)^2 dx ds, \end{aligned} \tag{44}$$

note that $V \hookrightarrow L^q(\Omega), 1 \leq q \leq \infty$, it follows that all the other term in (43) convergent to the limit, passing to the limit in (43), then, we deduce there is a positive constant c_1 independent of T such that

$$\begin{aligned} & \int_\Omega \hat{\rho}(T)|\hat{u}(T)|^2 dx + \alpha a + \int_0^t e^{-\frac{\bar{\mu}}{\alpha}(T-s)} \int_\Omega (\hat{\mu}(\hat{\rho} - \frac{\bar{\mu}}{2})(\partial_i \hat{u}_j + \partial_j \hat{u}_i)^2 dx ds \\ & \leq c_1 e^{-\frac{\bar{\mu}}{\alpha}T} + \int_0^t e^{-\frac{\bar{\mu}}{\alpha}(T-s)} (\int_\Omega \frac{\bar{\mu}}{\alpha} \hat{\rho} |\hat{u}|^2 dx + 2 \int_0^t < \hat{\rho} f, \hat{u} >) ds. \end{aligned} \tag{45}$$

Note that $(\hat{\rho}, \hat{u})$ also satisfies (34), comparing it with the last inequality we conclude that

$$\alpha a \leq c_2 e^{-\frac{\bar{\mu}}{\alpha}T} + \alpha \|\nabla w\|^2 \tag{46}$$

for some constants c_2 independent of T. Since T is arbitrary , then

$$a \leq \|\nabla w\|^2, \tag{47}$$

and we find a contradiction, and then u_n is asymptotic compact. Finally, the asymptotic compactness follows from the compactness of ρ^n and the asymptotic compactness of u^n and the theorem was proved by using an abstract result of J.Ball [10](theorem 3.3) for the existence of the global attractor.

4. Upper semi-continuity of the attractor

In this section, we prove theorem 1.4, we begin by proving an abstract result for the upper semi-continuity of the attractor for generalized semi-flows which is independently interesting.

Lemma 4.1. *Let Λ be a topology space(of parameter), $X_\lambda, \lambda \in \Lambda$ and X be matrix spaces, G_λ be a generalized semi-flow on X_λ, if*
i) $X_\lambda \hookrightarrow X, \ \forall \lambda \in \Lambda$.
ii) G_λ has a global attractor A_λ on X_λ.
iii) $\cup A_\lambda \subset B$, B is a bounded in X.
iv) For any given $T > 0$, and any sequence of λ_n with $\lambda_n \rightarrow \lambda_0$ in Λ when $n \rightarrow \infty$, and $z_n \in A_{\lambda_n}$ and $\phi_n(t) \in G_{\lambda_n}$ with $\phi_n = z_n$, there exists a subsequence $\{n'\}$ of $\{n\}$, and $\phi_0 \in G_{\lambda_0}$ such that

$$\phi_{n'}(t) \rightarrow \phi_0(t) \ \in X, \ \forall\, t \in [0, T]. \tag{48}$$

Then, the attractor is upper semi-continuous at $\lambda = \lambda_0$ namely,

$$\lim_{\lambda \rightarrow \lambda_0} dist_X(A_\lambda, A_{\lambda_0}) = 0. \tag{49}$$

Proof: First of all, under the assumption above we have the following property:

(P:) For any given number $\delta > 0$, and any sequence of λ_n with $\lambda_n \rightarrow \lambda_0$ in Λ when $n \rightarrow \infty$, and $z_n \in A_{\lambda_n}$ and a sequence of corresponding complete orbits $\psi_n(t) \in G_{\lambda_n}$, there exists a subsequence $\{n'\}$ of $\{n\}$ such that $\psi_{n'}(0)$ convergent to $\bar{x}$ in X and $\bar{x} \in N_\delta(A_{\lambda_0})$.

Indeed, from assumptions i),ii) and iii), it follows that there is time $T > 0$ such that all the solutions of G_{λ_0} with initial data belong to B are contained in $N_\delta(A_{\lambda_0})$ when $t \geq T$ since A_{λ_0} is attractive. Note that all the attractors $A_\lambda, \lambda \in \Lambda$ are invariant, using assupption iv) for this T, λ_n and $z_n = \psi_n(-T)$ and $\phi_n(t) = \psi(-T+t)$, it follows that $\psi_{n'}(-T+t) \rightarrow \phi_0(t)$ for some $\phi \in G_{\lambda_0}$ and then $\psi_{n'}(0) \rightarrow \phi_0(T) = \bar{x}$. Note that $\phi(0) \in B$, then

$\bar{x} = \phi(T) \in N_\delta(A_{\lambda_0})$, and the property (P:) hold.

Next, if lemma 4.1 is note true, then there exists a positive number δ and a sequence of parameter λ_n with $\lambda_n \rightarrow \lambda_0$and a corresponding sequence of points $x_n \in A_{\lambda_n}$ such that

$$dist_X(x_n, A_{\lambda_0}) \geq 2\delta. \tag{50}$$

Also by the invariance of the attractors, there is a sequence of complete orbit ψ_n with $\psi_n(0) = x_n$ which contradicts property (P), and the lemma was proved.

We now apply this result to prove theorem 1.4.

Proof of theorem 1.4:

Obviously, conditions i),ii) and iii) follows from lemma 2.2 and theorem 1.3. The rest is to clarify condition iv). Let $T > 0$, λn with $\lambda_n \rightarrow \lambda_0$ in Λ when $n \rightarrow \infty$, and $z_n = (\rho_0^n, u_0^n) \in A_{\lambda_n}$ and $\phi_n(t) = (\rho^n(t), u^n(t)) \in G_{\lambda_n}$ with $\phi_n = z_n$ be given. First of all, we claim that z_n has convergent subsequence. Obviously, $\rho_0^n \rightarrow a$; Since A_λ is invariant, we have $\bar{z_n} \in A_{\lambda_n}$ and $\psi_n \in G_{\lambda_n}$ with $\psi_n(0) = \bar{z_n}$ such that $z_n = \psi_n(n)$, the similar argument of proof of theorem 1.3 shows that u_0^n has convergent subsequence. Next, we assume $z_n = (\rho_0^n, u_0^n) \rightarrow (a, u_0)$, a similar argument of proof of theorem 1.3 shows that there exists a subsequence $\{n'\}$ and a $\phi = (\rho, u) \in G_{\lambda_0}$such that

$$\phi_{n'}(t) = (\rho^n(t), u^n(t)) \rightarrow \phi_0(t) = (\rho(t), u(t)) = (a, u(t)) \in X, \; \forall t \in [0, T]. \tag{51}$$

the proof was completed.

References

[1] A.V. Kazhkov,(1974) Resolution of bounded value problem for nonhomogeneous viscus fluids, Dokl. Akad. Nauh. 216(1974),pp. 1008-1010..

[2] A.V. Kazhkov and S.H. Smagulov, (1977)The correctness of bounded value problem in a diffusion model of an inhomogeneous liquid, Sov. Phys. Dokl., 22(1977),pp. 249-259.

[3] J. Simon, (1978)Ecoulement d'un fluid non homogeneous avec une densite initiale s'ananulant, C. R. Acad. Sci. Paris, 15 (1978), pp.1009-1012.

[4] J. Simon, (1989)Sur les fluid visqueux incompressible et non homogeneous, C. R. Acad. Sci. Paris, 309(1989), pp.447-452.

[5] J. Simon, (1990) Non-homogeneous viscus incompressible fluids; existence of velocity desity and pressure, SIAM J. Mathyh. Anal., 21 (1990).

[6] R.J. Diperna and P.L. Lions,(1989)Ordinary differential equations, Sobolev space and transport theory, Invent. Math. 98(1989), pp. 511- 547.

[7] R.J. Diperna and P.L. Lions,(1989)Equations differentielles ordinaries et equations de transport avec des coefficients irreguliers, In Seminaire EDP 1988-1989, Ecole Polytechnique, Palaiseau,(1989).

[8] P.L. Lions,(1996)Mathematical topics in fluid mechanics, Vol. 1, Incompressible models, Clarendon press Oxford (1996).

[9] Ting, T. W.,(1963)Certain non-steady flows of second order fluids, Arch. Rational mech. anal., 14(1963), pp. 1-26.

[10] Ball J.,(1997)Continuity properties and global attractors of generalized semi-flows and the Navier-Stokes equations, J. Nonlinear Sci. 7 (1997),pp. 475-502.

[11] Moise, I., Rosa, R. and Wang, X.,(1998)Attractors for non-compact semigroups via energy equations, Nonlinearity 11(1998), pp. 1369-1393.

[12] Hale, J.K. and Raugel, G.,(1988)Upper semicontinuity of the attractor for a singularly perturbed hyperbolic equation, J. Differential Eqns 73(1988), pp.197-214.

[13] Temam R.,(1997)Infinite Dimensional dynamical systems in mechanics and physics, 2nd ed. Belin Springer-Verlag, (1997).

[14] Temam R.,(1984)Navier-Stokes equations, Theory and numerical analysis, 3nd ed North-Holland, Amsterdam, 1984.

A CASCADIC MULTIGRID METHOD FOR SOLVING OBSTACLE PROBLEMS *

J. P. Zeng, S. Z. Zhou, J. T. Ma
Department of Applied Mathematics, Hunan University,
Changsha, Hunan, 410082, P.R. China

Abstract In this paper, we propose a cascadic multigrid method for the solution of the Lagrange finite element discretization for elliptic obstacle problems. We prove the convergence and obtain an error estimate of the method for the model obstacle problem. We also give some numerical results showing that the effectiveness of the proposed method.

Keywords: Cascadic multigrid method, obstacle problem, finite element, convergence

1. Introduction

Obstacle problems play an important role in the mathematical modeling of a variety of free boundary problems, arising for instance in porous media flow, device simulation or nonlinear mechanics. In the last century, various numerical iteration methods have been developed to solve obstacle problems. Among these iteration methods, it has been found that multigrid methods are effective (see, e.g., [3, 6, 8, 10, 16] and the references therein). Recently, a new kind of multigrid method, called cascadic multigrid method, was proposed to solve partial differential equations. Compared with traditional multigrid methods [2, 4], cascadic multigrid method is simple since it never goes back to coarse grid for correction in the iterations. Moreover, it has some good theoretical properties and can produce fast and accurate numerical solutions (see, e.g., [1, 7, 13, 14] and the references therein).

In this paper, we show how the cascadic multigrid method can be adopted to solve the Lagrange finite element discretization of a kind of elliptic variational inequality and analyze the convergence of the method. The method can be essentially considered as the PFMG scheme of Brandt and Cryer in [3]. We

*The work is supported by NNSF#10071017 of P. R. China.

also give some numerical results to show that cascadic multigrid method is competent with traditional multigrid methods.

Let $\Omega \subset R^d(d=1 \text{ or } 2)$ be a polygonal domain, $f(x) \in L^2(\Omega)$ be a given function, and K be defined as follows

$$K = \{v \in H_0^1(\Omega) | v \geq \phi\}, \tag{1.1}$$

where $\phi \in H^2(\Omega)$ satisfying $\phi|_{\partial\Omega} \leq 0$. We Consider the following veriational inequality of finding a $u \in K$ such that

$$a(u, v-u) \geq (f, v-u), \quad \forall v \in K, \tag{1.2}$$

where $a(v,w) = \int_\Omega \nabla v \nabla w dx + \int_\Omega qvwdx$ for some scale $q \geq 0$. Suppose that $T^{h_l}(l = 1,2,\ldots,L)$ are nested quasi-uniform triangulations and the corresponding linear conforming finite element spaces are $V_l(l = 1,2,\ldots,L)$. For simplicity, we assume that $h_l \cong 2^{-l}h_0$. Then the corresponding finite element approximation problem on level l is finding a $u_l \in K_l$, such that

$$a(u_l, v_l - u_l) \geq (f, v_l - u_l), \quad \forall v_l \in K_l, \tag{1.3}$$

where $K_l = \{v_l \in V_l | v_l(P) \geq \phi(P) \text{ for any node } P \text{ of } T^{h_l}\}$.

For one-dimensional case, we have the finite element error estimate [11, 12]

$$\|u - u_l\|_{L^2(\Omega)} \preceq h_l^2, \tag{1.4}$$

where u is the solution of (1.2), "$p_1 \preceq p_2$" means "$p_1 \leq Cp_2$" holds for some constant C independent of the mesh level and meshsize. However, as $d = 2$, it is still an open problem whether (1.4) holds. So, we give the following assumption.

Assumption 1.1. *For $d = 2$, the following estimate holds.*

$$\|u - u_l\|_{L^2(\Omega)} \preceq h_l^2, \tag{1.5}$$

where u is the solution of (1.2).

Now we can present a cascadic multigrid method as follows.

Algorithm 1.1

Step 1 Let u_0 be the finite element solution on level $l = 0$, and

$$u_0^0 = \tilde{u}_0 = u_0.$$

Step 2 For $l = 1,2,\ldots,L$, Let

$$u_l^0 = \tilde{u}_{l-1}, \qquad u_l^{m_l} = S_{l,m_l}(u_l^0), \qquad \tilde{u}_l = u_l^{m_l},$$

where $S_{l,m_l}(v)$ denotes the m_l steps basic iteration solution on level l with the initial v. Generally, we can use project Richardson, project Jacobi, or project symmetric Gauss-Seidel (SGS) as basic iterations.

Let $\{\varphi_l^{(i)}\}_{i=1}^{N_l}$ be the finite element basic functions on level l. Then we have that

$$u_l^j = \sum_{i=1}^{N_l}(x_l^j)_i\varphi_l^{(i)}, \quad u_l = \sum_{i=1}^{N_l}(x_l)_i\varphi_l^{(i)}, \quad \tilde{u}_l = \sum_{i=1}^{N_l}(\tilde{x}_l)_i\varphi_l^{(i)}.$$

Then, (1.3) can be reformulated as the following algebraic problem of finding an $x_l \in R^{N_l}$, such that

$$A_l x_l \geq b_l, \quad x_l \geq c_l, \quad (x_l - c_l)^{\mathrm{T}}(A_l x_l - b_l) = 0, \tag{1.6}$$

where the stiff matrix A_l is an S-matrix (see [5]). And therefore, algorithm 1.1 is equivalent to the following algorithm.

Algorithm 1.2

Step 1 Let $u_0 = \sum_{i=1}^{N_l}(x_0)_i\varphi_0^{(i)}$ be the finite element solution on level $l = 0$, and

$$x_0^0 = \tilde{x}_0 = x_0.$$

Step 2 For $l = 1, 2, \ldots, L$, Let

$$x_l^0 = I_l\tilde{x}_{l-1}, \qquad x_l^{m_l} = B_{l,m_l}(x_l^0), \qquad \tilde{x}_l = x_l^{m_l},$$

where I_l is some interpolation operator from $R^{N_{l-1}}$ to R^{N_l}, $B_{l,m_l}(x_l^0)$ is the m_l steps basic iteration solution for problem (1.6) with initial x_l^0.

2. Convergence of cascadic multigrid method

In the sequel, we will analyze the convergence of algorithm 1.1 for $d = 1$. Similar convergence results can be obtained for $d = 2$ under assumption 1.1 but we will omit it here. We refer to [9] for details.

Lemma 2.1. *Let us use the basic project iterations to solve the following problem of finding an $x \in R^n$, such that*

$$x \geq c \qquad A_l x \geq b, \qquad (x - c)^T(A_l x - b) = 0, \tag{2.1}$$

where $A_l \in R^{N_l \times N_l}$ is the stiff matrix on level l, $b, c \in R^{N_l}$. Denote the iteration error by ϵ^k, i.e. $\epsilon^k = x^k - x$. Then we have $|\epsilon^{k+1}| \leq G_l|\epsilon^k|$, where $|\epsilon| = (|\epsilon_j|)$ and $G_l \geq 0$ are as follows,

$$\begin{cases} G_l = I_l - \frac{A_l}{\sigma}, & for\ project\ Richardson\ iteration, \\ G_l = I_l - D_l^{-1}A_l, & for\ project\ Jacobi\ iteration, \\ G_l = (I_l - U_l)^{-1}L_l(I_l - L_l)^{-1}U_l, & for\ project\ SGS\ iteration, \end{cases}$$

where $A_l = D_l(I_l - L_l - U_l)$, matrices D_l, L_l and U_l are diagonal, strictly lower triangular and strictly upper triangular respectively, σ is some positive constant satisfying $\sigma I_l \geq D_l$.

Proof We give a simple proof for project Richardson iteration only. For project Richardson iteration, we have

$$x^{k+1/2} = x^k - \sigma^{-1}(A_l x^k - b),$$

and

$$x^{k+1} = \max\{x^{k+1/2}, c\} = \max\{x^k - \sigma^{-1}(A_l x^k - b), c\}.$$

Since the solution x of (2.1) satisfies

$$x = \max\{x - \sigma^{-1}(A_l x - b), c\},$$

it is easy to verify that

$$|\epsilon^{k+1}| \leq |I_l - \sigma^{-1} A_l||\epsilon^k| = (I_l - \sigma^{-1} A_l)|\epsilon^k|. \quad \square$$

From Lemma 2.1, we see that the $|\epsilon^{k+1}|$ can be estimated by the matrix G_l, while G_l is just the iteration matrix of correspondent basic iteration for linear equations $A_l x = b$. For such matrix G_l, we can introduce the following result.

Lemma 2.2. *[5, 7, 15] For the basic iterations, such as Richardson, Jacobi or Gauss-Seidel iteration, of linear equations on lever l, we have*

$$\|G_l^{m_l} y\|_2 \preceq \frac{h_l^{-1}}{m_l^{1/2}}\|y\|_2, \qquad \|G_l^{m_l} y\|_2 \leq \|y\|_2, \qquad \forall y \in R^{N_l}, \tag{2.2}$$

where G_l is the iteration matrix of Richardson, Jacobi or Gauss-Seidel iteration corresponding to linear equations $A_l x = b$.

The following lemma can be derived by careful calculation.

Lemma 2.3. *Let $v_l = \sum_{i=1}^{N_l} (y)_i \varphi_l^{(i)}$. Then*

$$\|v\|_{L^2} \preceq h_l^{\frac{1}{2}}\|y\|_2 \preceq \|v\|_{L^2}. \tag{2.3}$$

Based on above basic lemmas, we can obtain the main convergence result and the estimate of the total number of operations as follows.

Theorem 2.1. *Let $m_l = [\beta^{L-l} m_L]$. Then we have*

$$\|\tilde{u}_L - u_L\|_{L^2} \preceq \begin{cases} \dfrac{1}{(1 - \frac{2}{\sqrt{\beta}})m_L^{1/2}} \cdot h_L, & for \beta > 4, \\ \dfrac{L}{m_L^{1/2}} \cdot h_L, & for \beta = 4. \end{cases} \tag{2.4}$$

Proof By Lemma 2.1, we have that

$$|\tilde{x}_l - x_l| \leq G_l^{m_l}|x_l^0 - x_l| \leq G_l^{m_l}|I_l x_{l-1} - x_l| + G_l^{m_l}|I_l(\tilde{x}_{l-1} - x_{l-1})|.$$

Then, by lemma 2.2, we have that

$$\begin{aligned} \|\tilde{x}_l - x_l\|_2 &\leq \|G_l^{m_l}|I_l x_{l-1} - x_l|\|_2 + \||I_l(\tilde{x}_{l-1} - x_{l-1})|\|_2 \\ &\leq \|G_l^{m_l}|I_l x_{l-1} - x_l|\|_2 + \sqrt{2}\|\tilde{x}_{l-1} - x_{l-1}\|_2. \end{aligned} \tag{2.5}$$

By multiplying (2.5) by $(\sqrt{2})^{L-l}$ and taking the sum for $l = 1, 2, \ldots, L$, we get

$$\begin{aligned} \|\tilde{x}_L - x_L\|_2 &\preceq \sum_{l=1}^{L} \frac{(\sqrt{2})^{L-l} h_l^{-1}}{m_l^{1/2}} \|I_l x_{l-1} - x_l\|_2 \\ &\preceq \sum_{l=1}^{L} \frac{(\sqrt{2})^{L-l} h_l^{-3/2}}{m_l^{1/2}} \|u_{l-1} - u_l\|_{L^2} \\ &\preceq \sum_{l=1}^{L} \frac{(\sqrt{2})^{L-l} h_l^{1/2}}{m_l^{1/2}}. \end{aligned}$$

Noting that $h_l \cong 2^{-l} h_0$ and $m_l = [\beta^{L-l} m_L]$, we get immediately

$$\|\tilde{x}_L - x_L\|_2 \preceq \frac{h_L^{1/2}}{m_L^{1/2}} \sum_{l=1}^{L} (\frac{2}{\sqrt{\beta}})^{L-l},$$

and then

$$\|\tilde{u}_L - u_L\|_{L^2} \preceq \frac{h_L}{m_L^{1/2}} \sum_{l=1}^{L} (\frac{2}{\sqrt{\beta}})^{L-l}.$$

Therefore, for $\beta > 4$ we have

$$\|\tilde{u}_L - u_L\|_{L^2} \preceq \frac{h_L}{m_L^{1/2}} \sum_{l=1}^{\infty} (\frac{2}{\sqrt{\beta}})^{L-l} = \frac{1}{(1 - \frac{2}{\sqrt{\beta}}) m_L^{1/2}} \cdot h_L,$$

and for $\beta = 4$ we have

$$\|\tilde{u}_L - u_L\|_{L^2} \preceq \frac{h_L}{m_L^{1/2}} \sum_{l=1}^{L} 1 = \frac{L}{m_L^{1/2}} \cdot h_L. \quad \square$$

Theorem 2.2. *Under the same conditions of theorem 2.1, we have the following estimate of the total number of operations.*

$$W_L = \sum_{l=1}^{L} m_l N_l \preceq \begin{cases} \frac{1}{1 - \beta/2} \cdot m_L \cdot N_L, & \beta < 2; \\ m_L \cdot N_L \cdot L, & \beta = 2; \\ m_L \cdot N_L \cdot (\frac{\beta}{2})^L, & \beta > 2, \end{cases} \tag{2.6}$$

and then

$$W_L \preceq \begin{cases} \frac{1}{1-\beta/2} \cdot m_L \cdot N_L, & \beta < 2; \\ m_L \cdot N_L \cdot \log(N_L+1), & \beta = 2; \\ m_L \cdot N_L \cdot (N_L+1)^{(\log\beta-1)}, & \beta > 2. \end{cases} \tag{2.7}$$

Proof It is easy to get the estimate for $\beta \leq 2$. For $\beta > 2$, we have

$$\begin{aligned} W_L &= \sum_{l=1}^{L} m_l N_l \leq \sum_{l=1}^{L} \beta^{L-l} \cdot m_L \cdot \frac{N_L}{2^{L-l}} \\ &= m_L \cdot N_L \sum_{l=1}^{L} (\frac{\beta}{2})^{L-l} = m_L \cdot N_L \frac{1-(\frac{\beta}{2})^L}{1-\frac{\beta}{2}} \\ &\preceq m_L \cdot N_L \cdot (\frac{\beta}{2})^L \preceq m_L \cdot N_L \cdot (\frac{\beta}{2})^{\log(N_L+1)} \\ &= m_L \cdot N_L \cdot \frac{\beta^{\log(N_L+1)}}{N_L+1} \\ &= m_L \cdot N_L \cdot (N_L+1)^{(\log\beta-1)}. \qquad \square \end{aligned}$$

Corollary 2.1. *If we choose $\beta = 4$, $m_l = [\beta^{L-l} m_L]$, and $m_L = [m^* \cdot L^2]$, then for algorithm 1.1, we have the following estimates of iterative error and total number of operations.*

$$\|\tilde{u}_L - u_L\|_{L^2} \preceq h_L,$$

$$W_L \preceq L^2 \cdot 2^L \cdot N_L,$$

where m^* is some positive constant.

3. Numerical Examples

In this section, we give some numerical tests to show that the cascadic multigrid method proposed in the paper is effective for solving obstacle problems.

Example 3.1. *Find a $u \in K = \{v \in H_0^1(\Omega) | v \geq \phi\}$, such that*

$$a(u, v-u) \geq 0, \qquad \forall v \in K, \tag{3.1}$$

where $\Omega = (-2, 2)$, $\phi(t) = 1 - t^2$ and $a(w, v) = \int_{-2}^{2} (w'v' + 2wv)dt$.

We choose the finite element inner nodes $t_j = -2 + h_l j (j = 1, 2, \ldots, N_l)$, where $N_l + 1 = 4h_l^{-1}$, $h_l = 2^{-l} h_0$ with $h_0 = 2^{-3}$. Using Lagrange linear finite element method to discrete the problem (3.1), we get corresponding discrete problem (1.3). In algorithm 1.1, or equivalently algorithm 1.2, we choose $m_l = [4.2^{L-l}]$. In two-grid V-cycle algorithm (see e.g.[4]), we let μ_1, and μ_2 be the project iteration numbers of pre-smoother (performed before the coarse-grid correction) and post-smoother (performed after the coarse-grid correction) respectively, μ be the project iteration number on the coarse grids, that is the exact solution of the problem on the coarse-grid is replaced by μ steps of coarse-grid project iteration solution. precision= $|| \min\{A_L\tilde{x}_L - b_L, \tilde{x}_l - c_l\}||_\infty$. Table 1 and Table 2 show the computing results of the algorithms with project Jacobi smoother and project SGS smoother respectively. Table 3 shows the corresponding results of two-grid V-cycle algorithm.

Table 1. Numerical results for project Jacobi smoother

L	N_L	W_L	precision	time(second)
4	511	8252	1.1309×10^{-4}	0.44
5	1023	36393	3.0121×10^{-5}	1.32
6	2047	155471	7.2565×10^{-6}	5.05
7	4095	658462	1.7277×10^{-6}	19.06

Table 2. Numerical results for project SGS smoother

L	N_L	W_L	precision	time(second)
4	511	16504	3.6565×10^{-4}	1.04
5	1023	72786	1.0544×10^{-4}	4.39
6	2047	310942	2.3650×10^{-5}	18.84
7	4095	1316924	6.6756×10^{-6}	85.63

Table 3. Numerical results for two-grid V-cycle algorithm

$\mu_1/\mu_2/\mu$	N_L	cycle number	precision	time(second)
3/3/1	511	3	2.400×10^{-3}	0.55
3/3/1	1023	4	1.100×10^{-3}	2.14
3/3/1	2047	3	6.2469×10^{-4}	5.44
3/3/1	4095	3	3.1358×10^{-4}	20.76

Example 3.2. *Find a* $u \in K = \{v \in H_0^1(\Omega)|v \geq \phi\}$, *such that*

$$a(u, v - u) \geq 0, \qquad \forall v \in K, \tag{3.2}$$

where $\Omega = (0,1) \times (0,1)$, $\phi(t,s) = 0.35 - [(0.5-t)^2 + (0.5-s)^2]^{1/2}$ *and* $a(w,v) = \int_\Omega (\nabla w \nabla v) dtds$.

We choose the finite element inner nodes $(t_i, s_j) = (h_l i, h_l j)(i, j = 1, 2, \dots, \sqrt{N_l})$, where $N_l = (h_l^{-1} - 1)^2$, $h_l = 2^{-l}h_0$ with $h_0 = 2^{-2}$. Using Lagrange linear finite element method to discrete the problem (3.2), we get corresponding discrete problem (1.3). In algorithm 1.1, or equivalently algorithm 1.2, we choose $m_l = [4^{L-l}m_L]$, $m_L = m^* L^2$. Table 4 and Table 5 show the computing results of cascadic multigrid method and two-grid V-cycle algorithm respectively.

Table 4. Cascadic multigrid method with project Jacobi smoother

m^*	L	N_L	precision	time(second)
0.3	3	961	0.0195	0.38
0.15	4	3969	0.0093	4.40
0.1	5	16129	0.0046	77.55

Table 5. two-grid V-cycle algorithm

$\mu_1/\mu_2/\mu$	N_L	cycle number	precision	time(second)
2/2/1	961	2	0.0073	0.44
2/2/1	3969	2	0.0034	4.4
2/2/1	16129	2	0.0022	75.0

Remark 3.1 It is well known that multigrid methods are much faster than the traditional relaxation iteration methods. We see here from table 1 $\sim$ table 5 that the cascadic multigrid method is competent with two-grid V-cycle multigrid algorithm for both cases of $d = 1$ and $d = 2$.

Remark 3.2 Unlike the PDE case, the estimate (2.4) does not have optimal convergence order. The following Table gives a comparison of the cascadic method, for solving (3.1) and the corresponding boundary value problem, i.e. finding $u \in H_0^1(\Omega)$, such that

$$a(u, v - u) = 1, \qquad \forall v \in H_0^1(\Omega). \tag{3.3}$$

Table 6. Comparison with equation case

L	N_L	$\|\|u - \tilde{u}_L\|\|_{L_2}$	$\|\|w - \tilde{w}_L\|\|_{L_2}$
4	511	3.0242×10^{-4}	2.0509×10^{-3}
5	1023	2.6261×10^{-4}	1.3657×10^{-3}
6	2047	1.9290×10^{-4}	5.7183×10^{-4}
7	4095	1.2054×10^{-4}	3.3371×10^{-4}

In above table, $||u - \tilde{u}_L||_{L_2}$ and $||w - \tilde{w}_L||_{L_2}$ denote the L_2 error between the analytical and computed solutions of problem (3.1) and (3.3) respectively. From Table 6, we see that with the increase of L, $||u - \tilde{u}_L||_{L_2}$ decrease more slowly than $||w - \tilde{w}_L||_{L_2}$. However, the difference is not large, which implies that estimate better than (2.4) is possible.

References

[1] F. A. Bornemann and P. Deuflhard, The cascadic multigrid method for elliptic problems, Numer. Math., 75 (1996), pp. 135–152.

[2] A. Brandt, Multilevel adaptive solutions to boundary-value problems, Math. Comp., 31 (1977), pp. 333–409.

[3] A. Brandt and C. W. Cryer, Multigrid algorithms for the solution of linear complementarity problems arising from free boundary problems, SIAM J. Stat. Comput., 4 (1983), pp. 655–681.

[4] W. Hackbush, Multi-grid Methods and Applications, Springer Verlag, Berlin - Heidelberg - New York, 1992.

[5] W. Hackbusch, Iterative Solution of Large Sparse Systems of Equations, Springer Verlag, New York - Berlin - Heidelberg, 1994.

[6] R. H. W. Hoppe, Multigrid algorithms for variational inequalities, SIAM J. Numer. Anal., 24 (1987), pp. 1046–1065.

[7] Y. Huang, Z. Shi, T. Tang and W. Xue, Multilevel successive iteration methods for elliptic problems, to appear.

[8] R. Kornhuber, Monotone multigrid methods for elliptic variational inequalities I, Numer Math., 69 (1994), pp. 167–185.

[9] J. Ma, Z. Shi, J. Zeng, Multilevel successive iteration method for variational inequality problems, to appear.

[10] J. Mandel, On multigrid algorithm for variational inequalities, Appl. Math. Optim., 11 (1984), pp.77–95.

[11] U. Mosco, Error estimates for some variational inequalities, Lecture Notes in Math. 606, Springer Verlag, Berlin, (1977), 224–236.

[12] F. Natterer, Optimale L_2-konvergenz finiten elemente bei variationsungleichungen, Bonn Math. Schr., 89 (1976), pp. 1–12.

[13] Z. Shi and X. Xu, Cascadic multigrid method for elliptic problems, East-West J. Numer. Math., 7 (1999), pp. 199–222.

[14] Z. Shi and X. Xu, Cascadic multigrid method for the plate bending problem, East-West J. Numer. Math., 6 (1998), pp. 137–153.

[15] J. Xu, Iterative methods by space decomposition and subspace correction, SIAM Review, 34 (1992), pp. 581–613.

[16] J. P. Zeng, The convergence of multigrid methods for nonsymmetric elliptic variational inequalities, J. Comput. Math., 11 (1993), pp. 73–76.

THE FUNDAMENTAL EQUATION OF TWO-DIMENSIONAL LAYER FLOWS OF THE MELT FEEDSTOCK IN THE POWDER INJECTION MOLDING PROCESS *

Zhoushun Zheng
Central South University, Changsha,410083, P. R. China

Xuanhui Qu
Central South University, Changsha,410083, P. R. China
University of Science and Technology Beijing, Beijing,100083, P. R. China

Abstract The PIM mold filling analysis is based on two-dimensional layer flows assumptions for thin cavity filling under non-isothermal conditions. The mathematical model that described PIM filling process is created, and the formula that calculated the flow conductance is deduced. The pressure equation for the liquid phase is obtained as

$$\frac{\partial}{\partial x}\left(S\frac{\partial P}{\partial x}\right)+\frac{\partial}{\partial y}\left(S\frac{\partial P}{\partial y}\right)=0$$

which is a non-linear elliptic partial differential equation. It calculates for the model possible and lays a mathematical foundation for further analyzing the PIM mold filling flow.

Keywords: Powder Injection Molding, Rheology, Flow Conductance, Mathematical model

1. Introduction

Powder Injection Molding (PIM) is a new powder metallurgy near-net shape forming technology that has been evolved from the conventional injection molding process of plastics. PIM process offers many technological and economical advantages over the conventional powder products, such as near-net shaping, complex shape production, high performance and productivity etc., so PIM

*Projected supported by The National 973 Program, Natural Science Foundation of China and The State Ministry of Education

process has evoked many considerable researches. It is called "The technology in the greatest demand nowadays to manufacture parts". Figure 1 shows schematically the production process during powder injection molding. In a first step, the "granulates feedstock" is blended from the metal powder and a 35-55Vol% polymer binder system. In the second step, the melt feedstock is molded into shape under high pressure on an injection molding machine in the state of temperature ($100^0C - 180^0C$). Then, the polymer binder is extracted by a thermal or chemical process and the powder component is sintered to final density.

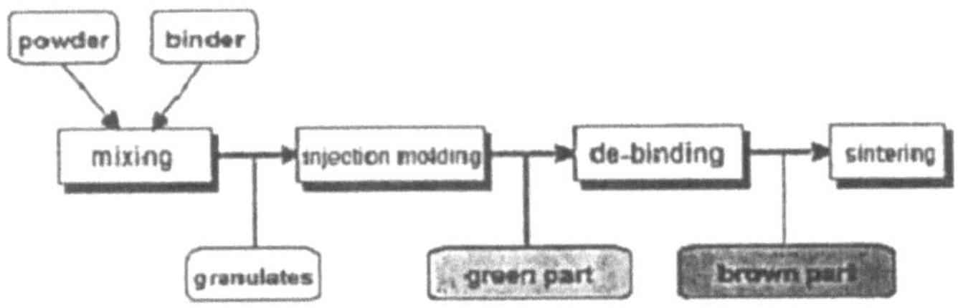

Figure 1. PIM - Powder Injection Molding

While there are many variables that control the PIM process, mold filling is the most critical phase of component manufacture. Defects, such as voids, sink marks, weld lines and density variation can result if the molding parameters and tool features are not properly specified, and these defects can't be compensated in the subsequent debinding or sintering process. Traditionally, the design of a PIM process involves iterative changes in mold variables and tool features until success has been achieved. Therefore, the design has been an art, rather than a science, involving costly and time consuming procedures. To bring this industry to a more scientific basis, the design process should be integrated with scientific analysis based upon fluid mechanics, heat transfer and stress analysis. We should create the partial differential equations describing the fluid flow and heat transfer in PIM processes, should solve the numerical solvation for the partial differential equations and simulate the flow front locations, distribution of velocities, temperature and pressure. Then we can analyze the possible defects and the region in which the defects maybe occur during the PIM process, provide the useful information for the design of a PIM process. This has become possible by numerical techniques implemented on computers.

2. The equations describing the feedstock mold filling flow

In this paper, we use the Euler Method. To describe the physical capacity of the melt fluid particle (every small parts getting from unlimitedly partition.) and their changing conditions, the velocity, pressure etc. of the fluid particle which is of every instant in space's certain place are studied. In order to study the melt feedstock mold filling flow, we consider that the melt fluid is even continuous medium, and powder's influence is summed to rheological behaviors and other material parameter's changing. So the rheological field of the melt fluid flow should obey the conservation law of mass, momentum and energy of continuous medium mechanics. According to Kwon [3], in a control volume,

The continuity equation is,

$$\nabla \cdot \vec{V} = \frac{\partial u_j}{\partial x_j} = 0 \tag{2.1}$$

where $\vec{V}$ is the velocity vector, $\vec{V} = u_1\vec{e_1} + u_2\vec{e_2} + u_3\vec{e_3}$, $\vec{e_1}$, $\vec{e_2}$ and $\vec{e_3}$ are the unit vectors in the direction of the 1, 2 and 3 axes.

The momentum transport equation is,

$$\rho\frac{\partial u_i}{\partial t} + \rho u_j\frac{\partial u_i}{\partial x_j} + \frac{\partial \tau_{ij}}{\partial x_j} + \frac{\partial P}{\partial x_i} + \rho g_i = 0 \tag{2.2}$$

where ρ is the density of feedstock, τ_{ij} is the deviatoric portion of the general stress tensor, P is the pressure, g_i is the gravitational acceleration vector in the direction of the i axes, $\vec{g} = g_1\vec{e_1} + g_2\vec{e_2} + g_3\vec{e_3}$.

The heat transfer equation of the melt is,

$$\rho C_p\frac{\partial T}{\partial t} + \rho C_p\vec{V} \cdot \nabla T - \rho\varphi_{mass}\triangle H_b\nabla f_s + \nabla\vec{q} - (-\tau : \nabla\vec{V}) - \dot{Q} = 0 \tag{2.3}$$

where T is the temperature, C_p is the specific heat, and $\vec{q}$ is the conductive heat flux, is given by $\vec{q} = q_1\vec{e_1} + q_2\vec{e_2} + q_3\vec{e_3} = q_i\vec{e_i}$, $q_i = -\Gamma\frac{\partial T}{\partial x_i}$ for conductive heat transfer, where Γ is the thermal conductivity. $\dot{Q}$ is the rate at which latent heat is released during freezing of the melt in unit volume,

$$\dot{Q} = \rho_b\varphi_v\triangle H_b\frac{\partial f_s}{\partial t} = \rho\varphi_{mass}\triangle H_b\frac{\partial f_s}{\partial t},$$

where ρ_b and $\triangle H_b$ are respectively the density and latent heat of the binder material, f_s is the solid volume fraction of the binder, and φ_v and φ_{mass} are the volume fraction and weight fraction, respectively, of the binder material in the melt. $-\tau : \nabla\vec{V}$ represent the term of viscous heating is determined by $-\tau : \nabla\vec{V} = \eta\dot{\gamma}^2$, where $\dot{\gamma}$ is the shear rate, η is the flow viscosity, $\eta = \eta(\dot{\gamma}, T)$ for a given feedstock..

On the basis of that many factors are considered and assume the melt fluid is well-distributed and satisfied the power law, the rheological equation $\eta = \eta(\dot{\gamma}, T)$ of the feedstock mold filling is given by German [2]. In general, the rheological equation $\eta = \eta(\dot{\gamma}, T)$ is simplified as

$$\eta = m_0 \, exp\left(T_\alpha / T \right) \dot{\gamma}^{m-1} \tag{2.4}$$

where, m_0 , T_α, are material constants.

3. The mathematical model of two-dimensional layer flows of the melt feedstock in the Powder Injection Molding process

The flow in the PIM process is a complicated problem with viscoelasticity effect, non-stability, non-isotherm. In addition, considering the complexity molding geometry shape. It is extremely difficult to describe the whole process of the flows exactly. What we have offered in the last paragraph is a three-dimensional flow with little feasibility. Actually, in many circumstances, the process of PIM feedstock mold filling flow could be simplified as a two-dimensional flow. Compared to viscosity, inertial force and gravity could be neglected. Most injection molded parts have a flat thin cavity. We therefore confine the analysis to a relatively thin part such that the flow can be considered to be two-dimensional layer flows. The following assumptions are made to simplify the formulation.

1 Powder and binder mix well without any gas hole, and the mixture never separates during the flow. The feedstock melt fluid flow is considered as even continuous medium non-Newtonian fluid flow, and the effect of heat expansion and the latent heat are neglected.

2 Heat conduction plays greatest role on cavity wall, and the convective transmit heat in cavity thickness z direction is neglected. While the convective transmit heat plays greatest role in cavity, and heat conduction in the streamwise x, y direction in cavity is neglected.

3 In the cavity, we only consider viscosity, and inertial force, elasticity and gravity are neglected. The pressure is assumed to be constant in the thickness z direction.

4 At the gate of cavity, the injecting temperature T0, the rate of volume Q and the temperature Tw of cavity wall are constants during the flow. The solidification is neglected and the no-slip boundary is assumed.

Then the governing equation for the continuity, momentum and energy balance for the melt fluid are described as

$$\frac{\partial}{\partial x}(b\bar{u}) + \frac{\partial}{\partial y}(b\bar{v}) = 0 \tag{3.1}$$

where $\bar{u}, \bar{v}$ are the average velocity vectors of the thickness in the direction of the x, y axes. $2b$ is the thickness of cavity.

$$\frac{\partial}{\partial z}\left(\eta\frac{\partial u}{\partial z}\right) - \frac{\partial P}{\partial x} = 0, \ \frac{\partial}{\partial z}\left(\eta\frac{\partial v}{\partial z}\right) - \frac{\partial P}{\partial y} = 0 \tag{3.2}$$

$$\rho C_p\left(\frac{\partial T}{\partial t} + u\frac{\partial T}{\partial x} + v\frac{\partial T}{\partial y}\right) = K_{th}\frac{\partial^2 T}{\partial z^2} + \eta\dot{\gamma}^2 \tag{3.3}$$

where u, v are the velocity vectors in the direction of the x, y axes. K_{th} is the coefficient of heat conduction. And η is given by

$$\eta = m_0 exp\left(T_\alpha/T\right)\dot{\gamma}^{n-1} \tag{3.4}$$

where $\dot{\gamma}$ is determined by

$$\dot{\gamma} = \left[\left(\frac{\partial u}{\partial z}\right)^2 + \left(\frac{\partial v}{\partial z}\right)^2\right]^{\frac{1}{2}} \tag{3.5}$$

The boundary conditions are

- On the cavity wall:

$$u(x, y, b, t) = v(x, y, b, t) = 0 \tag{3.6}$$
$$T(x, y, b, t) = T_w \tag{3.7}$$

- On the central line of cavity:

$$\frac{\partial u}{\partial z}(x, y, 0, t) = \frac{\partial v}{\partial z}(x, y, 0, t) = 0 \tag{3.8}$$
$$\frac{\partial T}{\partial z}(x, y, 0, t) = 0 \tag{3.9}$$

4. The flow conductance and the pressure equation for the melt fluid

For simplification, let

$$\Lambda_x = -\frac{\partial P}{\partial x}(x, y, t), \qquad \Lambda_y = -\frac{\partial P}{\partial y}(x, y, t). \tag{4.1}$$

By integrating the equation (3.2) and using the boundary condition (3.6), we obtain

$$\eta \frac{\partial u}{\partial z} = -\Lambda_x Z, \qquad \eta \frac{\partial v}{\partial z} = -\Lambda_y Z. \tag{4.2}$$

From equations (4.2) and (3.5), we deduce that

$$\dot{\gamma} = \frac{1}{\eta} \Lambda_z \tag{4.3}$$

where $\Lambda_z = \sqrt{\Lambda_x^2 + \Lambda_y^2}$. By integrating the equation (4.2) and using the boundary condition (3.6), we obtain

$$u = \Lambda_x \int_z^b \frac{z}{\eta}\, dz, \qquad v = \Lambda_y \int_z^b \frac{z}{\eta}\, dz \tag{4.4}$$

Therefore,

$$\begin{aligned} \bar{u} &= \frac{1}{b} \int_0^b \Lambda_x \left(\int_z^b \frac{z}{\eta}\, dz \right) dz = \frac{1}{b} \Lambda_x \int_0^b dz \int_z^b \frac{t}{\eta}\, dt \\ &= \frac{1}{b} \Lambda_x \int_0^b dt \int_0^t \frac{t}{\eta}\, dz = \frac{1}{b} \Lambda_x \int_0^b \frac{t^2}{\eta}\, dt \\ &= \frac{1}{b} \Lambda_x \int_0^b \frac{z^2}{\eta}\, dz \end{aligned}$$

Similarly,

$$\bar{v} = \frac{1}{b} \int_0^b \Lambda_y \left(\int_z^b \frac{z}{\eta}\, dz \right) dz = \frac{1}{b} \Lambda_y \int_0^b \frac{z^2}{\eta}\, dz.$$

Let

$$S = \int_0^b \frac{z^2}{\eta}\, dz \tag{4.5}$$

Equation (4.6) is the formula for calculating the flow conductance. Hence, we have the average velocity $\bar{u}, \bar{v}$,

$$\bar{u} = \frac{1}{b}\Lambda_x S, \qquad \bar{v} = \frac{1}{b}\Lambda_y S \tag{4.6}$$

By integrating the momentum equation, making use of the non-slip and symmetric boundary conditions for velocity as well as the continuity equation, we obtain from (3.1) and (4.6) the pressure equation for the melt liquid

$$\frac{\partial}{\partial x}\left(S\frac{\partial P}{\partial x}\right) + \frac{\partial}{\partial y}\left(S\frac{\partial P}{\partial y}\right) = 0 \tag{4.7}$$

which is a nonlinear elliptic partial differential equation. It lays a mathematical foundation for further analysis for the PIM flow.

References

[1] Rosof B. H.,"The Metal Injection Molding Process Comes of Age", J.of Mater., 41 (1989), pp. 13-16.

[2] R. M. German, "Powder Injection Molding, Metal Powder Industries Federation", Princeton, N.J., 1991.

[3] T. H. Kwon, "Numerical Simulation of Powder Injection Molding Filling Process for Three-Dimensional Complicated Cavity Geometries", Advances in Powder Metallurgy & Partiallar Materials, 5 (1996), pp. 19-79.

[4] X. H. Qu,"Numerical Simulation of Feedstock Melt Filling in a Cylindrical Cavity with Solidification in Powder Injection Molding", Trans. Of NMSC, 8 (1998), pp. 544-549.

[5] C. J. Hwang,"Computer-Aided Engineering System for Powder Injection Molding Filling Process", PIM 2000 World Congress.

Index

MIX
Papier aus verantwortungsvollen Quellen
Paper from responsible sources
FSC® C105338

If you have any concerns about our products,
you can contact us on
ProductSafety@springernature.com

In case Publisher is established outside the EU,
the EU authorized representative is:
Springer Nature Customer Service Center GmbH
Europaplatz 3, 69115 Heidelberg, Germany

Printed by Libri Plureos GmbH
in Hamburg, Germany